Burkhard König (Ed.)

Chemical Photocatalysis

Also of Interest

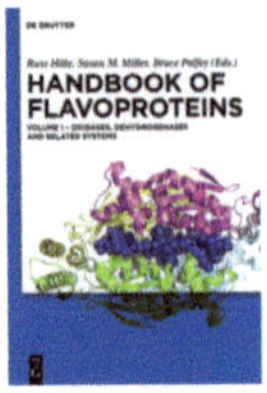

Handbook of Flavoproteins
Volume 1 Oxidases, Dehydrogenases and Related Systems
Russ Hille, Susan Miller, Bruce Palfey (Eds.), 2012
ISBN 978-3-11-026842-3, e-ISBN 978-3-11-026842-3

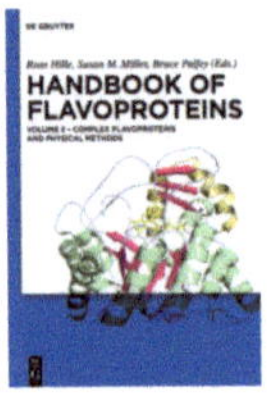

Handbook of Flavoproteins
Volume 2 Complex Flavoproteins, Dehydrogenase and Physical Methods
Russ Hille, Susan Miller, Bruce Palfey (Eds.), 2013
ISBN 978-3-11-029828-4, e-ISBN 978-3-11-029834-5

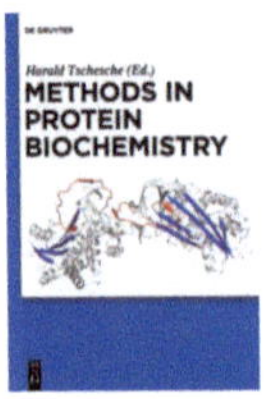

Methods in Protein Biochemistry
Harald Tschesche (Ed.), 2011
ISBN 978-3-11-025233-0, e-ISBN 978-3-11-025236-1

Inorganic Micro- and Nanomaterials
Synthesis and Characterization
Angela Dibendetto, Michele Aresta (Eds.), 2013
ISBN 978-3-11-030666-8, e-ISBN 978-3-11-030687-3

Chemical Photocatalysis

Edited by Burkhard König

DE GRUYTER

Editor
Prof. Dr. Burkhard König
Universität Regensburg
Institut für Organische Chemie
Universitätsstr. 31
93053 Regensburg
Germany
burkhard.koenig@chemie.uni-regensburg.de

ISBN 978-3-11-026916-1
e-ISBN 978-3-11-026924-6

Library of Congress Cataloging-in-Publication Data
A CIP catalog record for this book has been applied for at the Library of Congress.

Bibliographic information published by the Deutsche Nationalbibliothek
The Deutsche Nationalbibliothek lists this publication in the Deutsche Nationalbibliografie; detailed bibliographic data are available in the Internet at http://dnb.dnb.de.

Typesetting: le-tex publishing services GmbH, Leipzig
Printing and binding: Hubert & Co. GmbH & Co. KG, Göttingen
♾ Printed on acid-free paper

Printed in Germany

www.degruyter.com

List of contributing authors

Rafael Alonso Ruiz
Lehrstuhl für Organische Chemie I
Technische Universität München
Garching, Germany
Chapter 5

Iuliana Atodiresei
Institut für Organische Chemie
RWTH Aachen
Aachen, Germany
Chapter 10

Thorsten Bach
Lehrstuhl für Organische Chemie I
Technische Universität München
Garching, Germany
thorsten.bach@ch.tum.de
Chapter 1

Andreas Bauer
Lehrstuhl für Organische Chemie I
Technische Universität München
Garching, Germany
andreas.bauer@tum.de
Chapter 5

Maria Cherevatskaya
Institut für Organische Chemie
Universität Regensburg
Regensburg, Germany
Chapter 12

Radek Cibulka
Department of Organic Chemistry
Institute of Chemical Technology Prague
Czech Republic
cibulkar@vscht.cz
Chapter 4

Stephan Dankesreiter
Institut für Anorganische Chemie
Universität Regensburg
Regensburg, Germany
Chapter 12

Bernhard Dick
Institut für Physikalische und
Theoretische Chemie
Universität Regensburg
Regensburg, Germany
bernhard.dick@chemie.uni-regensburg.de
Chapters 3, 15

Anna Eisenhofer
Institut für Organische Chemie
Universität Regensburg
Regensburg, Germany
Chapter 12

Georgiy Kachkovskyi
Institut für Organische Chemie
Universität Regensburg
Regensburg, Germany
Chapter 8

Viktor Kais
Institut für Organische Chemie
Universität Regensburg
Regensburg, Germany
Chapter 8

Katharina Kastner
Institut für Anorganische Chemie II
Friedrich-Alexander-Universität
Erlangen-Nürnberg
Erlangen, Germany
Chapter 13

Uwe Kensy
Institut für Physikalische und
Theoretische Chemie
Universität Regensburg
Regensburg, Germany
Chapter 15

Burkhard König
Institut für Organische Chemie
Universität Regensburg
Regensburg, Germany
burkhard.koenig@chemie.uni-regensburg.de
Chapters 1, 4

Rene M. Königs
Institut für Organische Chemie
RWTH Aachen
Aachen, Germany
Chapter 10

Paul Kohls
Institut für Organische Chemie
Universität Regensburg
Regensburg, Germany
Chapter 8

Susanne Kümmel
Institut für Organische Chemie
Universität Regensburg
Regensburg, Germany
Chapter 4

Roger-Jan Kutta
Institut für Physikalische und
Theoretische Chemie
Universität Regensburg
Regensburg, Germany
Chapter 15

Thomas Merz
Institut für Physikalische und
Theoretische Chemie
Universität Regensburg
Regensburg, Germany
Chapter 14

Matthias Neumann
Institut für Organische Chemie
Universität Regensburg
Regensburg, Germany
Chapter 9

Suva Paria
Institut für Organische Chemie
Universität Regensburg
Regensburg, Germany
Chapter 8

Alexander Penner
Institut für Organische Chemie
Karlsruher Institut für Technologie (KIT)
Karlsruhe, Germany
Chapter 6

Michael G. Pfeffer
Anorganische Chemie I
Universität Ulm
Ulm, Germany
Chapter 11

Arno Pfitzner
Institut für Anorganische Chemie
Universität Regensburg
Regensburg, Germany
arno.pfitzner@chemie.uni-regensburg.de
Chapter 12

Michael Pirtsch
Institut für Organische Chemie
Universität Regensburg
Regensburg, Germany
Chapter 8

Daniel Rackl
Institut für Organische Chemie
Universität Regensburg
Regensburg, Germany
Chapter 8

Sven Rau
Anorganische Chemie I
Universiät Ulm
Ulm, Germany
sven.rau@uni-ulm.de
Chapter 11

Oliver Reiser
Institut für Organische Chemie
Universität Regensburg
Regensburg, Germany
oliver.reiser@chemie.uni-regensburg.de
Chapter 8

Eberhard Riedle
Lehrstuhl für BioMolekulare Optik
Ludwig-Maximilians-Universität
München, Germany
riedle@physik.uni-muenchen.de
Chapter 16

Magnus Rüping
Institut für Organische Chemie
RWTH Aachen
Aachen, Germany
magnus.rueping@rwth-aachen.de
Chapter 10

Peter Schroll
Institut für Organische Chemie
Universität Regensburg
Regensburg, Germany
Chapter 2

Martin Schütz
Institut für Physikalische und
Theoretische Chemie
Universität Regensburg
Regensburg, Germany
martin.schuetz@chemie.uni-regensburg.de
Chapter 14

Hana Seo
Institut für Organische Chemie
Universität Regensburg
Regensburg, Germany
Chapter 8

Robert Staehle
Lehrstuhl für Anorganische und
Allgemeine Chemie
Friedrich-Alexander-Universität
Erlangen-Nürnberg
Erlangen, Germany
Chapter 11

Carsten Streb
Institut für Anorganische Chemie II
Friedrich-Alexander-Universität
Erlangen-Nürnberg
Erlangen, Germany
carsten.streb@chemie.uni-erlangen.de
Chapter 13

Filip Teplý
Helquat Chemistry and Biology
Institute of Organic Chemistry and Biochemistry, v.v.i.
Academy of Sciences of the Czech Republic
Prague 6, Czech Republic
teply@uochb.cas.cz
Chapter 7

Johannes Tucher
Institut für Anorganische Chemie II
Friedrich-Alexander-Universität Erlangen-Nürnberg
Erlangen, Germany
Chapter 13

Hans-Achim Wagenknecht
Institut für Organische Chemie
Karlsruher Institut für Technologie (KIT)
Karlsruhe, Germany
wagenknecht@kit.edu
Chapter 6

Michael Weinberger
Institut für Organische Chemie
Karlsruher Institut für Technologie (KIT)
Karlsruhe, Germany
Chapter 6

Matthias Wenninger
Lehrstuhl für BioMolekulare Optik
Ludwig-Maximilians-Universität
München, Germany
Chapter 16

Kirsten Zeitler
Institut für Organische Chemie
Universität Regensburg
Regensburg, Germany
kzeitler@uni-leipzig.de
Chapter 9

Contents

Burkhard König, Thorsten Bach

1 Introduction

The only sustainable energy source on earth is sunlight. Within one hour the current annual energy demand of the world reaches the surface of earth as sunlight. Many concepts to harvest, use and store this energy directly or from secondary processes, such as wind or waves, are now under development to finally reduce our dependency on fossil energy sources. While the conversion of light into electrical energy has reached a high level of efficiency and is already at the stage of commercial applications, the use of visible light to drive chemical reactions is much less developed. This is surprising looking at the omnipresent biological photosynthesis, which harvests and stores energy from sunlight in chemical products as the basis for life on earth.

Light-induced reactions of molecules are the topic of a well established, classic area of chemistry, which is called photochemistry. However, most of the work in the past focused on direct excitation of molecules – and as most organic compounds do not absorb in the visible part of the spectrum – the excitation requires short wavelength (λ) ultraviolet irradiation, frequently in a region of $\lambda = 250–300$ nm. On earth, solar irradiation is unsuited to induce these reactions because its short wavelength contribution is luckily blocked by the ozone layer. The highest intensity of light in the solar spectrum is found in the blue, green, and red region at $\lambda = 400–650$ nm. To utilize this light energy for the conversion of molecules that do not absorb above 300 nm, sensitizers or photocatalysts are required. In principle, the concept of photocatalysis is well known, but less developed with respect to organic-chemical reactions: Visible light excites the sensitizer chromophore which transfers energy or an electron to the substrate to be converted. The most prominent example is the generation of reactive singlet oxygen from triplet oxygen by sensitizers such as methylene blue or porphyrins. Although the energy content of visible light is significantly lower compared to UV irradiation, it can still be sufficient to overcome activation barriers, cleave weaker bonds and generate reactive intermediates for synthesis. The energy of green light at $\lambda = 530$ nm corresponds to 222 kJ/mol, and blue light at $\lambda = 400$ nm to 300 kJ/mol.

In the last years the number of reports using visible ($\lambda = 400–800$ nm) and UV A ($\lambda = 320–400$ nm) light in chemical synthesis has dramatically increased. While one area of interest is the photocatalytic generation of dihydrogen as an energy source, other work focused on applications in synthesis mediating functional group transformations or bond formations. High power LEDs have become cheaply available as a perfect source of intense light in a defined wavelength range and make, in combination with photomicroreactor and microflow technologies, exploratory chemistry with visible light technical simple and affordable. Selectivity issues can now be addressed in a more sophisticated fashion. Catalytic cycles can be designed, which guarentee an efficient turnover and high chemoselectivity. A number of regio-, diastereo-, and even enantioselective reactions have become feasible in recent years. The idea to use long wavelength light for chemical processes is by no means new: More than 100 years ago Giacomo Ciamician promoted the use of visible light for chemical transformations as sustainable technology for the future. In the context of post Industrial Revolution with steam engines and combustion cars his ideas were not taken up widely, but now the time to consider photochemical reactions for a sustainable future has come.

We summarize in this book the chemical and physical principles of photocatalysis and discuss spectroscopic methods to determine the underlying reaction mechanisms. Recent examples of photocatalytic reactions are presented in various chapters, classified by the type of photocatalysts (e. g. heterogeneous or metal complex) or by the catalyzed reaction (photoorganoredox, arylation, oxidation, dihydrogen generation). The background information and the specific examples will give the reader an overview of the state of the art in the field and its future perspectives. The ongoing research presented in this book has been and continues to be supported by generous funding from the *Deutsche Forschungsgemeinschaft*, which established in 2010 a graduate college on Chemical Photocatalysis (GRK 1626) at the University of Regensburg (with collaborating groups in Munich and Erlangen).

Regensburg, Munich, August 2012
Burkhard König, Thorsten Bach

Peter Schroll

2 Early pioneers of organic photochemistry

"When oil will have been all burned in our prodigal industries, it may become necessary, even on social grounds, to come to exploit solar energy." [1] This foresightful statement fits perfectly into modern debates at the beginning of the twenty-first century about the necessity of using renewable energies, but is in fact more than one hundred years old, written by Giacomo Ciamician, an early pioneer of organic photochemistry. He was a visionary chemist who was invaluable in developing photochemistry, an upcoming science at the beginning of the twentieth century. Apart from his credits to discover several new photochemical reactions, he was interested in using these reactions on an industrial scale. Nevertheless, chemical transformations under the influence of light were known long before. In 1790, **Joseph Priestley (1733–1800)** exposed partially filled vials of "spirit of nitre" (nitric acid) to sunlight and observed a reddish color, which was attributed to the formation of nitrogen dioxide (Figure 2.1). This first photochemical reaction in the gas phase marks the beginning of photochemistry. [2] Moreover, Priestley is given credit for discovering basic principles of photosynthesis, the most important photochemical process for living organisms on earth. He "fully ascertained the influence of light in the production of dephlogisticated air (oxygen) in water by means of a green substance," which was identified as tiny plants. [2, 3] After **Nicholas Theodore de Saussure (1767–1845)** had shown in 1804 that the influence of light causes plants to consume water and carbon dioxide and to generate oxygen, a fundamental understanding of the process of photosynthesis in green plants had been discovered. [2, 4]

The English chemist **Sir Humphry Davy (1778–1829)** used sunlight to produce phosgene out of a mixture of chlorine and carbon dioxide in 1812. [5] Moreover,

Figure 2.1: Joseph Priestley (1733–1800), English philosopher, chemist and physicist.

Figure 2.2: Carl Theodor Liebermann (1842–1914), a pioneer of solid-state photochemistry.

during his studies of halogens and their salts he discovered the light sensitivity of silver iodine, which created the basis for the process of photography. For decades, photography was the most important application of photochemical processes.

As early as 1831, **Johann Wolfgang Döbereiner (1780–1849)** reported the light-induced reduction of metal ions by oxalic acid. [2, 6] In the case of an aqueous solution of oxalic acid and iron(III) oxide irradiated by sunlight, he obtained carbon dioxide and humboldtite, a basic iron(II) oxide. Similarly, he reduced salts of platinum, silver and iridium. In each case, the result was checked by control experiments in the dark. Unfortunately, he was not able to report the first photoreaction of a ruthenium compound, whose complexes today serve as powerful photocatalysts, since ruthenium was discovered thirteen years later in 1844.

The following decades were characterized by accidental discoveries of photochemical reactions. Several researchers contributed to the new field of photochemistry by discovering single reactions. Among the first reactions were [2+2]-cycloadditions, geometric isomerizations, light-induced halogenations, photoreductions of carbonyl compounds and photodimerizations.

Carl Julius Fritzsche (1808–1871) observed the light sensitivity of anthracene in 1867: "When a cold, saturated solution [of anthracene] is exposed to sunlight, microscopic crystals begin to precipitate." [7] However, it took another twenty-five years to recognize the photoproduct, being the dimer of anthracene, as a result of a [4+4]-cycloaddition. [8] In this case, the photoreaction was discovered due to the fact that the photoproduct was less soluble than the starting material.

In 1877, **Carl Theodor Liebermann (1842–1914)**, a German chemist working in Berlin, observed the conversion of yellow crystals of thymoquinone (2-isopropyl-5-methylbenzoquinone) producing a white photoproduct under irradiation with sunlight (Figure 2.2). This was the first example of a [2+2]-cycloaddition. [9] It is worth mentioning that he was the first to test artificial light sources since early photochemists used sunlight as the only and most important source of radiation at that

Figure 2.3: William Henry Perkin examined the first light-induced geometric isomerizations.

time. [10] Liebermann is today known as a pioneer of solid-state photochemistry due to his examinations concerning the dimerization of quinones and styrene derivatives. [11] As far as the latter are concerned, Bertram and Kürsten, two chemists working in industry, were a few months faster in publishing the light-mediated dimerization of cinnamic acid and are therefore given credit for the discovery of the dimerization of styrene derivatives. [12] However, Liebermann was the most important photochemist of the nineteenth century.

The first geometric isomerization of olefins was reported by **William Henry Perkin (1838–1907)** in 1881 (Figure 2.3). He irradiated 2-alkoxy-cinnamic acid and obtained the "β-isomer." [13] Liebermann found out that iodine accelerates the *cis-trans* isomerization of aromatic unsaturated acids such as δ-phenylpentadienoic acid. [10] **Johannes Wislicenus (1835–1902)** was able to extend the method to nonaromatic systems. Amongst others, when he irradiated maleic acid in an aqueous solution of bromine, fumaric acid was obtained. [2, 14]

Halogens also played an important role in photochemical experiments of **Julian Schramm (1852–1926)** in Lemberg. Between 1884 and 1888 he studied the light-induced bromination of alkylbenzenes and found out "that light and darkness work in the same way as elevated and low temperatures, respectively." [15] At that time it was known that the selectivity of a halogenation reaction is temperature-dependent: at low temperatures, the aromatic ring is halogenated while at elevated temperatures a substitution reaction occurs at the alkyl side chain. The new aspect of his work was the finding that the halogenation reaction can be carried out with light and that the selectivity can be controlled using sunlight irradiation or darkness. Moreover, Schramm was the first to realize the potential of this type of reaction for industrial application. Today, photochlorination is an important tool in industrial synthesis.

An impressive example of an accidental discovery of a new reaction is given by **Heinrich Klinger (1853–1945).** During his attempts to convert benzil to isobenzil in an aqueous solution of ether, he observed the unexpected formation of crystals. A close examination showed that the outcome of the reaction was not reproducible. Looking for the unknown parameter, he realized that "some of the tubes were exposed to sunlight in the morning hours." [16] The crystals were identified as a molecular complex of benzil and benzoin. Hence, in 1886, Klinger had observed the first light-mediated reduction of a keto group to the corresponding alcohol. A subsequent study of different ketones and the replacement of ether by an aldehyde led to interesting results. The intermolecular photoreaction of benzoquinone with benzaldehyde, for example, resulted in the formation of a new carbon-carbon bond in the photoproduct 2,5-dihydroxybenzophenone. [17] He observed a "synthetic effect of light" which he compared to photosynthesis in living plants. [18] This reaction is considered to be the first photoreaction for synthetic purposes.

After all these preliminary reactions had been reported and the knowledge about photochemical reactions had increased, the time was ready for a systematic study of light-driven reactions. Italy, as it turned out, would become a center of research for photochemical reactions. The development of photochemistry in Italy is closely related with the photoreaction of santonin **1** (Figure 2.6). This is the longest known photoreaction of an organic compound. Santonin was formerly used as an anthelmintic drug, but is today replaced by less toxic agents. In 1834, **Hermann Trommsdorff (1811–1884)**, a pharmacist in Erfurt (Germany), observed a color change of white santonin crystals to yellow ones followed by a crystal burst after exposure to sunlight. [19] Moreover, he examined the wavelength dependency of the reaction by means of a prism. He found out that the yellow color of the crystals appears after irradiation with direct sunlight as well as with violet and blue rays, but not with green, yellow or red beams. Looking back, this was a remarkable result for the first half of the nineteenth century, since not much was known about the nature of light at that time. Later scientists continued at this stage. Perkin and Klinger used colored solutions such as sulfate of quinine, cuprous ammonium sulfate and potassium dichromate as filters. Both researchers attributed the photochemical reactivity of their systems to the blue, violet and ultraviolet rays. Further investigations of Trommsdorff led to the result that santonin and its photoproduct were isomers. However, both structures could not be determined exactly with the methods of that time. The problem remained unsolved.

Several years later, **Fausto Alessandro Sestini (1839–1904)**, an Italian chemist working in Rome, was faced with the problem of santonin and its photoreaction (Figure 2.4). He identified the photoproduct as photosantonic acid. Sestini is also considered to be the founder of photochemistry in Italy. Cooperating with **Stanislao Cannizzaro (1826–1910)** in Rome (Figure 2.5), with Sestini introducing Cannizzaro to photochemistry, both scientists continued to work together on this topic. [20]

Figure 2.4: Fausto Alessandro Sestini, the originator of photochemistry in Italy, worked on the elucidation of the photorearrangement of santonin.

In almost twenty years of examination of the santonin photoreaction, Cannizzaro discovered an additional intermediate of the reaction which they called isophotosantonic acid. [21, 22] The final structure of photosantonic acid **5** was not described correctly until 1958, that is, 124 years after Trommsdorff first observed this photoreaction. [23, 24] Today we know that santonin **1** and photosantonic acid **5** have different crystal structures. This is the explanation for the observation that sunlight causes crystals of santonin to burst. The pathway for the rearrangement is given in Figure 2.6.

In his laboratory in Rome, Cannizzaro introduced Giacomo Ciamician and Paul Silber to photochemistry, two researchers, who had great influence in the development of this new branch of chemistry. **Giacomo Ciamician (1857–1922)** was an Italian chemist of Armenian descent who joined the Cannizzaro group in 1881 where

Figure 2.5: Stanislao Cannizzaro introduced Ciamician and Silber to photochemistry.

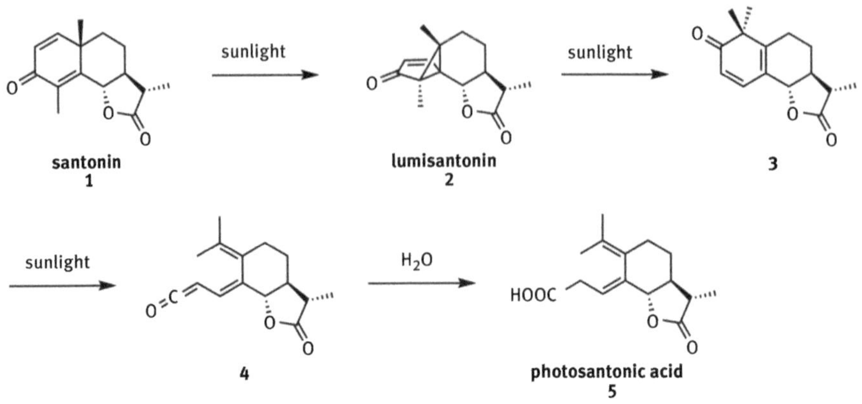

Figure 2.6: Photorearrangement of santonin to photosantonic acid.

he, too, was faced with the problem of the santonin reaction (Figure 2.7). In 1885, he started the first photochemical examinations and soon the German **Paul Silber (1851–1932)** joined his projects. [25] They discovered the photoreduction of nitrobenzene to aniline and 2-methylquinoline in an alcoholic solution and found out that a solution of benzoquinone in ethanol is converted to hydroquinone and acetaldehyde after five months of solar irradiation. [26] Although the work had been carried out independently from Klinger's photoreduction of ketones in Bonn (Germany), the latter was several months faster in publishing his results and claimed this field for his own research, which was possible in the nineteenth century. Thus, Ciamician's and Silber's first excursion in the field of organic photochemistry had come to a sudden end. They did not continue their work in photochemistry until the

Figure 2.7: Giacomo Ciamician's (left) and Paul Sailber's (right) work surpassed any previous effort in photochemistry.

year 1900. Ciamician moved to Padua in 1887 and then to Bologna in 1889 where he was accompanied by Paul Silber who became a co-worker for the next 25 years, until World War I ended their partnership. [27] The skills of both researchers, with "Ciamician as the genial and critical research director" and "Silber as the skilled and patient experimenter", being able to isolate and identify even labile photoproducts, made this collaboration "a profitable partnership". [28] Restarting in 1900, Ciamician and Silber conducted the first systematic study of the behavior of organic compounds toward light. This was the most complete study of the effect of light at that time. Ciamician was known for his enthusiastic dedication to teaching characterized by an interdisciplinary character. [29] In a lecture before the French Chemical Society in 1908, he summarized and classified the photoreactions discovered by him and other researchers. [1] Among the reported reactions were photochemical reactions of carbonyl groups such as oxidation, reduction and hydrogen transfer as well as α- and β-cleavage, reduction of nitro compounds, condensation reactions such as cycloaddition, dimerization or photopinacolization forming a new carbon-carbon bond, autoxidation, and geometric isomerization of several unsaturated compounds. [27, 30]

Ciamician's earliest photochemical experiment, the photochemistry of quinones, is a light-mediated redox reaction. An alcoholic solution of 1,4-benzoquinone was converted to hydroquinone and acetaldehyde under irradiation with sunlight. Both processes, the oxidation as well as the reduction, were useful reactions. As far as the oxidative side is concerned, a mild and selective oxidation of primary alcohols to aldehydes was found. This is particularly interesting since not too many synthetic tools are available for such a transformation. One might think of the Swern oxidation or the use of toxic chromium reagents, for example. With sunlight, the oxidation takes place under very mild conditions. Secondary alcohols were found to be oxidized to ketones, e. g. isopropanol to acetone. Furthermore, polyalcohols were oxidized to sugars, as shown by the reactions of glycerol to glyceraldehyde, erythrol **6** to erythrose **7**, and mannitol to mannose (Figure 2.8). [31–33] All these reactions benefit from the photoactivated oxidizing ability of 1,4-benzoquinone.

Taking the reducing side, not only was benzoquinone transformed to hydroquinone, but also aromatic ketones were found to undergo photopinacolization, since aromatic ketones underwent clean bimolecular reduction to pinacols in the presence of an alcohol as a reducing agent. Alloxane **8** was converted to alloxantin **9** (Figure 2.8) and aldehydes gave hydrobenzoins. These reactions were of preparative interest. The mechanistic background is the high reactivity of the triplet state of carbonyls and other $n\pi^*$ states with regard to fast hydrogen abstraction generating C-centered radicals.

In a similar way, the reaction was used for the light-driven reductive alkylation of ketones. Aromatic ketones such as benzophenone in the presence of benzylalcohol gave the alkylated product **10** (Figure 2.8). The reaction was considered to be similar to the aldol reaction.

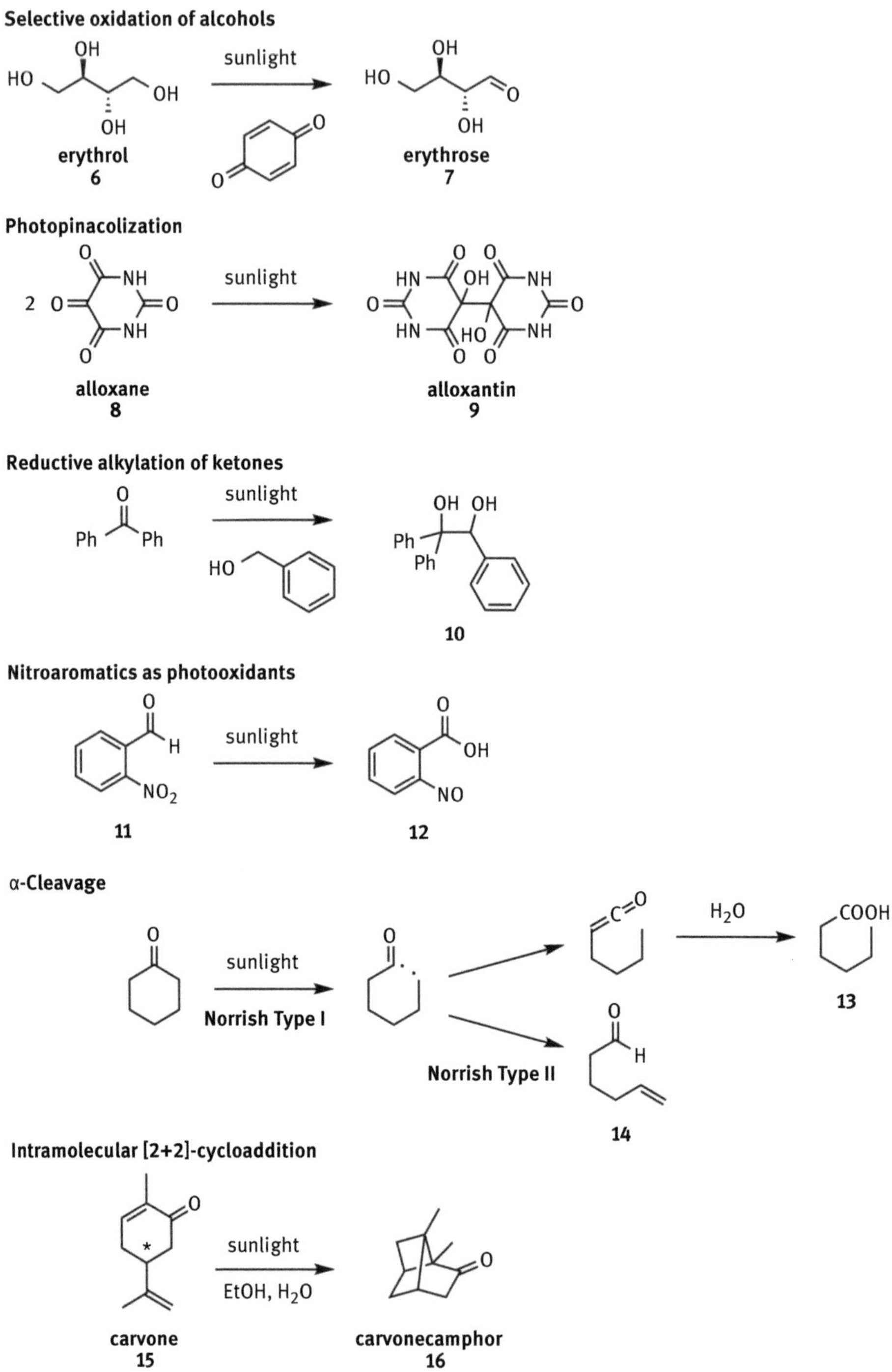

Figure 2.8: Types of photochemical reactions discovered by Ciamician and Silber.

Another group of photooxidants is that of nitroaromatics. They were reduced to the corresponding anilines upon irradiation in alcohols. [34, 35] Of particular importance was the intramolecular reaction of *ortho*-nitrobenzaldehyde **11** to give *ortho*-nitrosobenzoic acid **12** (Figure 2.8). This process was by far faster than other conversions, which were typically in the time scale of weeks or months. [36] Nowadays, intramolecular hydrogen abstraction by a nitro group is widely used in terms of the *ortho*-nitrobenzyl moiety as a photoremovable protecting group.

Besides oxidation reactions with benzoquinone or nitro compounds as oxidants, Ciamician and Silber discovered oxidation reactions involving molecular oxygen. Stilbene was converted to benzaldehyde in the presence of light and oxygen. The mechanism was explained by addition of the oxygen molecule to the double bond of stilbene forming a dioxetane intermediate followed by cleavage of the four-membered ring.

Ketones such as acetone were oxidatively cleaved giving acetic and formic acid under aerobic conditions. Furthermore, fragmentation reactions were found to occur in aqueous media. Irradiation of acetone gave acetic acid and methane – a formal hydrolysis of the carbon-carbon bond in α-position to the carbonyl group – as a result of *α*-cleavage. This reaction is today known as the Norrish Type I reaction. The name was established after detailed examinations in the gas phase by the English chemist **Ronald George Wreyford Norrish (1897–1978)** in the 1930s. [37, 38] Cyclic ketones were noticed to give two different products: open-chain saturated acids and unsaturated aldehydes. For example, the cleavage of cyclohexanone yielded hexanoic acid **13** and ω-hexenal **14**. The latter product was obtained by a Norrish Type II reaction, as it was called later (Figure 2.8). The identification of all products must be attributed to the extraordinary experimental skills of the scientists since it was very difficult to characterize aliphatic aldehydes at that time.

Apart from photochemical reactions including carbonyl compounds as reactants, Ciamician and Silber also used the reactivity of olefins. Geometric isomerizations were observed confirming the work of Liebermann and Wislicenus. Additionally, Ciamician and Silber studied the *syn-anti* isomerization of oximes and the isomerization of cyanoazo derivatives. [30] Reactions forming new bonds were summarized as "polymerization and condensation reactions". Cinnamic acid was found to undergo dimerization to truxilic acid in the solid state, confirming previous work of Bertram and Kürsten. The dimerization of stilbene in solution was reported after two years of irradiation and a coumarin dimer was found whose *syn* head-to-head structure was finally proven in 1962. [39] Besides such intermolecular [2+2]-cycloadditions, the first intramolecular reaction was found in case of the monoterpene ketone carvone **15**, which gave an isomeric product in a clean reaction, whose structure could not be determined exactly by Ciamician. He proposed the tricyclic product carvonecamphor **16** to be formed. The proposal was found to be correct in 1957. [40] A similar reaction including the [2+2]-cycloaddition of an olefin and a carbonyl group giving oxetanes was discovered by another important Italian pioneer of photochemistry, **Emanuele Paternò (1847–1935)**, in Rome in 1909. [41] The reaction was improved in 1954 by the Swiss-American chemist **George Hermann Büchi (1921–1998)** and is today known as the Paternò–Büchi reaction. [42]

The value of Ciamician's and Silber's work is not only described by the numerous new photoreactions they had discovered, but also by Ciamician's attempt to establish photochemistry as a science of its own by summarizing all the work known so far and by drawing a visionary picture about future applications of light-driven reactions in organic synthesis as well as in industry to produce solar fuels. He always emphasized the interdisciplinary character of his work and found parallels with physics and biology. Ciamician was looking for new synthetic methods and he was convinced that light was a novel reagent for synthesis under very mild conditions. He was somehow dissatisfied with the fact that many chemical reactions needed very aggressive reagents. In contrast, he realized that green plants were able to effectively use light. "For plants, light is the source of energy. Through the intervention of chlorophyll, green plants accumulate solar energy and transform it into chemical energy ... Chemistry in the laboratory differs from chemistry of organisms not in the materials, but in the reagents used. It is thus apparent that the further advancement of biology requires that all of the compounds present in nature can be produced by using only reagents present in nature, rather than agents that do not belong to the living world." [1, 27] Photochemistry was considered as a "better" chemistry, inspired by the mild reactions in green plants. According to Ciamician, molecular photochemistry could be a method to imitate the mild conditions of biochemical reactions. An understanding of chemical reactions in organisms was seen to be of high importance for the progress of man-made chemistry, which he encouraged should be biomimetic chemistry. [29] Throughout his studies of photochemical reactions, he found several examples where light served as an agent to achieve conversions under milder conditions. One example is the above-mentioned selective oxidation of alcohols to aldehydes by benzoquinone. This photoinduced oxidation was seen as a substitute for strong agents like bromine or nitric acid. The cross-coupling reaction of an aliphatic ketone with an aliphatic alcohol to give the corresponding pinacol was considered to be similar to the aldol reaction with the difference that no base was required. The oxidative cleavage of stilbene by means of light was found to be able to replace ozone. And the mild cleavage of ketones could avoid harsh conditions such as cleavage with permanganate. In a way, this kind of chemistry refers to what is today called "green" or "sustainable chemistry," although the terms were proposed later.

Ciamician realized that fossil fuels were not inexhaustible and stood up for the idea that industry should not wait for a possible shortage of fossil fuels, but begin immediately to take advantage of photochemistry. Photochemical processes for the production of basic chemicals like sulfur trioxide or ammonia on an industrial scale were proposed. One should not forget that these were times where coal was the backbone of the industry. Furthermore, for the first time in history, man-made materials began to get predominant over natural ones. Artificial products were easily available by synthetic methods. People were no longer dependent on laborious isolation from natural sources. There was a strong belief in technical progress. That is why his request to exploit solar energy as the only lasting form of energy came too early. The visionary picture of "The Photochemistry of the Future" from 1912 [43] did not find its breakthrough in industry, but was shared by a few other scientists. The German chemist **Hans Stobbe (1860–1938)** for example, who is best known for his pioneering studies on the photochromism of fulgides, was also campaigning for a bigger role for photochemistry. [44, 45]

The beginning of the twentieth century remains as a time of great progress for photochemistry since many types of photoreactions were already known. Besides Ciamician's summary of photochemical reactions, several books appeared. In 1912, **Alfred Benrath (1871–1969)** from Königsberg (Germany) published a "Lehrbuch der Photochemie" (Textbook of Photochemistry). [46] **Ivan Plotnikov (1878–1955)**, a Russian photochemist, wrote several books, for example, "Photochemische Versuchstechnik" (1911), "Grundriss der Photochemie", and "Lehrbuch der allgemeinen Photochemie." [47] He gave a summary of the development of photochemistry where he noticed a "pre-scientific period up to about 1850" and a "second period of lively advancement" mainly combined with the development of photography. According to him, no earlier than at the beginning of the twentieth century had photochemistry began its full development. [30]

World War I marked a break in many respects. A number of research activities had come to a sudden end. Later on, many researchers continued in the field of organic photochemistry. Among them should be named only a few: **Günther Otto Schenck (1913–2003)** did not only discover the Schenck reaction, which is a photosensitized diastereoselective ene reaction of alkenes with singlet molecular dioxygen that affords allylic hydroperoxides, [48] but is also known for his work on the theory of photosensitization and chemical engineering in the field of solar photochemistry. [49] In 1944, he used chlorophyll from spinach leaves as photosensitizers for the selective oxidation of α-terpinene to produce ascaridol via a cycloaddition of oxygen to the diene (Figure 2.9). [50] This drug was urgently needed at that time as an anthelmintic to fight ascaris infections in humans. To address this, Schenck constructed an open-air solar irradiation pilot plant for the methylene blue-sensitized production of ascaridol. Later, he became involved in the technical development of water disinfection by UV irradiation. [49]

The area of solid-state photochemistry that Liebermann had opened was enlarged by **Gerhard Martin Julius Schmidt (1919–1971)**, a German scientist working at the Weizmann Institute in Israel who was familiar with the fields of organic chemistry, crystallography, and spectroscopy (Figure 2.10). This constellation of knowledge led to the discovery of "topochemistry." The term "topochemistry" means that solid-state reactions are determined by the geometry of the reactant

Figure 2.9: Günther Otto Schenck and his pilot plant for the production of ascaridol in his garden in Heidelberg (1949).

lattice or, in other words, that reactions in the solid state occur with a minimum amount of atomic or molecular movement. [51–53]

Another scientist of international significance is the German **Alexander Schönberg (1892–1985)**. [54] He emigrated to Egypt in 1937 due to unsafe political circumstances in Germany (Figure 2.11). In Egypt, his work on solar photochemistry was favored by reason of an annual availability of 3500 hours of sunshine. [55] Besides many photochemical reactions he described the photochemical epoxidation of a carbonyl group in methanol [56] and the addition of olefins to *ortho*-quinones to give a six-membered ring, which is today known as "Schönberg addition." [57] But

Figure 2.10: Gerhard M. J. Schmidt, a pioneer of solid-state photochemistry.

Figure 2.11: Alexander Schönberg developed photochemistry in Egypt.

he is most famous for his book “Preparative Organic Photochemistry”. [58] This book is the last portrait of photochemistry as it was before the theory of electronically excited states came up. [30] It uses an order different from that of today’s textbooks. Nowadays, textbooks show a theoretically based classification. It is an interesting fact that many photochemical reactions were discovered although the underlying physical principles of all these processes were unexplored. Early photochemists did not know about the nature of light and its effect on organic molecules to create electronically excited states. The development of physical chemistry and spectroscopy, particularly time-resolved spectroscopy, made concepts like quantization of energy, electronically excited singlet and triplet states and deactivation of excited states via electron transfer available, which could explain the mechanisms of many photoreactions. [59, 60]

All pioneers of photochemistry have created the basis for today’s research concerning light-mediated chemical transformations. Today, the need of an improved disposal of solar energy is reflected in the discussion about “green chemistry” and the implementation of renewable energies. Maybe someday Ciamician’s hundred-year-old vision of the exploitation of solar energy will come true: “And if in a distant future the supply of coal will become completely exhausted, civilization will not be checked by that, for life and civilization will continue as long as the sun shines.” [43]

2.1 References

[1] G. Ciamician, (1908). Sur Les Actions Chimique De La Lumière. Bull. Soc. Chim. Fr. [4] 3, i.
[2] H. D. Roth, (1989). The Beginnings of Organic Photochemistry. Angew. Chem. Int. Ed. 28 (9), 1193–1207.

[3] J. Priestley, (1790). Experiments and Observations on Different Kinds of Air, Birmingham.
[4] A. J. Ihde, (1964). The Development of Modern Chemistry, New York.
[5] H. Remane, W. Girnus, (2012). Meilensteine der Chemie 2012. Nachrichten aus der Chemie 60 (1), 11–21.
[6] J. F. Döbereiner, S. J., (1831). Pharm. Centralbl. 2, 383–385.
[7] Fritzsche, (1867). Über die festen Kohlenwasserstoffe des Steinkohlentheers. J. Prakt. Chemie 101 (1), 333–343.
[8] K. Elbs, (1891). Über Paranthracen. J. Prakt. Chemie 44 (1), 467–469.
[9] C. Liebermann, (1877). Über Polythymochinon. Ber. Dtsch. Chem. Ges. 10 (2), 2177–2179.
[10] C. Liebermann, (1895). Über die Umlagerung der Allofurfurakrylsäure und Allocinnamylidenessigsäure im Sonnenlicht. Ber. Dtsch. Chem. Ges. 28 (2), 1443–1448.
[11] C. Liebermann, (1895). Über Cinnamylidenmalonsäure und die stereoisomeren Cinnamylidenessigsäuren (Phenylpentadiensäuren). Ber. Dtsch. Chem. Ges. 28 (2), 1438–1443.
[12] J. Bertram, R. Kürsten, (1895). Über das Vorkommen des Orthocumaraldehyd-methyläthers im Cassiaöl. J. Prakt. Chemie 51 (1), 316–325.
[13] W. H. Perkin, (1881). On the isomeric acids obtained from coumarin and the ethers of hydride of salicyl. J. Chem. Soc., Transactions 39, 409–452.
[14] J. Wislicenus, (1895). Ber. Verh. K. Saechs. Ges. Wiss. Leipzig, Math. Phys. Kl., 489–493.
[15] J. Schramm, (1888). Über den Einfluss des Lichtes auf den Verlauf chemischer Reactionen bei der Einwirkung der Halogene auf aromatische Verbindungen. Monatshefte für Chemie 9 (1), 842–853.
[16] H. Klinger, (1886). Über das Isobenzil und die Einwirkung des Sonnenlichts auf einige organische Substanzen. Ber. Dtsch. Chem. Ges. 19 (2), 1862–1870.
[17] H. Klinger, O. Standke, (1891). Über die Einwirkung des Sonnenlichts auf organische Verbindungen. Ber. Dtsch. Chem. Ges. 24 (1), 1340–1346.
[18] H. Klinger, (1888). Über die Einwirkung des Sonnenlichts auf organische Verbindungen. Justus Liebigs Annalen der Chemie 249 (1), 137–146.
[19] H. Trommsdorff, (1834). Über Santonin. Ann. Chem. Pharm. 11 (2), 190–207.
[20] S. Cannizzaro, S. Fausto, (1873). Gazz. Chim. Ital. 3, 241–251.
[21] S. Cannizzaro, (1876). Gazz. Chim. Ital. 6, 355–356.
[22] S. Cannizzaro, G. Fabris, (1886). Über eine neue vom Santonin abstammende Säure. (Isophotosantonsäure.). Ber. Dtsch. Chem. Ges. 19 (2), 2260–2265.
[23] E. E. van Tamelen, S. H. Levin, G. Brenner, J. Wolinsky, P. Aldrich, (1958). The Structure Of Photosantonic Acid. J. Am. Chem. Soc. 80 (2), 501–502.
[24] E. E. van Tamelen, S. H. Levin, G. Brenner, J. Wolinsky, P. E. Aldrich, (1959). The Structure of Photosantonic Acid1. J. Am. Chem. Soc. 81 (7), 1666–1678.
[25] G. Ciamician, P. Silber, (1886). Über die Einwirkung des Lichtes auf eine alkoholische Nitrobenzollösung. Ber. Dtsch. Chem. Ges. 19 (2), 2899–2900.
[26] G. Ciamician, (1886). Gazz. Chim. Ital. 16, 111–112.
[27] A. Albini, M. Fagnoni, (2004). Green chemistry and photochemistry were born at the same time. Green Chem. 6 (1), 1.
[28] N. D. Heindel, M. A. Pfau, (1965). A profitable partnership: Giacomo Ciamician and Paul Silber. J. Chem. Educ. 42 (7), 383.
[29] A. Albini, M. Fagnoni, (2008). 1908: Giacomo Ciamician and the concept of green chemistry. ChemSusChem 1 (1–2), 63–66.
[30] A. Albini, V. Dichiarante, (2009). The "belle époque" of photochemistry. Photochemical & Photobiological Sciences 8 (2), 248.
[31] G. Ciamician, P. Silber, (1900). Chemische Lichtwirkungen. Ber. Dtsch. Chem. Ges. 33, 2911.
[32] G. Ciamician, P. Silber, (1901). Chemische Lichtwirkungen. Ber. Dtsch. Chem. Ges. 34, 1530.

[33] G. Ciamician, P. Silber, (1911). Chemische Lichtwirkungen. Ber. Dtsch. Chem. Ges. 44, 1280.
[34] G. Ciamician, P. Silber, (1905). Chemische Lichtwirkungen. Ber. Dtsch. Chem. Ges. 38, 3813.
[35] G. Ciamician, P. Silber, (1906). Chemische Lichtwirkungen. Ber. Dtsch. Chem. Ges. 39, 4343.
[36] G. Ciamician, P. Silber, (1901). Chemische Lichtwirkungen. Ber. Dtsch. Chem. Ges. 34, 2040.
[37] R. G. W. Norrish, F. W. Kirkbride, (1932). Primary photochemical processes. Part I. The decomposition of formaldehyde. J. Chem. Soc. (Resumed), 1518–1530.
[38] R. G. W. Norrish, C. H. Bamford, (1937). Photo-decomposition of Aldehydes and Ketones. Nature 140 (3535), 195–196.
[39] R. Anet, (1962). The Photodimers of Coumarin and Related Compounds. Canadian Journal of Chemistry 40 (7), 1249–1257.
[40] G. Büchi, I. M. Goldman, (1957). The Intramolecular Cyclization of Carvone to Carvonecamphor. J. Am. Chem. Soc. 79 (17), 4741–4748.
[41] E. Paternò, G. Chieffi, (1909). Gazz. Chim. Ital. 39(I), 341.
[42] G. Büchi, C. G. Inman, E. S. Lipinsky, (1954). The Reaction of Carbonyl Compounds with 2-Methyl-2-butene in the Presence of Ultraviolet Light. J. Am. Chem. Soc. 76 (17), 4327–4331.
[43] G. Ciamician, (1912). The Photochemistry of the Future. Science 36 (926), 385–394.
[44] G. Reddelien, (1930). Hans Stobbe zum 70. Geburtstag. Angew. Chem. 43 (26), 599–600.
[45] H. Stobbe, (1908). Die Photochemie organischer Verbindungen. Zeitsch. Elektroch. 33, 173.
[46] A. Benrath, (1912). Lehrbuch der Photochemie, Heidelberg.
[47] G. S. Whitby, (1938). Prof. J. Plotnikow. Nature 142 (3605), 987–987.
[48] G. O. Schenck, (1955). Vol. Deutsches Patent Nr. DE-933925.
[49] K. Schaffner, (2003). Günther Otto Schenck (1913–2003): A Pioneer of Radiation Chemistry. Angew. Chem. Int. Ed. 42 (26), 2932–2933.
[50] G. O. Schenck, K. Ziegler, (1944). Die Synthese des Ascaridols. Naturwiss. 32, 157.
[51] D. Ginsburg, (1972). Gerhard M. J. Schmidt. Israel Journal of Chemistry 10, 59–72.
[52] G. Lehmann, (1977). Angew. Chem. 1 (89), 64.
[53] (1971). Wer ist's? G. M. J. Schmidt. Nachr. Chem. Techn. 23 (19), 435–436.
[54] E. Singer, (1987). Alexander Schönberg. Chem. Ber. (120), I–XIX.
[55] A. A. Nada, (1983). Professor A. Schönberg and the history of photochemistry in Egypt. J. Chem. Ed. 60 (6), 451.
[56] A. Schönberg, E. Singer, P. Eckert, (1980). Über die photochemische Epoxidierung einer Carbonylgruppe mit Methanol. Chem. Ber. 113 (9), 3094–3097.
[57] A. Schönberg, A. Mustafa, (1944). Reaction of ethylenes with phenanthraquinone. J. Chem. Soc. (Resumed), 387–387.
[58] W. Schroth, (1968). Preparative Organic Photochemistry Von A. Schönberg. 2., völlig überarb. Aufl. in Zusammenarbeit mit G. O. Schenck und O.-A. Neumüller., Vol. 9, Wiley-VCH Verlag GmbH & Co. KGaA, Springer-Verlag, Berlin, Heidelberg, New York.
[59] G. S. Hammond, N. J. Turro, (1963). Organic Photochemistry. Science 142, 1541–1553.
[60] H. D. Roth, (2001). Twentieth century developments in photochemistry. Brief historical sketches. Pure Appl. Chem. 73 (3), 395–403.

Bernhard Dick

3 Photophysics of Photocatalysts

A comprehensive description of the principles of the photophysics and photochemistry of organic compounds requires a whole book, and is hence far beyond the scope of this chapter. There exist excellent textbooks on this subject, for example the recent book by Turro, Ramamurthy, and Scaiano [1]. Here we can only provide a short summary of some aspects of the photophysics of organic molecules that are of particular interest in the context of photocatalytic applications.

A photochemical reaction is a chemical reaction that is initialized by the absorption of light and involves an electronically excited state in at least one reaction step. In the simplest case, the reactant A is promoted to one of its electronically excited states A* by absorption of a photon. The chemical properties of A* can be quite different from those of the electronic ground state A, i. e. A* can be regarded as a different chemical species. In many cases A* will undergo a unimolecular decomposition or rearrangement. Examples are the elimination of molecular nitrogen from azocompounds or diazonium salts, or *cis*/*trans* isomerization of double bonds. However, when A* encounters another molecule B while still in the excited state, a bimolecular reaction is also possible. A common example is hydrogen abstraction by excited ketones. Figure 3.1 presents a scheme of these two reaction types.

The energy diagram in Figure 3.1c illustrates the advantages of a photochemical reaction. Through electronic excitation, the reactant A is promoted in energy by a substantial amount. If the reaction from A to the product P involves a high barrier or is endothermic, the reaction from A* to this product P might be exothermic with almost no barrier. In order to achieve this in a thermal reaction, a very high temperature would be required. The photochemical synthesis strategy is, however, limited to reactants that are at the same time also good light absorbers.

The basic concept of photocatalysis is the separation of the light-absorbing step from the product-forming reaction. Figure 3.2 presents a schematic picture of the photocatalytic reaction.

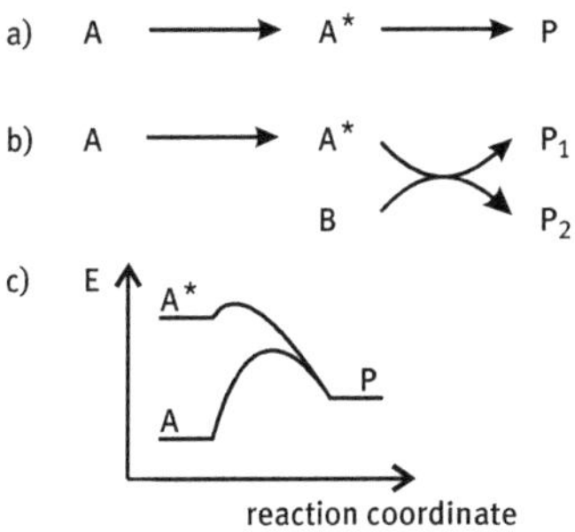

Figure 3.1: Reaction scheme for a unimolecular (a) and bimolecular (b) photochemical reaction. (c) Schematic reaction energy profile for a reaction from ground state A and electronically excited A* to a product P.

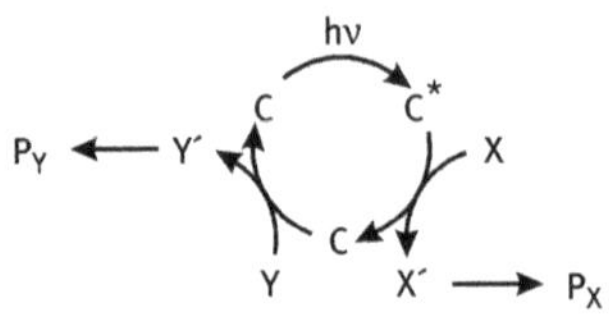

Figure 3.2: Reaction scheme for a photocatalytic reaction.

The catalyst C contains the chromophore that absorbs the light and is promoted to an excited state C*. Reaction of C* with a compound X will lead to two reaction products, one being a modified catalyst C′. If this primary reaction is transfer of the excitation energy from C* to X, this will bring the catalyst directly back to the ground state, i. e. C′ = C and form the excited state X*. This process is commonly known as sensitization. Here we consider a true chemical reaction step, which is in most cases electron transfer or proton transfer. In order to close the catalytic cycle, C must then be recovered by reaction of C′ with another compound Y. The reaction products X′ and Y′ are often reactive intermediates that undergo subsequent reactions in the electronic ground state before they end in stable products. Usually, the desired product is P_X, i. e. the product resulting from the bimolecular photochemical reaction of C* with X. Y then should be a cheap compound that is lost by the reaction. In the ideal case, both P_X and P_Y are useful products. In the overwhelming majority of reactions studied so far in the research training group, the two reaction steps $C^* + X \rightarrow C' + X'$ and $C' + Y \rightarrow C + Y'$ are one-electron transfer reactions. When C is electrically neutral, C′ is then either the radical anion $C^{-\bullet}$ or the radical cation $C^{+\bullet}$. In this chapter we shall describe the photophysical and photochemical principles that govern the two reaction steps $C \rightarrow C^*$ and $C^* \rightarrow C'$.

This chapter has been organized into three sections. The first reviews the fundamental concepts that allow us to think of molecules in terms of a structure, i. e. nuclei arranged at well-defined relative positions that can be defined by bond lengths and angles. The key concept here is the Born–Oppenheimer (BO) approximation. This also allows us to consider the total energy of a molecule as a sum of electronic, vibrational, and rotational energy. Thus, we arrive at the concept of electronic potential energy surfaces each accommodating a manifold of vibrational states. The second section describes those processes that are most relevant for photochemical reactions and photocatalysis. This is done from an experimentalist's point of view with a minimum of theoretical language. The key element here is the famous Jablonsky diagram, which is first discussed for the isolated catalyst, and subsequently for the interacting pair of catalyst and reactant. The third section presents a few considerations from a theoretician's point of view. These provide to the interested reader a more detailed picture of the consequences of the Born–Oppenheimer approximation in particular for transitions between vibronic states. It should also help in avoiding some pitfalls that can result from a too simplified quasi-classical treatment of quantum mechanical concepts.

3.1 Setting the Frame

A molecule consists of nuclei and electrons, and the rules of quantum mechanics apply to both classes of particles. We are used to considering electrons as being delocalized, as an electron cloud, around the nuclei. But the same delocalization then should apply to the nuclei. Chemists (and some physicists) will consider this idea disturbing, but it is true: the wavefunction for any molecule in its lowest rotational state is always completely spherically symmetric. Conversely, we rightly consider molecules as systems with a defined structure, i. e. well-defined bond lengths, bond angles, and so on. How can the idea of the wave-nature of the nuclei be reconciled with the notion of molecular structure? The key to this understanding is the Born–Oppenheimer approximation. Since this idea is so central to our ability to imagine molecular structure, it sets the frame for all discussions of (photo)chemical reactions, and is hence briefly summarized here.

The complete information of a molecule in a particular stationary state is contained in the corresponding wavefunction $\Theta(r, R)$ which is a function of the coordinates r of all electrons and the coordinates R of all nuclei. This wavefunction is an eigenfunction of the total molecular Hamiltonian

$$(T_n + T_e + V_{ee} + V_{en} + V_{nn})\Theta(r, R) = E\Theta(r, R). \tag{3.1}$$

In this equation, T and V refer to the operators of kinetic energy and potential energy, respectively. The indices e and n indicate electrons and nuclei. The operators T_e and T_n involve second derivatives with respect to the electron coordinates and the nuclear coordinates, respectively. The potential terms are the Coulomb repulsion between the electrons (V_{ee}) and the nuclei (V_{nn}), and the Coulomb attraction between electrons and nuclei (V_{en}). Even for the simplest case of two nuclei and a single electron Equation 3.1 can not be solved analytically. An often successful strategy to solve differential equations in several coordinates is to try separation of the variables. In the present case one would try to separate the nuclei from the electrons. The ansatz proposed by Born and Oppenheimer is

$$\Theta(r, R) = \Psi(R; r)\,\Phi(R), \tag{3.2}$$

where $\Psi(R; r)$ is the electronic wavefunction, and $\Phi(R)$ is a nuclear wavefunction. Note that the electronic wavefunction is a function of the electron coordinates, but also depends on the positions of the nuclei. This expresses the assumption that the electrons move much faster than the nuclei. That is the electronic wavefunction will instantaneously re-adjust when the nuclei change their positions. The electronic Schrödinger equation

$$(T_e + V_{ee} + V_{en})\Psi_a(R; r) = E_a(R)\Psi_a(R; r) \tag{3.3}$$

is then solved for a fixed arrangement R of the nuclei. This equation has many solutions, which are labeled by the index a. The solution with the lowest energy is called the electronic ground state, the others are electronically excited states. The corresponding energy eigenvalue $E_a(R)$ depends on the nuclear arrangement, and the set of all eigenvalues for the same electronic state for all nuclear arrangements form a hypersurface in configuration space.

In the second step of the Born–Oppenheimer approximation, the operators in the original Schrödinger equation involving electronic coordinates are replaced by their expectation value integrated over the electronic wavefunction. The equation involves only nuclear coordinates and is solved for the nuclear wavefunction $\Phi(R)$

$$(T_n + E_a(R) + V_{nn})\Phi_{av}(R) = E_{av}\Phi_{av}(R). \tag{3.4}$$

In this equation the sum of $E_a(R) + V_{nn}$ acts as an effective potential $V_a(R)$ for the movement of the nuclei. This is called the potential energy surface (PES) of the corresponding electronic state *a*. There is a different vibrational Schrödinger equation for each electronic state, and each equation has many solutions labeled with a vibrational quantum number *v*. The wavefunctions of the lowest total energies E_{av} are localized in the minima of the PES V_a. Hence, each minimum of the PES corresponds to a molecular structure.

Figure 3.3 presents a schematic sketch of two potential energy curves (PEC). A full PES for a molecule with N atoms is a function of $3N-6$ coordinates, $V_a(R_1, R_2, \cdots, R_{3N-6})$. These coordinates can be Cartesian coordinates after projection of those linear combinations that relate to the movement of the center of mass and to overall rotation of the molecule. However, internal coordinates (bond lengths and angles) or normal coordinates of vibrations or linear combinations thereof can also be used. The space defined by all possible values of these coordinates is called the configuration space of the molecule. A PEC is a plot of a PES along a single coordinate. Although this is – except for a diatomic molecule – a dramatic reduction in the dimen-

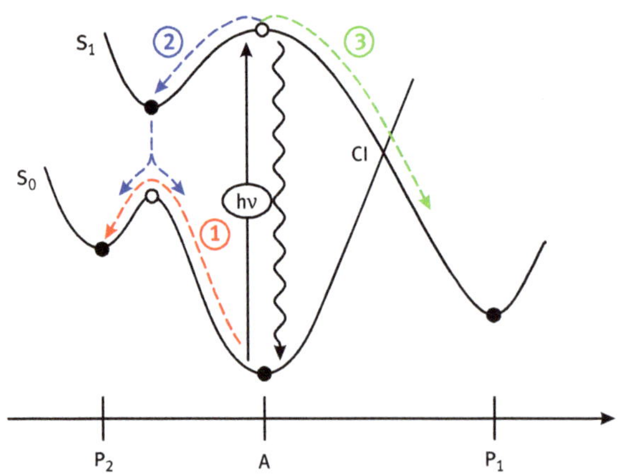

Figure 3.3: Potential energy curves of an electronic ground state and the first electronically excited state. M indicates minima, TS transition states, and CI a conical intersection.

sionality of the system, it is in many cases sufficient for the discussion of thermal reactions in the electronic ground state. This is due to the fact that all stable structures of the molecule are minima (i. e. zero-dimensional objects) in configuration space. At a minimum, the gradient of the energy with respect to all coordinates vanishes, and the matrix of the second derivatives (the Hessian matrix) has only positive eigenvalues. The square roots of these are the vibrational frequencies. The most likely path from one minimum to another minimum passes through a saddle point, which is like the pass between two mountains: it is that path which requires the smallest ascent in energy. Chemists refer to such a saddle point as a transition state. Mathematically it is characterized by a vanishing gradient, and a Hessian matrix with exactly one negative eigenvalue. In summary, chemical reactions in the electronic ground state can be represented by a one-dimensional path on the ground state PES which passes through three points, from one minimum over the saddle point to another minimum.

In contrast to this, photochemical reactions proceed on two PESs. Since electronic excitation might drastically change the bonding properties, the minima on the excited state PES are usually at quite different geometries than those of the electronic ground state. At the end of a photochemical reaction the molecule has to be back in the electronic ground state. It is plausible to assume (and supported by more sophisticated theoretical considerations) that the change from one PES to the other is more likely to occur at positions where the two PESs are close together. It is thus tempting to draw a reaction path like in the left part of Figure 3.3: the system propagates on the upper PES until it reaches a minimum on that PES, then jumps to the lower PES, and proceeds from there as it would in a ground state reaction. If this is the case, the minimum in the excited state is equivalent to another transition state, and the whole reaction can be described by a one-dimensional path in configuration space. However, in polyatomic molecules the PESs of the ground and excited electronic states may cross. In contrast to minima and transition states, these crossings do not occur in a single point but form a subspace in configuration space of dimension $3N-8$. That is only along two coordinates is the degeneracy lifted (the so-called branching space), whereas along the other $3N-8$ coordinates the degeneracy of the two electronic states is preserved. Along the two coordinates of the branching space the energies of both states increase linearly with the distance from the crossing, giving the PESs the shape of two cones that touch at their tips (Figure 3.4), hence the name "conical intersection" for this situation.

When a photochemical reaction proceeds via such a conical intersection, a single coordinate is not sufficient to describe the reaction. Instead, both coordinates of the branching space must be considered. As a consequence, two or more products might be formed by the same reaction. This is schematically shown in Figure 3.4: after excitation of a compound A to the upper electronic state, the system will evolve to the conical intersection. It can leave the degeneracy along any linear combination of the two coordinates (indicated as q_1 and q_2) leading for example to two different

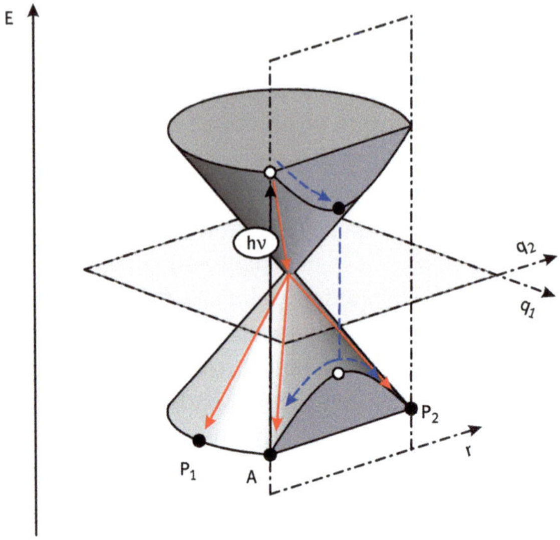

Figure 3.4: Potential energy surfaces near a conical intersection, shown as a function of the coordinates of the branching space. A projection along only one coordinate, as indicated by the vertical plane, would (erroneously) indicate an avoided crossing.

products P_1 and P_2 or relax back to A. Considering only one coordinate is equivalent to considering only a slice cut off the conical intersection, as schematically indicated by the vertical plane in Figure 3.4. This would lead to the erroneous conclusion that the reaction proceeds to a minimum on the excited state PES and then jumps to the ground state A or product P_1 only. This one-dimensional picture corresponds to the perspective of the Woodward–Hoffmann rules. The more comprehensive picture involving the conical intersection explains why the Woodward–Hoffmann rules are often less strict for excited state reactions than for ground state reactions.

3.2 The Experimentalist's Perspective

We shall now focus on those properties of electronically excited states that are most relevant for the application of a compound as a photocatalyst. A photocatalyst should absorb light and use this energy for the chemical reaction, but should emerge from the reaction sequence unchanged. Hence it must be an efficient light absorber, and it must keep the energy long enough to perform the reaction with the substrate molecule. This reaction should lead to only a small modification of its structure, so that the original state can be easily recovered by a second reaction on the electronic ground state. As a consequence, the photocatalyst should be photo-

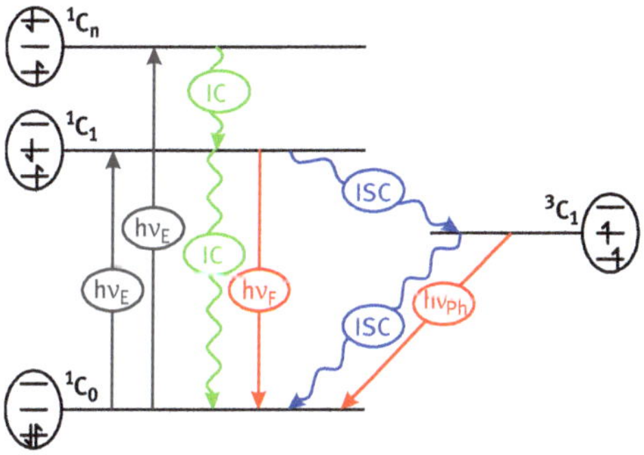

Figure 3.5: Jablonsky diagram for the photocatalyst.

stable in the absence of the substrate. Therefore, we first discuss the photophysics of the photocatalyst without the substrate, and then turn to the photochemistry of the photocatalyst–substrate pair.

The major photophysical processes can be illustrated by the popular Jablonsky diagram shown in Figure 3.5. In this diagram every electronic state is represented by a single energy, i. e. any dependence of this energy on nuclear coordinates is neglected. This is a valid approximation when the geometries of the minima in the various PESs are close to each other in configuration space. In the Jablonsky diagram states of different spin multiplicity are arranged in different columns. It is common practice to label singlet and triplet states with the letters S and T, respectively, and count the states with increasing energy by an index attached to the multiplicity symbol. A closed-shell ground state is thus labeled S_0, the excited singlet and triplet states are labeled S_n and T_n with $n = 1,2,3,\cdots$. We use a different notation ^MX_n here, where the symbol X represents the molecule, M is the spin multiplicity, and n is the state count. Thus, if we abbreviate the catalyst with the letter C, then $^1C_0$ is the singlet ground state of the catalyst, $^1C_2$ is the second excited singlet state, and $^3C_1$ the lowest triplet state. This nomenclature allows extension, for example $^2C_1^+$ and $^2C_1^-$ for the ground states of the radical cation and anion of the catalyst, or $^3R_1$ for the lowest triplet state of a reactant.

Transitions between different states in the Jablonsky diagram are either induced by absorption or emission of radiation (radiative transitions), or occur radiationless. The key quantity for radiative transitions is the transition dipole moment μ_{ab} between the two electronic states. Radiationless transitions can be described as the consequence of those terms in the Hamiltonian that are neglected in the Born–Oppenheimer approximation. As discussed in the last (i. e. theoretical) section of this chapter, in a first-order approximation both types of transitions are only allowed between states of the same spin multiplicity. The spin-forbidden transitions between singlet and triplet states occur only in a higher-order treatment.

We shall first discuss three types of radiative transitions shown in the Jablonsky diagram as curly lines:

- **Absorption:** Absorption of light will excite the photocatalyst from the ground state $^1C_0$ to any of the excited singlet states ^1C_n. The photon energy $h\nu$ must be large enough to provide the excitation energy for the first excited singlet state. Since relaxation from higher excited singlet states to $^1C_1$ is very fast, the energy of $^1C_1$ is the maximum energy available for the photoreaction. Chemists characterize the strength of the absorption usually by the molar extinction coefficient ϵ (units $1\,\text{Mol}^{-1}\text{cm}^{-1}$), physicists are more familiar with the absorption cross-section σ (units cm^2). These quantities are related by

$$\sigma = \epsilon \frac{\ln 10}{N_A} \times \frac{1000\ \text{cm}^3}{1}, \tag{3.5}$$

where N_A is Avogadro's constant. The cross section is related to the transition dipole moment by

$$\mu_{ab}^2 = \frac{3\hbar cn}{4\pi^2} \int \sigma(\nu) \frac{d\nu}{\nu}, \tag{3.6}$$

where c is the speed of light, n is the refractive index, and μ is in electrostatic units. A large value of the extinction coefficient results in strong light absorption according to the law of Lambert and Beer:

$$\frac{I(l)}{I(0)} = \exp(-\sigma N l) = 10^{-\epsilon\, cl}. \tag{3.7}$$

In this expression, c is the concentration in $\text{Mol}\,\text{l}^{-1}$, and N is the corresponding particle density in cm^{-3}. In both formulas the path length l is measured in cm. The extinction $E = \epsilon\, cl$ is also called the optical density. Optical densities of 1 or 2 correspond to 90 and 99 % light absorption, respectively. If the concentration of the photocatalyst is high, the sample will have a large optical density already for a small path length l. The average penetration depth of the light is

$$\langle l \rangle = \frac{\int_0^\infty l\, I(l)\, dl}{\int_0^\infty I(l)\, dl} = (\sigma N)^{-1} = (\epsilon\, c \ln 10)^{-1}. \tag{3.8}$$

For example, a concentration of 1 mM of a photocatalyst with an extinction coefficient of $10^{-4}\,\text{Mol}^{-1}\text{cm}^{-1}$ has an effective penetration depth of the light of $\langle l \rangle = 0.04$ cm only. Under these conditions only a thin film of ca. 400 µm thickness of the sample will contribute to the photoreaction. Obviously, the concentration of the photocatalyst must be carefully chosen in order to allow a substantial fraction of the reaction mixture to contribute.

- **Fluorescence:** Spin-allowed light emission from an excited singlet state is called fluorescence. The rate constant for this process is related to the extinction coefficient by the relations between the Einstein coefficients A and B for spontaneous emission and absorption, respectively,

$$k_r = \frac{1}{\tau_r} = A = \frac{8\pi h\nu^3}{c^3} B, \tag{3.9}$$

and B is proportional to the square of the transition dipole (see Equation 3.6). Equation 3.9 was derived for a two-level system. Considering the vibrational substructure of absorption and emission spectra, and the spectral shift between them due to vibrational relaxation, a more realistic form is given by the Strickler–Berg relation

$$k_r = \frac{8\pi c n^2 2303}{N_A} \left\langle \nu_F^3 \right\rangle \int \epsilon(\nu) \frac{d\nu}{\nu}. \tag{3.10}$$

We conclude that the rate constant of fluorescence emission increases with the area under the absorption spectrum. Typical values for strongly absorbing compounds are in the range $k_r = 10^7 - 10^9\,\mathrm{s}^{-1}$ corresponding to radiative lifetimes of 1–100 ns. Since in almost all compounds relaxation from higher excited singlet states to the first excited one occurs on a ps timescale, strong fluorescence is only observed for the transition from the first excited singlet to the ground state. Fluorescence has to compete with radiationless deactivation processes, and the observed decay time is the inverse of the sum of both rate constants,

$$\frac{1}{\tau_F} = k_r + k_{nr} \tag{3.11}$$

The fluorescence quantum yield, i. e. the number of fluorescence quanta divided by the number of molecules excited to the $^1C_1$ state, is

$$\Phi_F = \frac{k_r}{k_r + k_{nr}} = k_r \tau_F. \tag{3.12}$$

This relation can be used to determine k_r from the experimentally available values of Φ_F and τ_F. If the result strongly deviates from the value obtained from the Strickler–Berg relation, then it is very likely that the fluorescence does not derive from the state that is responsible for the absorption band. A well-known example is the phantom state in longer polyenes (with three or more conjugated double bonds): The strong absorption band is caused by a symmetry-allowed transition to the second excited singlet state. Optical excitation of the first excited singlet state in linear polyenes with three or more conjugated double bonds is symmetry forbidden, but this is the state that determines the photochemistry.

If a photocatalyst shows strong fluorescence in the absence of a substrate, this indicates that the excited singlet state is long-lived and hence available for photochemistry. If fluorescence is weak, a fast deactivation channel exists. This

does not necessarily mean that this compound can not be a good photocatalyst. Deactivation could lead to a lower lying state for which the optical fluorescence transition is forbidden. The existence of such a state could be verified by transient absorption measurements.
- **Phosphorescence:** Emission from the triplet state to the singlet ground state is spin forbidden and weak in organic molecules. It can, however, become comparable in intensity to fluorescence transitions of intermediate strength in coordination compounds that contain a heavy metal atom if the transition has some charge transfer character between the metal and the organic ligand. In almost all organic photocatalysts at room temperature the radiationless transitions from the triplet state are much faster than the radiative transition, and phosphorescence is not observed unless the sample is cooled to liquid nitrogen temperatures.

Nonradiative transitions are indicated by straight lines in the Jablonsky diagram of Figure 3.5. Several steps can be distinguished:
- **Internal vibrational redistribution (IVR):** Immediately after a transition to a different electronic state the molecule will be in one of the vibrational states with the largest transition probability, which is given by the Franck–Condon factor (see Section 3.3). In polyatomic molecules the density of vibrational states will increase strongly with increasing energy. For example, at ca. 10000 cm^{-1} above the potential minimum this density can be of the order of several million states in every 1-cm^{-1} energy interval. Even very small anharmonicities in the PES will lead to a fast redistribution of the vibrational energy onto these states within a ps timescale. Although the vibrational energy is still in the molecule, i. e. the distribution of total energy is still rather narrow, the probability of finding an excitation in a particular normal mode can be well described by a Boltzmann-like distribution. Hence these systems are customarily described as "hot", and even a temperature can be assigned (from the slope of the Boltzmann plot), although the system is not in thermal equilibrium.
- **Vibrational cooling:** This is usually the step following IVR in condensed media. Through collisions with the solvent molecules the vibrationally hot chromophore dissipates vibrational energy into the solvent until it is in thermal equilibrium. In liquids at room temperature this process typically takes ca. 10 ps. After this relaxation, the chromophore is energetically near the bottom of the minimum of the PES of its actual electronic state.
- **Internal conversion (IC):** Through this process an electronically excited molecule moves from one electronic PES to a lower one of the same multiplicity. Similar as in IVR, the total energy in the molecule is conserved, i. e. the electronic excitation energy is converted to vibrational energy. The situation is schematically depicted in Figure 3.6.

  The initial state is a vibrational state of the excited electronic state. When the step is preceded by vibrational cooling, this is usually the vibrational ground

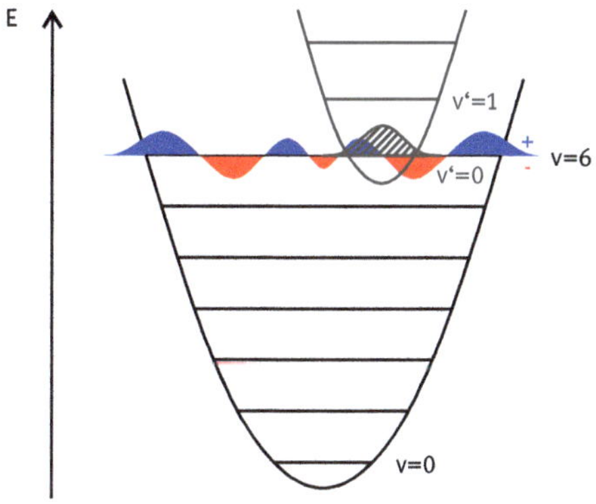

Figure 3.6: Franck–Condon factors in nonradiative processes. The decisive factor is the overlap between the two vibrational wavefunctions at the same total energy.

state ($v = 0$). The final state of the transition is a vibrationally excited state of the lower electronic state with the same total energy as the initial state (here $v = 6$). The wavefunction of this state alternates between positive and negative values (indicated by blue and red in the figure), and so does the product with the other wavefunction. Positive and negative contributions cancel in the integral, which will become smaller as the number of nodes in the wavefunction increases.

– **Intersystem crossing (ISC):** A transition from one electronic state to another one with a different spin multiplicity is called intersystem crossing. The situation is similar to IC, but an additional spin factor arises in the transition probability. For typical organic compounds ISC is slower by approximately a factor of 10^6 compared to IC when the energy gap is similar. For IC and ISC the transition probability is proportional to the Franck–Condon factor between the two vibrational wavefunctions involved. It decreases strongly with increasing number of nodes of the vibrational wavefunction of the final state. The most efficient vibrational modes will be those that can absorb the largest amount of energy with the smallest number of quanta, i. e. those modes with the largest frequencies. In organic molecules these involve stretch vibrations of C–H or O–H bonds. Substitution of the hydrogen atoms by deuterium will reduce the vibrational frequencies by approximately a factor of $\sqrt{2}$. For example, a transition that requires seven vibrational quanta of a C–H stretch will require ten vibrational quanta of the corresponding C–D vibration. The integrand in the Franck–Condon factor has three more nodes, making the factor much smaller. As a consequence, deuterated compounds show much longer triplet lifetimes at low temperatures compared to the nondeuterated compound. For example, the triplet lifetime of phenanthrene-h_8 at liquid nitrogen temperature is ca. 3 s, for phenanthrene-d_8 it is in the order of minutes.

A good photocatalyst must have a sufficiently large yield of the excited state species that performs the photoreaction, which can be either the lowest excited singlet or triplet state. The lifetime of this state in the absence of a substrate should be long

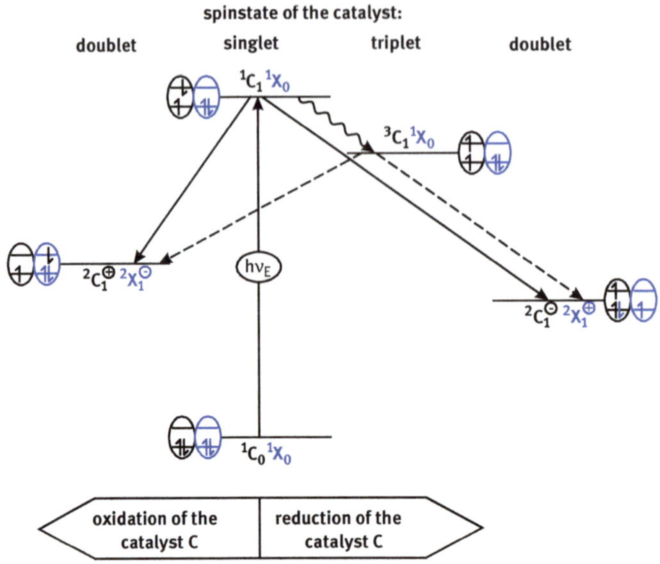

Figure 3.7: Jablonsky diagram for the complex of photocatalyst C and a reactant molecule X.

enough to allow the catalyst and substrate to encounter each other efficiently. The yield and lifetime of the excited singlet state can be obtained from fluorescence measurements. The corresponding data for the triplet state are available from transient absorption measurements.

In order to discuss the initial step of the photoreaction between the catalyst and a substrate molecule we have to extend the Jablonsky diagram as shown in Figure 3.7. In the overwhelming majority of situations studied so far this step is the transfer of an electron either from the reactant to the catalyst or in the opposite direction. The extended Jablonsky diagram thus contains, in addition to the ground state ${}^1C_0{}^1X_0$ and the locally excited states ${}^1C_1{}^1X_0$ and ${}^3C_1{}^1X_0$, also the charge transfer states ${}^2C_1^+\,{}^2X_1^-$ and ${}^2C_1^-\,{}^2X_1^+$. These latter two radical ion pairs could be produced either from the singlet state or from the triplet state. Depending on the redox properties of the excited state, only one of the two possibilities (oxidation or reduction) will occur with substantial yield. It is, however, possible that initially the radical ion pair has a well-defined total spin state, i. e. there may be consequences for subsequent reactions that are dependant on whether the ion pair state was formed from the excited singlet or the triplet state. An example is given in chapter 16.

For the reaction between the isolated molecules in the gas phase the energy balance for the electron transfer reaction is

$$\Delta E = I_P(D) - E_A(A) - E^* - \frac{e_0}{4\pi\varepsilon_0 R}, \tag{3.13}$$

where $I_P(D)$ is the ionization potential of the electron donor, $E_A(A)$ is the electron affinity of the electron acceptor, $E^\star$ is the energy of the excited state (which might be either the donor or the acceptor), and the last term is the Coulomb energy of the ion pair at distance R in units of eV. A negative value of ΔE indicates that the reaction is energetically possible, and solving $\Delta E = 0$ for R yields the largest distance over which electron transfer can occur. Rehm and Weller proposed a similar equation for reactions in solution, i. e.

$$\Delta G = E_{ox}(D) - E_{red}(A) - E^\star + C. \tag{3.14}$$

Here, ΔG is the change in the Gibbs free enthalpy, and the reaction is possible when $\Delta G < 0$. The ionization potential and electron affinity have been replaced by the oxidation and reduction potentials that can be obtained from electrochemical measurements. The excitation energy stored in one of the two reactants is $E^\star$, and the Coulombic interaction between the two radical ions is represented in C. The latter term is, however, solvent dependent. It can have negative values, but, in particular in unpolar solvents, positive values are also possible. The oxidation potential is frequently associated with the energy of the highest molecular orbital that is occupied by an electron. This is either the HOMO of the donor in the ground state, or the LUMO of the electronically excited donor. Similarly, the reduction potential is associated with the energy of the lowest orbital that can accommodate an electron. This is either the LUMO of the acceptor in the electronic ground state, or the HOMO of the electronically excited acceptor. This is, however, a rather simplified picture which can, for example, not account for the differences in singlet and triplet states. Nevertheless, Equation 3.14 provides a useful estimate of whether a photoinduced electron transfer is feasible or not. The oxidation and reduction potentials are often taken from the electrochemical measurements on the ground state donor and acceptor, respectively. A positive oxidation potential and a negative reduction potential both indicate a positive energy that is required to perform the respective charge transfer step. This positive energy must be overcome by the excitation energy in order to make ΔG negative. The Coulomb term is in most cases only a small correction, for example ca. -0.1 eV in polar solvents, and ca. $+0.5$ eV in unpolar solvents.

3.3 The Theoreticians' Perspective: A Closer Look

In the previous sections we have tried to use only a minimum of mathematical formalism. For a deeper understanding, however, some equations and formulae are unavoidable. This section will take a closer look at some of the important concepts discussed earlier. First, we will discuss the transition probabilities between various states of the photocatalyst or the reactive pair formed by the photocatalyst and the reactant. Within the framework of the Born–Oppenheimer approximation these transition probabilities can be factored, and the various terms can be interpreted in terms of several selection rules. A second concept concerns the representation of the molecular electronic states in terms of molecular orbitals (MOs). The concept of MOs is very popular and in many cases also useful, but there are also many situations where a simple MO consideration can be misleading. We will hence give a short introduction as to what MOs are and what they are not. This will not only permit a deeper understanding, but will hopefully make the reader also aware of certain popular concepts which can lead to erroneous conclusions when stressed beyond their limited range of validity.

3.3.1 Transition probabilities

The true eigenstates of a molecule are stationary, i. e. they do not evolve in time. The corresponding experimental situation would be the isolated molecule in the gas phase. Even then the Born–Oppenheimer (BO) wavefunctions are only an approximation and not the true eigenstates. In solution at elevated temperature the molecule is also subject to interactions with the solvent or the reactant, and in a photochemical experiment it also interacts with the radiation field. Hence we must recognize that the BO states are only a framework in which we may describe the time-dependent processes occurring in a photochemical reaction in solution. Conversely, only these BO states can be understood in the sense of a chemical structure. Hence it has become customary to describe the initial and the final state of a reaction as BO states. We are then left with the problem of describing how a system can evolve from an initial BO state to the final BO state. We will briefly sketch the time-dependent perturbation theory method: we consider the set of all possible BO states

$$|A\rangle = |a,v,\sigma\rangle = \Psi_a(R;r)\,\Phi_v(R)\,\Sigma_\sigma, \tag{3.15}$$

where A is short-hand notation for the set of electronic quantum number a, vibrational quantum number v, and spin quantum number σ. More precisely, v corresponds to the whole set of vibrational quantum numbers $v = \{v_1, v_2, \cdots, v_{3N-6}\}$, one for each normal mode. We also remember that the ket notation is the short-hand notation for the wavefunctions, as shown on the right hand side of Equation 3.15. We remind the reader that the electronic wavefunction $\Psi_a(R;r)$ depends not only on the coordinates of the electrons r, but also on the positions of the nuclei R, whereas the vibrational wavefunction $\Phi_v(R)$ depends only on the nuclear coordinates R.

The Hamiltonian H of the system is split into the sum of the BO Hamiltonian $H^{(BO)}$ and a term $V(t)$ which contains all contributions to the energy of the system that are neglected in the BO approximation. The BO states are eigenstates of the BO Hamiltonian,

$$H = H^{(BO)} + V(t) = H^{(BO)} + W\cos(\omega_0 t) \tag{3.16}$$

$$H^{(BO)}\left|A\right\rangle = E_A\left|A\right\rangle \tag{3.17}$$

$$E_A = E_{a,\sigma} + E_{vib} = E_{a,\sigma} + \sum_{j=1}^{N-6}\left(v_j + \frac{1}{2}\right)\hbar\omega_j. \tag{3.18}$$

In the second part of Equation 3.16 the interaction $V(t)$ is modeled as the product of a time-independent operator W and a cosine function oscillating with frequency ω_0. When the interaction is due to a monochromatic light wave, $W = -\mu\mathbf{E}$ is the product of the dipole operator with the electric field amplitude, and $\omega_0 = 2\pi\nu_0$ is the light frequency. Any polychromatic light field can be described as a superposition of such monochromatic terms. Other interactions can also be written in this form. For example, the terms neglected in the BO approximation are time-independent, hence $\omega_0 = 0$ in this case.

The BO states form a complete system that can be used to represent the true wavefunction $\left|\psi\right\rangle$ of the time-dependent system as a linear combination

$$\left|\psi(t)\right\rangle = \sum_J \left|J\right\rangle c_J(t). \tag{3.19}$$

The square modulus of the time-dependent coefficients $c_J(t)$ corresponds to the probability of observing the BO state $\left|J\right\rangle$ in a measurement performed at time t. Inserting the ansatz 3.19 into the time-dependent Schrödinger equation results in a system of coupled linear differential equations for the coefficients. For the perturbative solution of this equation system the reader is referred to textbooks on quantum mechanics. It is assumed that the system is initially in the BO state $\left|A\right\rangle$, and one asks for the probability of finding it at some later time in the final state $\left|B\right\rangle$. The result is usually presented as a rate constant k for this transition, the famous golden rule of Fermi:

$$k_{A\to B} = \frac{\pi}{\hbar}\left|W_{AB}\right|^2 g(E_B - E_A - \hbar\omega_0). \tag{3.20}$$

The line-shape function $g(E)$ is sharply peaked at $E = 0$, i. e. a transition from $\left|A\right\rangle$ to $\left|B\right\rangle$ will only occur when the condition

$$E_B - E_A = \hbar\omega_0 \tag{3.21}$$

is fulfilled. For a monochromatic light field this is just the resonance condition of Einstein. When the perturbation is time-independent, a transition occurs only when $E_A = E_B$. Thus, the lineshape function ensures conservation of the energy. The matrix element W_{AB} then determines the amplitude of the rate constant and is the source of the selection rules for transitions. After inserting the product form of the BO states the integrations along the spin, electronic, and nuclear coordinates can be performed consecutively.

We consider first the case where the operator W does not act on the spin coordinates,

$$\begin{aligned} W_{AB} &= \langle a, \nu_A, \sigma_A | W | b, \nu_B, \sigma_B \rangle \\ &= \langle \sigma_A | \sigma_B \rangle \langle a, \nu_A | W | b, \nu_B \rangle \end{aligned} . \tag{3.22}$$

Obviously the transition has zero probability when the spin wavefunctions of the two states are different. A spin-free operator W like the interaction with an electromagnetic field or the BO corrections will not allow transitions that change the multiplicity of the molecule. A direct excitation of a triplet state from the singlet ground state, phosphorescence, and intersystem crossing are not allowed at this level of approximation. These processes require an additional perturbation with an operator that acts on the spins. Such an operator will arise when relativistic contributions to the Hamiltonian of the molecule are considered. The first correction is called spin-orbit coupling (SOC) and is represented by an operator which is the scalar product of the operators for the electron spin and the electron angular momentum. This operator can have nonzero matrix elements for transitions in which the spin and the angular momentum change simultaneously. The most important contributions to the corresponding integrals in the angular momentum part have the form $\langle \chi_\mu | l_j | \chi_\nu \rangle$, where χ_μ and χ_ν are atomic orbitals *at the same atom* and l_j is a component of the angular momentum operator. These integrals can be nonzero only when both orbitals belong to the same angular momentum but have different m_l quantum numbers. An example would be $\langle p_x | l_y | p_z \rangle$. As an example, this integral will couple a $^1(n\pi^*)$ state to a $^3(\pi\pi^*)$ state. The integral of the SOC operator between these two states is given by the integral over the two orbitals in which these two states differ, i. e. $\langle n | l_j | \pi \rangle$. Expanding the molecular orbitals in the atomic orbital basis will yield a contribution of the form $\langle p_x | l_y | p_z \rangle$ when the n-MO involves a p_x orbital and the π-MO involves a p_z orbital at the same atom. This occurs, for example, in carbonyl compounds when the oxygen atom contributes both to the n-MO and to a π-MO. Intersystem crossing can then be very fast, for example in benzophenone it occurs on a ps timescale. SOC is a relativistic effect and becomes stronger when the electrons involved become faster. This is especially true for the inner electrons of so-called heavy atoms, i. e. atoms with a large nuclear charge Z. Integration over these electrons yields a prefactor to the SOC coupling matrix elements which increases with the fourth power of Z. Hence ISC can be increased in organic molecules by substitution of hydrogen atoms with heavy atoms like Br or J. The same effect is the reason for the fact that most coordination compounds show strong ISC when the lowest excited state is at least partially localized on the central metal atom.

The second factor in Equation 3.22 can be simplified when we first perform the integration over the electronic coordinates

$$\begin{aligned} \langle a, \nu_A | W | b, \nu_B \rangle &= \langle \nu_A | W_{ab}(R) | \nu_B \rangle \\ W_{ab}(R) &= \langle a | W | b \rangle \end{aligned} . \tag{3.23}$$

The second term in Equation 3.23 is called the electronic transition moment. Note that, although the electronic coordinates are already integrated over, the expression is still a function of the nuclear coordinates since the electronic wavefunctions depend on the positions of the nuclei. It should also be mentioned that only those terms in W that depend on the electronic coordinates enter this integral when $|a\rangle$ and $|b\rangle$ are different electronic states. All other terms, in particular the nuclear contributions to the dipole moment, cancel due to the orthogonality of the electronic wavefunctions.

In order to perform the integral over the nuclear coordinates in the first line of Equation 3.23 we need to know the functional form of $W_{ab}(R)$. Often this function varies only slowly with the nuclear coordinates and may be expanded in a Taylor series around the equilibrium structure R_0

$$W_{ab}(R) \approx W_{ab}^{(0)} + W_{ab}^{(1)}(R - R_0) + \ldots \tag{3.24}$$

When the first term on the right hand side of Equation 3.24 is nonzero, the transition is called electronically allowed. The second term is then neglected, and the transition rate is proportional to

$$k_{A\to B} \sim \left|\langle \sigma_A | \sigma_B \rangle\right|^2 \left|W_{ab}^{(0)}\right|^2 \left|\langle v_A | v_B \rangle\right|^2. \tag{3.25}$$

The last term in this expression is the famous Franck–Condon factor. It is the multidimensional projection of the vibrational wavefunctions in the initial electronic state onto the vibrational wavefunctions in the final electronic state. This factor determines the vibrational structure in absorption and emission spectra. When the transition starts from a molecule in the vibrational ground state, only totally symmetric vibrations contribute. The Franck–Condon factor also appears in the formula for the rate of nonradiative transitions. The only condition is that the first term in Equation 3.24 does not vanish. It could vanish for symmetry reasons. As an example we discuss radiative transitions induced by the electric field of a light wave. In this case

$$W_{ab}^{(0)} = -\mathbf{E}\langle a|\mu|b\rangle\Big|_{R=R_0}. \tag{3.26}$$

This integral is performed at the equilibrium structure of the molecule. It vanishes when the product of the irreducible representations of the three factors in the integral, i. e. the electronic wavefunctions of the initial and the final state and the dipole operator, does not contain the totally symmetric representation of the molecular point group. Frequently, one of the two states is totally symmetric. In this case the irreducible representation of the other state must be contained in the representation of the dipole operator.

If $W_{ab}^{(0)}$ vanishes the second term $W_{ab}^{(1)}$ must be considered. This term involves derivatives of the electronic wavefunctions with respect to the nuclear coordinates. A treatment introduced by Teller and Herzberg calculates this by expanding these derivatives in a series in all electronic states along displacements from the equilibrium structure. As a result, the intensity of these transitions is called "vibrationally induced". In distinction to the totally symmetric "Franck–Condon" modes involved in allowed transitions, the vibrational modes involved in the electronically forbidden or weak transitions are called "Herzberg–Teller" modes. Typical examples of such transitions are the first electronic transition in benzene, naphthalene, and pyrene.

3.3.2 Orbitals

The electronic wavefunction of a molecule is a function of the coordinates of all electrons. For example, the wavefunction of the water molecule depends on 30 coordinates (for fixed positions of the nuclei). A graphical representation of this wavefunction would be a hypersurface in a 30-dimensional space – which is quite beyond the imaginative capacity of most humans. This is one of the reasons why multielectron wavefunctions are constructed from orbitals. An orbital is a one-electron wavefunction and can hence be visualized as a surface in three-dimensional space. Orbitals centered around a single nucleus are called atomic orbitals $\chi_\mu(r)$, orbitals extending over two or more nuclei are called molecular orbitals $\phi_j(r)$. In the following we represent orbitals by small Greek letters, using Greek indices for atomic orbitals (AOs) and Latin indices for molecular orbitals (MOs). Multielectron wavefunctions are represented by capital Greek letters and capital Latin indices.

The simplest way of constructing a multielectron wavefunction from orbitals is to choose an orbital for each electron and multiply the orbitals. In doing so we also have to assign a spin to each electron. The example 5-electron wavefunction

$$\Psi(1,2,3,4,5)=\phi_a(1)\alpha(1)\phi_b(2)\alpha(2)\phi_c(3)\alpha(3)\phi_d(4)\beta(4)\phi_e(5)\beta(5) \tag{3.27}$$

then corresponds to electron 1, 2, and 3 being in orbitals a, b, and c with alpha spin, and electrons 4 and 5 in orbitals d and e with beta spin. However, electrons are indistinguishable. Hence any wavefunction in which the electron labels are permuted describes the same situation, and the correct wavefunction must be a linear combination of all these permutations (which are 5! = 120 in the present case). Electrons are Fermi-particles (having half-integer spin), hence the wavefunction must change sign for each pairwise exchange of electron labels. The linear combination that fulfills this requirement can be written in compact form as a so-called Slater determinant

$$\Psi=\begin{Vmatrix} a(1) & b(1) & c(1) & \bar{d}(1) & \bar{e}(1) \\ a(2) & b(2) & c(2) & \bar{d}(2) & \bar{e}(2) \\ a(3) & b(3) & c(3) & \bar{d}(3) & \bar{e}(3) \\ a(4) & b(4) & c(4) & \bar{d}(4) & \bar{e}(4) \\ a(5) & b(5) & c(5) & \bar{d}(5) & \bar{e}(5) \end{Vmatrix}=\left\| abc\bar{d}\bar{e} \right\|. \tag{3.28}$$

Every orbital occupies one column of the determinant, every electron one row. In Equation 3.28 we have abbreviated the orbital with alpha spin by its index, the orbital with beta spin by the index with an overbar. A theorem in quantum mechanics states that the set of all Slater determinants Ψ_J constructed from a complete basis of orbitals is itself a complete basis for the N-electron problem. Hence, any wavefunction of this molecule can be exactly written as a linear combination of these Slater determinants,

$$\Psi_{exact} = \sum_J \Psi_J \, c_J. \tag{3.29}$$

A particular distribution of electrons among orbitals is called configuration, and calculations based on expansion 3.29 are known as configuration interaction (CI). The problem is that a truly complete basis has an infinite number of orbitals. That is for practical calculations the orbital basis has to be truncated. But even with a truncated basis the number of possible Slater determinants grows dramatically with increasing number of orbitals and electrons. When a set of M orbitals $\{\phi_1, \phi_2, \cdots, \phi_M\}$ is available for N electrons, a total of

$$L_{det} = \binom{2M}{N} \tag{3.30}$$

Slater determinants can be constructed. For example, for water ($N = 10$) in the minimal basis of valence orbitals ($M = 7$) we obtain $L = 1001$. For benzene with the same basis we obtain $L = 118264581564861424$ even if we fix the 1 s-electrons of the carbon atoms. As a consequence, even with an orbital basis of moderate size, the number of Slater determinants has to be restricted. Clever strategies have been developed that combine Slater determinants to eigenfunctions of the spin operators S^2 and S_z. Calculations for a state of particular multiplicity can be of much smaller size when performed in the basis of these spin-adapted configurations (SAC).

So far we have used the orbitals only as a basis for constructing multielectron wavefunctions but did not specify them. When we mentioned valence orbitals we did, on purpose, not specify whether we mean atomic orbitals or molecular orbitals. The reason is that this does not matter when the full set of SACs for a particular multiplicity is used in the CI calculation. A full-CI calculation is invariant against a unitary transformation among the orbital basis. However, when the number of configurations considered is less than the full set, the choice of orbitals matters. One can then ask for those orbitals that yield the best approximation of the wavefunction with a given number of configurations. The extreme situation is to restrict expansion 3.29 to a single Slater determinant, or a single SAC. Optimizing the expectation value

$$\langle E_J \rangle = \langle \Psi_J | H^{(BO)} | \Psi_J \rangle \tag{3.31}$$

is solved by a procedure named after the inventors Hartree and Fock (HF). The effective one-particle Hamiltonian in this procedure depends on the orbitals. Equation 3.31 is solved iteratively until the energy and the electron density no longer change, hence the name self-consistent field (SCF) method. Usually the SCF-orbitals are expanded as a linear combination of atomic orbitals (LCAO),

$$\phi_j = \sum_{\nu} \chi_\nu c_{\nu j} . \tag{3.32}$$

The atomic orbitals χ_ν are taken as fixed functions, optimization is performed by varying the coefficients $c_{\nu j}$. Optimization is only possible when more AOs are available than MOs are incorporated in the SAC. The optimization is usually performed by finding the eigenfunctions of the effective one-electron Hamiltonian. The eigenvalues of these canonical Hartree–Fock orbitals are interpreted as orbital energies. It turns out that almost always the orbitals with the lowest eigenvalues are those involved in the optimized Slater determinant, i. e. the occupied orbitals. The number of canonical Hartree–Fock orbitals is the same as the number of AOs. Those canonical Hartree–Fock orbitals not appearing in the optimized Slater determinant are called unoccupied or virtual orbitals. The highest occupied molecular orbital (HOMO) and the lowest unoccupied molecular orbital (LUMO) are frequently used in discussions of the properties and reactivities of molecules. This has to be done with some caution, as will be explained in the following two paragraphs.

First of all, the canonical HF-orbitals are not the only solution of the HF-problem. If in a determinant one column is subtracted from another one, the value of the determinant does not change. More generally, if a unitary transformation is applied to a matrix, the determinant remains invariant. One of the consequences is the famous Pauli principle: if two columns in a Slater determinant are identical, i. e. the same orbital appears twice with the same spin, the determinant does not change when the second of these columns is replaced by zeros, making the whole determinant vanish. This property of the determinant has, however, also the consequence that the orbitals are not unique: any set of orbitals produced by applying a unitary transformation to the canonical HF orbitals is as good as the original set. To illustrate this fact we show in the left part of Figure 3.8 the canonical HF-orbitals ψ_σ and ψ_π of the ethene molecule. These are usually associated with the σ-bond and the π-bond between the two carbon atoms. Also, ψ_π is the HOMO. The right part of Figure 3.8 shows the orbitals

$$\psi_+ = \frac{1}{\sqrt{2}}\left(\psi_\sigma + \psi_\pi\right) \tag{3.33}$$

$$\psi_- = \frac{1}{\sqrt{2}}\left(\psi_\sigma - \psi_\pi\right) . \tag{3.34}$$

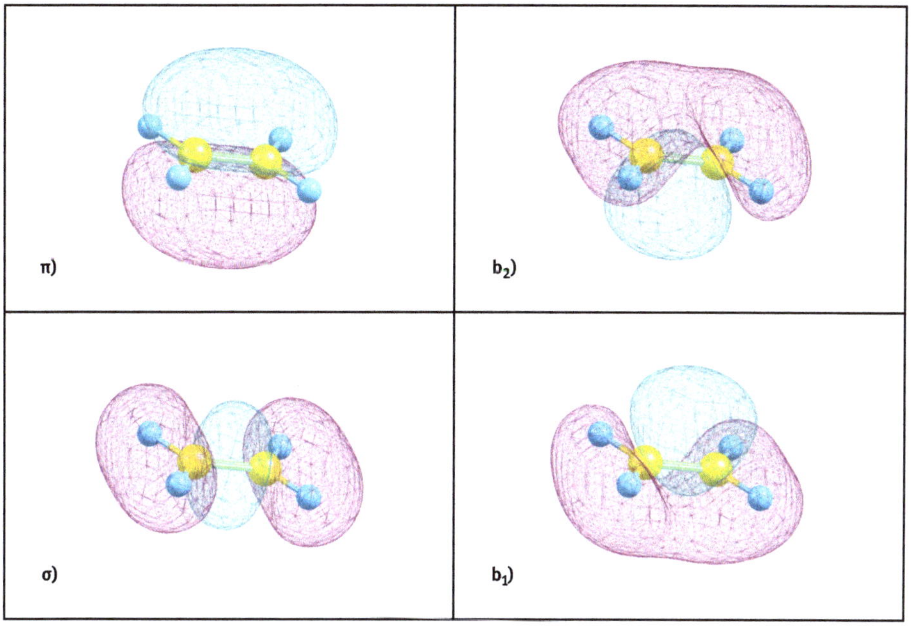

Figure 3.8: The multielectron wavefunction produced by the Slater determinant with doubly occupied σ and π orbitals (left side) is completely identical with the Slater determinant containing the doubly occupied "banana" bond orbitals (right side).

The Slater determinant formed with these orbitals is identical to the Slater determinant formed with the canonical HF-orbitals,

$$\left\| \psi_\sigma \bar{\psi}_\sigma \psi_\pi \bar{\psi}_\pi \right\| = \left\| \psi_+ \bar{\psi}_+ \psi_- \bar{\psi}_- \right\|. \tag{3.35}$$

We see that a particular orbital inside a Slater determinant has no real significance – this is only contained in the whole subspace of occupied orbitals. There is obviously no HOMO when we use the set of orbitals $\{\psi_+, \psi_-\}$. Both orbitals must be energetically equivalent since they transform into each other through a rotation by π around the carbon–carbon axis. With the virtual orbitals the situation is even worse. In simple MO theory as is taught in undergraduate courses the unoccupied orbitals are constructed as the antibonding counterparts of the bonding MOs. In the example of ethene, the bonding π orbital is constructed from the in-phase linear combination of the two *p*-orbitals at the two carbon atoms that are oriented perpendicular to the molecular plane. This orbital turns out to be the HOMO, i. e. orbital number 8 when ranked according to the energy. The corresponding antibonding linear combination is then thought to be the LUMO. In fact, a quantum chemical calculation which considers only the valence atomic orbitals finds the orbital as number 9. This is even true for a medium-sized atomic basis like cc-pVDZ, which contains 50 basis functions. However, when a triple-zeta basis augmented with diffuse functions is used,

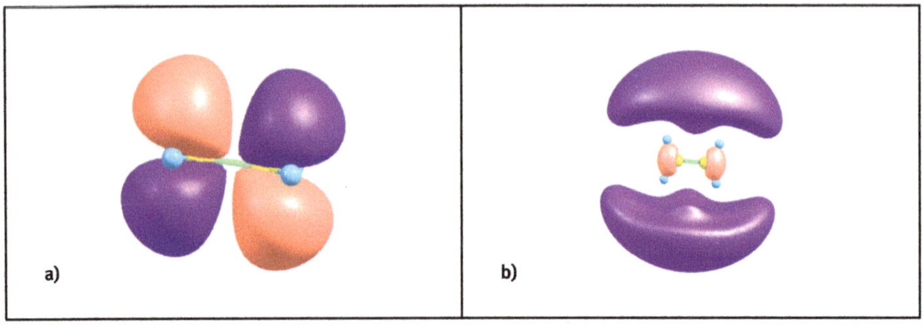

Figure 3.9: Plot of the lowest unoccupied orbital from a HF calculation of ethene with a cc-pVDZ basis (left) and an aug-cc-pVTZ basis (right). With the smaller cc-pVDZ basis this is the antibonding π^* orbital, but with the larger aug-cc-pVTZ basis this is a Rydberg orbital.

like aug-cc-pVTZ with 210 basis functions, orbital number 9 is a Rydberg orbital (Figure 3.9). The former LUMO is orbital number 21, i. e. it is now the thirteenth unoccupied orbital!

So why are canonical HF orbitals so popular among chemists? One reason has already been mentioned: only the canonical orbitals can be associated with an orbital energy. According to Koopmans' theorem these orbital energies correspond in a zero-order approximation to the ionization potentials (for the occupied orbitals) and the electron affinities (for the virtual orbitals) of the molecule. Hence it is tempting to use HOMO and LUMO as those orbitals that are involved in reduction, oxidation, and electronic excitation. Let us consider lumiflavin as a model for a photocatalyst. Lumiflavin has 134 electrons, i. e. orbitals 67 and 68 are the HOMO and LUMO, respectively. In the electronically excited singlet and triplet states both orbitals are singly occupied. In the radical cation only the HOMO is occupied by a single electron, whereas in the radical anion the HOMO is doubly filled and the LUMO contains one electron. This is indicated by the orbital diagram next to each molecular state in the Jablonski diagram. With the exception of the excited singlet state, all these states can be optimized with the ansatz of a single SAC. The two orbitals 67 and 68 obtained by each of these SCF calculations are shown in Figure 3.10.

The excited singlet state is the second state of this multiplicity. The variational principle requires that it must be optimized with the constraint that the wavefunction is orthogonal to that of the ground state. When the SAC from the HOMO–LUMO excitation is used as an ansatz, no optimization of the orbitals is permitted. Conversely, when configuration interaction with all singly excited configurations is performed, it is found that the HOMO–LUMO transition accounts for only 25 % of the norm of the excited state. Obviously the HOMO–LUMO excitation is a very crude approximation in this case. In order to obtain a guess for that pair of singly occupied orbitals which gives the best description of the first excited singlet state, we per-

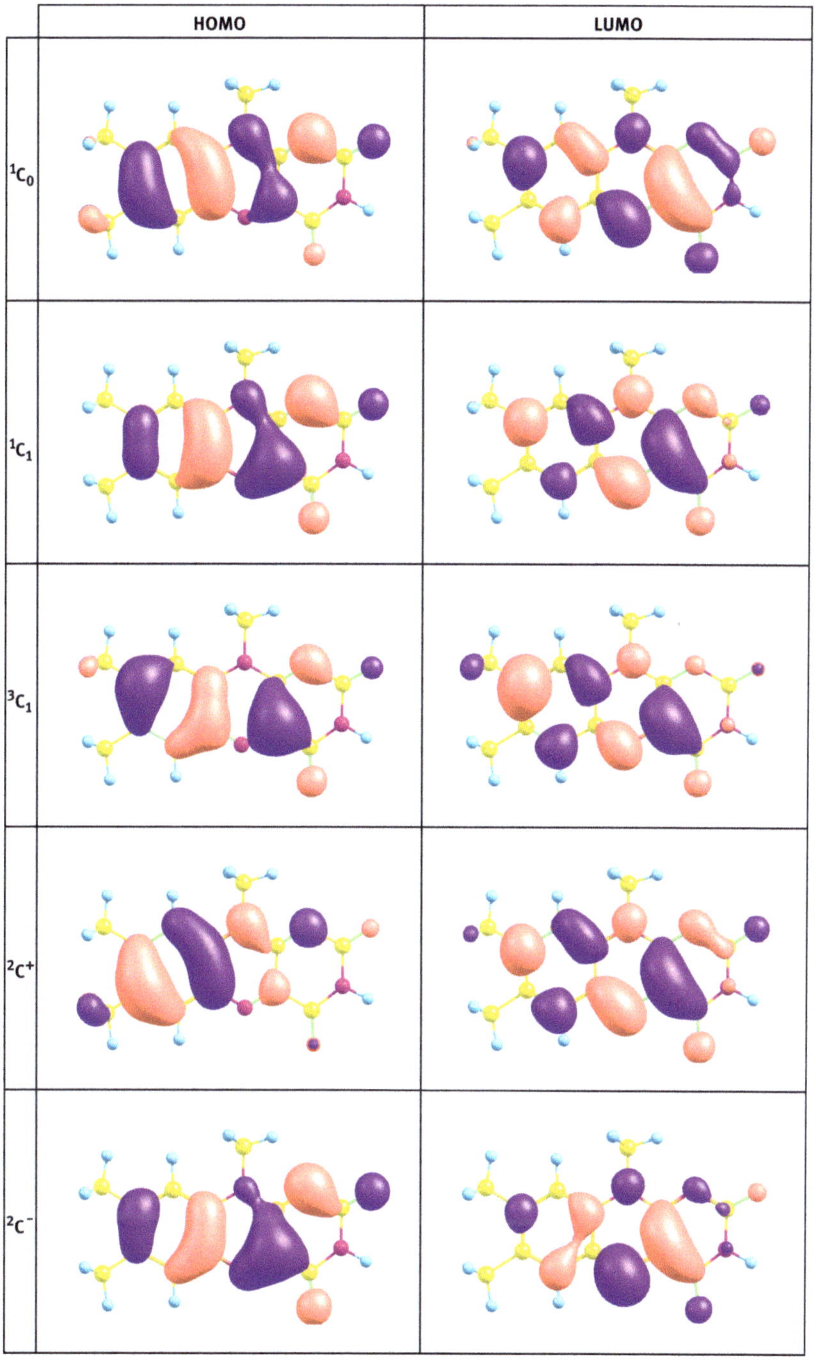

Figure 3.10: Orbitals of lumiflavin corresponding to the HOMO (left) and the LUMO (right) for neutral lumiflavin, the excited singlet and triplet states, the radical cation, and the radical anion (from top to bottom).

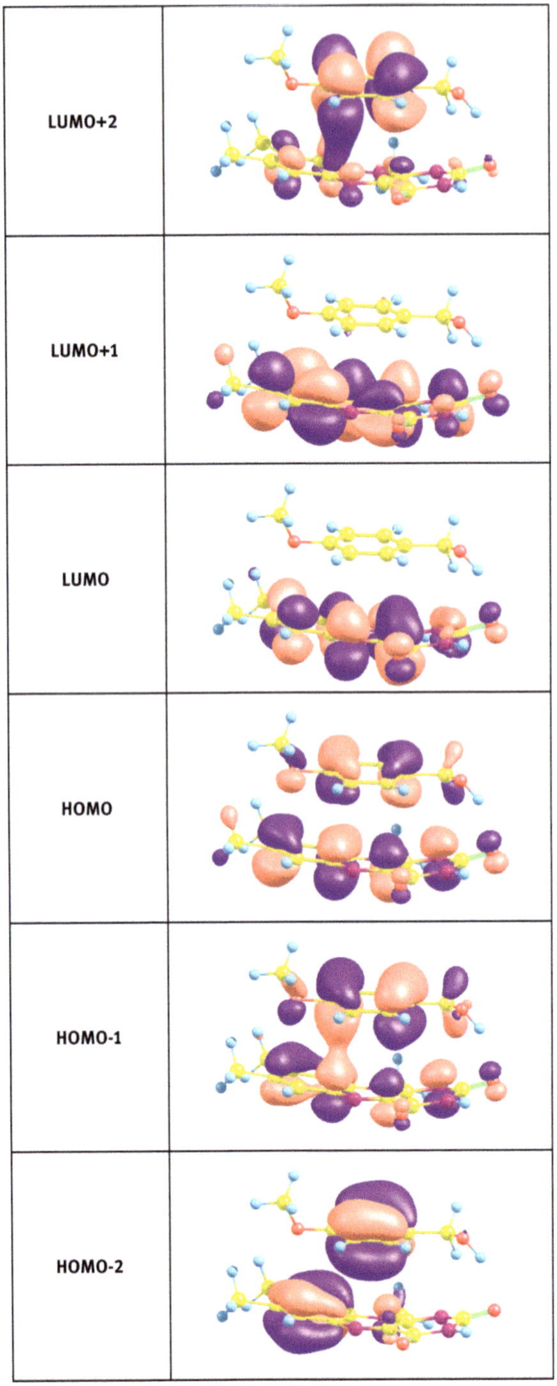

Figure 3.11: The three highest occupied and lowest virtual orbitals of the molecular complex of lumiflavin and MBA.

formed a multiconfiguration-SCF calculation with three SAC resulting from the distribution of two electrons among two orbitals and solved for the second state. This ensures orthogonality to the ground state, but results in noninteger occupation numbers of the orbitals. These orbitals are shown for the S_1 state in Figure 3.10. No orbital energies are associated with these orbitals, hence the assignment as HOMO or LUMO is based on the similarity with the SCF ground state orbitals.

Comparison of the orbitals collected in Figure 3.10 reveals that the HOMOs in these five systems are similar but not identical. The same applies to the LUMOs. The closest similarity exists between the HOMOs of the ground state and the cation, and between the HOMOs of the triplet and the anion. The LUMOs of the ground state and the cation are also rather similar, but these orbitals are unoccupied in both cases. We note that molecular orbitals are not a fixed property of a molecule but depend on the electronic state considered. Even worse, orbitals for the same state of a system can change when another molecule comes close. This is exemplified in Figure 3.11 displaying the three highest occupied orbitals and the three lowest virtual orbitals for lumiflavin in the presence of a molecule of p-methoxy benzyl alcohol (MBA). The two molecules are arranged in the optimized geometry of the ground state complex, calculated with the MP2 method. In this geometry HOMO and HOMO-1 are delocalized over both molecules; the LUMO is localized on the flavin. The calculated wavefunction of the first electronically excited state (at the CIS/cc-pVDZ level of theory) has a large contribution (80 %) from the HOMO–LUMO excitation. This corresponds to a partial charge transfer from the MBA molecule to the flavin molecule.

In conclusion: orbitals can often be useful for a discussion of mechanistic aspects, but this narrows the perspective to a one-particle picture. All electrons in a molecule are equal (i. e. indistinguishable), and orbitals are not observables in the strict quantum mechanical sense. In particular, in many molecules the first electronically excited state cannot be attributed to a HOMO–LUMO transition. For example, in flavin (but also in naphthalene or pyrene) it must be described by the linear combination of at least two SACs with similar weight. In an orbital picture, this corresponds to two orbitals occupied by 1.5 electrons, and two orbitals occupied by 0.5 electrons.

3.4 References

[1] N. J. Turro, V. Ramamurthy and J. C. Scaiano, *Modern Molecular Photochemistry of Organic Molecules*, University science books, Sausalito 2010.

Burkhard König, Susanne Kümmel, Radek Cibulka

4 Flavin photocatalysis

4.1 Introduction

Photocatalysis is a very common principle in nature: All plants and animals depend on sunlight and use it by means of photoreceptors. One of the most prominent examples of photoreceptor dyes is riboflavin (RF), also known as vitamin B_2, which takes part in biochemical redox reactions as a coenzyme. [1] Besides their role as photoreceptor dyes [2] the photoactivity of flavins is crucial for some other natural processes, e. g. the light generation by bacterial luciferase, [3] and DNA repair by photolyase. [4]

Riboflavin and its derivatives are yellow compounds and hence they can absorb visible light with a maximum absorption in the blue range. Upon excitation with blue light of 440 nm their redox power is dramatically increased by up to 2.48 eV (E_{00} of RF). [5] This value represents the maximal (theoretical) energy of light, which could be utilized in photocatalysis by flavins. Flavoenzyme models have been studied extensively including investigations of redox potential changes, [1, 6] photolyase [7] and luciferase models. [8] However, the use of flavins in chemical photocatalysis has not been well investigated. In this chapter we will give an overview of how these photophysical redox properties can be used in synthetic applications with flavins as photocatalysts. Due to the large amount of available literature, particularly from biochemical studies, the overview cannot be comprehensive. We principally discuss typical examples, including some light-independent redox reactions, to illustrate the potential of flavin photocatalysis.

Riboflavin was first described in 1879 as a bright yellow pigment isolated from cow milk, named lactochrome, [9] and found afterwards several times from different sources. [1] In 1934 Kuhn et al. developed the first synthesis and confirmed the molecular structure (see Figure 4.1). [10]

The name riboflavin was given due to the ribityl side chain (blue in Figure 4.1) and the bright yellow color (Latin: *flavus* – yellow), which is caused by the isoalloxazine unit (yellow in Figure 4.1). In nature it is mainly found as flavin adenine dinucleotide (FAD) or flavin adenine mononucleotide (FMN, riboflavin-5′-phosphate, see Figure 4.1) and today hundreds of flavoprotein enzymes containing predominantly noncovalently bound FAD or FMN are known. [1] In the case of flavoproteins, the most important domains are the LOV (light, oxygen and voltage) and the BLUF (blue-light using FAD) domains. LOV1 and LOV2 are found in the family of phototropins which undergo autophosphorylation upon blue light irradiation. [11] The BLUF domain was first discovered in a bacterium called *Rhodobacter sphaeroides* and was shown to control photosynthesis gene expression depending on blue light and oxygen. [12]

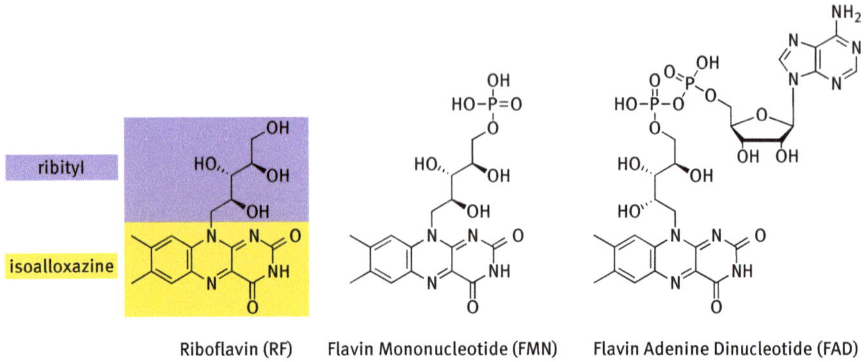

Figure 4.1: The most important flavins in nature: riboflavin (RF, also known as vitamin B_2, lactochrome, lactoflavin or ovoflavin), flavin mononucleotide (FMN) and flavin adenine dinucleotide (FAD).

Originally, the term flavin was used for 7,8-dimethyl-10-alkylisoalloxazines; later, all isoalloxazine derivatives have been called flavins. In this chapter we use the term for all 10-substituted isoalloxazines in general.

4.1.1 General properties

Flavins can exist in three redox states: oxidized, one-electron reduced (as semiquinone) or twofold reduced by two electrons, and each of these redox states has three different protonation states (see Figure 4.2). The redox properties of flavins, their

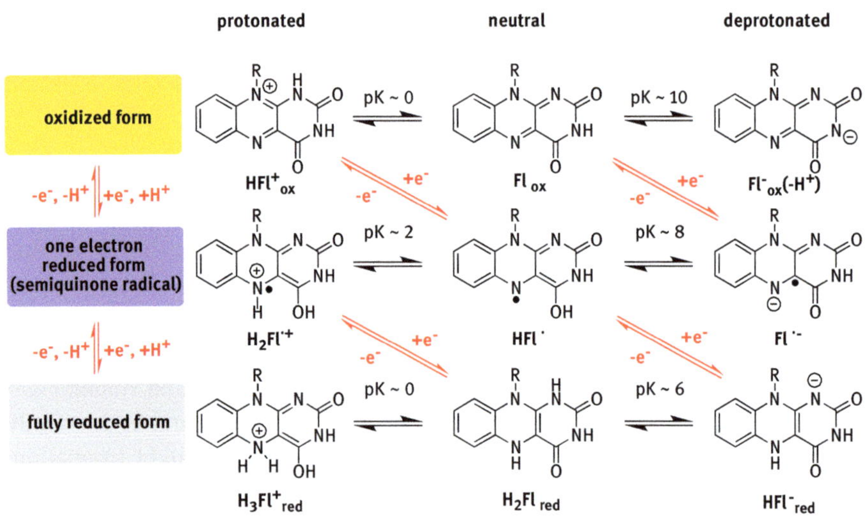

Figure 4.2: Different redox and protonation states of flavins (10-substituted isoalloxazines). [16]

absorption and the reactivity change with substitution, noncovalent interactions, such as hydrogen bonds, and the nature of the surrounding protein. [13] It is well known from electrochemical studies that in water and in protic organic solvents flavins are reduced reversibly by a two-electron process; the reduction potential of riboflavin (RF) is about −0.507 V (vs. ferrocene/ferrocenium (Fc/Fc^+)) in water. [14] In aprotic media, two consecutive one-electron reductions are observed. The reduction potentials of riboflavin tetraacetate (RFTA), −1.18 V and −1.87 V (vs. Fc/Fc^+) measured in acetonitrile, can be given as an example. [15]

The photocatalytic cycle in flavin photocatalysis starts with irradiation of the oxidized form of flavin (Fl_{ox}) with blue light exciting it into the singlet state ($^1Fl^*$). Then intersystem crossing (ISC) to the triplet state ($^3Fl^*$) occurs rapidly in a few nanoseconds (13.5 ns for RF, [17] 23.8 ns for RFTA [18]). The triplet state is the active species and key intermediate in catalysis. [19] It can then be reduced by an appropriate substrate to the radical anion ($Fl^{\cdot-}$), which is subsequently protonated (pK_a value of 8.3) [20] and further reduced to the flavohydroquinone anion (HFl^-_{red}). The different redox states are easily distinguished by UV/Vis spectroscopy. [13]

In principle the flavin redox cycle can be used for reductions as well as for oxidations (see Figure 4.3).

For the reduction of substrates (Figure 4.3, right) a sacrificial electron donor (e. g. typically triethylamine, EDTA or triethanol amine) [21] is needed while for oxidations, oxygen from the air is sufficient to reoxidize the flavin. [16]

The mechanism of the reoxidation by oxygen in catalytic oxidations was thoroughly investigated by Bruice et al. in 1979. [22] By this reaction the flavin cofactor in flavoenzymes is reoxidized and used either in an oxidative half-reaction or to generate flavin hydroperoxide, which oxygenates various substrates in flavin-dependent monooxygenases. Massey summarized the chemical and biological versatility of riboflavin in 2000 and described the possible reactions of the

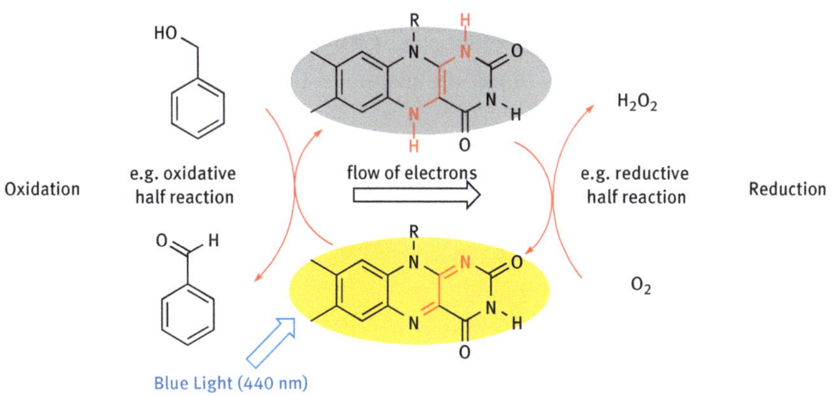

Figure 4.3: Catalytic cycle of flavin redox reactions. In the example of benzyl alcohol oxidation, air is used for the dark reoxidation of the catalyst.

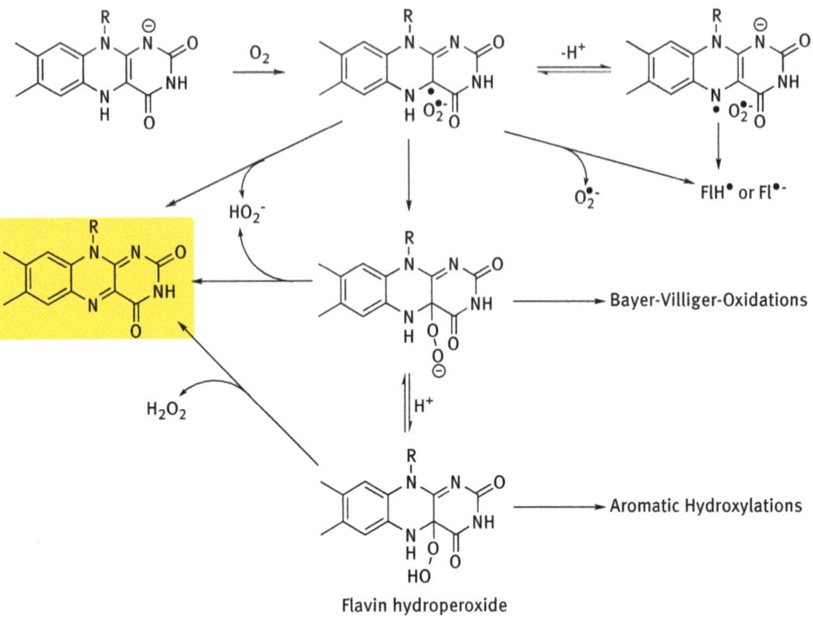

Figure 4.4: Reactions of reduced flavins with oxygen, the terminal oxidant in catalytic photo-oxidations. [23]

reduced flavin with oxygen (see Figure 4.4). [23] The process converts the reduced form of the flavin back to the oxidized form and activates oxygen for subsequent reactions.

The oxidized form of the flavin (yellow, on the right in Figure 4.4) is used in oxidation reactions when irradiated with blue light. The redox energy of the excited state can be estimated by the Rehm–Weller equation, [24] which shows that the potential is sufficient to activate substrates with low chemical reactivity, i. e. substrates with an oxidation potential below 1.9 V (vs. Fc/Fc^+). [25] In the next section we will discuss examples along the history of flavin chemistry.

4.2 Early examples of flavin photocatalysis

The first mechanistic investigations of the biochemical function of flavins as a coenzyme in dehydrogenases were reported in the 1930 s by several groups. [26] In 1939 Lipmann discussed the use of flavins in biochemistry as autoxidizable compounds for the catalysis of pyruvic acid oxidation via thiamine (vitamin B_1) oxidation. [27] This certainly inspired its use as a catalyst in chemical reactions.

The use of riboflavin as a photosensitizer in chemical applications was first mentioned in the literature in 1948 for the oxidation of ascorbic acid by light [25] and the

$$RF + h\nu \longrightarrow RF^*$$

$$\text{indole acetic acid} + RF^* \longrightarrow \text{oxidation product} + RFH_2$$

$$RFH_2 + \tfrac{1}{2}\,O_2 \longrightarrow RF + H_2O$$

Figure 4.5: One of the first reported chemical reactions with flavins: the oxidation of indole-acetic acid by riboflavin (RF). [28]

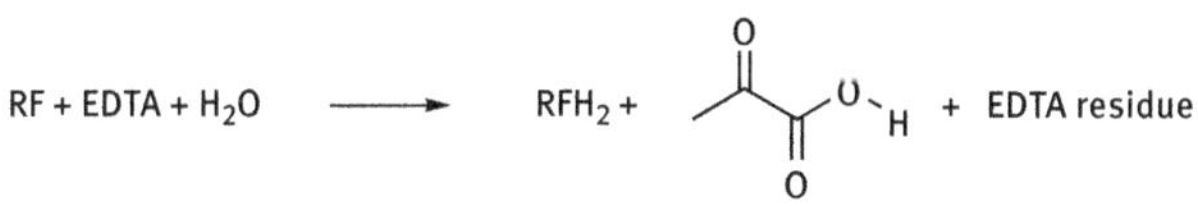

Figure 4.6: Stoichiometric reaction of EDTA and RF under anaerobic conditions. [29b]

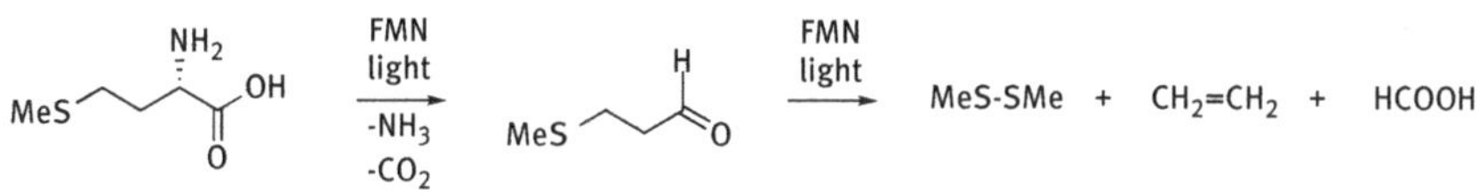

Figure 4.7: Stoichiometric photooxidation of methionine with flavin under anaerobic conditions.

riboflavin-sensitized photooxidation of indole-acetic acid investigated by Galston in 1949 (see Figure 4.5). [28] In these early days of flavin catalysis the side product of flavin reoxidation was thought to be water, instead of hydrogen peroxide.

Flavin-catalyzed photoreactions and their mechanisms have been intensively studied since then; most of the reports describe reactions with amino acids and amines. [29] Frisell et al. reported the oxidation of primary, secondary and tertiary amines and amino acids. The best results were obtained for tertiary amines such as EDTA, trimethylamine, dimethylglycine, but also with sarcosine and *N*-ethylglycine. Riboflavin was reported to be a better photocatalyst than FMN and FAD. [29a] Enns and Burgess described a stoichiometric reaction of riboflavin with methionine or EDTA under anaerobic conditions (see Figure 4.6) and could recover the riboflavin completely after oxidation with air. [29b] The first step of the mechanism of oxidation by flavins was discussed by several authors to be either hydride transfer [30] or one-electron radical mechanism. [29d, e, 31]

Later, methionine was reported to be oxidized to carbon dioxide, formic acid, ethylene, methyl disulfide and ammonia in the presence of FMN and light in aqueous solution under anaerobic conditions. Methional was suggested as the first intermediate of this transformation (Figure 4.7). [32]

In 1967 the debate about different mechanisms for different substrates was initiated by the results of Hemmerich, Massey et al. regarding the photoinduced decarboxylation of phenylacetate. [33]

They described the mechanism of the photodecarboxylation of phenylacetate, which results as a first step in a quantitative photoalkylation of the flavin (see

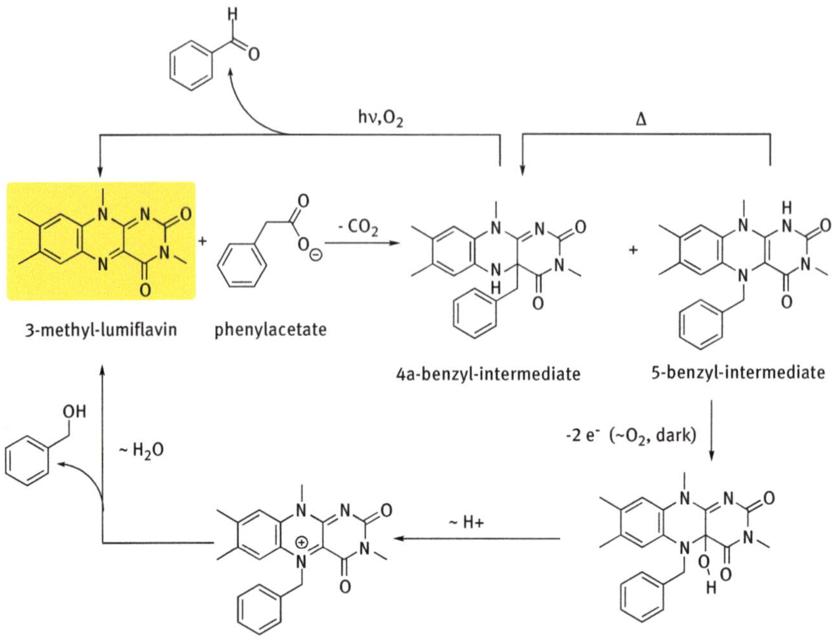

Figure 4.8: Mechanistic investigations on flavin-catalyzed oxidations. The phenylacetate is oxidized followed by decarboxylation and subsequently added to the flavin core. This is possible in two positions, 5 and 4a. The 4a-benzyl intermediate is thermally stable, but photolabile. It leads to an oxidized product (benzaldehyde in this case) while the 5-benzyl intermediate leads to benzyl alcohol formation via the purple oxidized 5-benzylflavinium cation. [33–34]

Figure 4.8). The structure shows the properties of leucoflavin derivatives. Depending on the conditions, the benzyl residue can either add to position 5 (nitrogen atom) or the bridge position 4a (carbon atom). Upon heating the benzyl residue migrates from N(5) to C(4a).

The flavinium salt (substituted in 5-position) is able to transfer the benzyl group to the solvent, i. e. water, to yield benzyl alcohol and the flavin starting material. The 4a-isomer decomposes under air atmosphere and irradiation to give benzaldehyde and the flavin starting material. It is not oxidized in the dark or in the absence of oxygen. [33b]

Hemmerich et al. compared the mechanism of oxidative photodecarboxylation of phenylacetic acids (i. e. group transfer) with the mechanism of "hydride" transfer, analyzing the spectra during the reaction. [34] They hypothesized that the latter mechanism might also be just a rapid sequence of group transfers. They found that for low pH values (<7), high temperature (>40 °C), with isoflavins (6-substituted flavins) [35] or with sterically demanding substrates the formation of 4a-benzyl-dihydroflavins is favored while at low temperature (<40 °C) and pH >7 the *N*5-substituted dihydroflavin is found, correspondingly. Then a dark rearrangement of the

N-substituted intermediate occurs, promoted by heat or acid leading to an equimolar mixture of both types of substitution. In the absence of light no reoxidized flavin can be found even in the presence of air. When irradiated again, the 4a-substituted reduced flavin is oxidized rapidly by air, while the 5-substituted derivative yields another product via a pathway involving a dark green radical as intermediate. This product can be hydrolyzed (under acidic conditions) to the oxidized flavin. [34]

Several other studies investigated the flavin reduction by carboxylic acids. [29e, 36] Penzer and Radda studied the photoreduction of flavins by amines and amino acids using the example of EDTA and DL-phenylglycine. [29e] They confirmed the better reactivity of FMN compared to FAD (reported by Frisell et al.) [29a] and found that the reaction with phenylglycine produces equimolar amounts of benzaldehyde, carbon dioxide and reduced flavin, but they observed differences in comparison with the reaction of phenylacetic acid. They proposed "that the first step is a one-electron process between the amino acid and the excited flavin. This is probably not the case for the reaction between excited flavins and phenylacetic acid." [29e] The flavin was not fully recovered after this decarboxylation reaction by air, which might be due to the different mechanism (like Hemmerich et al., [33, 37] see Figure 4.8). Carr and Weatherby reported the photooxidation of dihydrophtalates to benzoic acid or methylbenzoate. [38] They also proposed a mechanism with intermediate adducts.

These first examples for the chemical use of flavins were all performed with the intention of gaining insight into the biochemistry of flavoenzymes and not for the application in synthesis.

4.3 Flavin photocatalysis in synthesis application

In 1980 Yoneda et al. used 5-deazaflavin derivatives and analogues for the oxidation of alcohols and amines in the dark (Figure 4.9). [39]

Later, Fukuzumi et al. reported the efficient and substrate-selective photocatalytic oxidation of benzyl alcohol derivatives by oxygen using 3-methyl-10-phenylisoalloxazine and 10-phenyl-5-deazaflavine coordinated to Mg^{2+} and Zn^{2+} ions and protonated riboflavin tetraacetate. [40] Ever since, the photooxidation of substituted benzyl alcohols has became the most intensively studied photocatalytic reaction mediated by flavins.

In 1995 D'Souza et al. presented molecular orbital calculations on this photooxidation of substituted benzyl alcohol by riboflavin. [41] They conclude that N(1) of the flavin is preferentially protonated in the ground state and the excited states. The LUMO of the isoalloxazine is lowered by this protonation and hence the electron acceptance and therefore the oxidizing ability of the flavin are increased. The excitation results in energetically lowered corresponding SOMOs and the oxidation ability is significantly enhanced. [41] Interestingly, protonated flavin analogues were found

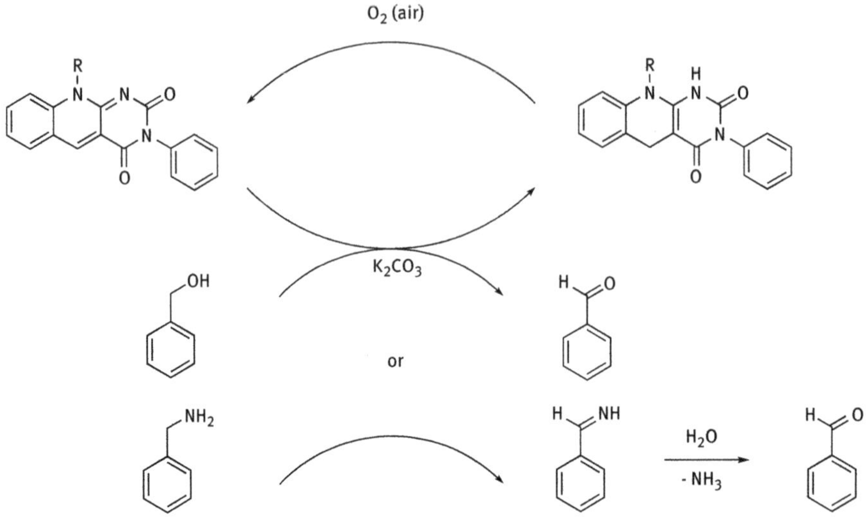

Figure 4.9: Synthetic application. Oxidation of benzyl alcohol or benzyl amine by 5-deaza-flavins in dark. [39a]

to be efficient photocatalysts for the oxidation of benzyl alcohols with oxidation potentials around 1.9 V vs. SCE (4-alkyl or 4-chloro substituted benzaldehydes) while no photooxidation has been observed in the case of benzyl alcohols with strongly withdrawing (NO_2 and CN) or electron-donating (OH or OMe) substituents. [42]

The effect of various metal ions on the oxidation power of photoexcited flavins has been studied by Fukuzumi in more detail. [43] It was shown that the effect of rare earth metal ions (Sc^{3+}, Lu^{3+} and La^{3+}) is more pronounced in comparison with the effect of magnesium or zinc ions. The metal ions interact with one or both carbonyl groups of riboflavin tetraacetate and this coordination shifts the reduction potential of the singlet excited state ($^1Fl^*$) positively by 390 mV for Mg^{2+} and even by 780 mV for Sc^{3+}, which causes an increase of the chemical quantum yield of up to 0.17 for the oxidation of 4-chlorbenzyl alcohol with the Sc^{3+} complex. The photooxidation proceeds via electron transfer from the benzyl alcohol to the singlet excited state of the flavin-metal ion complex [40b, 43] thus differing from the mechanism dominating when a nonmetal-ion-coordinated flavin is used as photocatalyst.

In 1997 Diederich et al. investigated a model system for pyruvate decarboxylase and reported a cooperative catalysis using a flavo-thiazolio-cyclophane as catalyst for the preparative electrooxidation of aromatic aldehydes to methyl esters or acids (Figure 4.10). [44]

Many attempts have been made to mimic the biologic properties of flavoenzymes obtaining better catalyst activities in chemical applications and understanding the enzyme catalysis. In 2003 D'Souza used modified cyclodextrins with flavin moieties as artificial enzymes (Figure 4.11). The oxidation of thiols and benzyl alco-

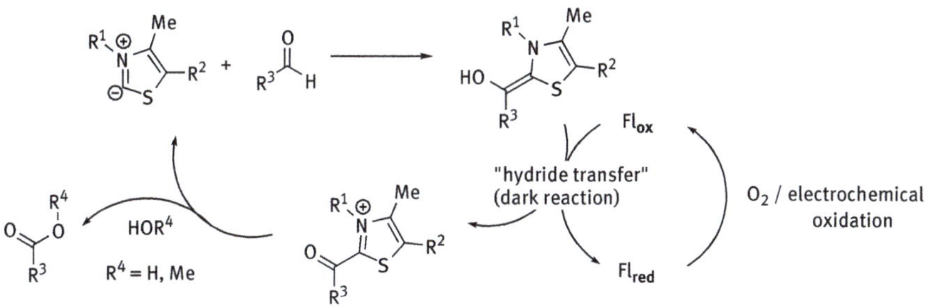

Figure 4.10: Principle of cooperative catalysis. Oxidation of aldehydes to acids with the help of a thiazolium ylide, Diederich et al. synthesized a flavo-thiazolio-cyclophane as catalyst for this reaction. [44]

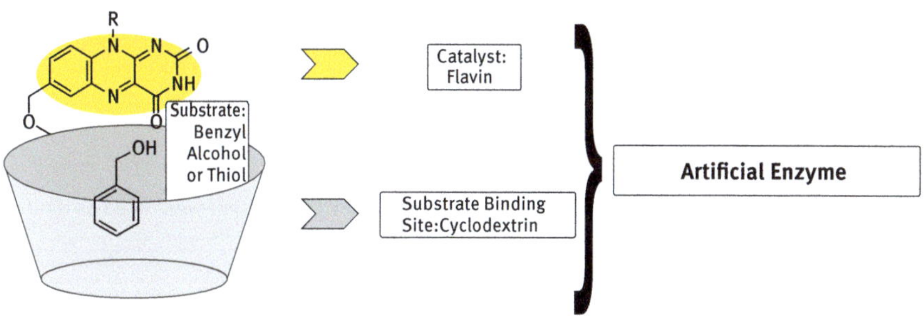

Figure 4.11: Artificial enzyme catalyzing the oxidation of benzyl alcohols or thiols more effective than riboflavin. [45]

hols turned out to be much better than with free riboflavin. [45] The photooxidation of benzyl alcohols showed a turnover number (TON) of 103, demonstrating an efficient recycling of the flavin-substituted cyclodextrin. Under the same conditions with riboflavin as catalyst a TON of only 6 could be obtained. [45]

Another attempt to design a supramolecular sensitizer containing a photoactive flavin unit used a zinc(II)-cyclen binding site. The assembly of a riboflavin-phenothiazine dyad based on a coordinative bond allowed an efficient intramolecular electron transfer between the electron-rich phenothiazine and the excited flavin (Figure 4.12), as well as the catalytic reductive photocleavage of thymine cyclobutane dimers. [15] The presence of the zinc(II)-cyclen unit covalently bound to the flavin also increases the efficiency of aerial 4-methoxybenzyl alcohol photooxidation significantly; the quantum yield was shown to be 30 times higher if the flavin-zinc(II)-complex (Figure 4.12) was used instead of the simple flavin sensitizer. The complex is water soluble and allows the photooxidation in an aqueous medium reaching a quantum yield of $\Phi = 0.4$. [46]

The double effect of the metal ion binding site has been shown in the light-mediated oxidative decarboxylation of mandelate salts by flavin possessing a crown ether moiety (Figure 4.12). It was found that potassium mandelate, having a cation

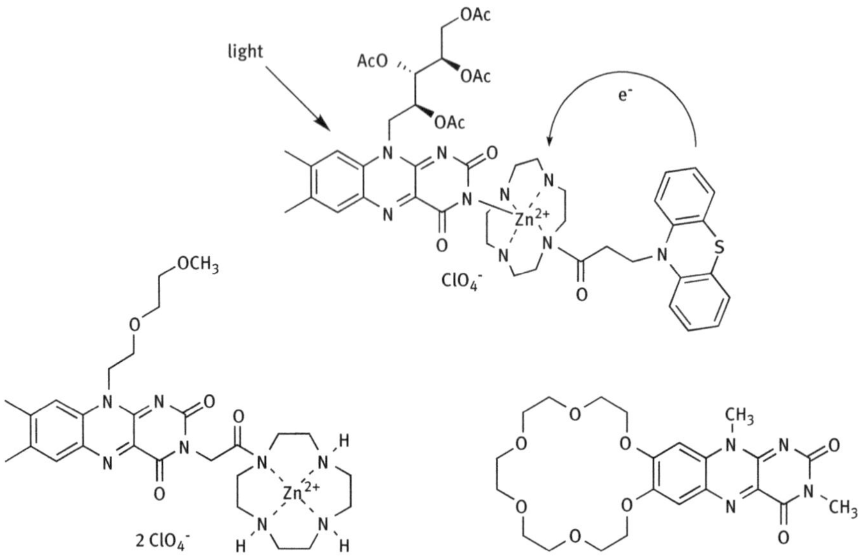

Figure 4.12: Supramolecular flavin catalysts: a zinc(II)-cyclen and a crown ether were used as binding sites.

with strong affinity to the 18-crown-6-ether host, is oxidized 110 times faster and with substantially higher quantum yield in comparison to the corresponding ammonium salt. [47]

The riboflavin tetraacetate photosensitized oxidation of substituted benzyl alcohols proceeds more efficiently in micelles of sodium dodecyl sulfate (SDS) than in acetonitrile solutions. The micelle-enhancing effect was attributed to the incorporation of a lipophilic flavin into an SDS micelle where the small volume of the micellar interior, the less polar medium and the negative charge of the micellar surface favors an electron transfer. [48]

In 2008, König et al. described the thiourea-enhanced flavin photooxidation of benzyl alcohol in acetonitrile solution. They report TONs up to 580 with their system; the presence of thiourea, either covalently bound to a flavin derivative or added stoichiometrically, led to a significant increase in the quantum yield of up to 30-fold. [49] They investigated the system further and discovered that the reaction works even better in aqueous solution where an addition of thiourea is not necessary. [50] A turnover frequency (TOF) of more than 800 h^{-1} and a TON of up to 68,800 were observed. The reaction worked also with heterogeneous photocatalysts with flavins being immobilized on solid supports, such as fluorous silica gel, reversed phase silica gel or PE pellets (see Figure 4.13). The catalytic activity decreased by a factor of 8–20 for the immobilization on silica gel and by a factor of 50 for the catalyst entrapment in polyethylene pellets compared to the reaction in homogeneous solution. [50]

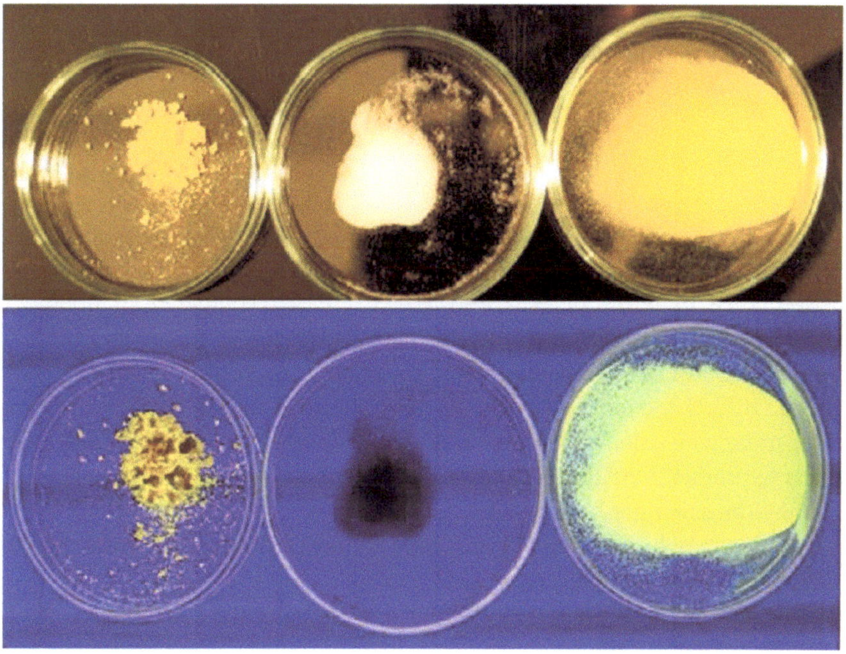

Figure 4.13: Immobilization of flavin catalysts on silica gel: normal light irradiation (top); UV blue light irradiation (bottom). Solid riboflavin tetraacetate (left), nonmodified silica gel (middle), immobilized catalyst (right). [50]

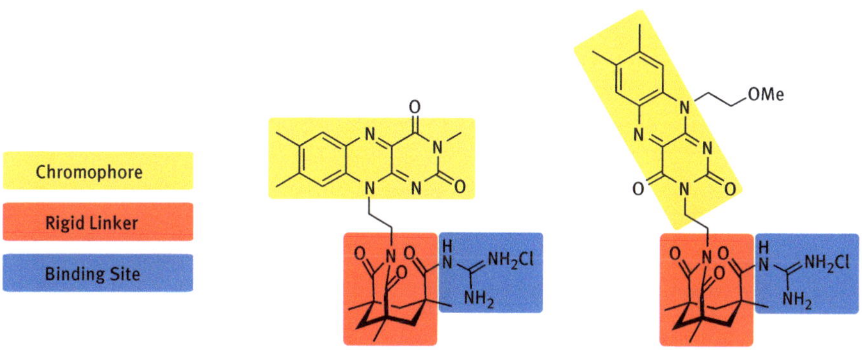

Figure 4.14: Flavin photocatalysts with guanidinium binding sites and Kemp's acid as rigid linker.

König et al. prepared new flavin derivatives with an acyl guanidinium group linked to the chromophore via a rigid Kemp's acid spacer (Figure 4.14). [51] This group was supposed to bind oxoanions, such as phosphates, via hydrogen bonds and was intended to position substrates in close proximity to the chromophore.

However, the expected benefit of the binding site for the photocatalytic activity was not observed as revealed by oxidative and reductive photocatalytic reactions.

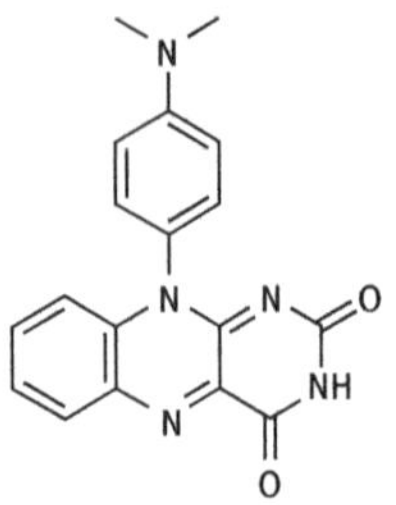

Figure 4.15: Structure of 10-(4-dimethylamino-phenyl)-isoalloxazine. A flavin derivative that is able to perform intramolecular charge transfer upon irradiation with light generating a very long-lived charge-separated state.

The performance in terms of TON and TOF decreased compared to riboflavin tetraacetate. Later, mechanistic studies of Riedle and Dick et al. rationalized this observation. [19] In the case of riboflavin tetraacetate the excited flavin has sufficient time in diffusion controlled reactions to reach the triplet state before colliding with a substrate molecule. It was shown, that only electron transfer from the triplet state leads to product formation. If the substrate is already positioned very close to the flavin by the substrate binding site, the electron transfer reaction occurs to the flavin singlet state, which will not produce the oxidation product due to very rapid back-electron transfer to the benzyl alcohol.

In 2010 Fukuzumi et al. reported an efficient intramolecular photoinduced electron transfer of flavin derivative DMA-Fl having an electron donor attached in 10-position (see Figure 4.15). They observed extremely small reorganization energies for electron self-exchange between DMA-Fl/DMA-Fl$^-$ resulting in a very long-lived charge-separation state (2.1 ms), which can do both: oxidize electron donors ($E_{ox} < 0.94$ V vs. SCE) and reduce electron acceptors ($E_{red} < -0.83$ V vs. SCE). [52]

In the same year several new applications of flavin photocatalysis with riboflavin tetraacetate (RFTA) were reported by König et al., for example the oxidation and deprotection of primary benzyl amines (see Figure 4.16). [53]

They reported the functionalization of toluene derivatives, the cleavage of styrenes and stilbenes (see Figure 4.17) and the direct oxidation of benzyl ethers to esters

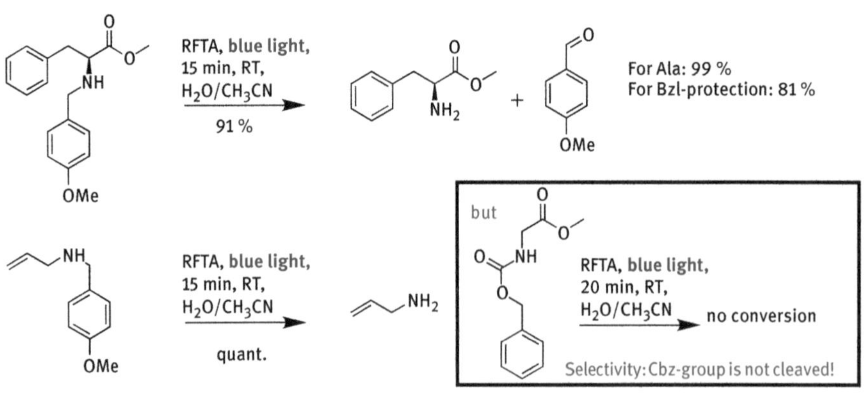

Figure 4.16: Deprotection of benzyl-protecting groups with flavin photocatalysis.

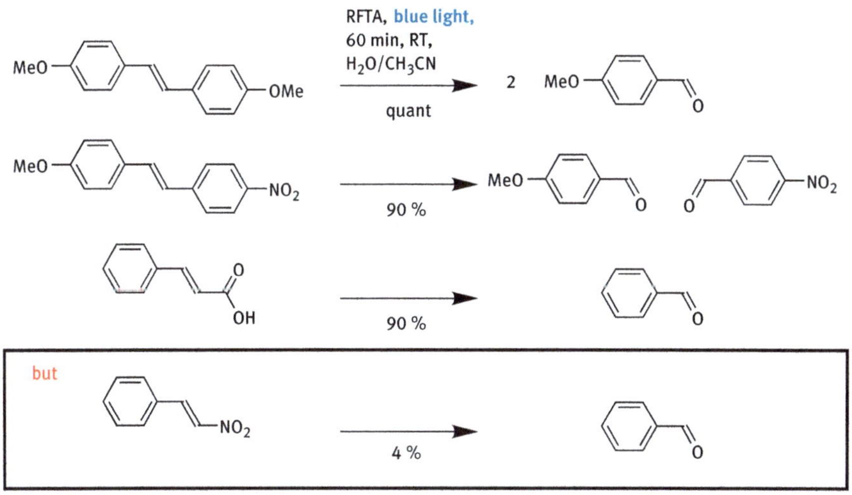

Figure 4.17: Cleavage of styrenes and stilbenes mediated by riboflavin tetraacetate (RFTA) and blue light.

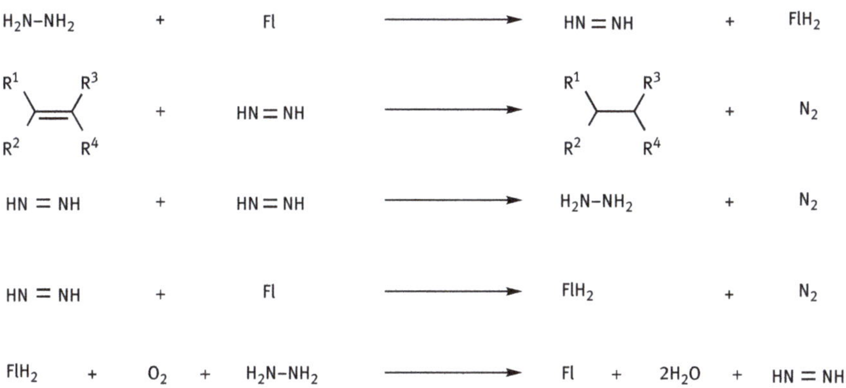

Figure 4.18: Indirect reductive use of flavins (Fl): hydrogenation of olefins via imide generation from hydrazine. The reduced flavin (FlH_2) is reoxidized by oxygen.

or benzyl amides to the corresponding imides. [54] These studies expand the possibilities of flavin applications significantly and show that not only benzyl alcohols are suitable substrates for flavin photooxidation. The details of the mechanism and kinetics of these oxidation reactions have been investigated by Riedle, Dick, König et al. in 2011 by transient absorption studies in the range of subpico to microseconds. [19]

Almost all applications of flavins as catalysts and photocatalysts in synthetic applications have been focused on oxidations. In 2010 Naota and Imada et al. used several flavins as organocatalysts for the aerobic hydrogenation of olefins via in situ generation of diimide from hydrazine producing water and nitrogen gas as the only waste products (Figure 4.18). [55] The reaction does not require irradiation.

4.4 Flavin-related compounds in photocatalysis

Additionally to flavins (isoalloxazines), the structurally related alloxazines and 5-deazaflavins occur in natural systems (see Figure 4.19). 8-Hydroxy-5-deazariboflavin acts as light harvesting cofactor in the class I photolyases. [56] Lumichrome is one of the photodegradation products of riboflavin. [57] Upon excitation of alloxazines without substitution in position N1 ($R^1 = H$, e. g. lumichrome), the N1-proton can be transferred to position N10 and thus the excited isoalloxazine form is created. [58] This phototautomerization can be catalyzed by proton donors or acceptors which are able to form hydrogen bonds with the alloxazine. [59]

Although the structure of alloxazines is very close to flavins, the photophysical properties of these classes of compounds differ. Particularly, fluorescence intensity and excited-state lifetimes are substantially lower for alloxazines than for flavins (see Table 4.1 for comparison). Nevertheless, there are some preliminary studies on the use of alloxazines in electron transfer processes. It was found that both singlet and triplet excited states of lumichrome are quenched with aliphatic and aromatic amines in methanol with similar rates as flavins. [60]

5-Ethylflavinium salts ($5EtFl^+ClO_4^-$, see Figure 4.20) have been applied mainly in catalysis of amine and sulfide oxidations and of Baeyer–Villiger oxidations with hydrogen peroxide or oxygen proceeding in the dark. [64] Quaternization of the flavin nitrogen N5 effects spectral and electrochemical properties of the flavin moiety significantly (Table 4.1). It increases the oxidation force of the flavin as evident from reduction potentials. On the other hand, the lifetime of flavinium salts excited states is short in comparison with neutral flavins. [63a] 5-Ethylflavinium salts easily add nucleophiles at position 4a (see Figure 4.20). [65] Water addition results in the formation of 5-ethyl-4a-hydroxyflavin (5EtFlOH) which is a suitable model compound for the investigation of the mechanism of bacterial bioluminescence. [8b, c, 66]

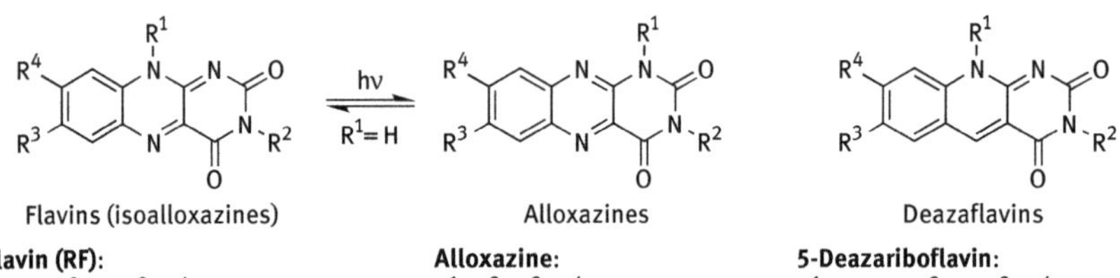

Riboflavin (RF):
R^1=ribityl, R^2=H, R^3=R^4=CH_3
Riboflavin tetraacetate (RFTA):
R^1=tetraacetylribityl, R^2=H, R^3=R^4=CH_3
3-Methyl riboflavin tetraacetate:
R^1=tetraacetylribityl, R^2=R^3=R^4=CH_3
Lumiflavin:
R^1=CH_3, R^2=H, R^3=R^4=CH_3

Alloxazine:
R^1=R^2=R^3=R^4=H
Lumichrome:
R^1=R^2= H, R^3=R^4=CH_3

5-Deazariboflavin:
R^1=ribityl, R^2=H, R^3=R^4=CH_3
8-Hydroxy-5-deazariboflavin:
R^1=ribityl, R^2=R^3=H, R^4=OH

Figure 4.19: Structures and names of important flavins (isoalloxazines), alloxazines and deazaflavins. Flavins can be transformed into alloxazines via phototautomerization and vice versa.

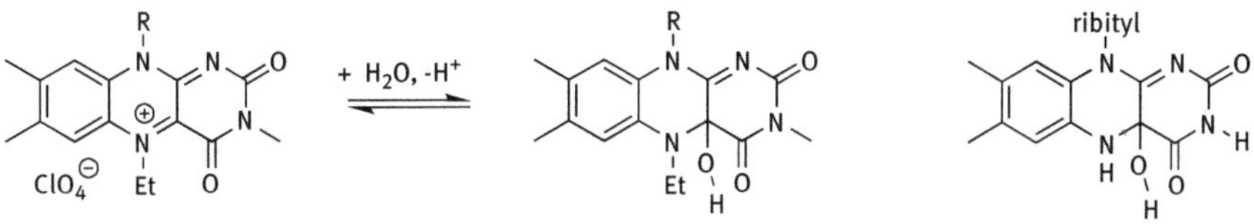

5-Ethylflavinium ($5EtFl^+ClO_4^-$)   4a-Hydroxy-5-ethylflavin (5EtFlOH)   4a-Hydroxyriboflavin

R = Et, Me

Figure 4.20: 5-Ethylflavinium, its water adduct and the natural analogue 4a-hydroxyriboflavin.

Table 4.1: Spectral and electrochemical characteristics of the selected flavin derivatives in acetonitrile.

Compound	λ_1 (nm)[a]	λ_2 (nm)[a]	λ_F (nm)[b]	Φ_F[c]	τ_F[d]	E^0 (V)[e]	Ref
Riboflavin tetraacetate	440	343	505	0.37	6.8 ns	−1.18(−1.87)	[15]
Lumiflavin	443	342	533	0.16	7.7 ns	−0.761	[55b,61]
5-Deazariboflavin	399	329	457	0.11	4.03 ns	–	[62]
Lumichrome	380	334	437	0.028	0.64 ns	−1.3	[60–61]
$5EtFl^+ClO_4^-$	557	414	661	–	590 ps	0.306 (−0.389)[f]	[63]
5EtFlOH	348	–	496	0.003	500 fs	–	[8c]

[a] λ_1 and λ_2 are the positions of the two lowest-energy bands in the absorption spectra; [b] maximum of the fluorescence emission spectrum; [c] fluorescence quantum yield; [d] fluorescence lifetime; [e] reversible redox potentials ($Fl \rightarrow Fl^{\bullet-}$ and $Fl^- \rightarrow Fl^{\bullet-}$ in the brackets) measured by CV using SCE as standard electrode; [f] value for 3,10-dimethyl-5-ethyl flavinium (R = Me in Figure 4.20).

In bacterial luciferase, excited 4a-hydroxyriboflavin is formed during the oxidation of a long-chain aldehyde (luciferin) to carboxylic acid and it returns to the ground state with emission of blue light (λ_{max} = 490 nm). [3b]

4.5 Photooxidations via singlet oxygen mechanism

Flavins and alloxazines are known to sensitize singlet oxygen production thus being able to participate on photooxidations proceeding by a singlet oxygen pathway (type II photooxidation). Singlet oxygen formation proceeds via energy transfer from triplet flavin to the ground state of triplet oxygen (Figure 4.21). [67]

Thus, the quantum yield of singlet oxygen production may be theoretically equal to the quantum yield of triplet flavin or alloxazine formation (≈ 0.7). In

$$^3Fl + {}^3O_2 \longrightarrow {}^1O_2 + Fl$$

Figure 4.21: General equation for singlet oxygen generation with flavins.

Table 4.2: Quantum yields of photosensitized production of singlet oxygen.[a]

Sensitizer	Solvent	$\Phi_\Delta (\lambda_{ex})$[b]	τ_F (μs)[c]	Ref.
Riboflavin tetraacetate	Acetonitrile	0.52 (355 nm)	–	[68]
Riboflavin tetraacetate	Ethanol	0.60 (355 nm)	–	[68]
3-Methyl riboflavin tetraacetate	Methanol	0.61 (355 nm)	10	[69]
Riboflavin	Methanol	0.51 (347 nm)	10	[70]
5-Deazariboflavin	Methanol	0.33 (355 nm)	10	[62]
Lumiflavin	Acetonitrile	0.85 (355 nm)	72	[67]
Lumiflavin	Methanol	0.48 (355 nm)	10	[61]
Lumiflavin	Water (pH 6)	0.31 (355 nm)	77	[71]
Lumichrome	Acetonitrile	0.73 (355 nm)	72	[67]
Lumichrome	Methanol	0.85 (355 nm)	10	[61]
Lumichrome	Water (pH 6)	0.36 (355 nm)	57	[71]

[a] For next data see also R. W. Rechmond, J. N. Gamlin, *Photochem. Photobiol.* **1999**, *70*, 391–475; [b] Quantum yields of singlet oxygen production (excitation wavelength); [c] singlet oxygen lifetime.

Table 4.2, quantum yields of singlet oxygen generation by various flavins as well as alloxazines in different solvents are compared.

Although singlet oxygen production by flavins is known for many decades, flavins have been utilized in type II photooxidations only rarely. This type of oxidation is rather described as a side reaction pathway to the electron transfer (type I) photooxidation, namely in oxidation of ascorbic acid, [72] tryptophan, [73] indole, [74] glucose, [75] and vitamin D, [76] however, electron transfer was described as being the dominant mechanism in these reactions. Flavin-sensitized photooxidation of esters of unsaturated fatty acids to the corresponding hydroperoxides were studied in more detail. [77] Both types of photooxidations were found to contribute to the formation of hydroperoxides from the esters of oleic, linoleic, linolenic and arachidonic acids. While the radical pathway results in conjugated hydroperoxides, singlet oxygen oxidation leads also to hydroperoxides with nonconjugated double bonds. [77a] Singlet oxygen becomes competitive to the free radical pathway with sufficient oxygen supply.

Riboflavin tetraacetate was found as an efficient sensitizer for the photooxidation of various types of sulfides to sulfoxides in alcohols (Figure 4.22). [68] The reaction is fastest in the presence of a small amount of water with the highest rates and quantum yields in 95 % ethanol (Φ up to 0.7). A dominant singlet oxygen mecha-

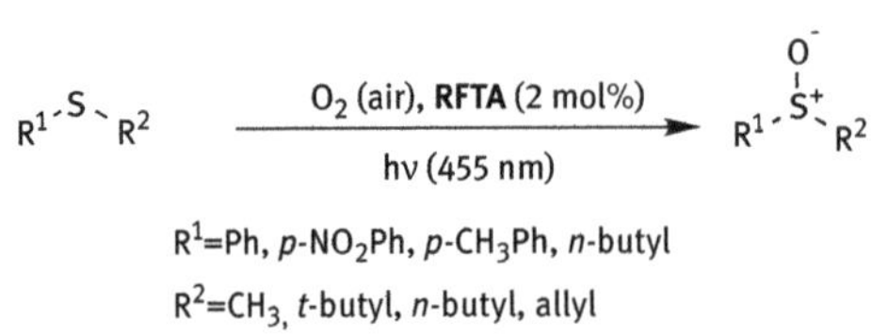

Figure 4.22: Oxidation of sulfides to sulfoxides: an example for the use of singlet oxygen produced by flavins.

nism was suggested based on significant differences of photooxidation rates in deuterated and nondeuterated solvents. It is advantageous that the reaction proceeds at low catalyst loading (2 mol%) and without side overoxidation to sulfones.

4.6 Conclusion

Flavin photocatalysis is a versatile and green method for several oxidation reactions in organic chemistry. The photocatalysts are easily accessible and possess high redox power in the excited state. In the last years some new reactions and catalysts were reported expanding the scope and applicability of the reaction. However, there are still problems to overcome: the photostability of the catalysts must be improved. Increasing the intersystem crossing rate to the triplet state of the oxidized form of flavin after excitation is necessary to use higher substrate concentrations and substrate binding sites. The application of reduced flavins as reduction reagents in organic synthesis is largely unexplored, but very promising as their potential can be further increased to very negative values upon irradiation with UV light (360 nm).

4.7 References

[1] Edwards, A. M. (2006). General Properties of Flavins. In: Flavins: Photochemistry and Photobiology, Silva, E., Edwards, A. M., Eds. (The Royal Society of Chemistry: Cambdrige), pp. 1–11.

[2] (a) Losi, A. and Gartner, W. (2008). Bacterial bilin- and flavin-binding photoreceptors. Photochem. Photobiol. Sci. 7, 1168–1178; (b) Losi, A. (2007). Flavin-based Blue-Light photosensors: a photobiophysics update. Photochem. Photobiol. 83, 1283–1300.

[3] (a) Baldwin, T. O., Christopher, J. A., Raushel, F. M., Sinclair, J. F., Ziegler, M. M., Fisher, A. J. and Rayment, I. (1995). Structure of bacterial luciferase. Curr. Opin. Struct. Biol. 5, 798–809; (b) Wilson, T. and Hastings, J. W. (1998). Bioluminescence. Annu. Rev. Cell Dev. Biol. 14, 197–230.

[4] (a) Heelis, P. F., Hartman, R. F. and Rose, S. D. (1995). Photoenzymic repair of UV-damaged DNA: a chemist's perspective. Chem. Soc. Rev. 24, 289; (b) Sancar, A. (2003). Structure and function of DNA photolyase and cryptochrome blue-light photoreceptors. Chem. Rev. 103, 2203–2237.

[5] Murov, S. L., Carmichael, I. and Hug, G. L. (1993). Handbook of Photochemistry (New York: CRC Press).

[6] (a) Fukuzumi, S. and Kojima, T. (2008). Control of redox reactivity of flavin and pterin coenzymes by metal ion coordination and hydrogen bonding. J. Biol. Inorg. Chem. 13, 321–333; (b) Jordan, B. J., Cooke, G., Garety, J. F., Pollier, M. A., Kryvokhyzha, N., Bayir, A., Rabani, G. and Rotello, V. M. (2007). Polymeric model systems for flavoenzyme activity: towards synthetic flavoenzymes. Chem. Commun. 1248–1250; (c) Mansoorabadi, S. O., Thibodeaux, C. J. and Liu, H. W. (2007). The diverse roles of flavin coenzymes – nature's most versatile thespians. J. Org. Chem. 72, 6329–6342; (d) Breinlinger, E., Niemz, A. and Rotello, V. M. (1995). Model Systems for Flavoenzyme Activity – Stabilization of the Flavin Radical-Anion through Specific Hydrogen-Bond Interactions. J. Am. Chem. Soc. 117, 5379–5380; (e) Cooke, G., Legrand, Y. M. and Rotello, V. M. (2004). Model systems for flavoenzyme activity: an electrochemically tuneable model of roseoflavin. Chem. Commun. 1088–1089; (f) Legrand, Y. M., Gray, M., Cooke, G. and Rotello, V.

M. (2003). Model systems for flavoenzyme activity: relationships between cofactor structure, binding and redox properties. J. Am. Chem. Soc. 125, 15789–15795.
[7] (a) Carell, T., Burgdorf, L., Butenandt, J., Epple, R. and Schwogler, A. (1999). DNA repair: from model compounds to artificial enzymes. Bioorg. Chem. 242–254; (b) Harrison, C. B., O'Neil, L. L. and Wiest, O. (2005). Computational studies of DNA photolyase. J. Phys. Chem. A 109, 7001–7012.
[8] (a) Kemal, C. and Bruice, T. C. (1976). Simple synthesis of a 4a-hydroperoxy adduct of a 1,5-dihydroflavine: preliminary studies of a model for bacterial luciferase. Proc. Natl. Acad. Sci. USA 73, 995–999; (b) Kemal, C. and Bruice, T. C. (1977). Chemiluminescence accompanying the decomposition of 4a-flavin alkyl peroxide. Model studies of bacterial luciferase. J. Am. Chem. Soc. 99, 7064–7067; (c) Zhou, D., Mirzakulova, E., Khatmullin, R., Schapiro, I., Olivucci, M. and Glusac, K. D. (2011). Fast excited-state deactivation in N(5)-ethyl-4a-hydroxyflavin pseudobase. J. Phys. Chem. B 115, 7136–7143.
[9] Blyth, A. W. (1879). LVI. – The composition of cows' milk in health and disease. J. Chem. Soc., Trans. 35, 530–539.
[10] Kuhn, R. and Weygand, F. (1934). Synthetisches Vitamin B2. Chem. Ber. 67, 2084–2085.
[11] Swartz, T. E., Corchnoy, S. B., Christie, J. M., Lewis, J. W., Szundi, I., Briggs, W. R. and Bogomolni, R. A. (2001). The photocycle of a flavin-binding domain of the blue light photoreceptor phototropin. J. Biol. Chem. 276, 36493–36500.
[12] Sadeghian, K., Bocola, M. and Schütz, M. (2008). A conclusive mechanism of the photoinduced reaction cascade in blue light using flavin photoreceptors. J. Am. Chem. Soc. 130, 12501–12513.
[13] Schmaderer, H., Svoboda, J. and König, B. (2009). Flavin Photocatalysts with Substrate Binding Sites. In: Activating Unreactive Substrates: The Role of Secondary Interactions, Bolm, C., Hahn, E., Eds. (Wiley-VCH: Weinheim), pp. 349–358.
[14] Ghisla, S., Kenney, W. C., Knappe, W. R., McIntire, W. and Singer, T. P. (1980). Chemical synthesis and some properties of 6-substituted flavins. Biochemistry 19, 2537–2544.
[15] König, B., Pelka, M., Zieg, H., Ritter, T., Bouas-Laurent, H., Bonneau, R. and Desvergne, J.-P. (1999). Photoinduced Electron Transfer in a Phenothiazine–Riboflavin Dyad Assembled by Zinc–Imide Coordination in Water. J. Am. Chem. Soc. 121, 1681–1687.
[16] Heelis, P. F. (1982). The photophysical and photochemical properties of flavins (isoalloxazines). Chem. Soc. Rev. 11, 15–39.
[17] Islam, S. D. M., Penzkofer, A. and Hegemann, P. (2003). Quantum yield of triplet formation of riboflavin in aqueous solution and of flavin mononucleotide bound to the LOV1 domain of Phot1 from Chlamydomonas reinhardtii. Chem. Phys. 291, 97–114.
[18] Kutta, R. J. (2012). PhD Thesis, Universität Regensburg.
[19] Megerle, U., Wenninger, M., Kutta, R. J., Lechner, R., Konig, B., Dick, B. and Riedle, E. (2011). Unraveling the flavin-catalyzed photooxidation of benzylic alcohol with transient absorption spectroscopy from sub-pico- to microseconds. Phys. Chem. Chem. Phys. 13, 8869–8880.
[20] Meisel, D. and Neta, P. (1975). One-electron reduction potential of riboflavine studied by pulse radiolysis. J. Phys. Chem. 79, 2459–2461.
[21] Amouyal, E. (1995). Photochemical Production of Hydrogen and Oxygen from Water – a Review and State-of-the-Art. Sol. Energy Mater. Sol. Cells 38, 249–276.
[22] (a) Miller, A. and Bruice, T. C. (1979). Oxidations by a 4a-hydroperoxyisoalloxazine hindered in the 9a and 10a positions. J. Chem. Soc., Chem. Commun. 896; (b) Eberlein, G. and Bruice, T. C. (1982). One and two electron reduction of oxygen by 1,5-dihydroflavins. J. Am. Chem. Soc. 104, 1449–1452; (c) Bruice, T. C. (1980). Mechanisms of flavin catalysis. Acc. Chem. Res. 13, 256–262.
[23] Massey, V. (2000). The Chemical and Biological Versatility of Riboflavin. Biochem. Soc. Trans. 28, 283–296.
[24] Rehm, D. and Weller, A. (1969). Kinetik und Mechanismus der Elektronübertragung bei der Fluoreszenzlöschung in Acetonitril. Ber. Bunsen-Ges. Phys. Chem. 73, 834–839.

[25] Julliard, M. and Chanon, M. (1983). Photoelectron-transfer catalysis: its connections with thermal and electrochemical analogs. Chem. Rev. (Washington, DC, U. S.) 83, 425–506.
[26] (a) Ogston, F. J. and Green, D. E. (1935). The mechanism of the reaction of substrates with molecular oxygen. II. Biochem. J. 29, 2005–2012; (b) Ogston, F. J. and Green, D. E. (1935). The Mechanism of the Reaction of Substrates with Molecular Oxygen. I. Biochem. J. 29, 1983–2004; (c) Kuhn, R., Rudy, H. and Weygand, F. (1936). Über die Bildung eines künstlichen Ferments aus 6.7-Dimethyl-9-l-araboflavin-5′-phosphorsäure. Chem. Ber. 69, 2034–2036; (d) Kuhn, R. and Rudy, H. (1936). Lactoflavin als Co-Ferment; Wirkstoff und Träger. Chem. Ber. 69, 2557–2567; (e) Green, D. E. (1936). α-Glycerophosphate Dehydrogenase. Biochem. J. 30, 629–644; (f) Das, B. N. (1936). Studies on Amino-Acid Dehydrogenase. II. Activator of Proline Dehydrogenase. Biochem. J. 30, 1617–1621; (g) Kuhn, R. and Ströbele, R. (1937). Synthese von Flavin-glucosiden. Chem. Ber. 70, 747–752; (h) Kuhn, R., Vetter, H. and Rzeppa, H. W. (1937). Zur Spezifität des lactoflavins; Ersatz der Methylgruppen durch den Tetramethylen- und Trimethylen-Ring. Chem. Ber. 70, 1302–1314; (i) Kuhn, R. and Ströbele, R. (1937). Über o-Nitranilin-glucoside. Chem. Ber. 70, 773–787; (j) Dewan, J. G. and Green, D. E. (1937). A New Oxidation Catalyst. Nature 140, 1097–1098; (k) Adler, E. and Euler, H. V. (1938). Lactoflavin in the Eyes of Fish. Nature 141, 790–791; (l) Dewan, J. G. and Green, D. E. (1938). Coenzyme Factor – A New Oxidation Catalyst. Biochem. J. 32, 626–639; (m) Corran, H. S., Green, D. E. and Straub, F. B. (1939). The catalytic function of heart flavoprotein. Biochem. J. 33, 793–801.
[27] Lipmann, F. (1939). Flavin Component of the Pyruvic Acid Oxidation System. Nature 143, 436–436.
[28] Galston, A. W. (1949). Riboflavin-Sensitized Photooxidation of Indoleacetic Acid and Related Compounds. Proc. Natl. Acad. Sci. USA 35, 10–17.
[29] (a) Frisell, W. R., Chung, C. W. and Mackenzie, C. G. (1959). Catalysis of Oxidation of Nitrogen Compounds by Flavin Coenzymes in the Presence of Light. J. Biol. Chem. 234, 1297–1302; (b) Enns, K. and Burgess, W. H. (1965). The Photochemical Oxidation of Ethylenediaminetetraacetic Acid and Methionine by Riboflavin. J. Am. Chem. Soc. 87, 5766–5770; (c) McCormick, D. B., Koster, J. F. and Veeger, C. (1967). On the Mechanisms of Photochemical Reductions of FAD and FAD-Dependent Flavoproteins. Eur. J. Biochem. 2, 387–391; (d) Byrom, P. and Turnbull, J. H. (1968). Excited States of Flavine Coenzymes-Tv. Kinetics of the Photoreduction of Lumiflavine by Methionine. Photochem. Photobiol. 8, 243–254; (e) Penzer, G. R. and Radda, G. K. (1968). The Chemistry of Flavines and Flavoproteins – Photoreduction of Flavins by Amino Acids. Biochem. J. 109, 259–268.
[30] (a) Suelter, C. H. and Metzler, D. E. (1960). The oxidation of a reduced pyridine nucleotide analog by flavins. Biochim. Biophys. Acta 44, 23–33; (b) Fox, J. L. and Tollin, G. (1966). Flavoenzyme Models. I. Flavin Free-Radical Formation in the Reduced Nicotinamide-Adenine Dinucleotide-Flavin Mononucleotide System. Biochemistry 5, 3865–3872.
[31] (a) Radda, G. K. and Calvin, M. (1964). Chemical and Photochemical Reductions of Flavin Nucleotides and Analogs. Biochemistry 3, 384–393; (b) Penzer, G. R. and Radda, G. K. (1967). The chemistry and biological function of isoalloxazines (flavines). Q. Rev. Chem. Soc. 21, 43; (c) Byrom, P. and Turnbull, J. H. (1967). Excited States of Flavine Coenzymes – II. Anaerobic Oxidation of Amino Acids by Excited Riboflavine Devrivatives. Photochem. Photobiol. 6, 125–131.
[32] Yang, S. F., Ku, H. S. and Pratt, H. K. (1967). Photochemical Production of Ethylene from Methionine and Its Analogues in the Presence of Flavin Mononucleotide. J. Biol. Chem. 242, 5274–5280.
[33] (a) Hemmerich, P., Massey, V. and Weber, G. (1967). Photo-induced Benzyl Substitution of Flavins by Phenylacetate: a Possible Model for Flavoprotein Catalysis. Nature 213, 728–730; (b) Walker, W. H., Hemmerich, P. and Massey, V. (1967). Reductive photoalkylation of flavin nuclei and flavin-catalyzed photodecarboxylation of phenylacetate. Helv. Chim. Acta 50, 2269–2279.

[34] Walker, W. H., Hemmerich, P. and Massey, V. (1970). Light-Induced Alkylation and Dealkylation of the Flavin Nucleus. Stable Dihydroflavins: Spectral Course and Mechanism of Formation. Eur. J. Biochem. 13, 258–266.
[35] Tishler, M., Pfister, K., Babson, R. D., Ladenburg, K. and Fleming, A. J. (1947). The Reaction between o-Aminoazo Compounds and Barbituric Acid. A New Synthesis of Riboflavin. J. Am. Chem. Soc. 69, 1487–1492.
[36] Bruestlein, M. and Hemmerich, P. (1968). Photoreduction of flavocoenzymes by pyruvic acid. FEBS Lett. 1, 335–338.
[37] Hemmerich, P., Nagelschneider, G. and Veeger, C. (1970). Chemistry and molecular biology of flavins and flavoproteins. FEBS Lett. 8, 69–83.
[38] Weatherby, G. D. and Carr, D. O. (1970). Riboflavine-catalyzed photooxidative decarboxylation of dihydrophthalates. Biochemistry 9, 344–350.
[39] (a) Yoneda, F., Mori, K., Ono, M., Kadokawa, Y., Nagao, E. and Yamaguchi, H. (1980). Syntheses of 2-deoxo-2-phenyl-5-deazaflavins and 3-phenyl-5-deazaflavins and their use in the oxidation of benzyl alcohol and benzylamine. Chem. Pharm. Bull. 28, 3514–3520; (b) Nagamatsu, T., Matsumoto, E. and Yoneda, F. (1982). Autorecycling oxidation of alcohol catalyzed by pyrimidopteridines as a flavin model. Chem. Lett. 1127–1130.
[40] (a) Fukuzumi, S., Tanii, K. and Tanaka, T. (1989). Protonated pteridine and flavin analogues acting as efficient and substrate-selective photocatalysts in the oxidation of benzyl alcohol derivatives by oxygen. J. Chem. Soc., Chem. Commun. 816; (b) Fukuzumi, S., Kuroda, S. and Tanaka, T. (1985). Flavin analog-metal ion complexes acting as efficient photocatalysts in the oxidation of p-methylbenzyl alcohol by oxygen under irradiation with visible light. J. Am. Chem. Soc. 107, 3020–3027.
[41] Tong, W., Ye, H., Zhu, H. and D'Souza, V. T. (1995). Photooxidation of substituted benzyl alcohol by riboflavin. J. Mol. Struct. THEOCHEM 333, 19–27.
[42] Fukuzumi, S. and Kuroda, S. (1999). Photooxidation of benzyl alcohol derivatives by oxygen, catalyzed by protonated flavin analogs. Res. Chem. Intermed. 25, 789–811.
[43] Fukuzumi, S., Yasui, K., Suenobu, T., Ohkubo, K., Fujitsuka, M. and Ito, O. (2001). Efficient Catalysis of Rare-Earth Metal Ions in Photoinduced Electron-Transfer Oxidation of Benzyl Alcohols by a Flavin Analogue. J. Phys. Chem. A 105, 10501–10510.
[44] Mattei, P. and Diederich, F. (1997). Catalytic Cyclophanes. Part XI. A flavo-thiazolio-cyclophane as a biomimetic catalyst for the preparative-scale electro-oxidation of aromatic aldehydes to methyl esters. Helv. Chim. Acta 80, 1555–1588.
[45] D'Souza, V. T. (2003). Modification of cyclodextrins for use as artificial enzymes. Supramol. Chem. 15, 221–229.
[46] Cibulka, R., Vasold, R. and König, B. (2004). Catalytic photooxidation of 4-methoxybenzyl alcohol with a flavin-zinc(II)-cyclen complex. Chem. Eur. J. 10, 6223–6231.
[47] (a) Shinkai, S., Nakao, H., Ueda, K. and Manabe, O. (1984). Light-mediated oxidation of alcohols and mandelate by flavin-metal complexes. Tetrahedron Lett. 25, 5295–5298; (b) Shinkai, S., Nakao, H., Ueda, K., Manabe, O. and Ohnishi, M. (1986). Selective photooxidation of alkali mandelates by a flavin bearing a crown ring as a metal recognition site. Bull. Chem. Soc. Jpn. 59, 1632–1634.
[48] Yasuda, M., Nakai, T., Kawahito, Y. and Shiragami, T. (2003). Micelle-Enhancing Effect on a Flavin-Photosensitized Reaction of Benzyl Alcohols in Aqueous Solution. Bull. Chem. Soc. Jpn. 76, 601–605.
[49] Svoboda, J., Schmaderer, H. and König, B. (2008). Thiourea-enhanced flavin photooxidation of benzyl alcohol. Chem. Eur. J. 14, 1854–1865.
[50] Schmaderer, H., Hilgers, P., Lechner, R. and König, B. (2009). Photooxidation of Benzyl Alcohols with Immobilized Flavins. Adv. Synth. Catal. 351, 163–174.

[51] Schmaderer, H., Bhuyan, M. and König, B. (2009). Synthesis of rigidified flavin-guanidinium ion conjugates and investigation of their photocatalytic properties. Beilstein J. Org. Chem. 5, 26.
[52] Murakami, M., Ohkubo, K. and Fukuzumi, S. (2010). Inter- and intramolecular photoinduced electron transfer of flavin derivatives with extremely small reorganization energies. Chem. Eur. J. 16, 7820–7832.
[53] Lechner, R. and König, B. (2010). Oxidation and Deprotection of Primary Benzylamines by Visible Light Flavin Photocatalysis. Synthesis 2010, 1712–1718.
[54] Lechner, R., Kümmel, S. and König, B. (2010). Visible light flavin photo-oxidation of methylbenzenes, styrenes and phenylacetic acids. Photochem. Photobiol. Sci. 9, 1367–1377.
[55] (a) Imada, Y., Kitagawa, T., Ohno, T., Iida, H. and Naota, T. (2010). Neutral flavins: green and robust organocatalysts for aerobic hydrogenation of olefins. Organic letters 12, 32–35; (b) Imada, Y., Iida, H., Kitagawa, T. and Naota, T. (2011). Aerobic reduction of olefins by in situ generation of diimide with synthetic flavin catalysts. Chem. Eur. J. 17, 5908–5920.
[56] (a) Kay, C. W. M., Bacher, A., Fischer, M., Richter, G., Schleicher, E. and Weber, S. (2006). Blue Light-Initiated DNA Repair by Photolyase. In: Flavins: Photochemistry and Photobiology, Silva, E., Edwards, A. M., Eds. (The Royal Society of Chemistry: Cambdrige), pp. 151–182; (b) Kim, S.-T. and Sancar, A. (1993). Photochemistry, Photophysics and Mechanism of Pyrimidine Dimer Repair by DNA Photolyase. Photochem. Photobiol. 57, 895–904; (c) Carell, T. and Epple, R. (1998). Repair of UV Light Induced DNA Lesions: A Comparative Study with Model Compounds. Eur. J. Org. Chem. 1998, 1245–1258.
[57] (a) Metzler, D. E. and Cairns, W. L. (1971). Photochemical degradation of flavines. VI. New photoproduct and its use in studying the photolytic mechanism. J. Am. Chem. Soc. 93, 2772–2777; (b) Kino, K., Kobayashi, T., Arima, E., Komori, R. and Miyazawa, H. (2009). Photoirradiation products of flavin derivatives, and the effects of photooxidation on guanine. Bioorg. Med. Chem. Lett. 19, 2070–2074.
[58] Song, P.-S., Sun, M., Koziolowa, A. and Koziol, J. (1974). Phototautomerism of lumichromes and alloxazines. J. Am. Chem. Soc. 96, 4319–4323.
[59] (a) Kozioł, J. (1969). Studies on Flavins in Organic Solvents – III. Spectral Behaviour of Lumifalvin. Photochem. Photobiol. 9, 45–53; (b) Moyon, N. S. and Mitra, S. (2011). Fluorescence solvatochromism in lumichrome and excited-state tautomerization: a combined experimental and DFT study. J. Phys. Chem. A 115, 2456–2464.
[60] Porcal, G., Bertolotti, S. G., Previtali, C. M. and Encinas, M. V. (2003). Electron transfer quenching of singlet and triplet excited states of flavins and lumichrome by aromatic and aliphatic electron donors. Phys. Chem. Chem. Phys. 5, 4123.
[61] Sikorska, E., Khmelinskii, I. V., Prukała, W., Williams, S. L., Patel, M., Worrall, D. R., Bourdelande, J. L., Koput, J. and Sikorski, M. (2004). Spectroscopy and Photophysics of Lumiflavins and Lumichromes. J. Phys. Chem. A 108, 1501–1508.
[62] Insińska-Rak, M., Sikorska, E., Bourdelande, J. L., Khmelinskii, I. V., Prukała, W., Dobek, K., Karolczak, J., Machado, I. F., Ferreira, L. F. V., Komasa, A., Worrall, D. R. and Sikorski, M. (2006). Spectroscopy and photophysics of flavin-related compounds: 5-deaza-riboflavin. J. Mol. Struct. 783, 184–190.
[63] (a) Sichula, V., Kucheryavy, P., Khatmullin, R., Hu, Y., Mirzakulova, E., Vyas, S., Manzer, S. F., Hadad, C. M. and Glusac, K. D. (2010). Electronic properties of N(5)-ethyl flavinium ion. J. Phys. Chem. A 114, 12138–12147; (b) Imada, Y., Iida, H., Ono, S., Masui, Y. and Murahashi, S. (2006). Flavin-catalyzed oxidation of amines and sulfides with molecular oxygen: biomimetic green oxidation. Chem. Asian J. 1, 136–147.
[64] (a) Gelalcha, F. G. (2007). Heterocyclic hydroperoxides in selective oxidations. Chem. Rev. 107, 3338–3361; (b) Imada, Y. and Naota, T. (2007). Flavins as organocatalysts for environmentally benign molecular transformations. Chem. Rec. 7, 354–361; (c) Murahashi, S., Oda, T. and Masui, Y.

(1989). Flavin-catalyzed oxidation of amines and sulfur compounds with hydrogen peroxide. J. Am. Chem. Soc. 111, 5002–5003; (d) Jurok, R., Cibulka, R., Dvořáková, H., Hampl, F. and Hodačová, J. (2010). Planar Chiral Flavinium Salts – Prospective Catalysts for Enantioselective Sulfoxidation Reactions. Eur. J. Org. Chem. 2010, 5217–5224; (e) Murahashi, S.-I., Ono, S. and Imada, Y. (2002). Asymmetric Baeyer-Villiger Reaction with Hydrogen Peroxide Catalyzed by a Novel Planar-Chiral Bisflavin. Angew. Chem. Int. Ed. 41, 2366–2368; (f) Baxová, L., Cibulka, R. and Hampl, F. (2007). Organocatalytic sulfoxidation in micellar systems containing amphiphilic flavinium salts using hydrogen peroxide as a terminal oxidant. J. Mol. Catal. A: Chem. 277, 53–60.

[65] (a) Kim, J. M., Bogdan, M. A. and Mariano, P. S. (1993). Mechanistic analysis of the 3-methyllumiflavin-promoted oxidative deamination of benzylamine. A potential model for monoamine oxidase catalysis. J. Am. Chem. Soc. 115, 10591–10595; (b) Ménová, P., Eigner, V., Čejka, J., Dvořáková, H., Šanda, M. and Cibulka, R. (2011). Synthesis and structural studies of flavin and alloxazine adducts with O-nucleophiles. J. Mol. Struct. 1004, 178–187.

[66] Lei, B., Ding, Q. and Tu, S. C. (2004). Identity of the emitter in the bacterial luciferase luminescence reaction: binding and fluorescence quantum yield studies of 5-decyl-4a-hydroxy-4a,5-dihydroriboflavin-5′-phosphate as a model. Biochemistry 43, 15975–15982.

[67] Sikorska, E., Sikorski, M., Steer, R. P., Wilkinson, F. and Worrall, D. R. (1998). Efficiency of singlet oxygen generation by alloxazines and isoalloxazines. J. Chem. Soc., Faraday Trans. 94, 2347–2353.

[68] Dad'ová, J., Svobodová, E., Sikorski, M., König, B. and Cibulka, R. (2012). Photooxidation of Sulfides to Sulfoxides Mediated by Tetra-O-Acetylriboflavin and Visible Light. ChemCatChem 4, 620–623.

[69] Insińska-Rak, M., Sikorska, E., Bourdelande, J. L., Khmelinskii, I. V., Prukała, W., Dobek, K., Karolczak, J., Machado, I. F., Ferreira, L. F. V., Dulewicz, E., Komasa, A., Worrall, D. R., Kubicki, M. and Sikorski, M. (2007). New photochemically stable riboflavin analogue-3-Methyl-riboflavin tetraacetate. J. Photochem. Photobiol., A 186, 14–23.

[70] Sikorska, E., Khmelinskii, I., Komasa, A., Koput, J., Ferreira, L. F. V., Herance, J. R., Bourdelande, J. L., Williams, S. L., Worrall, D. R., Insińska-Rak, M. and Sikorski, M. (2005). Spectroscopy and photophysics of flavin related compounds: Riboflavin and iso-(6,7)-riboflavin. Chem. Phys. 314, 239–247.

[71] Sikorski, M., Sikorska, E., Koziolowa, A., Gonzalez Moreno, R., Bourdelande, J. L., Steer, R. P. and Wilkinson, F. (2001). Photophysical properties of lumichromes in water. J. Photochem. Photobiol., B 60, 114–119.

[72] Sahbaz, F. and Somer, G. (1993). Photosensitized decomposition of ascorbic acid in the presence of riboflavin. Food Chem. 46, 177–182.

[73] Kanner, J. D. and Fennema, O. (1987). Photooxidation of tryptophan in the presence of riboflavin. J. Agric. Food Chem. 35, 71–76.

[74] Yoshimura, A. and Ohno, T. (1988). Lumiflavin-Sensitized Photooxygenation of Indole. Photochem. Photobiol. 48, 561–565.

[75] Silva, E., Edwards, A. M. a. and Pacheco, D. (1999). Visible light-induced photooxidation of glucose sensitized by riboflavin. J. Nutr. Biochem. 10, 181–185.

[76] King, J. M. and Min, D. B. (1998). Riboflavin Photosensitized Singlet Oxygen Oxidation of Vitamin D. J. Food Sci. 63, 31–34.

[77] (a) Fukuzumi, S., Tanii, K. and Tanaka, T. (1989). Flavin-sensitized photo-oxidation of unsaturated fatty acids. J. Chem. Soc., Perkin Trans. 2 2103; (b) Chacon, J. N., McLearie, J. and Sinclair, R. S. (1988). Singlet Oxygen Yields and Radical Contributions in the Dye-Sensitized Photooxidation in Mehtanol of Esters of Polyunsaturated Fatty Acids (Oleic, Linoleic, Linolenic and Arachidonic). Photochem. Photobiol. 47, 647–656; (c) Huvaere, K., Cardoso, D. R., Homemde-Mello, P., Westermann, S. and Skibsted, L. H. (2010). Light-induced oxidation of unsaturated lipids as sensitized by flavins. J. Phys. Chem. B 114, 5583–5593.

Andreas Bauer, Rafael Alonso Ruiz

5 Templated Enantioselective Photocatalysis

5.1 Introduction

Enantioselective photochemistry represents a specially challenging field of synthetic organic chemistry. In reactions of this kind, prochiral substrates are converted into enantiomerically pure or enriched products. These photoreactions can proceed in an enantioselective manner if enantiotopic faces or groups of the substrate are sufficiently differentiated by a chiral entity. The degree of enantioselectivity is measured by the enantiomeric excess (*ee*) of the corresponding product.

A number of photochemical approaches to prepare enantiomerically enriched compounds from prochiral substrates in solution can be found in the literature. [1] Many efforts have been focused on the use of circularly polarized light as a source of the chiral information to induce an effective differentiation of enantiotopic faces but, only recently, a few examples of excellent *ee* have been reported. [2] Photoreactions carried out using chiral solvents did not give rise to highly enantiopure products. [3] A different approach relies on a chiral solid state environment as the source of chirality. In this context, photochemical reactions were carried out within homochiral crystals, [4] employing the "ionic chiral auxiliary" approach, [5] in inclusion complexes, [6] and using chiral cavities such as chirally modified zeolites. [7]

Chiral auxiliaries covalently bound to one of the substrates have given great results in many cases [8] but, in general, it is desirable to transfer the chiral information by a noncovalent interaction, thus, avoiding additional steps for the introduction and the cleavage of the chiral moiety from the substrate. Furthermore, reactions taking place in solution greatly expand the scope of suitable substrates and also offer the possibility of performing intermolecular processes. Possible synthetic methods following this strategy are based on the stoichiometric use of chiral reagents or chiral complexing agents. Until recently, however, only a few examples of this kind of photoreactions using complexing agents have been performed. [9]

By definition, chiral complexing agents interact with prochiral substrates by noncovalent interactions in equilibrium. Apart from other supramolecular interactions, [10] hydrogen bonds are the most suitable noncovalent interactions to provide a chiral environment, [11] in which a photochemical reaction takes place. Furthermore, this approach appeared particularly promising for enantiocontrol in photochemical reactions, since light does not normally interfere with hydrogen bonds. Following this idea, templates, which exhibit hydrogen bond donor and hydrogen bond acceptor sites, have been designed.

The templates employed for this purpose were 1,5,7-trimethyl-3-azabicyclo[3.3.1]nonan-2-one derivatives which are available from *cis,cis*-1,3,5-trimethylcyclohexane-1,3,5-tricarboxylic acid (Kemp's triacid). [12] They act as rigid hosts which offer

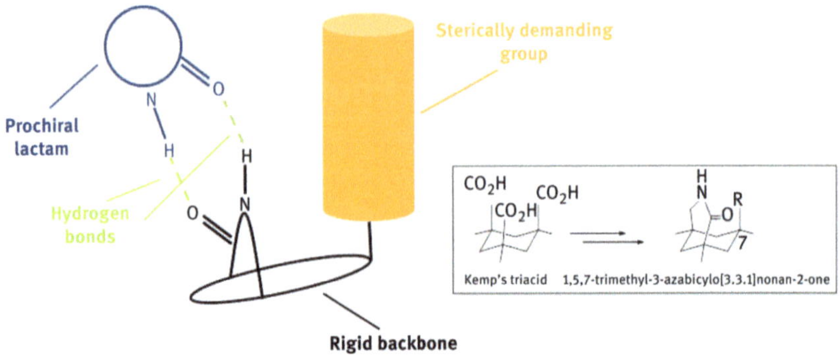

Figure 5.1: General mode of action of hydrogen bonding templates based on Kemp's triacid for the differentiation of enantiotopic faces in prochiral lactams.

an easily accessible binding site and which simultaneously exhibit an effective sterically demanding group that is able to protect one of the faces of the substrate bound to the complexing agent. The self-association of the enantiopure compounds is low due to steric hindrance of the substituent in the 7-position of the chiral backbone. Consequently, templates of this type can coordinate to prochiral amides or lactams through a pair of self-complementary hydrogen bonds between the respective carbonyl oxygen atoms as acceptors and the nitrogen-bound hydrogen atoms as donors (Figure 5.1). These chiral complexing agents have enabled both intra- and intermolecular photochemical reactions in solution to be performed with unprecedented yields and enantiomeric excesses. These compounds were also used as effective ^{1}H NMR shift reagents for the *ee* determination of chiral lactams, quinolones and oxazolidinones, and they were employed for the determination of the absolute configuration of new chiral templates and catalysts derived from Kemp's triacid. [13]

Performing enantioselective reactions in a catalytic fashion was another major step in this field. For this purpose chiral complexing agents were designed, which are able to absorb light and allow sensitization of the substrate by electron or energy transfer.

The most relevant results obtained in the field of template enantioselective photocatalysis during the last few years are reported herein. A number of photochemical processes performed with a high enantioselectivity are detailed below, namely Paternò–Büchi reactions, [4+4]-photocycloadditions of 2-pyridones, Norrish–Yang processes, [6π]-photocyclizations of anilides, Diels–Alder processes, formal [3+2]-photocycloadditions of 2-substituted naphthoquinones, radical reactions (irrespective of whether the radical chain process was initiated photolytically or chemically) and [2+2]-photocycloadditions, some of them performed in a catalytic photoinduced electron or energy transfer fashion.

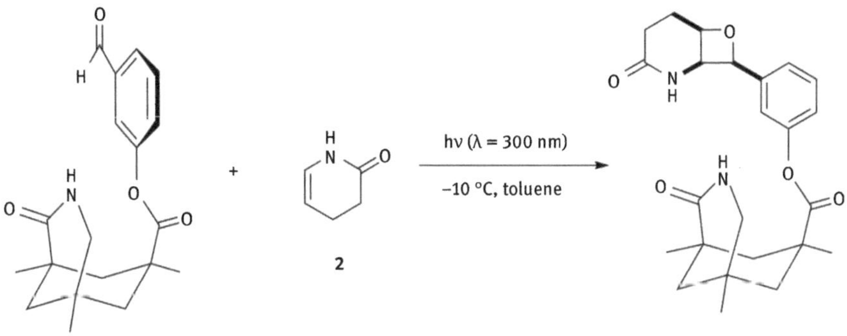

Figure 5.2: Paternò–Büchi cycloaddition reactions of the chiral aromatic aldehyde **1** and 3,4-dihydro-2-pyridone (**2**).

5.2 Early studies. Paternò–Büchi cycloadditions of a chiral aromatic aldehyde and cyclic enamines

Initial work in the field of template-mediated stereoselective photoprocesses was related to the use of hydrogen bonds in order to obtain facial diastereoselectivity for the well-known Paternò–Büchi reaction between an aldehyde and an olefin leading to oxetanes. For this purpose compound **1** derived from Kemp's triacid was synthesized. Reaction of **1** with 3,4-dihydro-2-pyridone (**2**) gave rise to the desired oxetane **3** (Figure 5.2).

Precoordination of compound **2** to lactam **1** provided a high facial diastereoselectivity for the reaction (up to 95 % *de*). The photoexcited aldehyde reacts predominantly within the association complex, thus approaching the bound compound **2** from a single enantioface. The best results were found when the reaction was carried out in a nonpolar solvent like toluene and at low temperature (−10 °C). The photocycloaddition of enantiomerically pure (+)-**1** to **2** gave the enantiomerically pure oxetane (−)-**3**. The bicyclic oxetano[2,3-*b*]piperidone fragment could be readily cleaved by transesterification. [14]

5.3 Enantioselective Norrish–Yang cyclization reaction of prochiral imidazolidinones

The Norrish–Yang reaction was performed for imidazolidinone **6** (see below). The reaction was carried out in toluene at low temperature and in the presence of the templates **4** and **5** derived from Kemp's triacid depicted in Figure 5.3. These diastereomeric complexing agents are expected to behave as if they were enantiomers.

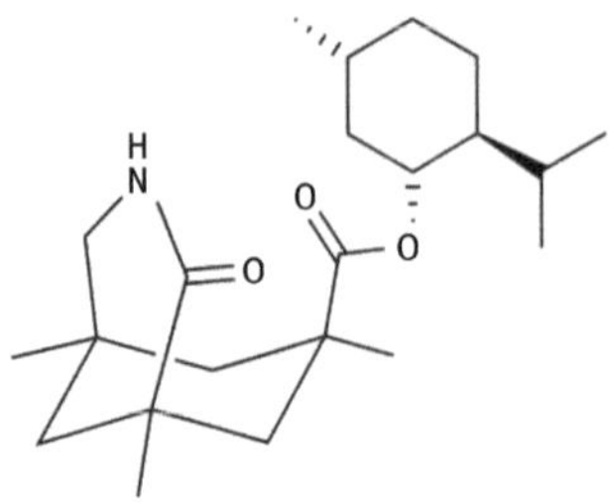

4

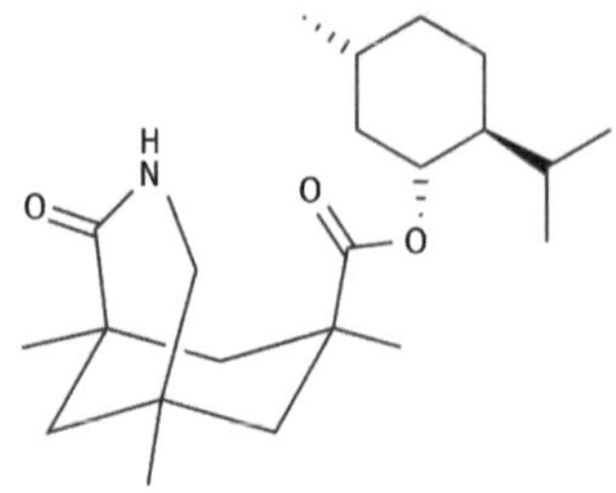

5

Figure 5.3: Structures of menthyl esters **4** and **5**.

Good yields (~70–80 %) and diastereoselectivities (80–90 % in favor of the *exo* isomer) were obtained for the reaction, but unfortunately, although both templates were found to bind to **6** through two hydrogen bonds and provide a chiral environment in which the Norrish–Yang reaction takes place, enantioselectivities were found to be no higher than 26 % *ee*.

In order to improve these results, template **7** and its enantiomer *ent*-**7**, presenting the planar group tetrahydronaphthalene as a sterically demanding shield, were prepared. When these templates were used for the Norrish–Yang reaction of imidazolidinone **6** the yield and the simple diastereoselectivity (*dr* = 80 % in favor of the *exo* compound) remained nearly unchanged compared to the results obtained with templates **4** and **5**, but a significant rise in the enantioselectivity was observed (up to 60 % *ee*) (Figure 5.4). A further increase in the enantiomeric excess could not be achieved, presumably due to insufficient binding of the imidazolone substrate.

When employing **7** as a template, the C–C bond formation step, which is the decisive step for the absolute configuration of products, takes place from the C-5 *Si* face of the biradical intermediate (Figure 5.5) due to the shielding of the *Re* face by the sterically demanding tetrahydronaphthalene group of the complexing agent

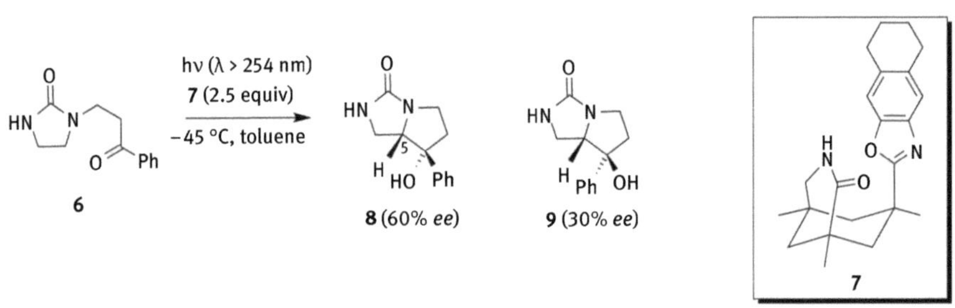

Figure 5.4: Enantioselective Norrish–Yang cyclization reaction of compound **6** to products **8** and **9** in the presence of template **7**.

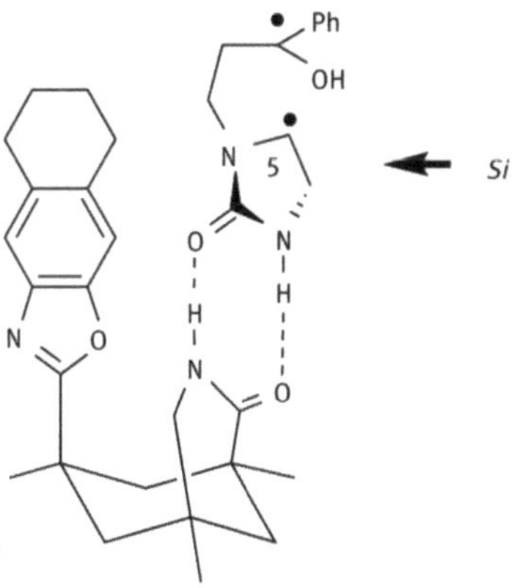

7

Figure 5.5: Differentiation of the enantiotopic faces of the biradical intermediate in the environment of chiral template 7.

(Figure 5.2). Thus, both major enantiomers of the *exo* (**8**) and *endo* (**9**) product show *R* configuration at C-5. [15]

5.4 Enantioselective photochemical [4+4]-cycloadditions and electrocyclic [4π]-ring closure of 2-pyridones

The [4+4]-photocycloaddition of 2-pyridone (**10**) and cyclopentadiene led to a 3:2 mixture of the *exo*- and *endo*-products. Performing the photoreaction in toluene at −50 °C and in the presence of 1.2 equivalents of the template **7** gave rise to two diastereoisomers **11** and **12**, both with high *ee* (84 and 87 % *ee*, respectively) (Figure 5.6).

Upon association to the template, the *Si*-face of carbon atom C-3 of the 2-pyridone is the only accessible face of the substrate, which explains the absolute configurations found for compounds **11** and **12**. The back of the 2-pyridone **10** is shielded by the tetrahydronaphthalene unit of the complexing agent.

The [4π]-cyclization of several substituted 2-pyridones to 3-oxo-2-azabicyclo [2.2.0]-5-hexenes (**13**) was also studied (Figure 5.7). This reaction – unlike most photochemical reactions – turned out to be very sensitive towards low temperature.

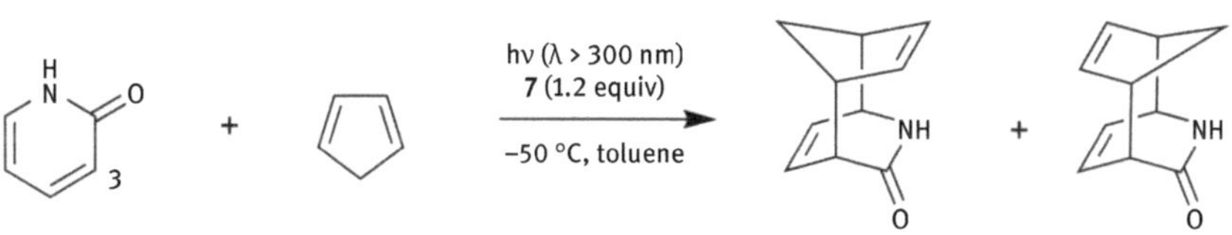

10

11 (84% *ee*)
exo

12 (87% *ee*)
endo

Figure 5.6: Enantioselective photochemical [4+4]-cycloaddition of 2-pyridone and cyclopenadiene in the presence of template 7.

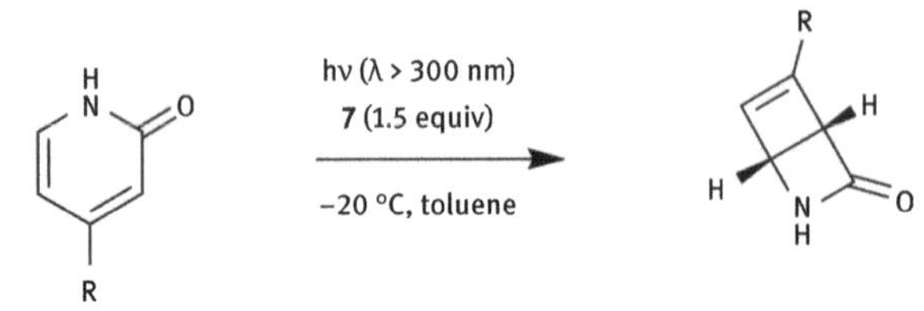

R= OMe, OBn

13 (20-23% *ee*)

Figure 5.7: Enantioselective electrocyclic [4π]-ring closure of 4-methoxy- and 4-benzyloxy-2-pyridone in the presence of template **7**.

Therefore, the reaction could only be conducted at temperatures above −20 °C and consequently, the association was not as good as in the previous examples. In addition, steric differences between the two possible cyclization ways are apparently not significant. These two facts prevent the reaction of occurring with high stereoselectivities and therefore, only enantioselectivities between 20 and 23 % *ee* were achieved (Figure 5.7). [16]

5.5 Enantioselective [6π]-photocyclization of acrylanilides

The [6π]-photocyclization of anilide **14** was studied under the previously optimized conditions, thus, employing the enantiomerically pure chiral templates **7** or *ent*-**7** at low temperature using toluene as the solvent. The *trans* product **15** was obtained in 57 % *ee*, and the minor diastereoisomer (*dr* = 73 / 27), *cis* product **16**, was obtained in 30 % *ee* (Figure 5.8). It is noteworthy that the diastereoselectivity of the process is changed in the presence of the template since the *cis* compound is the major product in its absence (*dr* = 46 / 54). While the enantiomeric enrichment, as expected, is larger at lower temperatures for *trans* product **15**, surprisingly, the *ee* of **16** reaches a maximum at −15 °C and decreases at lower temperature.

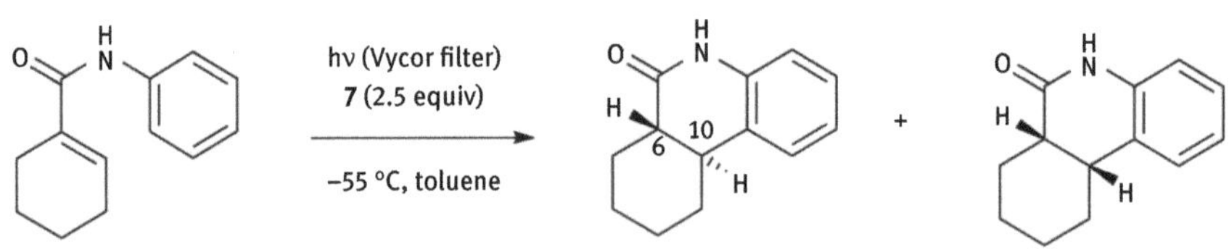

14

15 (57% *ee*)

16 (30% *ee*)

Figure 5.8: Enantioselective [6π]-photocyclization of acrylanilide **14** using compound **7** as chiral template.

The major enantiomers for both the *cis* and the *trans* products showed identical absolute configurations at the stereogenic carbon atom C-6 but the absolute configuration at C-10 is opposite for the two structures **15** and **16**. The latter is formed in the C–C bond-forming step and cannot be inverted in the further course of the reaction. The configuration at carbon atom C-6 is established in the protonation step. These individual steps have been studied using DFT calculations and deuteration experiments to understand how each stereocenter is formed in the course of this reaction. [17]

5.6 Enantioselective Diels–Alder reaction of a photochemically generated *ortho*-quinodimethane

Experiments on the [4+2]-photocycloaddition of *o*-quinodimethane **17**, obtained photochemically from lactam **18**, with different dienophiles (**19**) in the presence of the chiral template **7** were also performed (Figure 5.9). Only the *E*-diastereoisomer (**17**) of the two formed enols is able to participate in the cycloaddition reaction, while the *Z*-diastereoisomer (**20**) tautomerizes rapidly to the starting material. The *exo* product was obtained as the major diastereoisomer when acrylonitrile (R = CN, R′ = H) was employed as the dienophile, whereas methyl acrylate (R = COOMe, R′ = H) and dimethyl fumarate (R = COOMe, R′ = COOMe) led to the *endo* products.

In the presence of the chiral complexing agent **7** the enantiotopic faces of the photochemically generated *o*-quinodimethane were differentiated. The experiments showed a very effective face differentiation, perfect regioselectivity and good to excellent simple diastereoselectivity. The reaction was extraordinarily fast and the

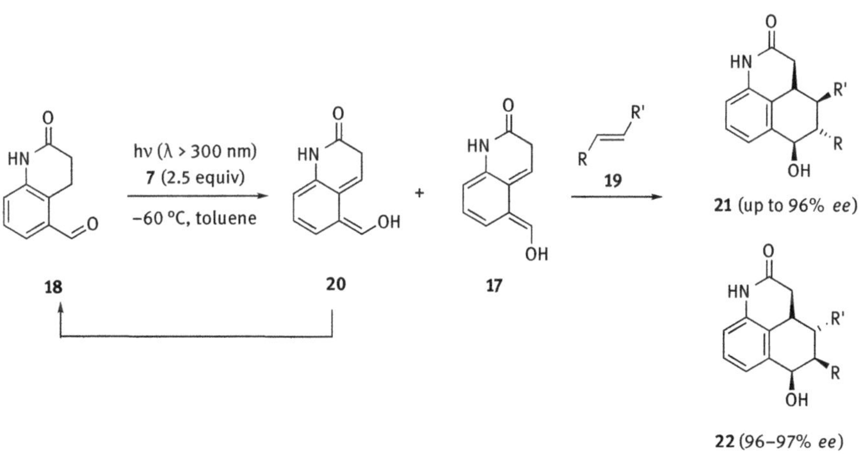

Figure 5.9: Template-mediated enantioselective Diels–Alder reaction of photochemically generated *ortho*-quinodimethane **17** with different dienophiles.

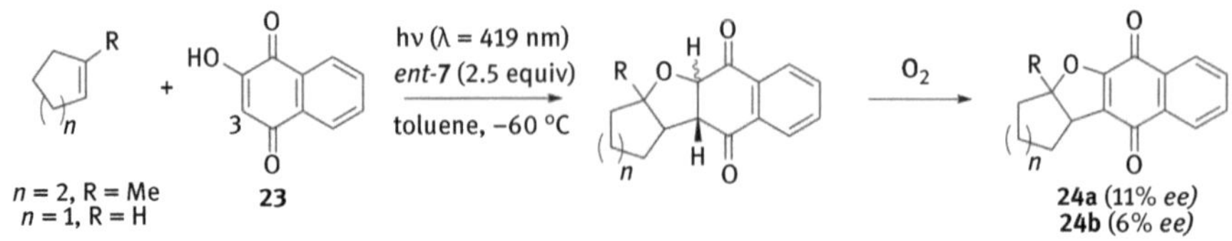

Figure 5.10: Variation of binding motif: formal [3+2]-photocycloaddition of naphthoquinone **23** with different cyclic alkenes resulting in low enantioselectivities.

enantioselectivities were high (96–97 % *ee* at −60 °C in toluene). The high enantioselectivities observed are partially due to the fact that the products **21** and **22** show a much lower association to the template than compound **18.** [18]

5.7 Formal [3+2]-photocycloadditions of 2-substituted naphthoquinones

The formal [3+2]-photocycloaddition of 2-substituted 1,4-naphthoquinone **23** to cyclic alkenes leading to tetracyclic compounds **24** was studied in the presence of template *ent*-**7**. The reactions were carried out at λ = 419 nm using 200 equivalents of the respective olefin. Although high yields and diastereoselectivities were achieved, the observed enantioselectivities were low (Figure 5.10). The reason for this is the fact that the stereogenic center formed in the initial photocycloaddition is lost during the oxidation to the final product. The primary reaction may be perfectly enantioselective regarding a C–C bond formation at C-3 of the 2-substituted naphthoquinone, but the relative configuration of the two stereogenic centers, between which the bond is formed, determines the absolute configuration of compounds **24**. Furthermore, in studies on the association complex equilibrium, no defined complex formation could be detected but rather mixtures of 1 : 1 and 2 : 1 substrate-template complexes are likely to be present under the reaction conditions. [19]

5.8 Intramolecular [2+2]-photocycloadditions of substituted 5,6-dihydro-1H-pyridin-2-ones

The 3-(ω′-Alkenyl)- and 3-(ω′-alkenyloxy)-5,6-dihydro-1*H*-pyridin-2-ones (**25**) undergo [2+2]-photocycloaddition reactions upon irradiation at λ = 254 nm. When *n* = 1 the crossed products **26** and **27** were obtained exclusively. Irradiation of longer chain compounds led to the straight products **29–30** (Figure 5.11).

High enantioselectivities were obtained (40–84 *ee*) when performing the photoreaction either employing the compound **7**, or the chiral template **31**, which presents a γ-lactam binding site (Figure 5.11). [20]

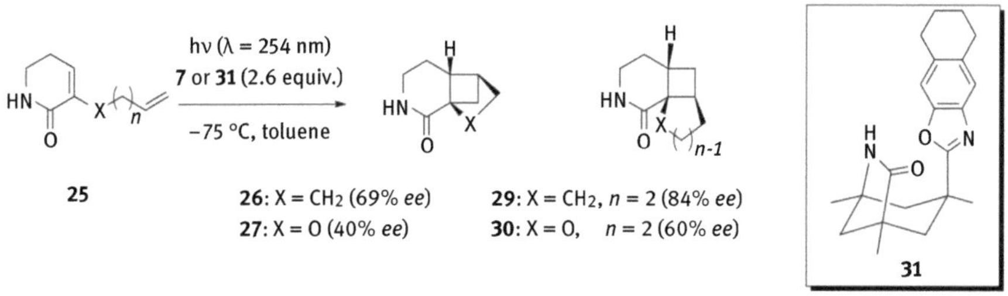

Figure 5.11: Intramolecular [2+2]-photocycloaddition reactions of substituted 5,6-dihydro-1H-pyridine-2-ones **25** in the presence of chiral templates **7** and **31** leading to enantiomerically enriched straight and crossed products.

5.9 Enantioselective radical cyclizations

The encouraging results achieved in Norrish–Yang cyclizations (cf. Chapter 3) led to an investigation on other enantioselective radical cyclization reactions. Radical reactions have been carried out enantioselectively without an auxiliary being attached covalently to the substrate, for example, employing chiral hydrogen-atom donors and Lewis acids coordinated to chiral ligands. [21] Using the previously described template **7** and its enantiomer a number of reductive radical cyclizations of α,β-unsaturated amide derivatives were successfully carried out in an enantioselective fashion.

5.9.1 Reductive radical cyclization reactions of 3-(ω-iodoalkylidene)-piperidin-2-ones

In the presence of an initiator and tributyltin hydride the alkenyl iodides **32** react in a 5- or 6-*exo-trig*-cyclization depending on the chain length (Figure 5.12). The intermediate radicals **33** possess a prostereogenic center in α-position to the carbonyl function, which is transformed by an intermolecular reaction with Bu_3SnH

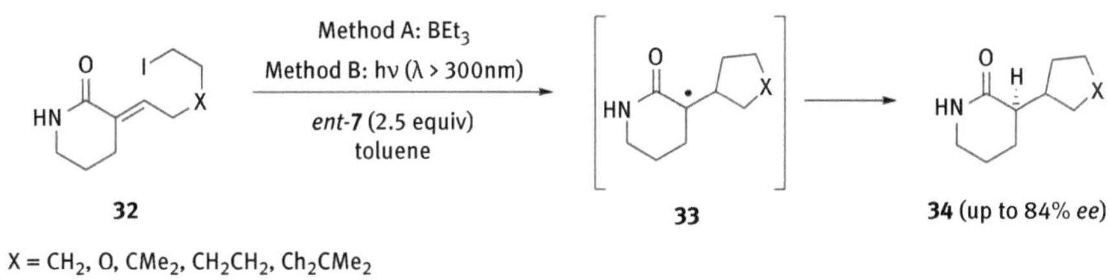

Figure 5.12: Reductive radical cyclization reactions of 3-(ω-iodoalkylidene)-piperidin-2-ones **32** to products **34**.

to a stereogenic carbon atom. The reaction can be initiated both, by addition of BEt_3 or by ultraviolet light. Performing the reaction in toluene at low temperature and in the presence of the template *ent*-**7** high enantioselectivities were achieved (up to 84 % *ee*).

Hydrogen transfer from Bu_3SnH to radical **33**, which is associated to template *ent*-**7** by means of hydrogen bonding, takes place from the *Si* face due to the fact that the *Re* face is efficiently blocked by the tetrahydronaphthalene moiety. Experiments employing substrates with X = O and X = CMe_2, for which both the radical cyclization and the hydrogen abstraction step led to a new stereocenter, did not show changes in the respective diastereoselectivities, which indicated that the cyclization step was not enantioselective. [22]

5.9.2 Reductive radical cyclization of 3-(3-iodopropoxy)propenoic acid derivatives

Since enantioselective radical cyclization reactions of acrylates are rare, [23] 5-*exo-trig* radical cyclizations of 3-(3-iodopropoxy)propenoic acid derivatives **35** were attempted in the presence of the chiral template **7** in order to control the conformation of the acrylate and to provide an efficient enantioface differentiation. The reaction was carried out in the presence of tributyltin hydride and an appropriate initiator (BEt_3, AIBN or ultraviolet light) leading to the corresponding tetrahydrofurans **36** with a new stereogenic center at C-2 (Figure 5.13).

The major enantiomer of the radical cyclization is formed by *Si*-face attack at the β-carbon atom. The enantioselectivity was significantly lower than in the previous studies. Cyclic ureas gave rise to the highest enantioselectivities (up to 59 % *ee*). From these results, it could be concluded that the association to the template increases for acrylic acid derivatives in the order hydrazide < amide < acid < cyclic urea. [24]

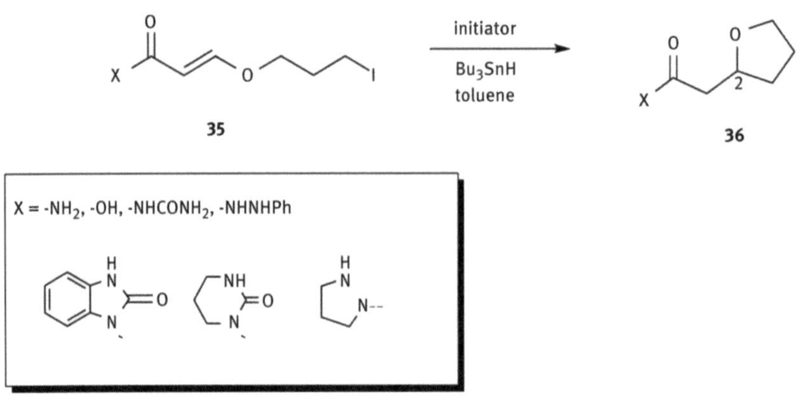

Figure 5.13: Reductive radical cyclization of 3-(3-iodopropoxy)propenoic acid derivatives **35** to tetrahydrofuran **36**.

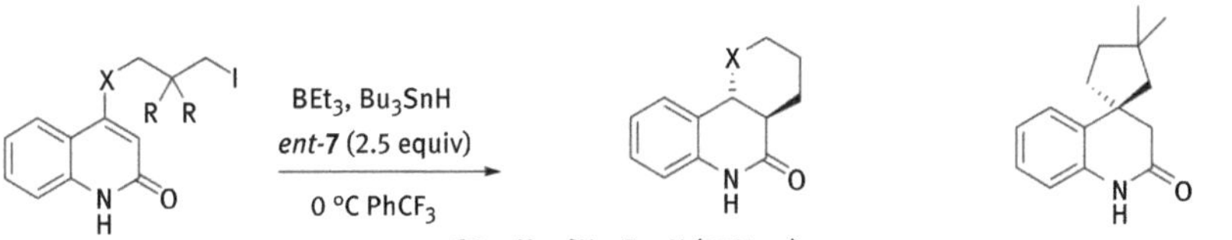

37

38a: X = CH_2, R = H (99% *ee*)
38b: X = S, R = H (99% *ee*)

39: X = CH_2, R = Me (94% *ee*)

Figure 5.14: Enantioselective synthesis of 6-*endo* (**38a, b**) and 5-*exo* (**39**) products by templated radical cyclizations of 4-substituted quinolones **37**.

5.9.3 Radical cyclization reactions of 4-substituted quinolones

Reductive radical cyclizations of quinolones **37** were carried out employing BEt_3/O_2 as initiator and either Bu_3SnH or TMS_3SiH as hydride source. 4-(4-Iodobutyl)-quinolone and 4-(3-iodopropylthio)-quinolone gave 6-*endo*-cyclization products in good yields, while 4-(3,3-dimethyl-4-iodobutyl)-quinolone reacted in a 5-*exo*-fashion. When the cyclization reactions were conducted in the presence of the chiral template *ent*-**7** high enantiomeric excesses were obtained (94–99 % *ee*) (Figure 5.14). Even better selectivities were obtained upon employing trifluorotoluene as the solvent. In this case, catalytic reactions could be conducted with 25 and 10 mol% of template with recovery yields of the template higher than 90 %. The effective chirality transfer and the chirality multiplication in the reaction were probably due to the low solubility of the substrates in trifluorotoluene. The reaction mixture remained heterogeneous throughout the reaction. It is likely that thc complexing agent dissolves the substrate and allows the radical reaction to take place under homogeneous conditions (phase-transfer catalysis).

The association behavior of substrate **37a** (X = CH_2, R = H) to *ent*-**7** was studied by NMR titration experiments. Association data obtained by these experiments confirm that the high enantioselectivities are in agreement with the degree of complexation. [25]

5.10 [2+2]-Photocycloaddition reactions of substituted isoquinolones

Intramolecular [2+2]-photocycloadditions of isoquinolones were also studied, taking into account that the isoquinoline moiety is a recurrent motif in natural products. [26] Thus, performing the intramolecular [2+2]-photocycloaddition of isoquinolones in a regio- and enantioselective way is potentially very useful.

Starting from easily accessible isoquinolones (**40–47**), a number of complex cyclobutane photoproducts (**48–56**) were obtained with high enantioselectivity (88–96 % *ee*) after irradiation at $\lambda = 366$ nm in the presence of the chiral template *ent*-**7** (2.6 equiv.) at –60 °C in toluene (Figure 5.15).

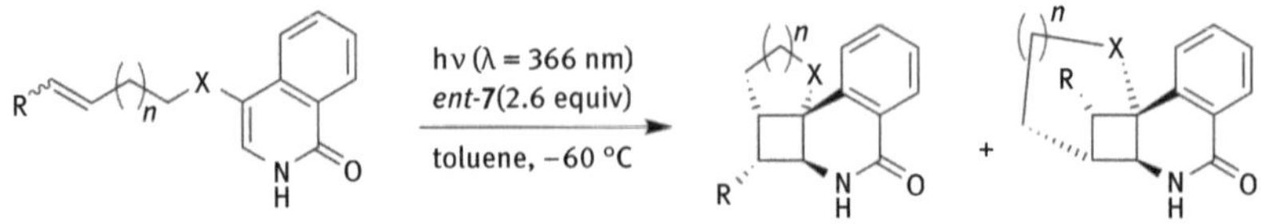

40 X = O, *n* = 1, R = H	**48** (93% *ee)*	**49** (96% *ee)*
41 X = CH2, *n* = 1, R = H	**50** (88% *ee)*	–
42 X = O, *n* = 0, R = H	–	**51** (89% *ee)*
43 X = O, *n* = 2, R = H	–	**52** (95% *ee)*
44a X = O, *n* = 2, R = Me, (*E*)-isomer **44b** X = O, *n* = 2, R = Me, (*Z*)-isomer	–	**53** (96% *ee)*

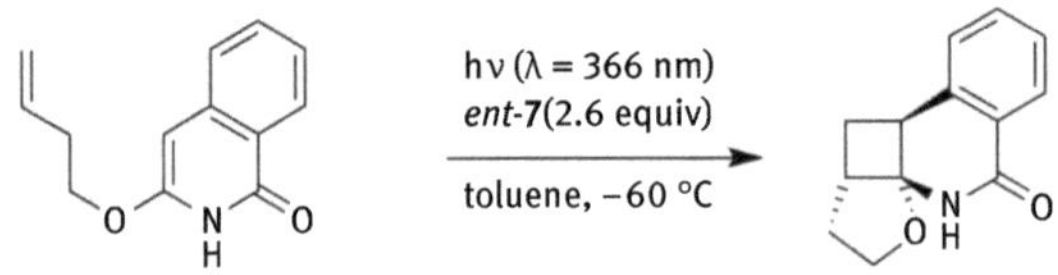

45 **54** (75% ee)

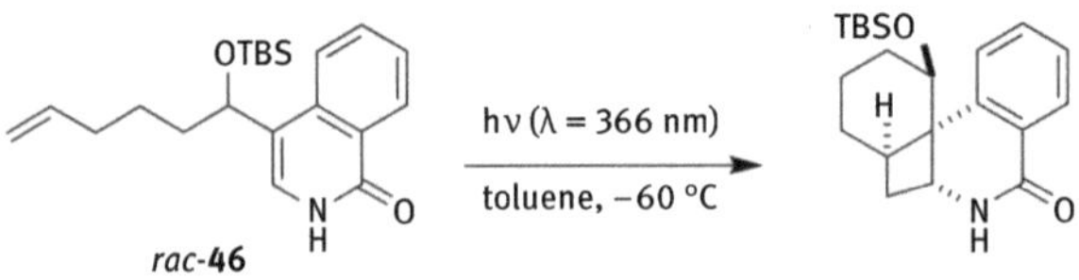

rac-**46** *rac*–**55** (isolated as single diastereoisomer*)*

Figure 5.15: Enantioselective intramolecular [2+2]-photocycloaddition reactions of substituted isoquinolones **40–46**.

Due to the high association of some isoquinolones to the template, enantiomorphic conformations in the 1 : 1 complex are favored. An application of this phenomenon to the kinetic resolution of racemic isoquinolone *rac*-**47** was successfully carried out (Figure 5.16). Compound **56** was obtained in more than 95 % *ee* at 2 % conversion although it decreased to 53 % *ee* when the starting material was consumed due to photochemical side reactions taking place during irradiation. The enrichment of compound *ent*-**47** was also observed (23 % *ee* at 40 % conversion). [27]

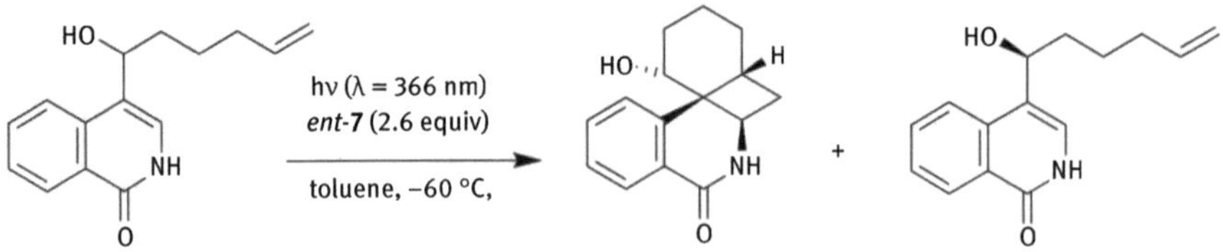

rac-**47** **56** *ent*-**47**

Figure 5.16: Kinetic resolution of racemic isoquinolone *rac*-**47** by employing template *ent*-**7**.

5.11 [2+2]-Photocycloaddition reactions of substituted quinolones

An in-depth study on the [2+2]-photocycloadditions of quinolones was carried out taking into account that such quinolones appeared to be perfect guests for chiral lactam complexing agents. Thus, inter- and intramolecular processes were performed and some of the results obtained were used in natural product total synthesis.

5.11.1 Intermolecular [2+2]-photocycloaddition reactions of quinolones

Template *ent*-**7** was used as a chiral complexing agent for the intermolecular [2+2]-photocycloaddition of quinolone **57** with several alkenes obtaining very high enantioselectivities for cycloadducts **58** (81–98 % *ee*). *Exo* isomers were obtained with high diastereoselectivity for R = $(CH_2)_3OH$, CH_2OAc and CO_2Me (Figure 5.17). The *endo* product was exclusively obtained for R = Ph while R = OAc gave rise to a mixture of both diastereomers (63 % *dr* in favor of the *exo* compound **58d**). In the absence of the template the simple diastereoselectivity was not significantly changed. As expected, the products with opposite configuration were obtained when the template **7** was employed instead of *ent*-**7**. The use of *ent*-**7** induces a *Re*-attack at carbon atom C-3 while **7** favors the reaction taking place through the *Si*-face. [28]

5.11.2 Intramolecular [2+2]-photocycloadditions of 4-(2′-aminoethyl)quinolones

Enantioselective [2+2]-photocycloadditions were carried out on 4-(2′-aminoethyl)-quinolones in solution employing complexing agent **7** as the source of chirality. The intermolecular reaction of the substrate **59** with different 2-alkyl-substituted acrylates was performed. *N*-Benzylic-protected photoproducts **60** were obtained in high yields and with high diastereoselectivity (up to 95/5 *dr*) and enantioselectivity

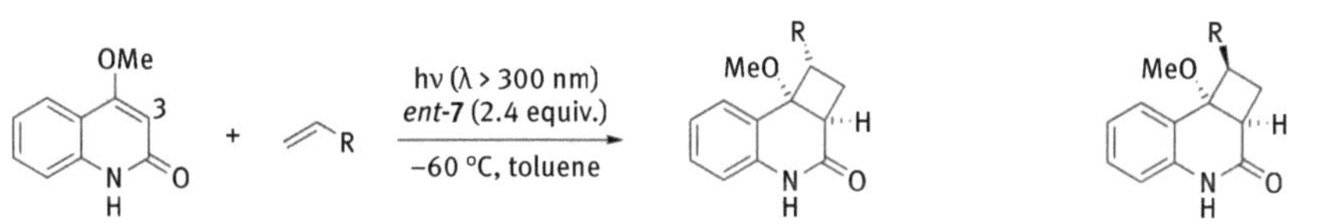

57

58a: R = $(CH_2)_3OH$ (81% ee)
58b: R = CH_2OAc (92% ee)
58c: R = CO_2Me (82% ee)
58d: R = OAc (93% ee)

exo

58e: R = OAc (98% ee)
58f: R = Ph (83% ee)

endo

Figure 5.17: Photochemical intermolecular [2+2]-cycloaddition reactions of 4-methoxy-2-quinolone **57** with several alkenes for the preparation of enantiomerically enriched cyclobutanes **58**.

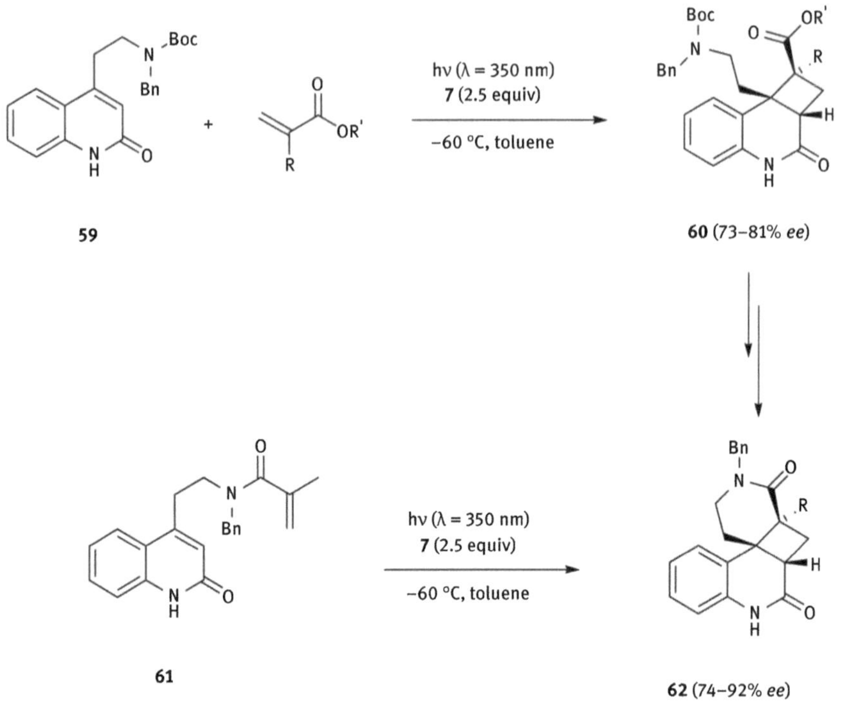

Figure 5.18: Synthesis of enantiomerically enriched compounds **62** by inter- and intramolecular photoreactions of 4-(2′-aminoethyl)quinolones **59** and **61** in the presence of chiral template **7**.

(73–81 % *ee*). The intramolecular [2+2]-photocycloaddition of acrylic acid amides **61** was carried out under the same conditions leading to photoproducts **62** (46–55 % yield, 74–92 % *ee*). Photoproducts **60** were converted to compounds **62** by a sequence of *Boc* deprotection and thermal lactamization (74–98 % yield) (Figure 5.18). Comparative investigation of both routes showed that quinolone dimerization was the single most decisive factor preventing a complete chirality transfer. [29]

The possibility of performing a 1,2-rearrangement reaction of compounds possessing the product **62** skeleton towards the enantioselective synthesis of *Melodinus* alkaloids was subsequently tested. In this context, the total synthesis of (+)-meloscine (**63**), which is the prototype of *Melodinus* alkaloids, was achieved in 15 steps in a 9 % overall yield. [30] The enantioselective [2+2]-photocycloaddition of **59** with methyl 2-(trimethylsilyloxy)acrylate [31] to compound **60a** in the presence of **7**, followed by a *retro*-benzilic acid rearrangement to compound **64**, was used as the key step for the synthesis of this alkaloid (Figure 5.19). This enantioselective synthesis of (+)-meloscine represented the first example of a natural product synthesis employing an enantioselective [2+2]-photocycloaddition as its key step.

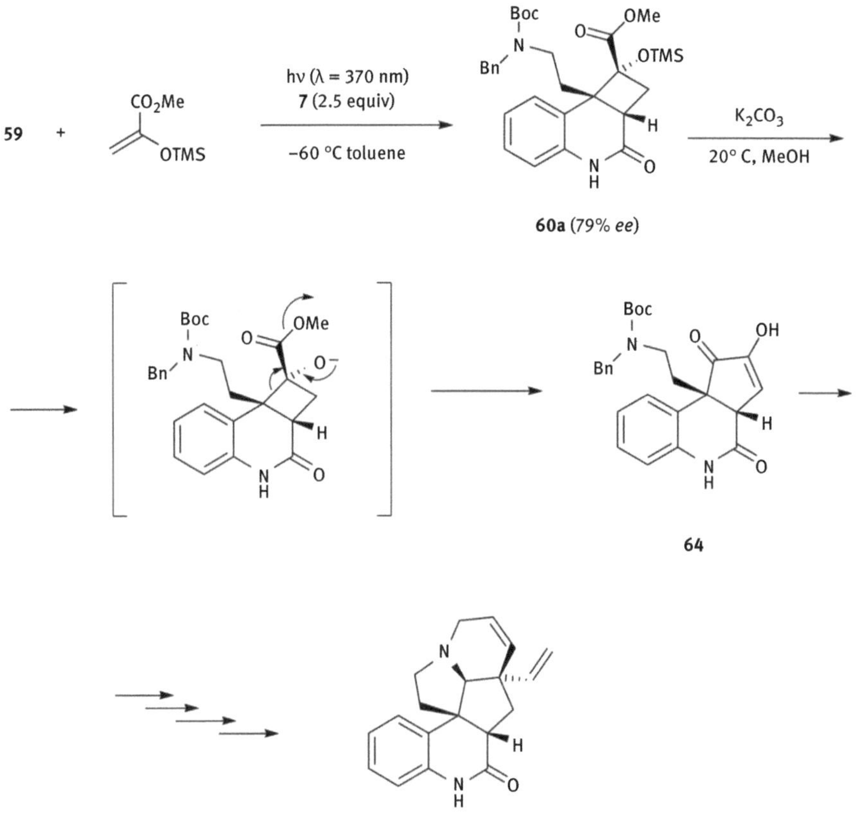

Figure 5.19: Total synthesis of (+)-meloscine (**63**) by enantioselective [2+2]-photocycloaddition of **59** with methyl 2-(trimethylsilyloxy)acrylate to compound **60a** in the presence of **7** as the key step.

5.11.3 Intramolecular [2+2]-photocycloadditions of 4-(ω-alkenyloxy)-quinol-2-ones

Irradiation of 2-quinolones **65** and **66** in the presence of **7** at −60 °C in toluene led to the corresponding [2+2]-cycloadducts **67** and **68** in high yields and with excellent enantioselectivities (Figure 5.20). In both cases only a single diastereomer was obtained. When template *ent*-**7** was employed the corresponding enantiomers *ent*-**67** and *ent*-**68** were formed. Upon binding to *ent*-**7** the *Si*-face is shielded by the tetrahydronaphthalene moiety, and the *Re*-face is accessible to the intramolecular attack. Further experiments performed in the presence of the templates derived from menthol **4** and **5** did not give rise to better results. [32]

The photoreaction of **65** leading to *ent*-**67** was also performed employing template *ent*-**7** immobilized on polymeric supports in toluene at −74 °C. Two new tem-

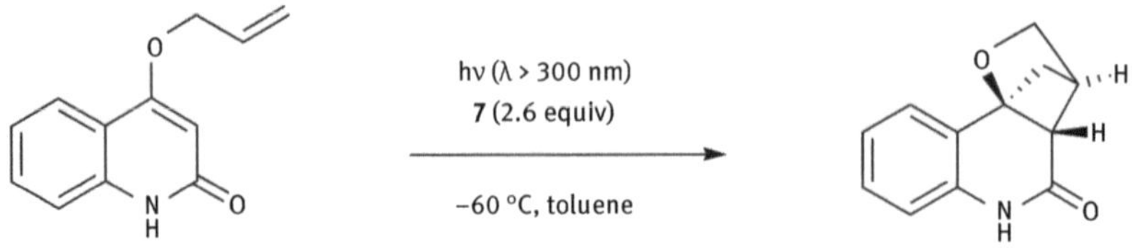

65 67 (93% *ee*)

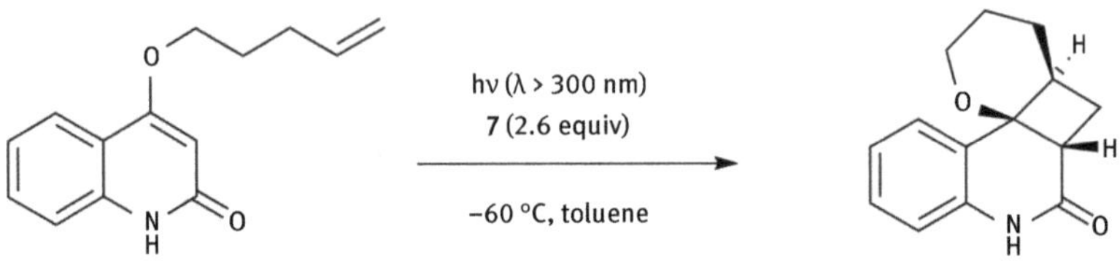

66 68 (>90% *ee*)

Figure 5.20: Templated intramolecular [2+2]-photocycloaddition reactions of 4-allyloxy-2-quinolinone (**65**) and 4-(4-pentenyloxy)-2-quinolinone (**66**) to the enantomerically enriched photoproducts **67** and **68**.

plates were prepared for this purpose: *ent*-**7** grafted on a Wang resin and *ent*-**7** bound to methoxypolyethylene glycol. In both cases high enantioselectivities were achieved and did not deteriorate even after the catalyst had been recycled four times. Experiments conducted with methoxypolyethylene glycol bound template gave rise to better conversions due to the transparency of the support while Wang resin was not transparent and the penetration depth of the light remained low. [33]

5.12 Light-induced enantioselective catalysis

Two main approaches are currently under investigation in the field of light-induced enantioselective catalysis. The first one relies on the interaction of a substrate with a catalyst during the photochemical key step of a reaction. This key step can be a sensitization by energy or electron transfer as well as other means to accelerate a catalytic enantioselective reaction over a racemic background reaction. The second approach is the combination of well-established enamine or iminium ion organocatalysis with photochemical reactions. This rapidly developing field of research is the topic of a separate chapter in this book. The catalysis of photochemical primary reactions such as [2+2]- or [4+4]-photocycloadditions remains difficult. A number of successful approaches in this area rely on the rational approach to combine chiral templates

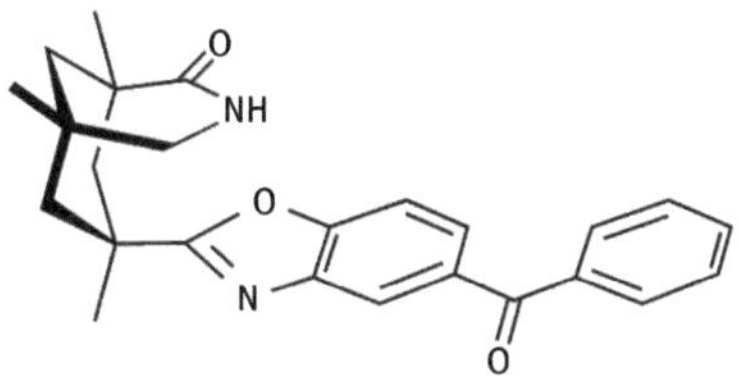

69

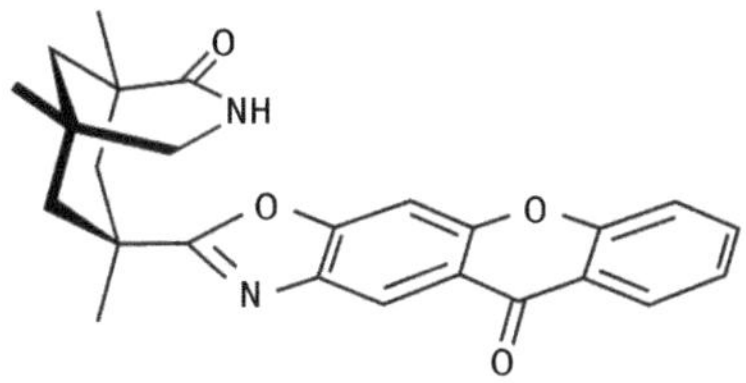

70

Figure 5.21: Structures of the chiral photocatalysts based on the benzophenone (**69**) and xanthone (**70**) chromophore.

with sensitizing units which are capable of acting as an antenna for light and which subsequently transfer the excitation energy by any means of sensitization to a bound substrate. The use of such established templates guarantees a high face differentiation, which is a prerequisite for a highly enantioselective reaction. The combination of cyclodextrins with sensitizing units such as biphenyl-, anthrancene or benzoate derivatives which are able to deliver the energy provided by UV irradiation by sensitization to the complexed substrates such as cyclic alkenes and 1,1-diphenylpropene leads to the respective products in enantioselectivities of up to 46 % *ee* applying 10 mol% of the cyclodextrin-sensitizer hybrid catalyst. [34] If the substrate is sensitized by triplet sensitization or photoelectron transfer, which act only over very short ranges the excitation of unbound substrate molecules is unlikely. If the sensitizer is the only absorbing species at the irradiation wavelength, only a bound substrate can undergo the photochemical transformation. The reaction should therefore take place exclusively within the chiral environment leading to high stereoselectivities. To avoid preliminary dissociation of excited substrate from the chiral environment leading to a reaction in the absence of the chiral information, low temperatures are used commonly for these reactions. Ideally, the complex dissociation is significantly slower than the rate of the desired reaction but still fast enough to allow efficient turnover in a catalytic reaction. The application of this rationale to the concept of template **7** led to the synthesis of catalysts **69** and **70**, depicted in Figure 5.21. These compounds, as the parent compound **7**, are able to bind amides and lactams by forming two hydrogen bonds and, in addition, they can act as sensitizers due to the presence of the benzophenone or xanthone chromophore.

Both benzophenone and xanthone were used frequently as sensitizers in triplet–triplet-energy transfer catalyzed photoreactions as well as in photoelectron transfer (PET) sensitized processes. The latter is also applicable to this activation concept as photoelectron transfer is similarly short-ranging and reactions initiated by reductive or oxidative photoelectron transfer should therefore occur also exclusively with bound substrates.

Benzophenone derivative **69** was used to sensitize electron transfer as well as energy transfer processes. Xanthone **70** was found to induce higher enantioselec-

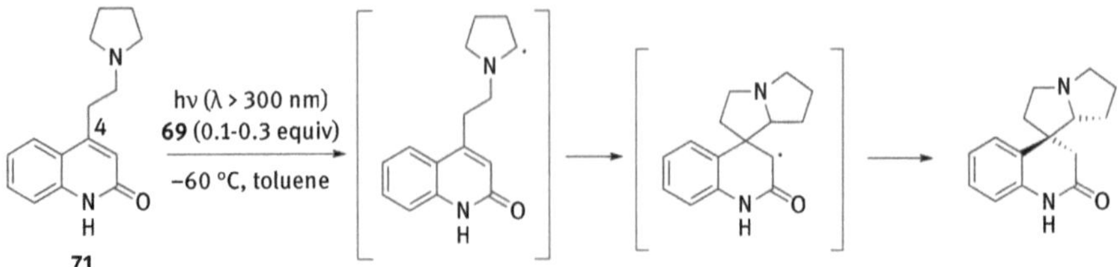

71

72 (70% *ee*)

Figure 5.22: Catalyzed enantioselective photoinduced electron transfer reaction of quinolone **71** in the presence of the benzophenone-based catalyst **69**.

tivity than compound **69** for energy transfer processes due to the xanthone chromophore exhibiting both higher triplet energy and significantly red-shifted absorption spectra.

5.12.1 Photoinduced electron transfer enantioselective catalytic reactions

The reaction of quinolone **71** leading to chiral spirocyclic pyrrolizidines **72** and *ent*-**72** (Figure 5.22) was chosen in order to study the enantioselective catalytic photoinduced electron transfer (PET) process. For this purpose, catalysts **69** and *ent*-**69** were employed. In the presence of a catalyst, ultraviolet light induces a PET from the amine to the photoexcited catalyst. Subsequent proton loss from the intermediate cation radical presumably leads to an α-aminoalkyl radical, which adds intramolecularly to carbon atom C-4 of the quinolone. After the radical addition reaction, a hydrogen atom transfer occurs, by either back electron transfer from the catalyst or by direct protonation from the suggested ketyl radical. The generated enolate is eventually protonated to yield the products.

As expected, the catalyst acted not only as a PET catalyst, but also as a stereocontrolling agent inducing the desired enantiofacial differentiation in the cyclization step. The reaction proceeded with considerable turnover and high enantioselectivity. Catalyst **69** gave rise to **72** as major product, while *ent*-**69** led to *ent*-**72**. The products were obtained in significant enantiomeric excess (up to 70 %) and in yields reaching 64 %. [35]

5.12.2 Catalyzed enantioselective [2+2]-photocycloadditions of 4-substituted quinolones

The reaction depicted in Figure 5.23 was selected in order to investigate energy transfer processes. High enantioselectivities (43–99 % *ee*) were achieved when performing the reaction in the presence of template *ent*-**7**.

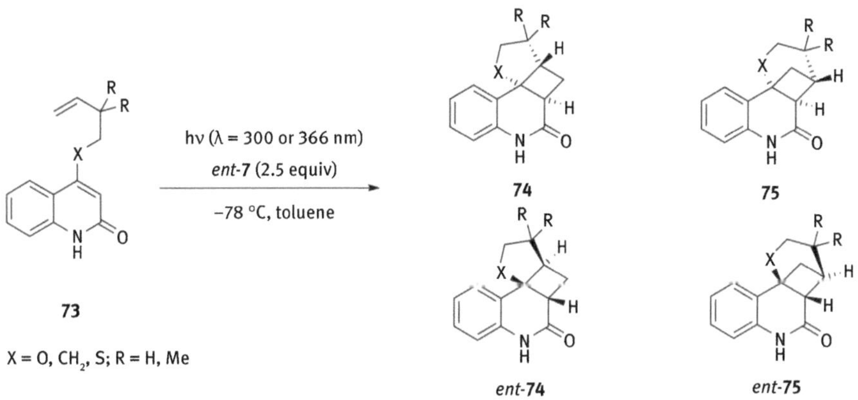

Figure 5.23: Photochemical [2+2]-photocycloaddition reactions of quinolones **73** using *ent*-**7** as chiral template to induce enantioselectivity.

Benzophenone catalyst **69** was employed since previous work from Krische et al. showed that a sensitization of this reaction is possible by a chiral benzophenone (19 % *ee* with 25 mol% catalyst) [36] but, disappointingly, its use did not lead to high enantioselectivities. However, when xanthone catalyst **70** was employed, much better results were obtained, achieving a significant rate acceleration and high enantioselectivities. These results are maybe due to the fact that the triplet energy of the xanthone is higher than the one of the benzophenone, which implies that the process is more effective when the former is used.

An extraordinarily high chirality turnover was achieved for the photoreaction of **73a** (X = O, R = H) with 0.10 equivalents of **70** with enantioselectivities of at least 90 % *ee* for both products **74a** and **75a** after one hour of irradiation in trifluorotoluene at −25 °C. When the reaction was performed in the presence of the parent compound xanthone conversion was found to be lower. This fact indicates that intramolecular sensitization within the substrate-catalyst complex is more efficient than intermolecular sensitization.

In a further experiment, a different behavior was observed comparing the reactivity of compounds **73a** and **66**. The reaction of **73a** leading to a five-membered ring was much more enantioselective than the reaction of **66** giving rise to a six-membered ring (*ent*-**68**) (Figure 5.24). [37]

Emission spectra and laser flash photolysis experiments showed that the photocycloaddition of **73a** was significantly faster. A much smaller phosphorescence quantum yield and a larger triplet state lifetime were found for compound **66**. This is in agreement with faster kinetics for the cyclization of **73a** compared to **66**. In the case of substrate **66**, the lower photocycloaddition rate (relative to **73a**) leads to an increase in cycloaddition reactions occurring after dissociation, which implies loss of enantioselectivity due to the process occurring in the absence of the chiral information. [38]

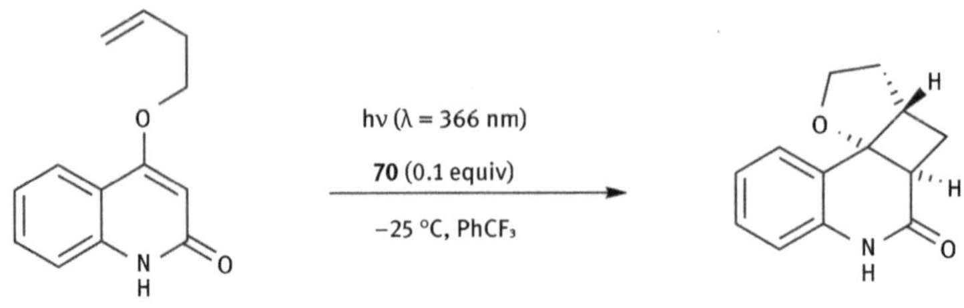

73a **74a** (92% *ee*)

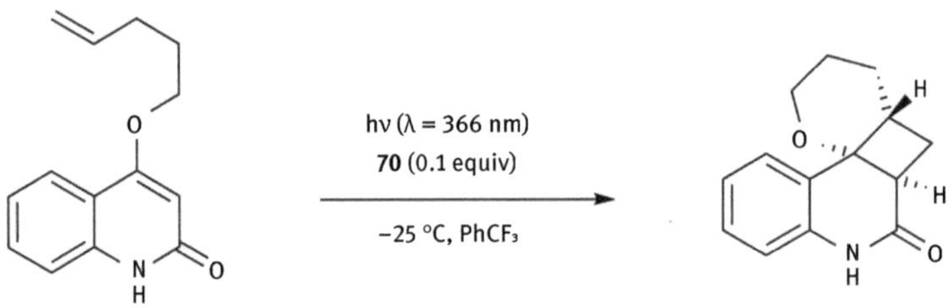

66 *ent*-**68** (27% *ee*)

Figure 5.24: Catalyzed [2+2]-photocycloaddition reactions of 4-(4-butenyloxy)-2-quinolone (**73a**) and 4-(4-pentenyloxy)-2-quinolone (**66**) performed in an enantioselective fashion employing chiral xanthone catalyst **70**.

5.13 Conclusion

Hydrogen bonds have been described as an ideal means to place a prochiral substrate into a chiral environment. It was shown that the general working mode of such complexing agents derived from Kemp's triacid can be applied to induce enantioselectivity in a wide range of photochemical reactions. Templates bearing the xanthone or the benzophenone chromophore are able to sensitize the prochiral substrates as well as to induce the desired enantioselectivity. Due to its robustness, which allows for its almost quantitative recovery and its reuse, templates of this kind are suitable for being used as catalysts in photochemical processes.

5.14 References

[1] a) Everitt, S.R.L., and Inoue, Y. (1999). Molecular and Supramolecular Photochemistry: Organic Molecular Photochemistry, Vol. 3 (CRC Press: New York), pp. 71–130; b) Inoue, Y. (1992). Asymmetric photochemical reactions in solution. Chem. Rev. 92, 741–770; c) Rau, H. (1983). Asymmetric photochemistry in solution. Chem. Rev. 83, 535–547.

[2] a) Kawasaki, T., Sato, M., Ishiguro, S., Saito, T., Morishita, Y., Sato, I., Nishino, H., Inoue, Y., and Soai, K. (2005). Enantioselective synthesis of near enantiopure compound by asymmetric autocatalysis triggered by asymmetric photolysis with circularly polarized light. J. Am. Chem. Soc. 127, 3274–3275; b) Sato, I., Sugie, R., Matsueda, Y., Furumura, Y., and Soai, K. (2004). Asymmetric synthesis utilizing circularly polarized light mediated by the photoequilibrium of chiral olefins in conjunction with asymmetric autocatalysis. Angew. Chem. Int. Ed. 43, 4490–4492; c) Nishino, H., Kosaka, A., Hembury, G.A., Aoki, F., Miyauchi, K., Shitomi, H., Onuki, H., and Inoue, Y. (2002). Absolute asymmetric photoreactions of aliphatic amino acids by circularly polarized synchrotron radiation: critically pH-dependent photobehavior. J. Am. Chem. Soc. 124, 11618–11627.

[3] a) Thomas, R., and Tamaoki, N. (2011). Chirality transfer from chiral solvents and its memory in an azobenzene derivative exhibiting photo-switchable racemization. Org. Biomol. Chem. 9, 5389–5393; b) Boyd, D.R., Campbell, R.M., Coulter, P.B., Grimshaw, J., Neill, D.C., and Jennings, W.B. (1985). Dynamic stereochemistry of imines and derivatives. Part 18. Photosynthesis and photoracemization of optically active oxaziridines. J. Chem. Soc., Perkin Trans. 1, 849–855.; c) Seebach, D., Oei, H-A., and Daum, H. (1977). Asymmetric synthesis in photo-, electro-, and alkali metal-pinacolizations of benzaldehyde and phenones in the chiral medium DDB. Chem. Ber. 110, 2316–2333.

[4] a) Sakamoto, M. (2006). Spontaneous chiral crystallization of achiral materials and absolute asymmetric photochemical transformation using the chiral crystalline environment. J. Photochem. Photobiol., C 7, 183–196; b) Sakamoto, M. (2004). Molecular and Supramolecular Photochemistry (Vol. 11): Chiral Photochemistry (Dekker: New York), pp. 415–461.

[5] a) Zhao, G., Yang, C., Chen, Q., Jin, J., Zhang, X., Zhao, L., and Xia, W. (2009). Photochemical studies on *exo*-bicyclo[2.1.1]hexyl and bicyclo[3.1.0]hexyl aryl ketones: two approaches for synthesis of enantiomerically enriched cyclopentene derivatives. Tetrahedron 65, 9952–9955; b) Yang, C., Xia, W., Chen, Q., Zhang, X., Li, B., and Gou, B. (2009). Photochemical studies on benzonorbornene derivatives. medium effects and asymmetric induction. Lett. Org. Chem. 6, 41–43; c) Yang, C., and Xia, W. (2009). Solid-state asymmetric photochemical studies using the ionic chiral auxiliary approach. Chem. Asian J. 4, 1774–1784; d) Xia, W., Scheffer, J.R., Botoshansky, M., and Kaftory, M. (2005). Photochemistry of 1-isopropylcycloalkyl aryl ketones: ring size effects, medium effects, and asymmetric induction. Org. Lett. 7, 1315–1318; e) Scheffer, J.R. (2001). Asymmetric induction in the photochemistry of crystalline ammonium carboxylate salts. Can. J. Chem. 79, 349–357.

[6] a) Fukuhara, G., Mori, T., and Inoue, Y. (2009). Competitive enantiodifferentiating anti-Markovnikov photoaddition of water and methanol to 1,1-diphenylpropene using a sensitizing cyclodextrin host. J. Org. Chem. 74, 6714–6727; b) Nakamura, A., and Inoue, Y. (2003). Supramolecular catalysis of the enantiodifferentiating [4+4] photocyclodimerization of 2-anthracenecarboxylate by γ-cyclodextrin. J. Am. Chem. Soc. 125, 966–972; c) Toda, F. (2001). Crystalline inclusion complexes as media of molecular recognitions and selective reactions. Aust. J. Chem. 54, 573–582.

[7] a) Sivasubramanian, K., Kaanumalle, L.S., Uppili, S., and Ramamurthy, V. (2007). Value of zeolites in asymmetric induction during photocyclization of pyridones, cyclohexadienones and naphthalenones. Org. Biomol. Chem. 5, 1569–1576; b) Shailaja, J., Kaanumalle, L.S., Sivasubramanian, K., Natarajan, A., Ponchot, K.J., Pradhan, A., and Ramamurthy, V. (2006). Asymmetric induction during electron transfer mediated photoreduction of carbonyl compounds: role of zeolites. Org. Biomol. Chem. 4, 1561–1571; c) Sivaguru, J., Natarajan, A., Kaanumalle, L.S., Shailaja, J., Uppili, S., Joy, A., and Ramamurthy, V. (2003). Asymmetric photoreactions within zeolites: role of confinement and alkali metal ions. Acc. Chem. Res. 36, 509–521.

[8] a) Natarajan, A., Mague, J.T., and Ramamurthy, V. (2005). Asymmetric induction during Yang cyclization of α-oxoamides: the power of a covalently linked chiral auxiliary is enhanced in the crystalline state. J. Am. Chem. Soc. 127, 3568–3576; b) Natarajan, A., Mague, J.T., and Ramamurthy, V. (2005). Viability of a covalent chiral auxiliary method to induce asymmetric induction in solid-state photoreactions explored. Cryst. Growth Design 5, 2348–2355; c) Bach, T. (1998). Stereoselective intermolecular [2+2] photocycloaddition reactions and their application in synthesis. Synthesis, 683–703.

[9] a) Kawanami, Y., Pace, T.C.S., Mizoguchi, J., Yanagi, T., Nishijima, M., Mori, T., Wada, T., Bohne, C., and Inoue, Y. (2009). Supramolecular complexation and enantiodifferentiating Photocyclodimerization of 2-anthracenecarboxylic acid with 4-aminoprolinol derivatives as chiral hydrogen-bonding templates. J. Org. Chem. 74, 7908–7921; b) Müller, C., and Bach, T. (2008). Chirality Control in Photochemical reactions: enantioselective formation of complex photoproducts in solution. Aust. J. Chem. 61, 557–564; c) Svoboda, J., and König, B (2006) Templated photochemistry: toward catalysts enhancing the efficiency and selectivity of photoreactions in homogeneous solutions. Chem. Rev. 106, 5413–5430; d) Mizoguchi, J., Kawanami, Y., Wada, T., Kodama, K., Anzai, K., Yanagi, T., and Inoue, Y. (2006). Enantiodifferentiating photocyclodimerization of 2-anthracenecarboxylic acid using a chiral *N*-(2-hydroxymethyl-4-pyrrolidinyl)-benzamide template. Org. Lett. 8, 6051–6054.

[10] a) Alonso, R., Jiménez, M.C., and Miranda M.A. (2011). Stereodifferentiation in the compartmentalized photooxidation of a protein-bound anthracene. Org. Lett. 13, 3860–3863
b) Fukuhara, G., Chiappe, C., Mele, A., Melai, B., Bellina, F., and Inoue, Y. (2010). Photochirogenesis in chiral ionic liquid: enantiodifferentiating [4+4] photocyclodimerization of 2-anthracenecarboxylic acid in (*R*)-1-methyl-3-(2,3-dihydroxypropyl)imidazolium bistriflimide. Chem. Commun. 46, 3472–3474. c) Ke, C.F., Yang, C., Mori, T., Wada, T., Liu, Y., and Inoue, Y. (2009). Catalytic enantiodifferentiating photocyclodimerization of 2-anthracenecarboxylic acid mediated by a non-sensitizing chiral metallosupramolecular host. Angew. Chem., Int. Ed. 48, 6675–6677. d) Yang, C., Mori, T., Origane, Y., Ko, Y.H., Selvapalam, N., Kim, K., and Inoue, Y. (2008). Highly stereoselective photocyclodimerization of α-cyclodextrin-appended anthracene mediated by γ-cyclodextrin and cucurbit[8]uril: a dramatic steric effect operating outside the binding site. J. Am. Chem. Soc. 130, 8574–8575.

[11] MacGillivray, L.R., Papaefstathiou, G.S., Friscic, T., Hamilton, T.D., Bucar, D.-K., Chu, Q., Varshney, D.B., and Georgiev, I.G. (2008). Supramolecular control of reactivity in the solid state: from templates to ladderanes to metal-organic frameworks. Acc. Chem. Res. 41, 280–291.

[12] Bach, T., Bergmann, H., Grosch, B., Harms, K., and Herdtweck, E. (2001). Synthesis of enantiomerically pure 1,5,7-trimethyl-3-azabicyclo[3.3.1]nonan-2-ones as chiral host compounds for enantioselective photochemical reactions in solution. Synthesis, 1395–1405.

[13] a) Bergmann, H., Grosch, B., Sitterberg, S., and Bach T. (2004). An enantiomerically pure 1,5,7-trimethyl-3-azabicyclo[3.3.1]nonan-2-one as ^{1}H NMR shift reagent for the *ee* determination of chiral lactams, quinolones, and oxazolidinones. J. Org. Chem. 69, 970–973;
b) Bauer, A., and Bach, T. (2004). Assignment of the absolute configuration of 7-substituted 3-azabicyclo[3.3.1]nonan-2-ones by NMR-titration experiments. Tetrahedron Asymmetry 15, 3799–3803.

[14] a) Bach, T., Bergmann, H., Brummerhop, H., Lewis, W., and Harms K. (2001). The [2+2]-photocycloaddition of aromatic aldehydes and ketones to 3,4-dihydro-2-pyridones: regioselectivity, diastereoselectivity, and reductive ring opening of the product oxetanes. Chem. Eur. J. 7, 4512–4521; b) Bach, T., Bergmann, H., and Harms, K. (1999). High facial diastereoselectivity in the photocycloaddition of a chiral aromatic aldehyde and an enamide induced by intermolecular hydrogen bonding. J. Am. Chem. Soc. 121, 10650–10651.

[15] a) Bach, T., Aechtner, T., and Neumüller, B. (2002). Enantioselective Norrish–Yang cyclization reactions of *N*-(ω-oxo-ω-phenylalkyl)-substituted imidazolidinones in solution and in the solid state. Chem. Eur. J. 8, 2464–2475; b) Bach, T., Aechtner, T., and Neumüller, B. (2001). Intermolecular hydrogen binding of a chiral host and a prochiral imidazolidinone: enantioselective Norrish–Yang cyclisation in solution. Chem. Commun., 607–608.
[16] Bach, T., Bergmann, H., and Harms, K. (2001). Enantioselective photochemical reactions of 2-pyridones in solution. Org. Lett. 3, 601–603.
[17] Bach, T., Grosch, B., Strassner, T., and Herdtweck E. (2003). Enantioselective [6π]-photocyclization reaction of an acrylanilide mediated by a chiral host. Interplay between enantioselective ring closure and enantioselective protonation. J. Org. Chem. 68, 1107–1116.
[18] a) Grosch, B., Orlebar, C.N., Herdtweck, E., Kaneda, M., Wada, T., Inoue, Y., and Bach, T. (2004). Enantioselective [4+2]-cycloaddition reaction of a photochemically generated *o*-quinodimethane: mechanistic details, association studies, and pressure effects. Chem. Eur. J. 10, 2179–2189; b) Grosch, B., Orlebar, C.N., Herdtweck, E., Massa, W., and Bach, T. (2003). Highly enantioselective Diels–Alder reaction of a photochemically generated *o*-quinodimethane with olefins. Angew. Chem. Int. Ed. 42, 3693–3696.
[19] Müller, C., Bauer, A., and Bach T. (2011). Chirogenic [3+2]-photocycloaddition reactions of 2-substituted naphthoquinones with cyclic alkenes. Photochem. Photobiol. Sci. 10, 1463–1468.
[20] Albrecht, D., Vogt, F., and Bach, T. (2010). Diastereo- and enantioselective intramolecular [2+2] photocycloaddition reactions of 3-(ω′-alkenyl)- and 3-(ω′-alkenyloxy)-substituted 5,6-dihydro-1H-pyridin-2-ones. Chem. Eur. J. 16, 4284–4296.
[21] Zimmerman J., and Sibi, M.P. (2006). Enantioselective Radical Reactions. Top. Curr. Chem. 263, 107–162.
[22] a) Dressel, M., Aechtner, T., and Bach, T. (2006). Enantioselectivity and diastereoselectivity in reductive radical cyclization reactions of 3-(ω-iodoalkylidene)-piperidin-2-ones. Synthesis, 2206–2214; b) Aechtner, T., Dressel, M., and Bach, T. (2004). Hydrogen bond mediated enantioselectivity of radical reactions. Angew. Chem. Int. Ed. 43, 5849–5851.
[23] Nishida, M., Hayashi, H., Nishida, A., and Kawahara, N. (1996). Enantioselective radical cyclization controlled by a chiral aluminium reagent. Chem. Commun., 579–580.
[24] Kapitán, P., and Bach, T. (2008). Template-induced enantioselectivity in the reductive radical cyclization of 3-(3-iodopropoxy)propenoic acid derivatives depending on the binding motif. Synthesis, 1559–1564.
[25] a) Bakowski, A., Dressel, M., Bauer, A., and Bach, T. (2011). Enantioselective radical cyclisation reactions of 4-substituted quinolones mediated by a chiral template. Org. Biomol. Chem. 9, 3516–3529; b) Dressel, M., and Bach, T. (2006). Chirality multiplication and efficient chirality transfer in *exo*- and *endo*-radical cyclization reactions of 4-(4′-iodobutyl)quinolones. Org. Lett. 8, 3145–3147.
[26] a) Bentley, K.W. (2006). β-Phenylethylamines and the isoquinoline alkaloids. Nat. Prod. Rep. 23, 444–463; b) Jin, Z. (2011). *Amaryllidaceae* and *Sceletium* alkaloids. Nat. Prod. Rep. 28, 1126–1142.
[27] Austin, K.A.B., Herdtweck, E., and Bach, T. (2011). Intramolecular [2+2] photocycloaddition of substituted isoquinolones: enantioselectivity and kinetic resolution induced by a chiral template. Angew. Chem. Int. Ed. 50, 8416–8419.
[28] a) Bach, T., Bergmann, H., Grosch, B., and Harms, K. (2002). Highly enantioselective intra- and intermolecular [2+2] photocycloaddition reactions of 2-quinolones mediated by a chiral lactam host: host-guest interactions, product configuration, and the origin of the stereoselectivity in solution. J. Am. Chem. Soc. 124, 7982–7990; b) Bach, T., and Bergmann, H. (2000). Enantioselective intermolecular [2+2]-photocycloaddition reactions of alkenes and a 2-quinolone in solution. J. Am. Chem. Soc. 122, 11525–11526.

[29] a) Selig, P., and Bach, T. (2006). Photochemistry of 4-(2′-aminoethyl)quinolones: enantioselective synthesis of tetracyclic tetrahydro-1a*H*-pyrido[4′,3′:2,3]-cyclobuta[1,2-*c*] quinoline-2,11(3*H*,8*H*)-diones by intra- and intermolecular [2+2]-photocycloaddition reactions in solution. J. Org. Chem. 71, 5662–5673; b) Brandes, S., Selig, P., and Bach, T. (2004). Stereoselective intra- and intermolecular [2+2] photocycloaddition reactions of 4-(2′-aminoethyl)quinolones. Synlett, 2588–2590.

[30] a) Selig, P., Herdtweck, E., and Bach, T. (2009). Total synthesis of Meloscine by a [2+2]-photocycloaddition/ring-expansion route. Chem. Eur. J. 15, 3509–3525; b) Selig, P., and Bach, T. (2008). Enantioselective total synthesis of the *Melodinus* alkaloid (+)-Meloscine. Angew. Chem. Int. Ed. 47, 5082–5084.

[31] Selig, P., and Bach, T. (2010). Methyl 2-(Trimethylsiloxy)Acrylate. Encyclop. Reag. Org. Synt. (John Wiley & Sons, Ltd).

[32] a) Bach, T., Bergmann, H., Grosch, B., and Harms, K. (2002). Highly enantioselective intra- and intermolecular [2+2] photocycloaddition reactions of 2-quinolones mediated by a chiral lactam host: host-guest interactions, product configuration, and the origin of the stereoselectivity in solution. J. Am. Chem. Soc. 124, 7982–7990; b) Bach, T., Bergmann, H., and Harms, K. (2000). Enantioselective intramolecular [2+2]-photocycloaddition reactions in solution. Angew. Chem. Int. Ed. 39, 2302–2304.

[33] Breitenlechner, S., and Bach, T. (2008). A polymer-bound chiral template for enantioselective photochemical reactions. Angew. Chem. Int. Ed. 47, 7957–7959.

[34] a) Lu, R., Yang, C., Cao, Y., Wang, Z., Wada, T., Jiao, W., Mori, T., and Inoue, Y. (2008). Supramolecular enantiodifferentiating photoisomerization of cyclooctene with modified β-cyclodextrins: critical control by a host structure. Chem. Comm. 374–376. b) Fukuhara, G., Mori, T., Wada, T and Inoue, Y. (2006). Entropy-controlled supramolecular photochirogenesis: enantiodifferentiating *Z-E* Photoisomerization of Cyclooctene Included and Sensitized by Permethylated 6-*O*-Modified β-cyclodextrins. J. Org. Chem. 71, 8233–8243. c) Asaoka, S., Wada, T., and Inoue, Y. (2003). Microenvironmental polarity control of electron-transfer photochirogenesis. Enantiodifferentiating polar addition of 1,1-diphenyl-1-alkenes photosensitized by saccharide naphthalenecarboxylates. J. Am. Chem. Soc. 125, 3008–3027.

[35] Bauer, A., Westkämper, F., Grimme, S., and Bach, T. (2005). Catalytic enantioselective reactions driven by photoinduced electron transfer. Nature, 436, 1139–1140.

[36] Cauble, D.F., Lynch, V., and Krische, M.J. (2003). Studies on the enantioselective catalysis of photochemically promoted transformations: “sensitizing receptors” as chiral catalysts. J. Org. Chem. 68, 15–21.

[37] Müller, C., Bauer, A., and Bach, T. (2009). Light-driven enantioselective organocatalysis. Angew. Chem. Int. Ed. 48, 6640–6642.

[38] Müller, C., Bauer, A., Maturi, M.M., Cuquerella, M.C., Miranda, M.A., and Bach, T. (2011). Enantioselective intramolecular [2+2]-photocycloaddition reactions of 4-substituted quinolones catalyzed by a chiral sensitizer with a hydrogen-bonding motif. J. Am. Chem. Soc. 133, 16689–16697.

Hans-Achim Wagenknecht, Michael Weinberger, Alexander Penner

6 Photocatalysis with nucleic acids and peptides

6.1 Introduction

Chemical photocatalysis in the organic-chemical context describes the use of a redox active chromophore that upon irradiation with light induces an electron transfer to the substrate. This photophysical process (charge separation) is coupled to a subsequent chemical reaction. The absorbed light energy contributes to the overall energy balance of the reaction and thereby increases its sustainability. Furthermore, oligonucleotides and oligopeptides offer the possibility to control regio- and stereoselectivity as catalysts of organic reactions. The combination of both general concepts, photochemistry and biooligomer catalysis, drives our search for new photocatalytic methods which can be accomplished by rethinking photochemical reactions that were discovered decades ago. The photocatalytic versions of these reactions are both biocompatible and suitable for the UV-A/Vis range in order to increase sustainability by both applying aqueous solutions as reaction media and high-power LEDs as a cheap and reliable source for light.

6.2 DNA-assisted enantioselective reactions

Nature often inspires chemists to use or adopt its systems to chemical reactions. Most often proteins (enzymes) are used as catalysts to build up defined molecular structures with high enantioselectivities. Since it is also known that RNA or DNA can catalyze chemical reactions, there have been several approaches to use this ability for organic reactions. Oligonucleotides are able to either fold in a variety of defined three-dimensional structures or to build up a very stable right-handed B-DNA helix. The use of DNA as a source of chirality in asymmetric organic reactions provides several advantages: First of all DNA is chemically more stable than RNA or proteins. For technical applications the lower price for DNA is also important, compared to other biomolecules. It is readily available in high amounts, in diverse sequences and lengths by synthetic procedures, molecular-biological techniques, or from natural sources. [1] While B-DNA is the predominant conformation, many more are possible, depending on the solvent, the ionic strength and the presence of DNA-binding molecules. [2] Besides A-DNA, left-handed Z-DNA or G-quadruplexes, more sophisticated two- and three-dimensional structures have been investigated. [3] In terms of green chemistry in asymmetric catalysis, DNA is a good choice of ligand due to its high solubility in water. [1]

In principle, there are two different ways of using DNA for organic catalysis, both with pros and cons. In the first approach the catalytic center is brought into the

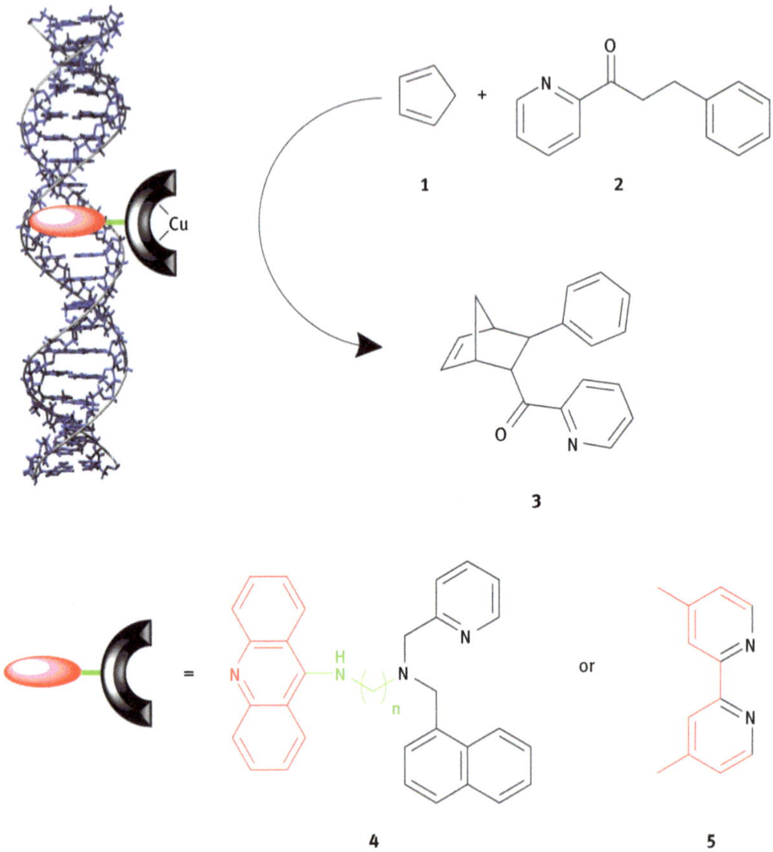

Figure 6.1: Diels–Alder reaction catalyzed by DNA-based Cu(II) complex.

vicinity by self-assembly of the catalytically active complex and the DNA. On the other hand, the oligonucleotide can be modified covalently, either postsynthetically or by using modified DNA building blocks. [2]

The first report on a DNA-based hybrid catalyst was in 2005 by Feringa and Roelfes. [4] A copper(II) complex was brought into the vicinity of DNA in order to catalyze a Diels–Alder reaction between aza-chalcone and cyclobutadiene in water (Figure 6.1). The nonchiral ligand consists of three parts: the copper binding part, a spacer and 9-aminoacridine as an intercalator for DNA and ee values of up to 48 % of the major (endo) product were obtained. The enantioselectivity was highly dependent on the design of the ligand. Interestingly, they observed different enantiomers in excess by variation of the ligand, but could not find general principles for these contradicting results. It is not surprising that the elongation of the spacer causes a dramatic drop in the enantiomeric excess. [2] To bring the catalytic center even closer to DNA a second generation of ligands was prepared, in which the spacer was completely removed. [5] The best choice was 4,4′-dimethyl-2,2′-bipyridine, which achieved up to 99 % regioselectivity and up to 99 % enantioselectivity. [1] The authors successfully transferred this principle to other types of reactions such as Michael additions and Friedel–Crafts alkylations. The use of α,β-unsaturated 2-acyl imidazoles as new substrates allows further synthetic modifications by removing the imidazole group. [2, 6] The Toru group applied the catalytic system for electrophilic fluorination reactions. [7]

All these reactions are highly sequence dependent, because of the different microenvironment of the Cu(II) provided by different oligonucleotides. In this supramolecular approach a scientific elucidation of the structure-activity relationship is quite difficult. To learn more about the system, the covalent approach seems to be very promising. To date only a few examples are published. The catalysis of allylic amination was described successfully by Jäschke et al. [8] They used a diene ligand complexing iridium. The observed ee values are modest; it is, however, a promising system since there is slight enantioselectivity. The covalent modification approach was also tested by Feringa and Roelfes for the Cu(II)-bipyridine system. Such a DNA catalyst was formed via postsynthetic modification and bore the bipyridine ligand in the middle of a double strand; ee values up to 93 % were obtained, slightly lower than the supramolecular assembly previously described. [2]

These different systems show clearly that DNA is a promising source of chirality in catalysis. Additionally, a covalently linked catalytic moiety allows control of the microenvironment of the reaction. [2]

6.2.1 Photocatalytically active DNA (PhotoDNAzymes)

Catalytically active nucleic acids are called ribozymes and DNAzymes. In 1994, the first artificial DNAzyme with the ability to cleave RNA was identified. [9] Since then, the majority of the newly reported DNAzymes showed catalytic activity only by interaction with RNA. There are only few reports on DNA cleavage, phosphorylation and capping. [10] We want to focus herein on PhotoDNAzymes in which the activity is directly linked to excitation by light. There are two fundamentally different approaches for how light is used for DNAzymes. First, light can be used to uncage DNA. [11] That means that a chemical modification of a nucleoside that interferes with the Watson–Crick hydrogen bonds is removed by light. Deiters et al. use, e. g. 6-nitropiperonyloxymethyl as a photocleavable group at the N-3 position of thymidine. After irradiation with 365 nm the normal hydrogen bonding pattern and thus most of the DNAzyme activity is restored. [10a] The second class of light interacting DNAzymes is using the excitation energy directly as a cofactor. Sen et al. reported in 2004 the first example for this category of DNAzymes (Figure 6.2). [12]

An *in vitro* selected oligonucleotide (termed UV1C) showed the ability to repair T-T-dimers by UV irradiation > 300 nm. The light harvesting component was found to be a G-quadruplex structure, hence the DNAzyme itself. [12, 13] The UV1C is a

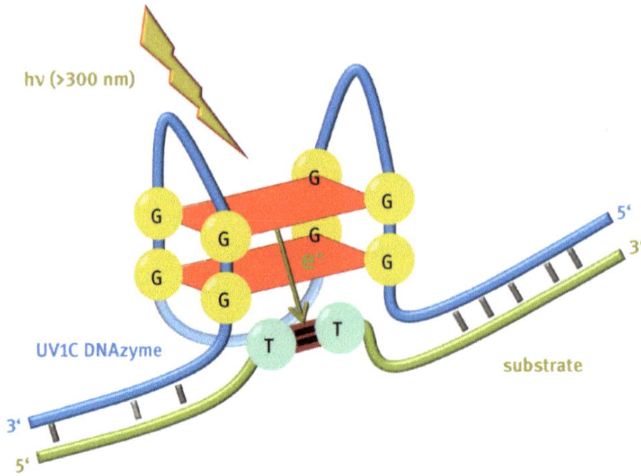

Figure 6.2: Schematic representation of DNAzyme UV1C with the T-T-dimer containing substrate.

good starting point for further research on PhotoDNAzymes, because it shows the ability of DNA to use light directly for catalytic purposes. [10a] The future goal must be to harvest and transfer visible light to DNA structures by using dyes as sensitizers. Moreover, there is no DNAzyme reported up to now that catalyzes C-C bond forming reactions. [10a]

6.2.2 Benzophenone as photosensitizer in DNA for the development of PhotoDNAzymes

Combining the concepts of DNA-assisted enantioselective catalysis (see above) and templated enantioselective photocatalysis (cf. Chapter 5) should yield promising PhotoDNAzymes. For this approach, the C-nucleoside **1α** [14] was chosen as DNA modification because the glycosidic C-C bond is chemically more stable compared to natural bases, and, moreover, the base surrogate is embedded into the base stacking. The defined microenvironment should potentially be useful for photocatalytic organic reactions with DNA. As shown in many examples described above, this reaction environment may lead to a chirality transfer and results in highly enantioselective reactions.

The synthesis of **1α** and **1β** (Figure 6.3) started with the protection of the carbonyl function of 4-bromobenzophenone (**2**) to the ethylene glycol ketal **3**. The sugar moiety was equipped with a reactive anomeric center according to the literature about Hoffer's chlorosugar **4**. [15] In a Grignard reaction, the two moieties were attached forming the C-C glycosidic bond in **5**. The two anomers ($\alpha:\beta=4:1$) were separated by flash chromatography and assigned based on NOE experiments. After deprotection, the desired C-nucleosides **1α** and **1β** were obtained. By reason of better

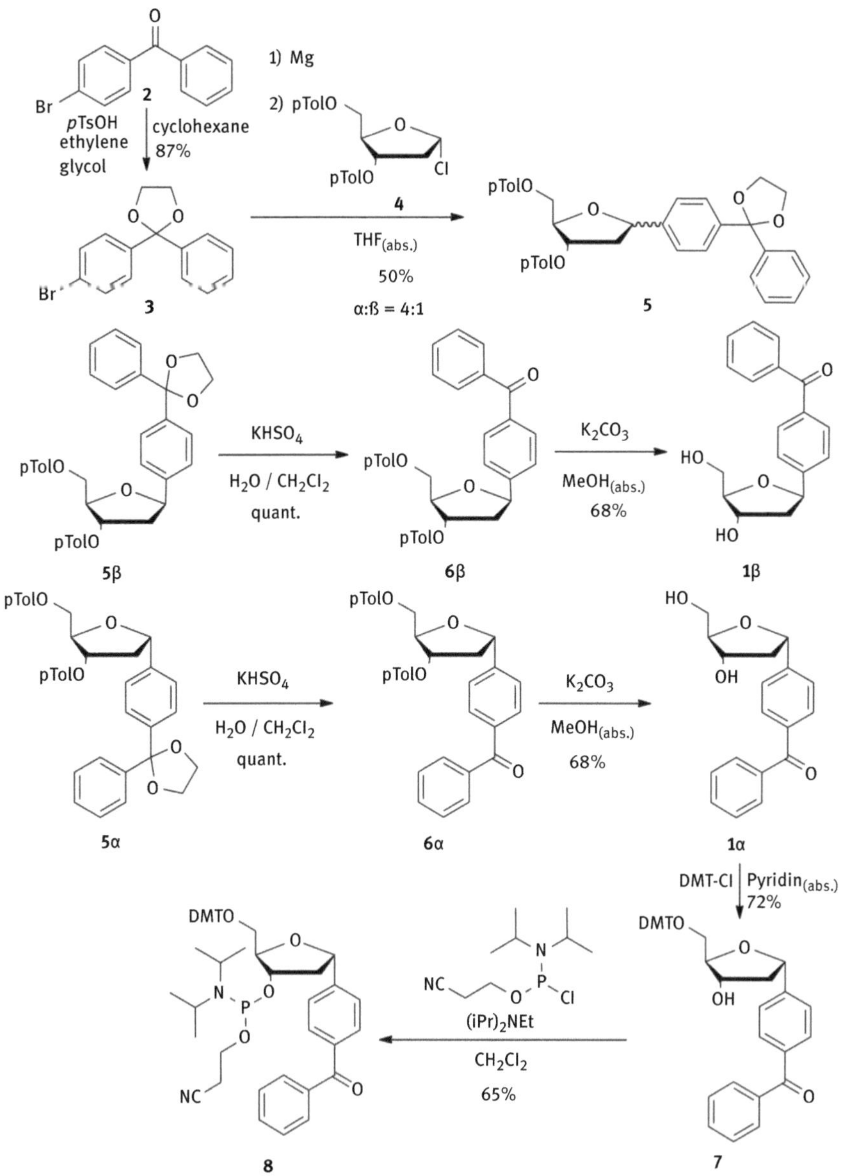

Figure 6.3: Synthesis of C-nucleosides **1α**, **1β** and DNA building block **8**.

synthetic accessibility, the α-anomer was representatively incorporated into oligonucleotides, which followed standard methods of phosphoramidite chemistry to the DNA building block **8**.

The benzophenone building block was incorporated into a representative oligonucleotide with random sequence to study the optical properties in DNA (Figure 6.4).

DNA1 5'–G –C –A –G –T –C –T –A –1α –A –T –C –A –C –T –G –A – 3'

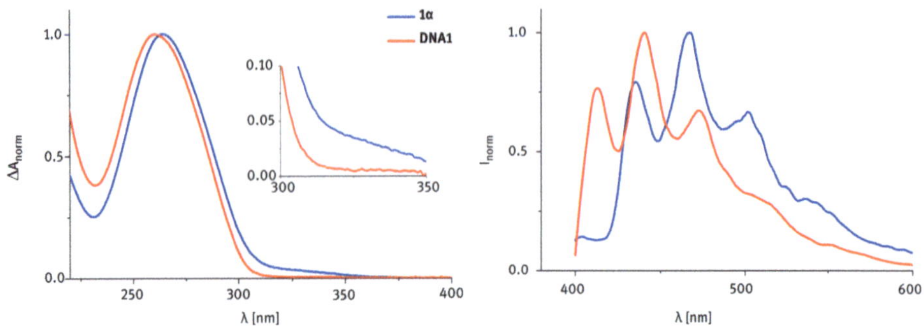

Figure 6.4: Sequence and optical properties of benzophenone building block **1α** and **DNA1**; normalized UV spectra at r.t. (left); normalized phosphorescence spectra at 77 K (right).

After 355 nm excitation at 77 K **1α** shows very similar phosphorescence to unmodified benzophenone. The modified oligonucleotide **DNA1** preserves the typical benzophenone pattern, although the maxima are shifted slightly. While **1α** has its main maxima at 467 nm, **DNA1** shows a blue shift to 440 nm. Remarkably, a very similar blue shift is also observed when **1α** is dissolved in methanol instead of water. This clearly indicates that the surrounding DNA structure of **1α** in **DNA1** behaves like a polar organic solvent for the benzophenone moiety. This is important for the photophysical behavior of **1α** embedded in oligonucleotides. Since the phosphorescence lifetime of **DNA1** did not show any decrease compared to **1α**, there is no predominant photoredox process from the triplet state at 77 K. This might be different when carrying out catalytic experiments, because it is performed under moderate temperatures.

To study the direct influence of each adjacent natural base to **1α**, four different dinucleotides **1X** were synthesized (Figure 6.5, top left), where **X** represents the natural bases A, T, G and C. We chose the well-known intramolecular [2+2]-cycloaddition of 4-(3-butenyloxy)quinolone-2(1*H*)-one **9** as benchmark reaction. [16] UV/Vis spectra were recorded to follow the reaction.

All catalytic experiments were carried out under the same conditions (100 μM **9**, 15 μM catalyst, 10 °C, 250 μL H_2O/MeCN 4:1, $\lambda_{exc} = 366$ nm). Obviously the desired reaction only works in the presence of benzophenone or one of the anomers **1α** and **1β**. With benzophenone-modified dinucleotides (representatively shown for **1C**) and oligonucleotides (representatively shown for **DNA1**) the photocatalysis failed. In order to explain this observation, a closer look at this system by ultrafast transient absorption spectroscopy (cf. Chapter 16) and highly sophisticated theoretical methods (cf. Chapter 14) was crucial. These studies revealed that electron transfer within the dinucleotides that interferes potentially with photocatalytic conversion of the substrate **9** is controlled by the dinucleotide conformation and hence by the solvent.

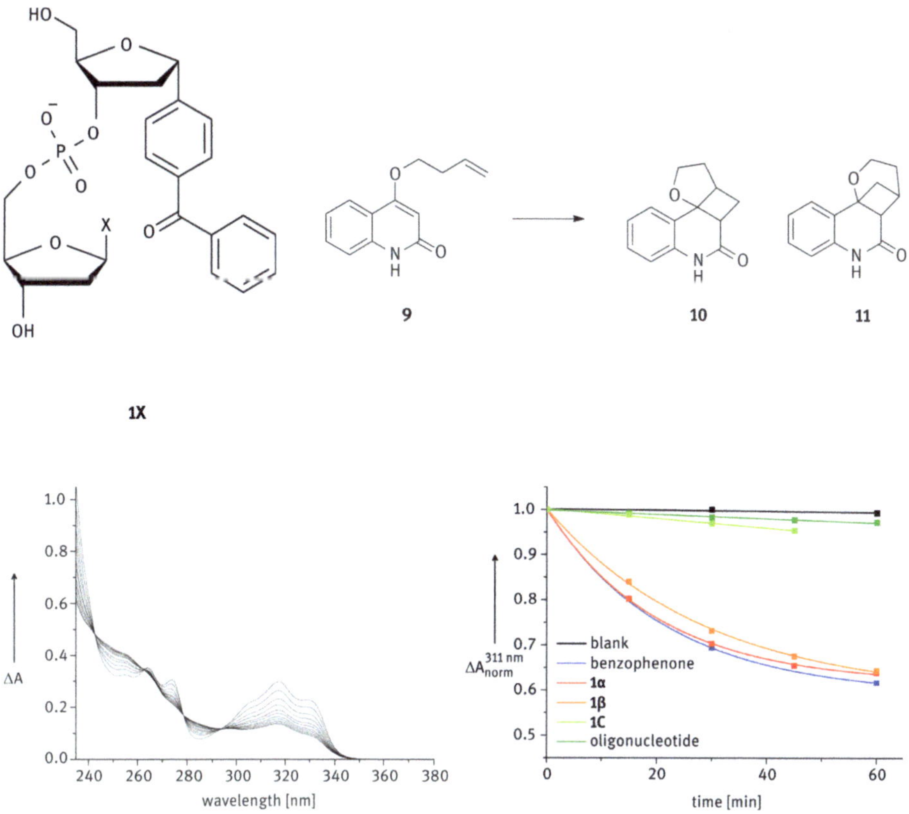

Figure 6.5: General structure of dinucleotides **1X** (top, left) and [2+2]-cycloaddition of 4-(3-butenyl-oxy)quinolone-2(1*H*)-one **9** (top, right). Representative UV/Vis measurements (bottom, left); kinetics of catalytic experiments followed by UV/Vis at 311 nm (bottom, right).

It turned out that the dinucleotides **1X** exist mainly in stacked conformations in water due to hydrophobic interactions between benzophenone and the DNA base. This conformation causes different decay pathways from the singlet and the triplet state of the benzophenone moiety, which are competitive to the photocatalysis. In contrast to water the conformation changes dramatically when methanol is used as a solvent. There is a predominant stretched conformation, which prohibits the mentioned decays. These competitive decay pathways could be diminished potentially by creating a binding site that brings the substrate closer to the sensitizer. There are two different strategies: First, in analogy to the DNA-based catalysts mentioned in Section 6.2, a metal ion can be used to bind the substrate during the photocatalytic circle. Alternatively, the DNA itself can fulfil the task and acts as an aptamer in analogy to Section 6.2.1. A corresponding DNAzyme could potentially recognize and bind the substrate prior to catalysis. Work including the SELEX method is in progress to further develop the PhotoDNAzymes.

6.3 Small peptides as organocatalysts

Peptides provide the following advantages in order to use them as organocatalysts. Their building blocks of amino acids are easily and cheaply accessible from natural products. Automated synthesis allows for designing catalysts specifically for desired reactions. Substrate binding motifs can be created by natural and artificial amino acids as building blocks to address substrates selectivity. In addition, amino acids typically bear a chiral center, which can induce enantioselectivity. There is the advantage of solvent variety. Peptides are soluble in water as well as in organic solvents by use of adequate protection and functionalization groups. Finally, some amino acids and their side chains are able to perform covalent organocatalysis. In principle, imine groups (e. g. methylated side chains of histidine) can perform, e. g. acyl-transfer reactions to nucleophiles by iminium catalysis. Secondary amines (esp. N-terminal proline) can perform enamine catalysis and create nucleophiles from carbonyl compounds, which undergo a reaction with electrophiles.

Peptides as organocatalysts are established between monomeric catalytic units like proline [17], MacMillan [18] or Jørgensen [19] catalyst on the one side and a large variety of polymeric catalysts in the field of enzymes [20, 21] on the other side. Although peptides are known for a large multitude of functions, it is surprising that their potential as catalysts has only been explored since the late 1970s. [22, 23] Their value as asymmetric catalysts was mainly investigated over the last decade by the groups of Miller and Wennemers. [24] In 1998, Miller and coworkers introduced asymmetric catalytic peptides which were able to catalyze acyl-transfer reactions via iminium catalysis. [25] Methylhistidine and methylimidazole-modified peptides showed a high nucleophilicity and performed catalytic activities exceeding those of 4-dimethylaminopyridine to acylate chiral alcohols with excellent enantioselectivities (Figure 6.6). [26] The essential structure motif of these peptidic catalysts was identified to be a β-turn induced by proline that has the ability to provide a chiral environment that brings the reactive moieties into precise arrangements.

The Miller group extended the application of the same type of peptidic catalysts to more challenging substrates, showing that they can discriminate between functional groups that are separated by a long distance. The catalyst was able to

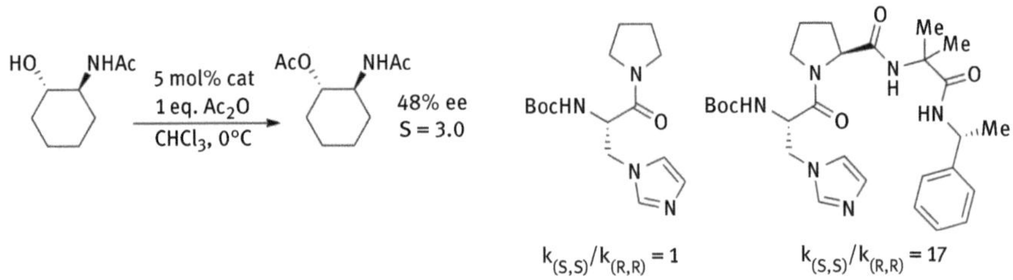

Figure 6.6: Example of Miller's initial stereoselective acylation reaction.

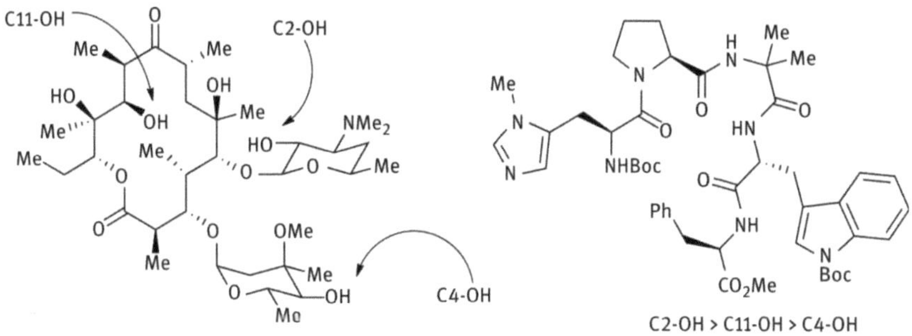

Figure 6.7: Example of side-selective functionalization of erythromycin A by a peptide catalyst.

desymmetrize a long-range bifunctionalized achiral *meso*-diol (~1 nm between enantiotropic hydroxyl groups) in 95 % enantiomeric purity. [27] Even more challenging selectivities could be demonstrated by discrimination of functional groups within complex molecules such as the polyol erythromycin A (Figure 6.7). [28] The best peptidic catalysts were also successfully applied on stereoselective phosphorylations and sulfonations of polyols. [29, 30]

An impressive development in the field of peptidic catalysts was the incorporation of a second and orthogonal reactive center. Schreiner and coworkers coupled a methylhistidine moiety as a nucleophilic center and a TEMPO residue as oxidative reagent to the contrary ends of a non-β-folded peptide catalyst (Figure 6.8). Hence, this was able to undergo cascade reactions with *meso*-diols. The nucleophilic side was used to perform kinetic resolution via acylation by iminium catalysis, as seen before in Miller's work followed by oxidation of the unreacted hydroxyl group to the ketone. [31] This is an impressive example of how the unique conformational features of peptides can be applied for arrangement of orthogonal reactive moieties in highly controlled reaction cascades.

The described peptidic catalysts carry a nucleophile methylhistidine or methylimidazole moiety to perform iminium catalysis and generate acyl electrophiles. In contrast, enamine catalysis requires the condensation of a secondary amine with a carbonylic compound to an enamine that can be regarded as a nucleophilic enolate equivalent. Based on pioneering reports about the single amino acid proline as enamine catalyst for aldol reactions, [32] peptides bearing N-terminal prolinyl resi-

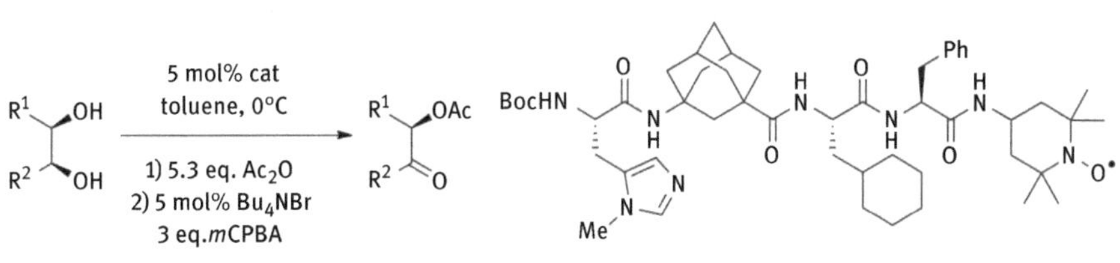

Figure 6.8: Dual catalysis (acylation and oxidation) performed by peptidic catalysts.

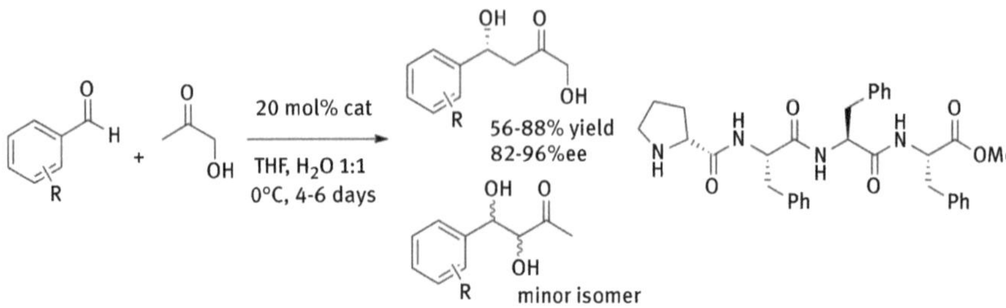

Figure 6.9: Regio- and stereoselective aldol reactions using a peptide organocatalyst.

dues were evaluated as enamine catalysts for aldol and conjugate addition reactions. [33] The groups of Reymond and List proved that peptides are catalytically active for stereoselective aldol reactions relying on enamine catalysis. [17b, 34] Gong and coworkers demonstrated furthermore that peptidic catalysts not only effect the stereoselectivity but can also change regioselectivity compared to simple proline. Proline provides the 1,2-diol as the major product of the aldol reaction between hydroxyacetone and aromatic aldehydes, whereas the peptidic catalyst H-Pro-$(Phe)_3$-OMe supports the formation of the 1,4-diol (Figure 6.9). [35]

The stereoselective induction is astonishingly high with respect to the rather flexible peptidic backbone. In contrast, Wennemers and coworkers used fixed β-folded tripeptides of the general type Pro-Pro-Xaa (Xaa stands for the acidic amino acids aspartate and glutamate) for aldol reactions as well as conjugate addition reactions between aldehydes and nitroolefins to afford even higher stereoselectivities. [36] These peptidic catalysts also provide astonishingly high reactivites as well as stability, and allow catalyst loadings of as little as ≤1 mol% whereas typical catalyst loadings for enamine catalysis are 10–30 mol%. The aptitude of the peptidic catalysts differs by the length of the side chain of the acidic amino acid. H-Pro-Pro-Asp-NH_2 was found by combinatorial screening to be the best catalyst for direct aldol reactions, [37] whereas rational design based on conformational studies illustrated H-Pro-Pro-Glu-NH_2 to be an excellent catalyst for conjugate reactions between aldehydes and nitroolefins (Figure 6.10). The better reaction yields synthetically useful γ-nitroolefins in ≥90 % yield (≥90 % ee), which represent precursors for other valuable compounds like chiral γ-amino acids or butyrolactams. [38]

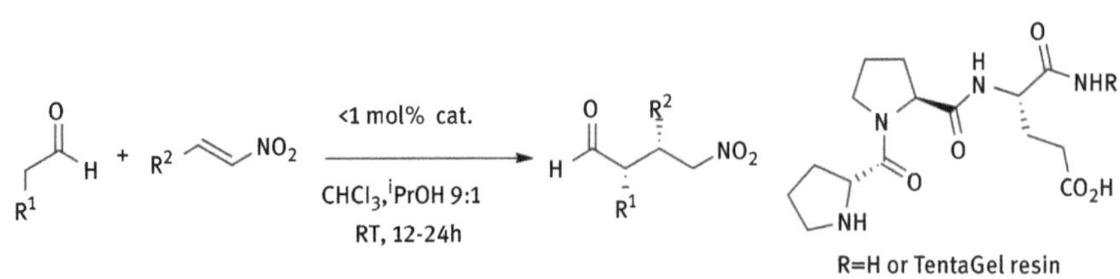

Figure 6.10: Conjugate addition reactions of aldehydes to nitroolefins.

It is remarkable that the catalyst containing glutamate shows high chemoselectivity as no by-products are formed, whereas the aspartate containing catalytic peptide represents a good catalyst for aldol reactions yielding the *homo*-aldol product in the same approach. This demonstrates that the chemoselectivity of peptidic catalysts of the type Pro-Pro-Xaa can be easily tuned by variation of acidic amino acid Xaa. Wennemers and coworkers extended the synthetic applicability of these catalytic peptides on a large scale by immobilizing them on TentaGel resin at the C-terminus without losing reactivity or stereoselectivity.

6.3.1 Development of peptides for photocatalytic addition to olefins

This project started from a known photoreaction from the literature (Figure 6.10) and followed the stepwise development of photocatalysts consisting of electron-rich chromophores and modified peptides. In the prototype reaction a nucleophile is added to an aryl-substituted ethylene **12** yielding Markovnikov addition products (Figure 6.11). In the original report, 1,4-dimethoxy-naphthlene was used as sensitizer and irradiated by a medium-pressure mercury lamp to undergo a reductive electron transfer catalytic cycle. [39] Products resulting from the nucleophiles H_2O, MeOH and CN^- were achieved.

The aim of the project was to design and synthesize chromophore-attached peptides to investigate their aptitude as peptidic photocatalysts. The modular synthetic approach is based on an azide-modified amino acid and its incorporation to a small variety of peptides. Acetylene-modified electron-rich organic chromophores should be attached to the peptides by "click"-chemistry. These peptides could potentially bind substrate molecules and thus bring the photocatalytic unit and the substrate in close proximity in order to enhance the electron transfer efficiency. Moreover, the peptide could also drive this reaction stereoselectively. 3-*L*-Azido-alanine (**16**, Aza) was chosen for the click-type cycloaddition, which can be synthesized from commercially available *L*-serine (**13**). This building block was incorporated via LPPS using the Boc-strategy into tripeptides (Figure 6.12).

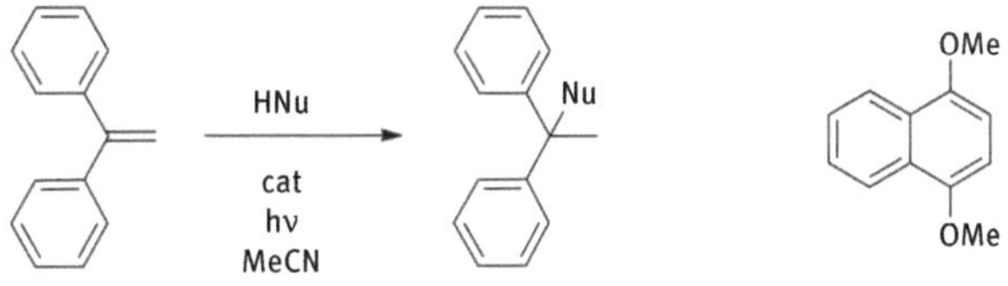

12

Figure 6.11: Photocatalytic test reaction and initial photocatalyst.

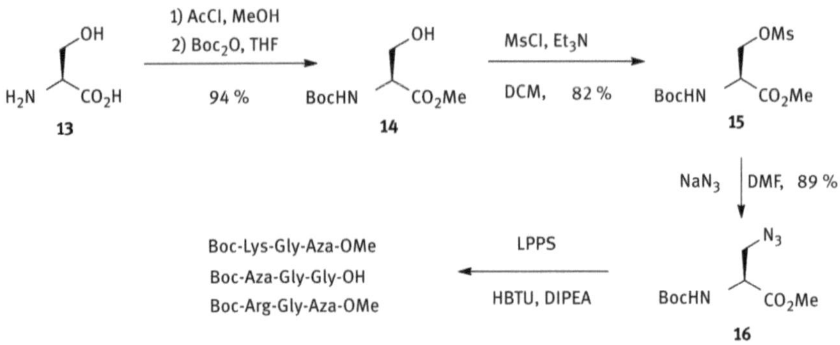

Figure 6.12: Synthesis of 3-azide-alanine (**16**, Aza) and azide-modified tripeptides.

Alternatively, 4-*L*-azido-proline (**20**, Azp) offers additional advantages. Besides the possibility to attach the photocatalyst via click-ligation at this position, the resulting monomeric catalyst can bind substrates via hydrogen bonds promoted by the free carboxylic acid group and it can generate C-nucleophiles via enamine catalysis. Starting from carbonyl derivatives, the free secondary amine group can undergo a reaction with the photocatalytically generated electrophile derived from the substrate. The peptide building block was synthesized by standard procedures (Figure 6.13).

To find good candidates for photocatalysis a screening of electron-rich chromophores was performed (Figure 6.14). The following served as potential candidates: naphthalene derivatives (to evaluate the substitution effects), anthracenes, pyrenes, perylene and two ruthenium(II)-complexes. These compounds were applied equimolar in the previously described model reaction. A 200 W Hg-Xe arc lamp was applied as a broad-band excitation light source. The unsubstituted and alkyl-substituted chromophores were not efficient whereas methoxy-, meththio- and dimethylamino-substituted chromophores showed good conversions. Multiple substituted chromophores showed the best catalytic activity, which also varied depending on the substitution pattern.

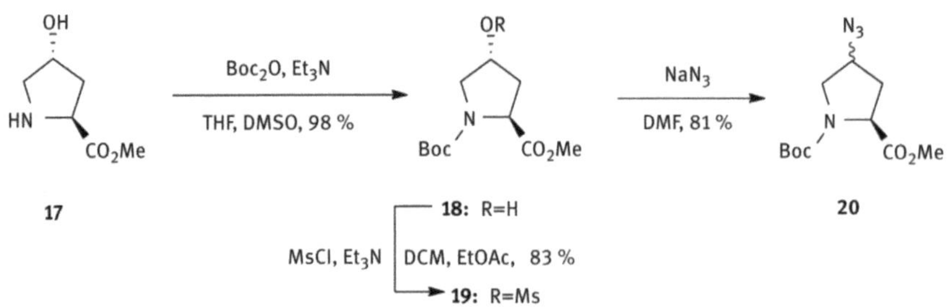

Figure 6.13: Synthesis of *trans*-4-azide-proline (**20**, Azp).

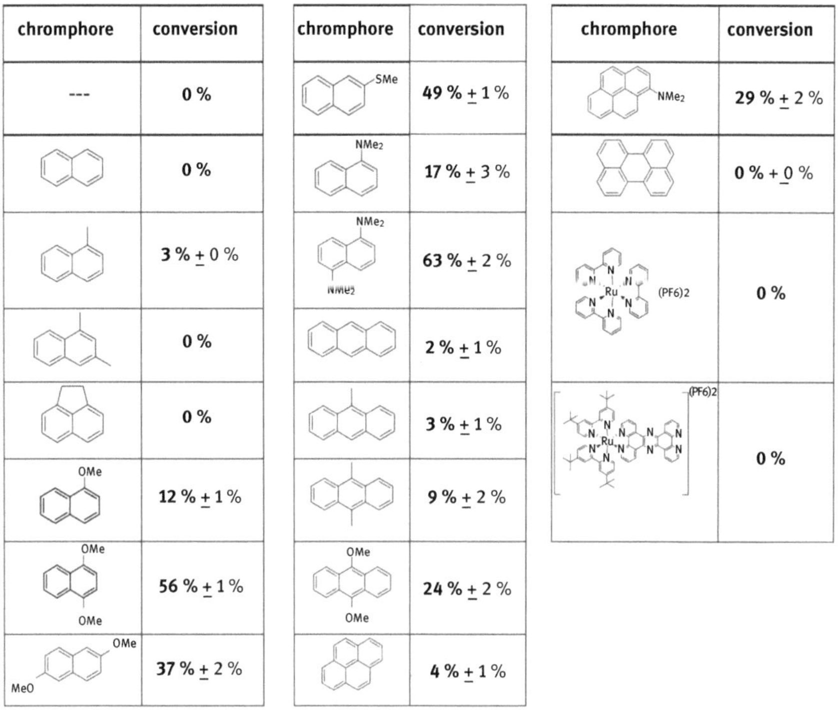

Figure 6.14: Photocatalytic screening for the addition of MeOH to **12**.

Naphthalenes proved to be very efficient photocatalysts but they have the drawback that they absorb mainly in the UV-B region. In order to push the photocatalytic absorption further to the UV-A/Vis range, larger aromatic systems are needed. Among those 9,10-dimethoxyanthracene and 1-(N,N-dimethylamino)-pyrene (DIMAP) were found to be the most efficient ones. Moreover, spectroelectrochemical data revealed spectroscopic signatures of oxidized chromophore radical cations (Figure 6.15). This represents an important prerequisite to follow the catalytic reaction via transient absorption spectroscopy.

An ethynyl-linked derivative of 9,10-dimethoxyanthracene (**25**) ready for "click"-ligation to the azide-modified tripeptides was prepared. The chromophore was then incorporated into the group of presynthesized tripeptides via "click"-chemistry (Figure 6.16).

The potential catalyst **3** was applied in submolar stoichiometries (10 mol%) and provided 23 % conversion under the same conditions as the chromophore screening. This preliminary result shows that the peptides could be developed further by substrate binding.

Furthermore, the catalytic properties of DIMAP were investigated more closely. An advantageous feature of this chromophore is its strong absorption band around

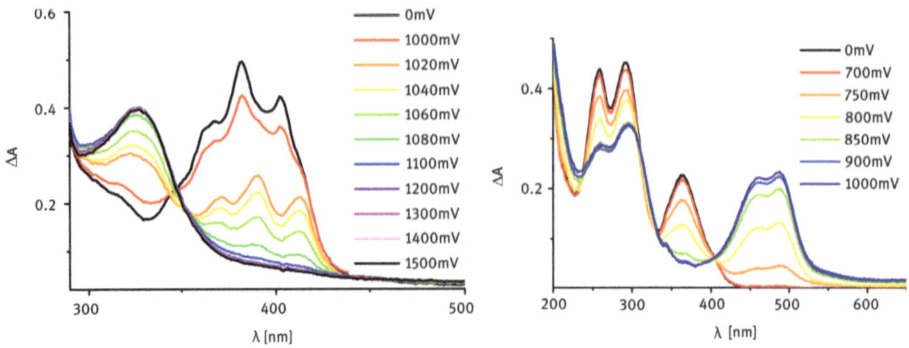

Figure 6.15: Spectroelectrochemical data of 9,10-dimethoxyanthracence (left) and 1-(N,N-dimethyl-amino)-pyrene (right).

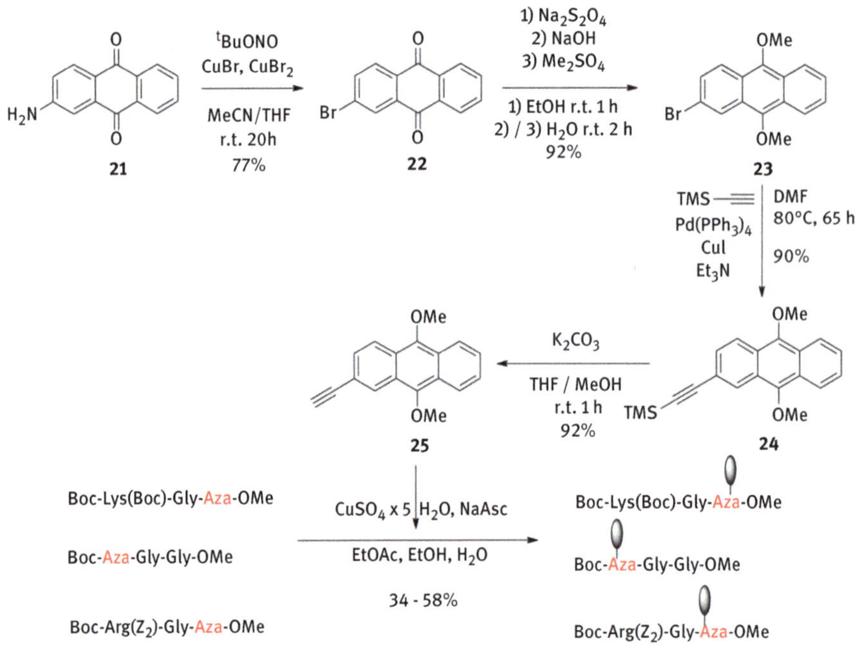

Figure 6.16: Synthesis of 2-ethynyl-9,10-dimethoxy-anthracene (**25**) and "click"-ligation of azide-modified tripeptides.

363 nm. It is possible to excite this catalyst with high-power LEDs emitting light at 366 nm. The catalytic model reaction with **12** under the same conditions as the screening experiment lead to 33 % conversion. Careful MS analysis revealed that the catalyst was destroyed during the experiment. It seemed likely to extend the DIMAP lifetime by addition of Et_3N as an additional electron donor. A 5 vol% Et_3N as addi-

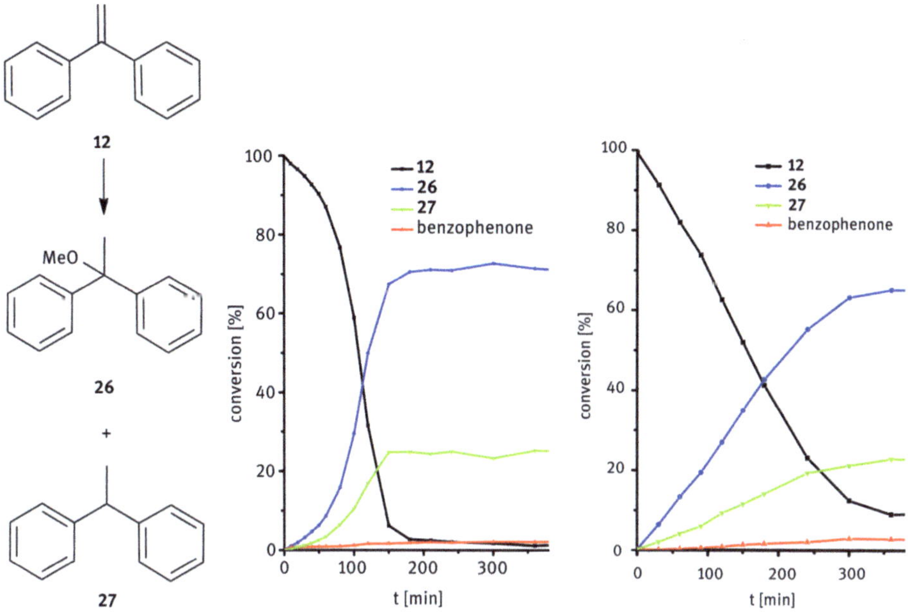

Figure 6.17: Kinetic conversion of substrate **12** to product **26** and side product **27** with 1 eq (left) and 0.1 eq (right) DIMAP as photocatalyst.

tive had a terrific impact on the result of the catalytic experiment. The conversion could be raised up to 75 % yield of **26** based on complete consumption of the starting material. Notably 1,1-diphenylethane (**27**) was formed as a by-product (25 %) in the latter reaction. The kinetic measurement revealed that the reaction is finished after 3 h (Figure 6.17). The by-product 1,1-diphenylethane (SMH_2) is formed concomitantly with the product. Benzophenone, which was the main by-product in the screening experiment, is formed in traces (< 3 %). This is due to the fact that irradiation with 366 nm LEDs forms less singlet oxygen. The addition of Et_3N in this reaction allows for the use of the photocatalyst in substoichiometric amounts. The 10 mol% of DIMAP was still enough to get a conversion of 42 %. In the kinetic measurement with 0.1 eq DIMAP the most noticeable difference is that the reaction is finished after 5 h of excitation instead of 3 h with 1 eq DIMAP. 1,1-dipheny-ethane is also produced concomitantly with the desired product. The conversion is slower and the yield slightly lower because the photocatalyst is bleached after about 5 h of excitation even in the presence of Et_3N.

Based on the publication by Arnold and Maroulis, an electron transfer mechanism is proposed for this nucleophilic addition (Figure 6.18). Stern–Volmer plots support this proposal since fluorescence quenching of DIPEA is observed in the presence of the substrate and in the absence of Et_3N.

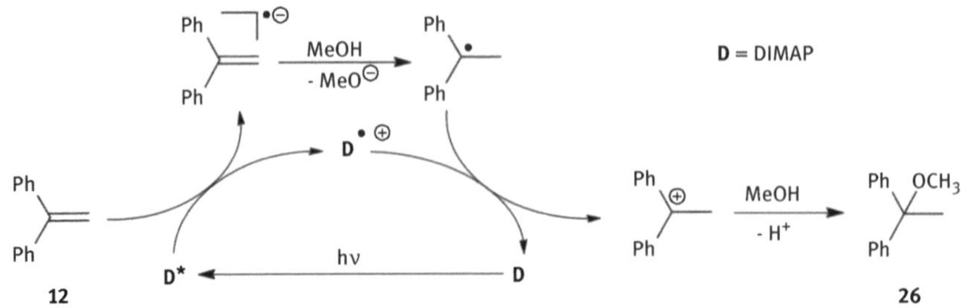

Figure 6.18: Proposed mechanism for the DIMAP-photocatalyzed addition of MeOH to **12** in the absence of Et_3N.

It was detected that the amount of DIMAP gets reduced during the photocatalytic cycle due to bleaching of the chromophore. Thus, the most critical step of the mechanistic cycle seems to be the back electron transfer from the protonated neutral diphenylethyl radical to DIMAP. Obviously, Et_3N supplies the electron to the DIMAP radical cation more efficiently than the substrate radical and the degradation of the photocatalyst is significantly reduced. Moreover, Et_3N increases the yield notably. This implies the idea that Et_3N closes the cycle and gets regenerated by oxidation of the diphenylethyl radical (Figure 6.19). It turned out that Et_3N serves as an additional electron shuttle and thereby improves back electron transfer from the intermediate substrate radical to the chromophore radical cation. Consequently, this proposal also gives an explanation for the formation of the side product 1,1-diphenylethane (**27**). The diphenylethyl radical cannot only be oxidized by the Et_3N radical cation but also reduced by Et_3N itself to the anionic species. Subsequent protonation of the diphenylethyl radical anion gives the undesired product 1-methoxy-1,1-diphenylethane.

Our results clearly show that it is worth developing photocatalytic versions for peptide catalysis. The irradiation in the UV-A/Vis range leads to more efficient processes by applying energy-efficient high-power LEDs as light source. This principal concept and type of methodology will certainly be applicable for other peptide-catalyzed reactions in the future.

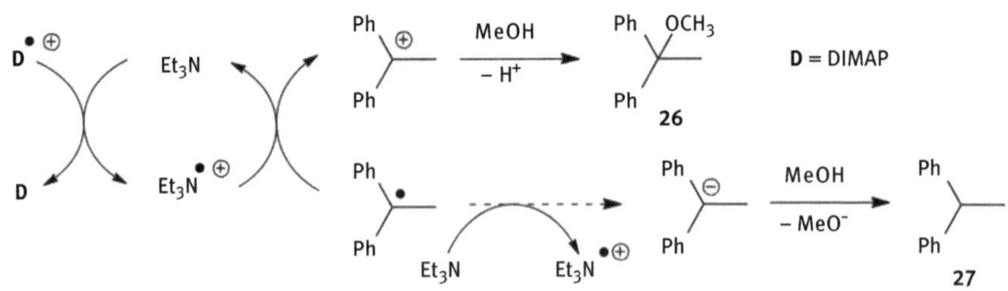

Figure 6.19: Proposed photocatalytic mechanism in the presence of NEt_3.

6.4 Conclusion

Although both parts of this project are not yet completed, our results already show the value in revisiting known photochemical reactions and in developing versions that are photocatalyzed by the assistance of nucleic acids or peptides. The irradiation in the UV-A/near-Vis range increases the sustainability significantly by applying energy-efficient high-power LEDs as a cheap and reliable source for light. The mild and nonacidic and nonbasic conditions make this type of photocatalysis a promising alternative. Oligonucleotides and oligopeptides offer the general possibility to control regio- and stereoselectivity of organic reactions.

6.5 References

[1] Park, S.; Sugiyama, H. (2010). DNA-Based Hybrid Catalysts for Asymmetric Organic Synthesis. Angew. Chem. Int. Ed., 49, 3870–3878.
[2] Boersma, A. J.; Megens, R. P.; Feringa, B. L.; Roelfes, G. (2010). DNA-based asymmetric catalysis. Chem. Soc. Rev., 39, 2083–2092.
[3] (a) Rothemund, P. W. K. (2006). Folding DNA to create nanoscale shapes and patterns. Nature, 440, 297–302; (b) Chen, J.; Seeman, N. C. (1991). Synthesis from DNA of a molecule with the connectivity of a cube. Nature, 350, 631–633.
[4] Roelfes, G.; Feringa, B. L. (2005). DNA-Based Asymmetric Catalysis. Angew. Chem. Int. Ed., 44, 3230–3232.
[5] Roelfes, G.; Boersma, A. L.; Feringa, B. L. (2006). Highly enantioselective DNA-based catalysis. Chem. Commun., 635–637.
[6] Evans, D. A.; Fandrick, K. R.; Song, H.-J. (2005). Enantioselective Friedel–Crafts Alkylations of α,β-Unsaturated 2-Acyl Imidazoles Catalyzed by Bis(oxazolinyl)pyridine–Scandium(III) Triflate Complexes. Journal of the American Chemical Society, 127, 8942–8943.
[7] Shibata, N.; Yasui, H.; Nakamura, S.; Toru, T. (2007). DNA-Mediated Enantioselective Carbon-Fluorine Bond Formation. Synlett, 1153–1157.
[8] Fournier, P.; Fiammengo, R.; Jäschke, A. (2009). Allylic Amination by a DNA-Diene-Iridium(I) Hybrid Catalyst. Angew. Chem. Int. Ed., 48, 4426–4429.
[9] Breaker, R. R.; Joyce, G. F. (1994). A DNA enzyme that cleaves RNA. Chemistry & Biology, 1, 223–229.
[10] (a) Höbartner, C.; Silverman, S. K. (2007). Recent advances in DNA catalysis. Biopolymers, 87, 279–292; (b) Pan, W.; Clawson, G. A. (2008). Catalytic DNAzymes: derivations and functions. Exp. Opin. Biol. Ther., 8, 1071–1085.
[11] (a) Lusic, H.; Young, D. D.; Lively, M. O.; Deiters, A. (2007). Photochemical DNA Activation. Org. Lett., 9, 1903–1906; (b) Young, D. D.; Lively, M. O.; Deiters, A. (2010). Activation and Deactivation of DNAzyme and Antisense Function with Light for the Photochemical Regulation of Gene Expression in Mammalian Cells. J. Am. Chem. Soc., 132, 6183–6193.
[12] Chinnapan, D. J.-F.; Sen, D. (2004). A desoxyribozyme that harnesses light to repair thymine dimers in DNA. Proc. Natl. Acad. Sci. USA, 101, 65–59.
[13] Chinnapen, D. J. F.; Sen, D. (2007). Towards Elucidation of the Mechanism of UV1 C, a Deoxyribozyme with Photolyase Activity. J. Mol. Biol., 365, 1326–1336.
[14] Weinberger, M.; Wagenknecht, H.-A. (2012). Synthesis of a Benzophenone C-Nucleoside as Potential Triplet Energy and Charge Donor in Nucleic Acids. Synthesis, 2012, 648,652.

[15] (a) Hoffer, M. (1960). α-Thymidin. Chem. Ber., 93, 2777–2781; (b) Rolland, V.; Kotera, M.; Lhomme, J. (1997). Convenient preparation of 2′-Deoxy-3,5-di-O-*p*-toluoyl-α-D-erythropentofuranosyl chloride. Synth. Commun., 27, 3505–3511; (c) Dhimitruka, I.; Jr., J. S. (2004). Efficient Preparation of 2-Deoxy-3,5-di-O-*p*-toluoyl-α-D-ribofuranosyl Chloride. Synlett, 335–337; (d) Chin, T.-M.; Huang, L.-K.; Kan, L.-S. (1997). Improved Syntheses of Halofuranose Derivatives with the Desired α-Configuration. J. Chin. Chem. Soc., 44, 413–416.

[16] (a) Bauer, A.; Westkämper, F.; Grimme, S.; Bach, T. (2005). Catalytic enantioselective reactions driven by photoinduced electron transfer. Nature, 436, 1139–1140; (b) Kaneko, C.; Suzuki, T.; Sato, M.; Naito, T. (1987). Cycloadditions in Syntheses. XXXII. Intramolecular Photocycloaddition of 4-(ω-Alkenyloxy)quinolin-2(1*H*)-one: Synthesis of 2-Substituted Cyclobuta[*c*]quinolin-3(4*H*)-ones. Chem. Pharm. Bull., 35, 112–123; (c) Cauble, D. C.; Lynch, V.; Krische, M. J. (2003). Studies on the Enantioselective Catalysis of Photochemically Promoted Transformations: "Sensitizing Receptors" as Chiral Catalysts. J. Org. Chem., 68, 15–21.

[17] (a) Hoang, L.; Bahmanyar, S.; Houk, K. N.; List, B. (2002). Kinetic and Stereochemical Evidence for the Involvement of Only One Proline Molecule in the Transition States of Proline-Catalyzed Intra- and Intermolecular Aldol Reactions. J. Am. Chem. Soc., 125, 16–17; (b) Mukherjee, S.; Yang, J. W.; Hoffmann, S.; List, B. (2007). Asymmetric Enamine Catalysis. Chem. Rev., 107, 5471–5569; (c) Seebach, D.; Beck, A. K.; Badine, D. M.; Limbach, M.; Eschenmoser, A.; Treasurywala, A. M.; Hobi, R.; Prikoszovich, W.; Linder, B. (2007). Are Oxazolidinones Really Unproductive, Parasitic Species in Proline Catalysis? – Thoughts and Experiments Pointing to an Alternative View. Helv. Chim. Acta, 90, 425–471.

[18] (a) Ahrendt, K. A.; Borths, C. J.; MacMillan, D. W. C. (2000). New Strategies for Organic Catalysis: The First Highly Enantioselective Organocatalytic Diels–Alder Reaction. J. Am. Chem. Soc., 122, 4243–4244; (b) Paras, N. A.; MacMillan, D. W. C. (2001). New Strategies in Organic Catalysis: The First Enantioselective Organocatalytic Friedel–Crafts Alkylation. J. Am. Chem. Soc., 123, 4370–4371; (c) Northrup, A. B.; MacMillan, D. W. C. (2002). The First General Enantioselective Catalytic Diels–Alder Reaction with Simple α,β-Unsaturated Ketones. J. Am. Chem. Soc., 124, 2458–2460.

[19] Dinér, P.; Kjærsgaard, A.; Lie, M. A.; Jørgensen, K. A. (2008). On the Origin of the Stereoselectivity in Organocatalysed Reactions with Trimethylsilyl-Protected Diarylprolinol. Chem. Eur. J., 14, 122–127.

[20] Drauz, K.; Waldmann, H., Enzyme Catalysis in Organic Synthesis. VCH: Weinheim, 1995.

[21] Dugas, H., Bioorganic Chemistry. A Chemical Approach to Enzyme Action. 3. ed.; Springer Verlag: Berlin, 1996.

[22] (a) Oku, J.-i.; Ito, N.; Inoue, S. (1979). Asymmetric cyanohydrin synthesis catalyzed by synthetic dipeptides, 1. Makromol. Chem., 180, 1089–1091; (b) Oku, J.-i.; Inoue, S. (1981). Asymmetric cyanohydrin synthesis catalysed by a synthetic cyclic dipeptide. Journal of the Chemical Society, Chem. Commun., 229–230.

[23] (a) Juliá, S.; Masana, J.; Vega, J. C. (1980). "Synthetic Enzymes". Highly Stereoselective Epoxidation of Chalcone in a Triphasic Toluene-Water-Poly[(S)-alanine] System. Angew. Chem. Int. Ed., 19, 929–931; (b) Julia, S.; Guixer, J.; Masana, J.; Rocas, J.; Colonna, S.; Annuziata, R.; Molinari, H. (1982). Synthetic enzymes. Part 2. Catalytic asymmetric epoxidation by means of polyamino-acids in a triphase system. J. Chem. Soc., Perkin Trans. 1, 1317–1324.

[24] Wennemers, H. (2011). Asymmetric catalysis with peptides. Chem. Commun., 47, 12036–12041.

[25] Miller, S. J.; Copeland, G. T.; Papaioannou, N.; Horstmann, T. E.; Ruel, E. M. (1998). Kinetic Resolution of Alcohols Catalyzed by Tripeptides Containing the N-Alkylimidazole Substructure. J. Am. Chem. Soc., 120, 1629–1630.

[26] Miller, S. J. (2004). In Search of Peptide-Based Catalysts for Asymmetric Organic Synthesis. Acc. Chem. Res., 37, 601–610.

[27] Lewis, C. A.; Gustafson, J. L.; Chiu, A.; Balsells, J.; Pollard, D.; Murry, J.; Reamer, R. A.; Hansen, K. B.; Miller, S. J. (2008). A Case of Remote Asymmetric Induction in the Peptide-Catalyzed Desymmetrization of a Bis(phenol). J. Am. Chem. Soc., 130, 16358–16365.
[28] Lewis, C. A.; Miller, S. J. (2006). Site-Selective Derivatization and Remodeling of Erythromycin A by Using Simple Peptide-Based Chiral Catalysts. Angew. Chem. Int. Ed., 45, 5616–5619.
[29] Sculimbrene, B. R.; Miller, S. J. (2001). Discovery of a Catalytic Asymmetric Phosphorylation through Selection of a Minimal Kinase Mimic: A Concise Total Synthesis of d-myo-Inositol-1-Phosphate. J. Am. Chem. Soc., 123, 10125–10126.
[30] Fiori, K. W.; Puchlopek, A. L. A.; Miller, S. J. (2009). Enantioselective sulfonylation reactions mediated by a tetrapeptide catalyst. Nat. Chem., 1, 630–634.
[31] Müller, C. E.; Zell, D.; Schreiner, P. R. (2009). One-Pot Desymmetrization of meso-1,2-Hydrocarbon Diols through Acylation and Oxidation. Chem. Eur. J., 15, 9647–9650.
[32] List, B.; Lerner, R. A.; Barbas, C. F. (2000). Proline-Catalyzed Direct Asymmetric Aldol Reactions. J. Am. Chem. Soc., 122, 2395–2396.
[33] (a) Tanaka, F.; Barbas Iii, C. F. (2001). Phage display selection of peptides possessing aldolase activity. Chem. Commun., 769–770; (b) Tsogoeva, Svetlana B.; Jagtap, Sunil B.; Ardemasova, Zoya A.; Kalikhevich, Victor N. (2004). Trends in Asymmetric Michael Reactions Catalysed by Tripeptides in Combination with an Achiral Additive in Different Solvents. Eur. J. Org. Chem., 2004, 4014–4019; (c) Andreae, M. R. M.; Davis, A. P. (2005). Heterogeneous catalysis of the asymmetric aldol reaction by solid-supported proline-terminated peptides. Tetrahedron: Asymmetry, 16, 2487–2492; (d) Zou, W.; Ibrahem, I.; Dziedzic, P.; Sunden, H.; Cordova, A. (2005). Small peptides as modular catalysts for the direct asymmetric aldol reaction: ancient peptides with aldolase enzyme activity. Chem. Commun., 4946–4948.
[34] (a) Martin, H. J.; List, B. (2003). Mining Sequence Space for Asymmetric Aminocatalysis: N-Terminal Prolyl-Peptides Efficiently Catalyze Enantioselective Aldol and Michael Reactions. Synlett, 1901,1902; (b) Kofoed, J.; Nielsen, J.; Reymond, J.-L. (2003). Discovery of new peptide-Based catalysts for the direct asymmetric aldol reaction. Bioorg. Med. Chem. Lett., 13, 2445–2447.
[35] Tang, Z.; Yang, Z.-H.; Cun, L.-F.; Gong, L.-Z.; Mi, A.-Q.; Jiang, Y.-Z. (2004). Small Peptides Catalyze Highly Enantioselective Direct Aldol Reactions of Aldehydes with Hydroxyacetone: Unprecedented Regiocontrol in Aqueous Media. Org. Lett., 6, 2285–2287.
[36] (a) Krattiger, P.; Kovasy, R.; Revell, J. D.; Ivan, S.; Wennemers, H. (2005). Increased Structural Complexity Leads to Higher Activity: Peptides as Efficient and Versatile Catalysts for Asymmetric Aldol Reactions. Org. Lett., 7, 1101–1103; (b) Krattiger, P.; Kovàsy, R.; Revell, J. D.; Wennemers, H. (2005). Using Catalyst–Substrate Coimmobilization for the Discovery of Catalysts for Asymmetric Aldol Reactions in Split-and-Mix Libraries. QSAR Comb. Sci., 24, 1158–1163; (c) Wiesner, M.; Revell, J. D.; Wennemers, H. (2008). Tripeptides as Efficient Asymmetric Catalysts for 1,4-Addition Reactions of Aldehydes to Nitroolefins – A Rational Approach. Angew. Chem. Int. Ed., 47, 1871–1874.
[37] (a) Revell, J. D.; Gantenbein, D.; Krattiger, P.; Wennemers, H. (2006). Solid-supported and pegylated H-Pro-Pro-Asp-NHR as catalysts for asymmetric aldol reactions. Pept. Sci., 84, 105–113; (b) Revell, J. D.; Wennemers, H. (2007). Functional group requirements within the peptide H-Pro-Pro-Asp-NH2 as a catalyst for aldol reactions. Tetrahedron, 63, 8420–8424; (c) Revell, J. D.; Wennemers, H. (2008). Investigating Sequence Space: How Important is the Spatial Arrangement of Functional Groups in the Asymmetric Aldol Reaction Catalyst H-Pro-Pro-Asp-NH2? Adv. Synth. Catal., 350, 1046–1052; (d) Messerer, M.; Wennemers, H. (2011). Reversing the Enantioselectivity of a Peptidic Catalyst by Changing the Solvent. Synlett, 2011, 499,502.

[38] (a) Wiesner, M.; Neuburger, M.; Wennemers, H. (2009). Tripeptides of the Type H-D-Pro-Pro-Xaa-NH2 as Catalysts for Asymmetric 1,4-Addition Reactions: Structural Requirements for High Catalytic Efficiency. Chem. Eur. J., 15, 10103–10109; (b) Wiesner, M.; Upert, G. g.; Angelici, G.; Wennemers, H. (2009). Enamine Catalysis with Low Catalyst Loadings – High Efficiency via Kinetic Studies. J. Am. Chem. Soc., 132, 6–7; (c) Wiesner, M.; Wennemers, H. (2010). Peptide-Catalyzed Conjugate Addition Reactions of Aldehydes to Nitroolefins. Synthesis, 2010, 1568,1571.
[39] Arnold, D. R.; Maroulis, A. J. (1977). Radical ions in photochemistry. 4. The 1,1-diphenylethylene anion radical by photosensitization (electron transfer). J. Am. Chem. Soc., 99, 7355–7356.

Filip Teplý

7 Visible light photoredox catalysis with $[Ru(bpy)_3]^{2+}$: General principles and the twentieth century roots

7.1 Introduction

Organic reactions effected by light have been pursued as an important branch of chemistry since the early twentieth century. [1–3] In 2012, one hundred years after Giacomo Ciamician's visionary address on the future of photochemistry, [1b] many chemists feel eager to explore transformations triggered by light in the visible region of the electromagnetic spectrum as amply documented in this book. The expanding interest is mainly curiosity-driven and is fueled by economical and ecological advantages of using visible light as an abundant source of energy.

Processes that can be induced by photoredox catalysis using a combination of $[Ru(bpy)_3]^{2+}$ catalyst and visible light have been considered as a tool for preparative organic chemistry since the 1970s. However, only a few research groups dealt with this topic until the beginning of the twenty-first century. In 2008 and 2009, the attractiveness of this bond-forming strategy for organic synthesis was highlighted due to the seminal studies by MacMillan, Yoon, and Stephenson groups. [4–7] Since then, organic chemistry has witnessed rejuvenated interest in organic transformations triggered by metal-based dyes, and recently also purely organic dyes, [8] as photoredox catalysts. This chapter turns to the roots of this branch of research and summarizes the pioneering examples from the twentieth century. The emphasis is on $[Ru(bpy)_3]^{2+}$ (Figure 7.1) and the transformations where this complex has been used as a photoredox catalyst. With permission of the Collection of Czechoslovak Chemical Communications, large parts of this chapter have been adopted from a review published in the last year. [4b]

7.2 $[Ru(bpy)_3]^{2+}$ and its photoredox properties

Salts of $[Ru(bpy)_3]^{2+}$ (**1**, bpy = 2,2′-bipyridine) were first reported by Burstall in 1936 (Figure 7.1). [9] Notably, Burstall also achieved resolution of the two enantiomers of **1** via diastereomeric tartrate salts. [9] At that time, it was shown that $[Ru(bpy)_3]^{2+}$ was not only chemically robust, but also a remarkably configurationally stable species. [10] Whereas chirality of this C_3-symmetric complex cation was recognized at the time of its first synthesis, its unique photochemical and electrochemical properties have attracted attention with a more than 30-year delay. [11, 12] However, at that time the research focused on the photoredox manifold of $[Ru(bpy)_3]^{2+}$ and its applications have

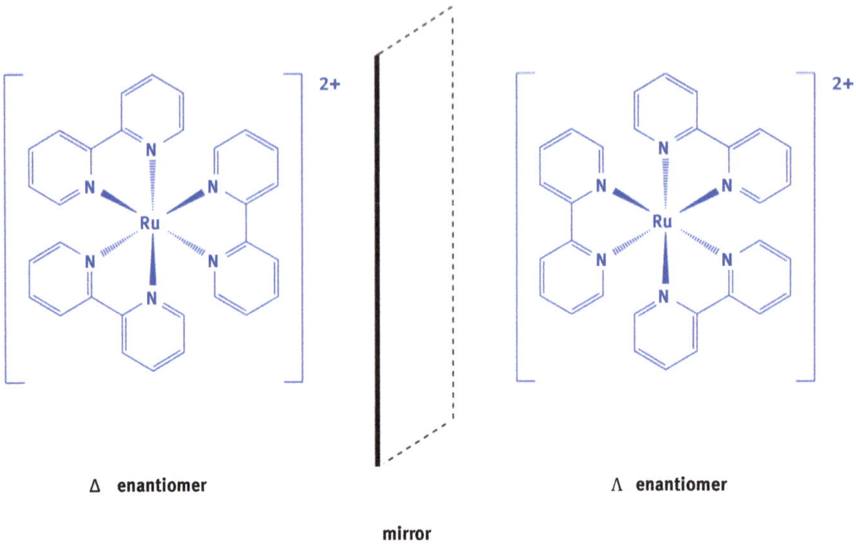

Figure 7.1: Structure of $[Ru(bpy)_3]^{2+}$ (**1**) and its chirality.

since experienced a massive expansion; especially, the conversion of solar radiation into other forms of energy has been studied vigorously. In particular, conversion of solar energy into electrical current [13] and into chemical energy of fuels by photoreduction of small molecules, such as water and CO_2, have received focused attention due to the practical implications of harnessing solar energy to meet the global energy needs. [14] Besides this, even at the beginning of the twenty-first century, $[Ru(bpy)_3]^{2+}$ and its rich spectrum of useful properties continues to inspire important conceptual advances in other areas, as recently discussed by Balzani et al. in the context of a novel molecular encoder–decoder for sensing, labeling, or molecular computing. [15]

Photoredox manifold of $[Ru(bpy)_3]^{2+}$ complex offers unique opportunities on how visible light can be channeled to trigger formation of chemical bonds that are of interest for preparative organic chemistry. [4] Since 2008, this promising potential has started to be explored by numerous research groups. Visible light matches well a broad absorption band of the $[Ru(bpy)_3]^{2+}$ complex (λ_{max} = 452 nm) leading to efficient excitation that gives the lowest singlet excited state ($^1Ru^{2+*}$ in Figure 7.2). This initially generated singlet state (^{1}MLCT state, MLCT = metal-to-ligand charge-transfer) then undergoes intersystem crossing yielding a long-lived luminescent triplet excited state $[Ru(bpy)_3]^{2+*}$ (^{3}MLCT, $^3Ru^{2+*}$ in Figure 7.2). The triplet-excited state is generated with highest quantum efficiency and is long-lived (~ 600 ns). This high-energy species can serve either as a single-electron oxidant or reductant depending on other chemical species present. If reductive quenching of the $[Ru(bpy)_3]^{2+*}$ takes place, the strongly reducing species $[Ru(bpy)_3]^+$ is produced (−1.33 V vs. SCE in CH_3CN), whereas oxidative quenching pathway generates $[Ru(bpy)_3]^{3+}$ that is a

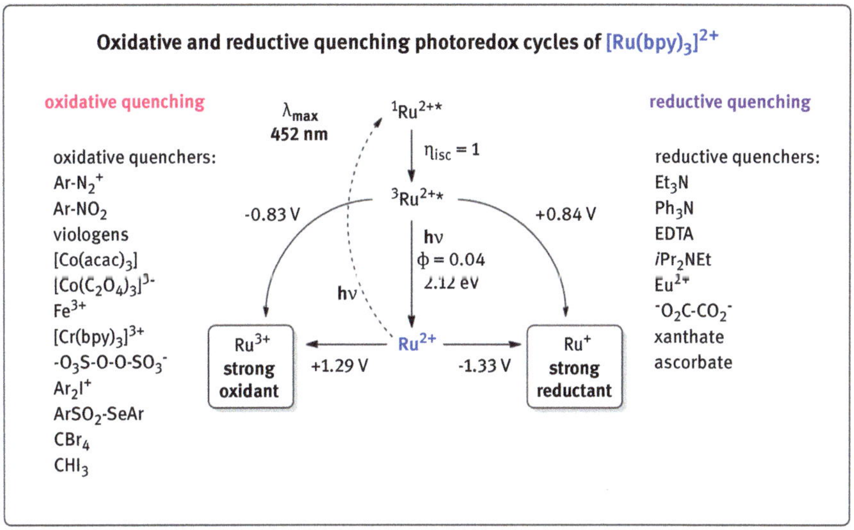

Figure 7.2: Photoredox manifold of $[Ru(bpy)_3]^{2+}$ and representative oxidative and reductive quenchers.

strong oxidant (+1.29 V *vs.* SCE in CH_3CN). Depending on the choice of suitable reductive or oxidative quencher, the $[Ru(bpy)_3]^{2+}$ catalyst can be used to trigger photoreduction or photooxidation, respectively (Figure 7.2). Recently, more details on $[Ru(bpy)_3]^{2+}$ as a prototypical photoredox catalyst have been summarized in an excellent overview in which photoelectronic properties of other photoredox catalysts are also discussed. [4a]

7.3 Application of $[Ru(bpy)_3]^{2+}$ as catalyst in the twentieth century

In the twentieth century, only a handful of research groups dealt with the application of photoredox properties of $[Ru(bpy)_3]^{2+}$ in organic synthesis.

In 1978, Kellogg's group reported reduction of phenacylsulphonium salts (e. g. **2**) by 1,4-dihydropyridine **3** (**2** + **3** → **4**, Figure 7.3). The reaction was found to be induced by irradiation with visible light and greatly accelerated in the presence of $[Ru(bpy)_3]^{2+}$ and other dyes, such as TPP (*meso*-tetraphenylporphine), or eosin disodium salt. [16] Importantly, the roles of light and dyes were carefully tested. Thus, a control reaction performed in the dark in the absence of sensitizing dye did not lead to any conversion at room temperature even after 72 h. However, the reaction reached complete conversion upon exposure of the same reaction mixture to incident room light for 48 h (neon fluorescent lamps at ca. 2 m distance). Addition of 1 mol% of $[Ru(bpy)_3]Cl_2$, TPP, or eosin disodium salt led to acceleration of the irradi-

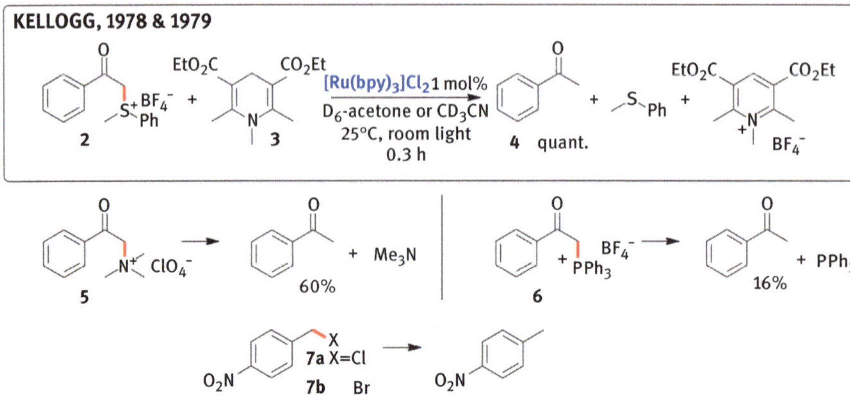

Figure 7.3: Dye-accelerated photoreduction of phenacyl sulfonium salt **2** and related substrates. [16, 17]

ated reactions resulting in complete conversions within significantly shorter reaction times of 0.3 h, 1 h, or 3 h, for the three dyes, respectively. The authors suggested that light-induced single-electron transfer steps are responsible for the observed sulfonium salt reduction and proposed that the large acceleration effect of $[Ru(bpy)_3]Cl_2$ might be due to the involvement of this photoredox catalyst in the single-electron transfers (SET). [16] Later on, results of a more detailed mechanistic study were disclosed. [17] Although the authors did not arrive at a definitive conclusion about the exact role of $[Ru(bpy)_3]^{2+}$ in this process, an impressively insightful discussion of the gathered body of evidence was presented. [17]

Reductions of other substrates such as ammonium, phosphonium salts, and 4-nitrobenzyl halides have also been achieved (**5**, **6**, **7a**, **b**, respectively, Figure 7.3). [16, 17] In 1985, the same group described reductions of phenacylbromide, bromomalonates, and related substrates (Figure 7.4) using 3-methyl-2,3-dihydrobenzothiazoles using $[Ru(bpy)_3]Cl_2$ or rose bengal as photocatalysts. [18] All these examples provided important early evidence that various dyes can function as remarkable accelerators of some light-induced reductions.

Another type of transformation photosensitized by $[Ru(bpy)_3]^{2+}$ has been independently discovered by the group of Pac (Figure 7.5). [19, 20] In their work Pac et al. studied reactions of various olefins with 1-benzyl-1,4-dihydronicotinamide (BNAH) as an NADH model. The outcome of these processes was shown to be substrate specific. For example, dimethyl maleate **8** afforded mainly dimethylsuccinate (**9**), whereas olefin **11** led to a product **12** containing dihydropyridine moiety.

To account for the observed results, the authors provided a mechanistic rationale depicted in Figure 7.6. Ruthenium catalyst was proposed to participate in this transformation via reductive quenching cycle. To this end, the photocatalyst's excited state $[Ru(bpy)_3]^{2+*}$ is reductively quenched by BNAH to form *in situ* $[Ru(bpy)_3]^+$

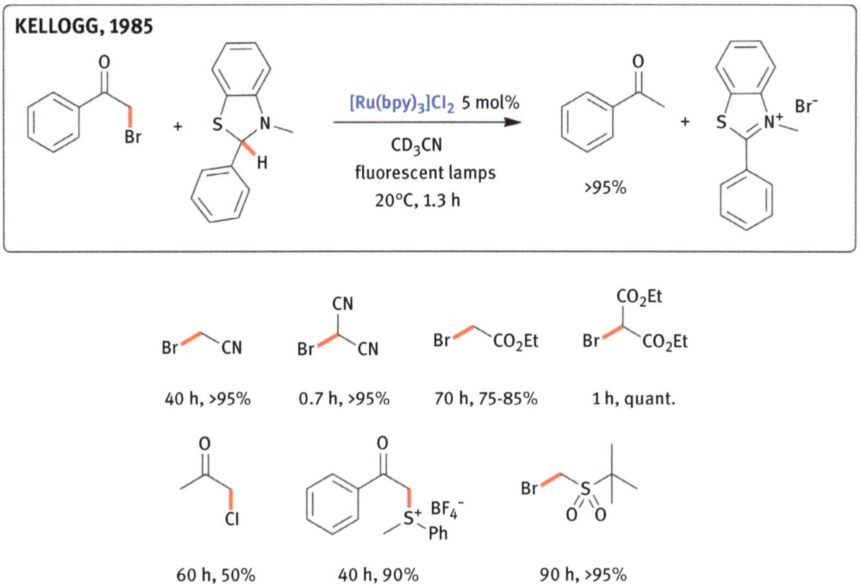

Figure 7.4: Dye-enhanced photoreductions at saturated carbon atoms using 3-methyl-2,3-dihydro-benzothiazole reductant. [18]

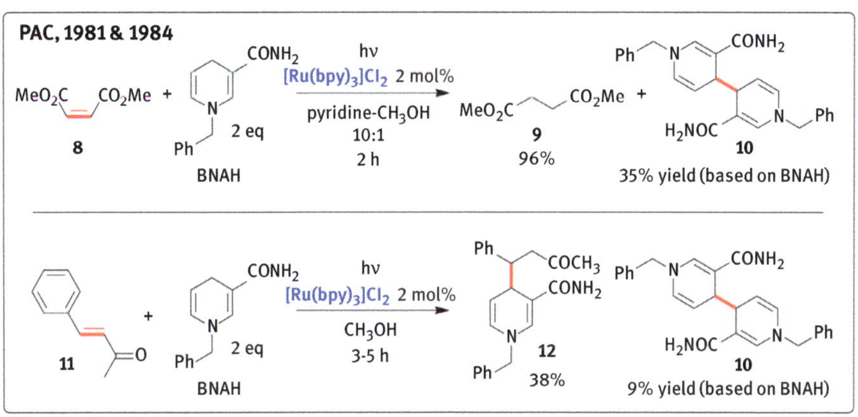

Figure 7.5: $[Ru(bpy)_3]^{2+}$-catalyzed photoreduction of olefins with 1-benzyl-1,4-dihydronicotinamide. [19, 20]

and $BNAH^{\bullet+}$. The latter species can subsequently lose proton leading to $BNA^{\bullet}$ that will result in production of the dimer **10.** Strongly reducing $[Ru(bpy)_3]^{+}$ was proposed to transfer its electron onto the olefin substrate to generate the radical anion **13** that is then protonated to give radical intermediate **14.** The resultant radical **14** can be either reduced to the product **15** or can alternatively lead to the structure **16**

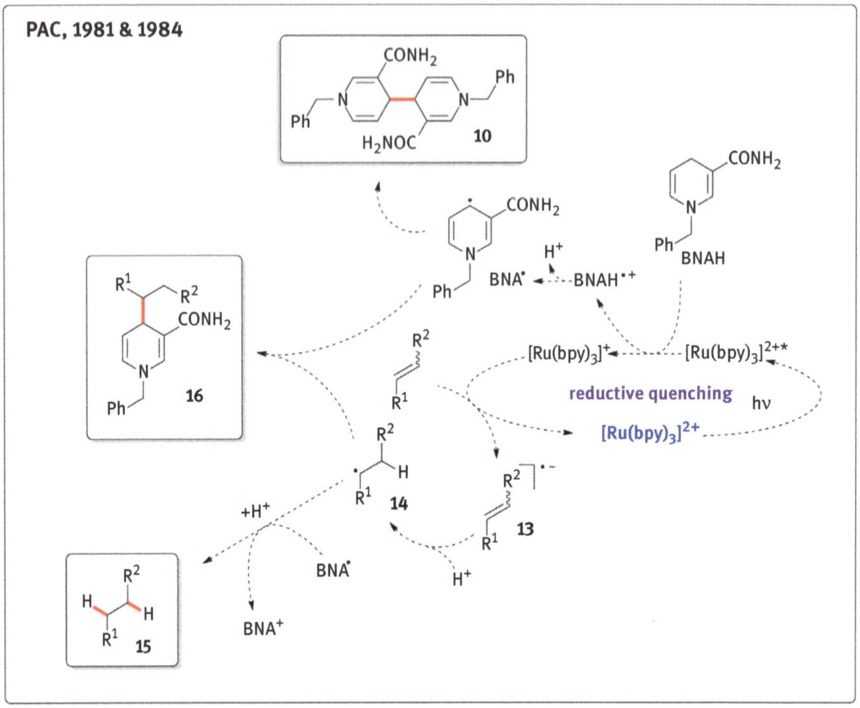

Figure 7.6: Mechanistic picture proposed for the transformations of olefins in Figure 7.5. [19, 20]

containing dihydropyridine moiety. Whether radical **14** undergoes electron transfer resulting in structure **15** or leads to radical coupling giving dihydropyridine product **16**, depends on steric and electronic properties of **14** that should be affected by the substituents at the radical center.

In a related work, Pac and coworkers disclosed the reactivity of aromatic carbonyl compounds with BNAH under similar conditions (Figure 7.7). [21, 22] The reaction course was found to be governed by the structural parameters of the used carbonyl compound. Thus, di-2-pyridylketone (**17**) led to the alcohol **18** together with a dimeric dihydronicotinamide structure **10**, whereas benzaldehyde was converted to a secondary alcohol **19** in high yield. It was reasoned that the exclusive formation of alcohol **18** from di-2-pyridylketone (**17**) is governed by the two pyridyl groups that have a dual role. Because of the electron-withdrawing nature of these groups, the one electron reduction of the radical HO-C$^{\bullet}$(2-py)$_2$ is favored (see single-electron transfer to **20** leading to **21** in Figure 7.8). In addition to that, radical coupling of the species **20** (leading to possible product **22**) is inhibited because of steric hindrance around the radical center in HO-C$^{\bullet}$(2-py)$_2$.

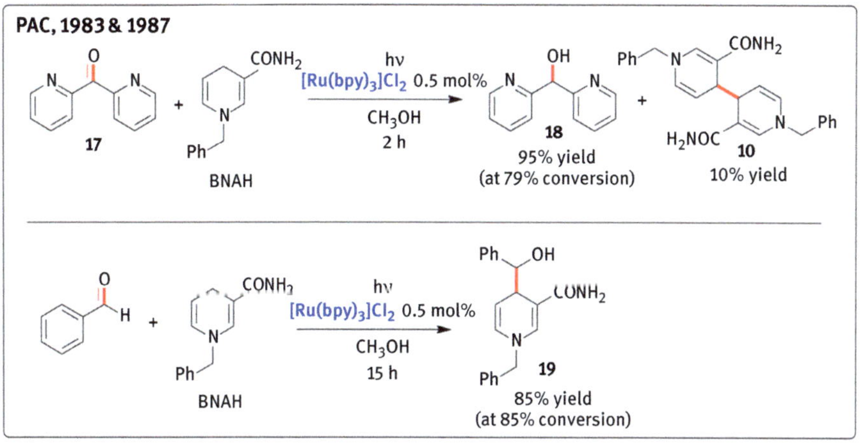

Figure 7.7: $[Ru(bpy)_3]^{2+}$-photosensitized reactions of carbonyl compounds with an NADH model. [21, 22]

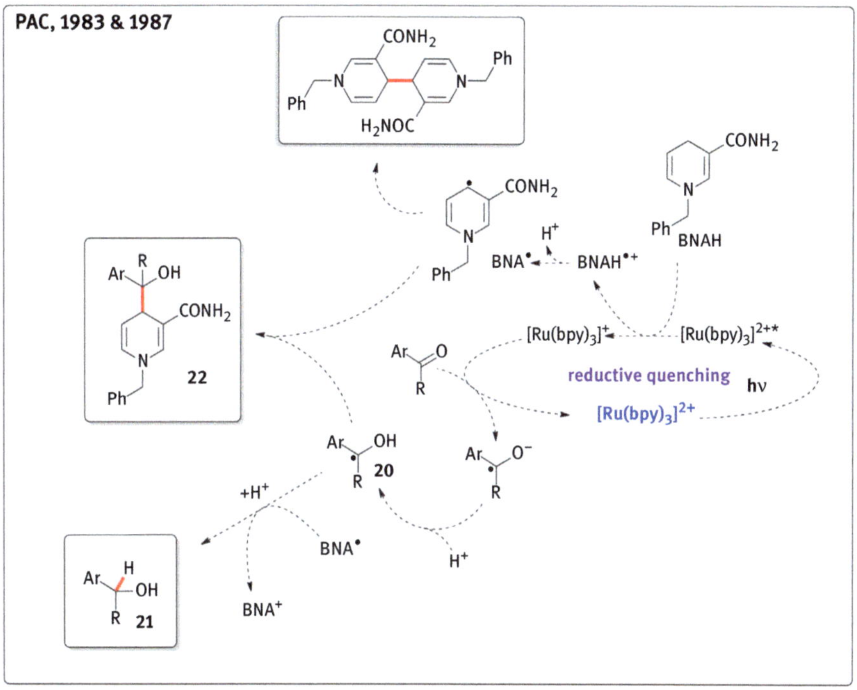

Figure 7.8: Mechanistic picture proposed for transformations of carbonyl compounds in Figure 7.7. [21, 22]

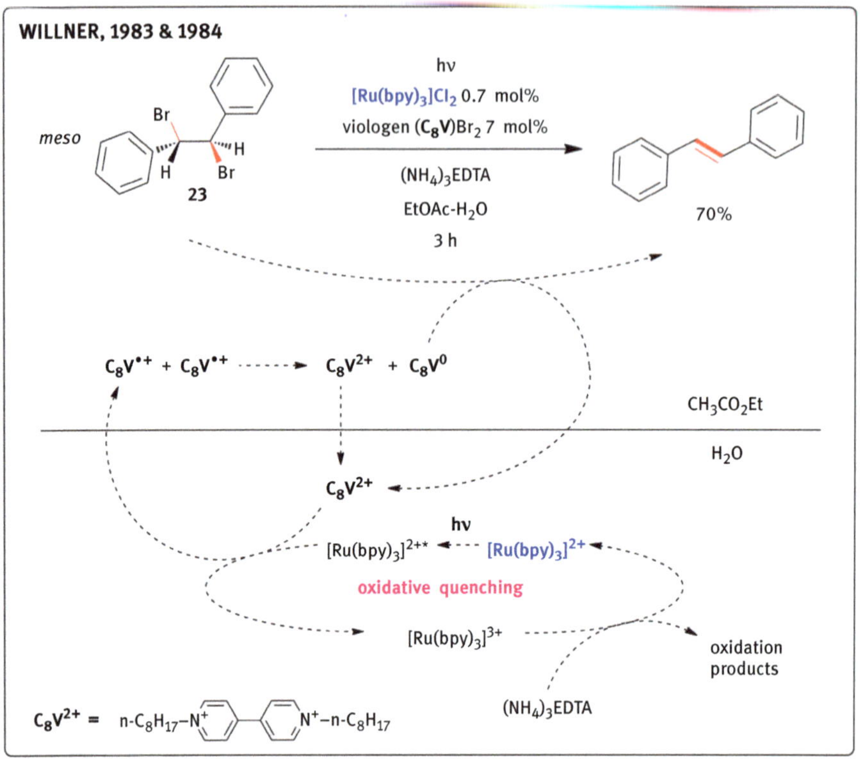

Figure 7.9: Photosensitized debromination of *meso*-1,2-dibromostilbene in a two-phase system. [23, 24]

Photoinduced electron transfer has been extensively studied because of solar energy conversion and storage. [12–14] In this context, the photoinduced production of 4,4′-bipyridinium radical cations from the corresponding dications (viologens) with visible light is very well known. Usually, $[Ru(bpy)_3]^{2+}$ or zinc porphyrins are used as sensitizers and triethanolamine, ethylenediaminetetraacetic acid (EDTA), or cysteine are introduced as electron donors. Inspired by this, Willner group studied several photochemical transformations involving photogenerated reductant $[Ru(bpy)_3]^{2+}*$. [23–25] Based on their research of viologen/$Na_2S_2O_4$ reduction system in ethyl acetate–water, they described a photocatalytic version of their debromination of *meso*-1,2-dibromostilbene (**23**) triggered by light/$[Ru(bpy)_3]^{2+}$ and di(n-octyl) viologen (C_8V^{2+}, Figure 7.9). [23, 24] Disproportionation of viologen radical cation $C_8V^{\bullet+}$ in the ethyl acetate–water biphasic system generates neutral species C_8V^0, which is believed to be the key reducing agent in the reported debrominations. Interestingly, the enzyme-catalyzed variant of this biphasic photodebromination reaction involves ethanol/alcohol dehydrogenase/NADH reduction system. [26] In this process, enzymatically generated NADH serves instead of EDTA or other amines to reduce

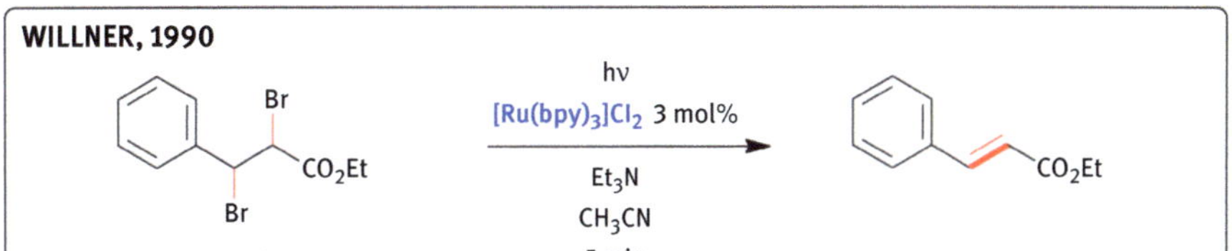

24

25
>80%

Figure 7.10: Photocatalyzed debromination of ethyl dibromocinnamate. [28]

$[Ru(bpy)_3]^{3+}$ to $[Ru(bpy)_3]^{2+}$. [27] Importantly, these reports from the Willner group provided early examples showing that the oxidative quenching cycle of $[Ru(bpy)_3]^{2+*}$ by using strong oxidants (e. g. viologens) can be taken advantage of in transformations interesting to organic synthesis.

In later studies, photocatalysis in the debromination reactions was also described to proceed with $[Ru(bpy)_3]Cl_2/Et_3N$ in acetonitrile solvent in the absence of viologen catalyst (e. g. **24**→**25**, Figure 7.10). [28] Under similar conditions, successful reduction of activated ketones has been demonstrated (Figure 7.11). In this way, ethyl benzoylformate (**26**) is reduced to ethyl mandelate (**27**) and benzil (**28**) to benzoin (**29**).

In 1986, Tomioka group reported examples of reduction of nitroalkenes to oximes using photoredox catalysis with $[Ru(bpy)_3]^{2+}$/viologen/Na_2EDTA system in biphasic medium (Figure 7.12). [29] The corresponding ketones were also isolated in small quantities. Viologen species serves as an electron phase transfer catalyst shut-

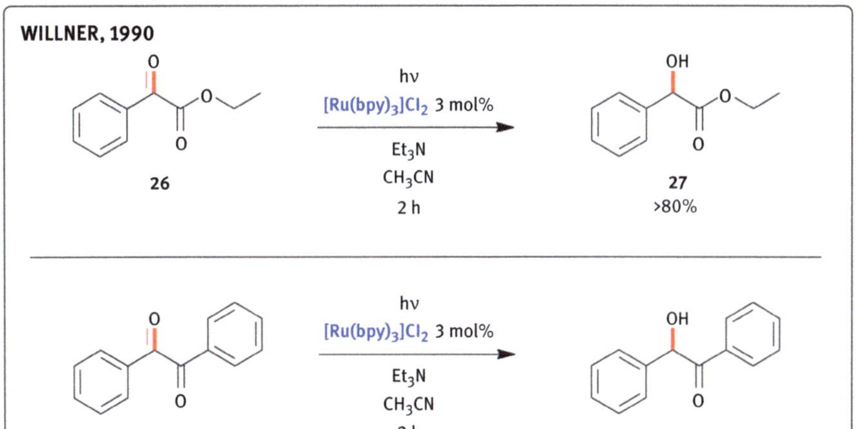

26

27
>80%

28

29
>80%

Figure 7.11: Photocatalyzed reduction of activated ketones triggered by $[Ru(bpy)_3]^{2+}/Et_3N$ system. [28]

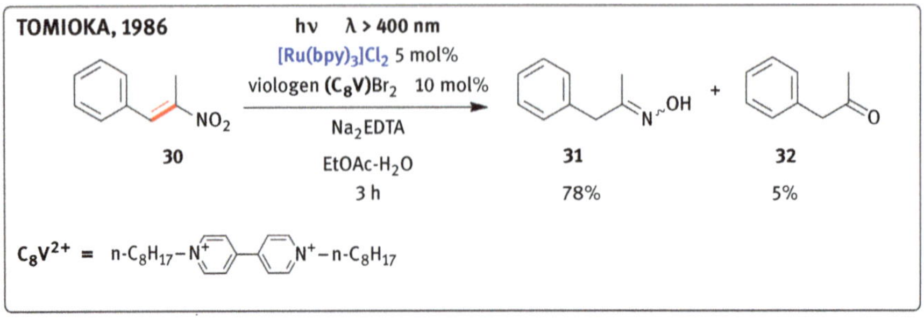

Figure 7.12: Photochemical reduction of nitroalkenes using viologen as an electron phase transfer catalyst. [29]

tling electrons between the aqueous and organic phase and, in this respect, the roles of the ruthenium photocatalyst, Na_2EDTA and viologen are similar to those in the debromination reaction mechanism depicted in Figure 7.9. In this case, the authors stated that the reducing species that directly transfers its electrons to the nitroalkene substrate is either the viologen radical cation $C_8V^{\bullet +}$, or the neutral species C_8V^0 generated by the disproportionation mechanism.

In 1984, Cano-Yelo and Deronzier described the photocatalytic Pschorr reaction converting aryldiazonium salts **33** to phenanthrene derivatives **34** in the presence of $[Ru(bpy)_3]^{2+}$ photoredox catalyst (Figure 7.13). [30] The authors proposed a mechanism involving oxidative quenching of photoexcited $[Ru(bpy)_3]^{2+\star}$ with aryldiazonium salt leading to aryl radical species **35** and $[Ru(bpy)_3]^{3+}$ which is a strong oxidant. [31] Radical cyclization of **35** gives intermediate **36.** In the next step, the radical intermediate **36** is oxidized by $[Ru(bpy)_3]^{3+}$ and subsequently restores aromaticity upon loss of a proton, affording phenanthrene skeletons **34.**

Importantly, when the aryldiazonium salts **33** are subjected to direct photolysis in the absence of a photoredox catalyst, the phenanthrene products **34** become minor components of the reaction mixture (10–20 %, Figure 7.14). The major products are acetanilides **37** isolated in 80 % yields. They result from interaction of a photogenerated aryl cation **38** with acetonitrile and subsequent hydrolysis of the intermediate **39** (Figure 7.14).

The oxidative quenching cycle of Ru-photocatalyst was also used for transformation of carbinols (e. g. **40,** Figure 7.15) to aldehydes with aryldiazonium salts as oxidants. [32, 33] Besides the aldehyde **41,** this process results in benzophenone **42** and fluorenone **43,** which is a product of the Pschorr cyclization pathway similar to that outlined in Figure 7.13. Oxidation of other primary [32] and secondary [33] alcohol substrates was investigated in the presence of various bases and the results were compared with the electrochemical variant of this reaction (electrochemical redox catalysis). [33]

In 1984, the Tanaka group reported a reaction of benzylbromide (**44**) with 1-benzyl-1,4-dihydronicotinamide (BNAH) photosensitized by $[Ru(bpy)_3]^{2+}$ resulting in

1,2-diphenylethane (**45**) as the main product (Figure 7.16). However, in the absence of $[Ru(bpy)_3]^{2+}$ sensitizer, irradiation of the absorption band of BNAH triggers a different photoinduced electron-transfer process leading to reduction of benzyl bromide (**44**) to yield toluene as the major product (Figure 7.16, bottom and Figure 7.17, right). [34] Reaction in the absence of $[Ru(bpy)_3]^{2+}$ proceeds by a radical chain mechanism involving benzyl radical, as the chain carrier, formed by a single-electron transfer from the excited state of BNAH to benzyl bromide (Figure 7.17, right). By contrast, in the presence of $[Ru(bpy)_3]^{2+}$, benzyl bromide is subject to two-electron reduction by $[Ru(bpy)_3]^{+}$ generated by reductive quenching of $[Ru(bpy)_3]^{2+}$* with BNAH (Figure 7.17, left). The authors noted there is some similarity between $[Ru(bpy)_3]^{+}$ reductant and sodium naphthalenide that can also, as a strong reductant, trigger reduction of benzyl halides to generate anionic species $PhCH_2^-$ (**46**) via two-electron transfer. [35a, b] In 1987, related photoassisted C-C coupling of benzyl halide substrates with $Cu(dap)_2^+$ photocatalyst (dap = 2,9-bis(p-anisyl)-1,10-phenanthroline) was described by Kern and Sauvage. [35c] The authors suggested that the oxidative quenching pathway of photoexcited $Cu(dap)_2^{+}$* with benzyl halide is operative. Radical coupling of the resulting benzyl radical, or alternatively, two-electron transfer pathway have been proposed as possible mechanistic scenarios.

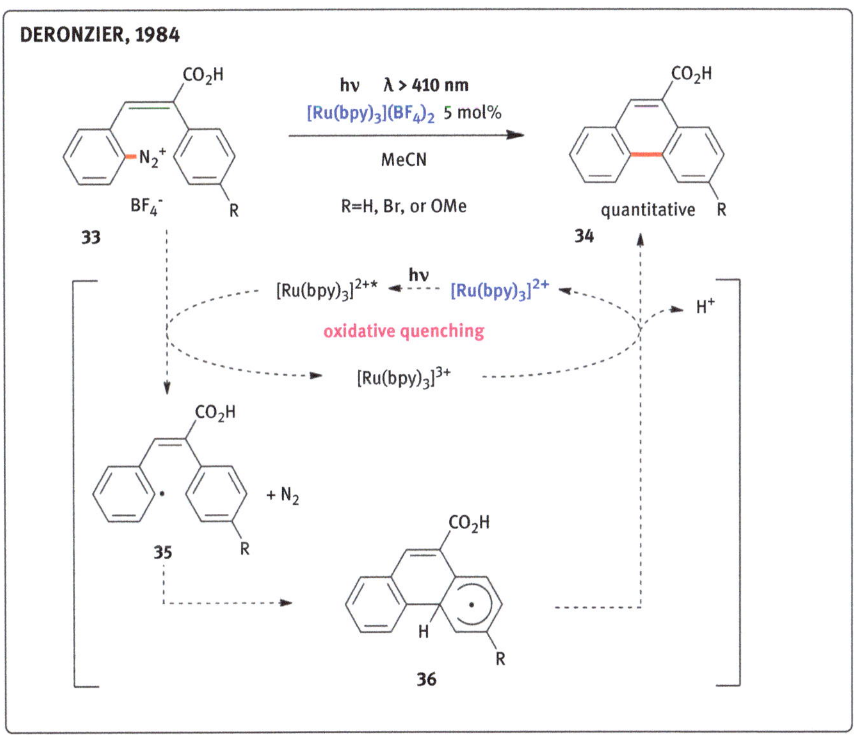

Figure 7.13: Pschorr reaction via visible light photoredox catalysis with $[Ru(bpy)_3]^{2+}$. [30]

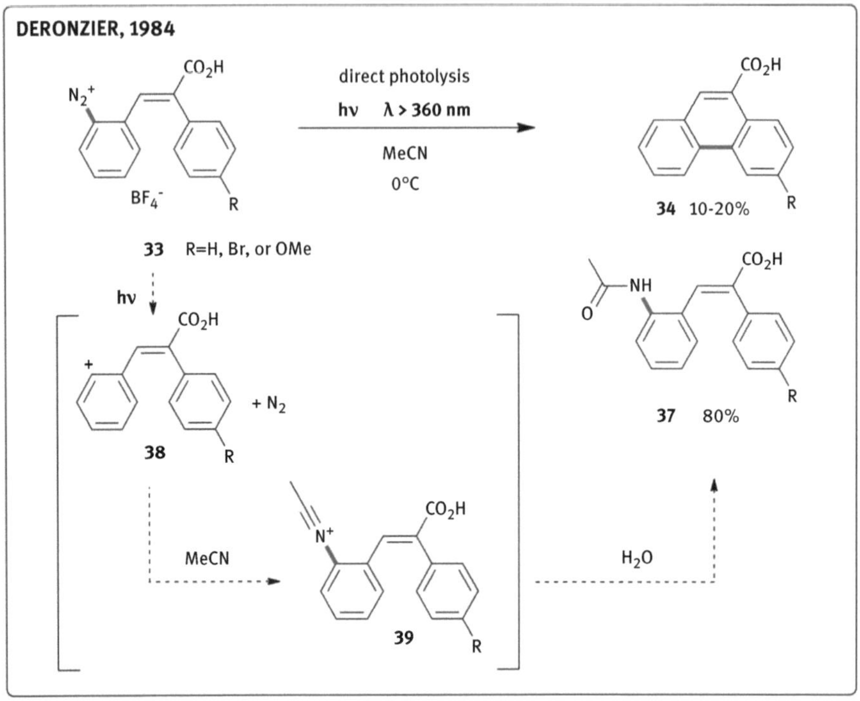

Figure 7.14: Direct photolysis of aryldiazonium salts **33** in the absence of a photoredox catalyst and the pathway leading to major acetanilide products **37** via aryl cations **38.** [30]

In 1990, Fukuzumi et al. reported reduction of phenacyl halides by dihydroacridine derivative **48** photocatalyzed by $[Ru(bpy)_3]^{2+}$ (Figure 7.18). [36] Remarkable influence of perchloric acid on the mechanistic pathway was investigated. [37] In the absence of acid, the reaction was found to proceed via reductive quenching of $[Ru(bpy)_3]^{2+\star}$ by **48** generating $[Ru(bpy)_3]^{+}$ reductant. Conversely, in the presence of perchloric acid, oxidative quenching of $[Ru(bpy)_3]^{2+\star}$ by phenacyl halides is operative producing $[Ru(bpy)_3]^{3+}$ oxidant *in situ*. The two proposed mechanistic pathways are depicted in Figure 7.19.

In 1991, Okada and Oda published a photocatalytic decarboxylative Michael addition of N-(acyloxy)phthalimides (Figure 7.20). [38] Reductive quenching of the photoexcited $[Ru(bpy)_3]^{2+\star}$ with electron-rich BNAH leads to strongly reducing $[Ru(bpy)_3]^{+}$ that transfers its electron to phthalimide moiety to generate species **52.** This leads to decarboxylation and generation of the key alkyl radical intermediate **54** and subsequent incorporation of Michael acceptors such as methyl vinyl ketone to give products **50** and **51.** In a related work, the key alkyl radical intermediates **54** generated, as depicted in Figure 7.20, can give rise to reduced species R-H when tBuSH reagent is used as a source of the hydrogen atom. [39] Alternatively the alkyl radicals **54** can be trapped with PhSeSePh to give phenyl selenides R-SePh. [40]

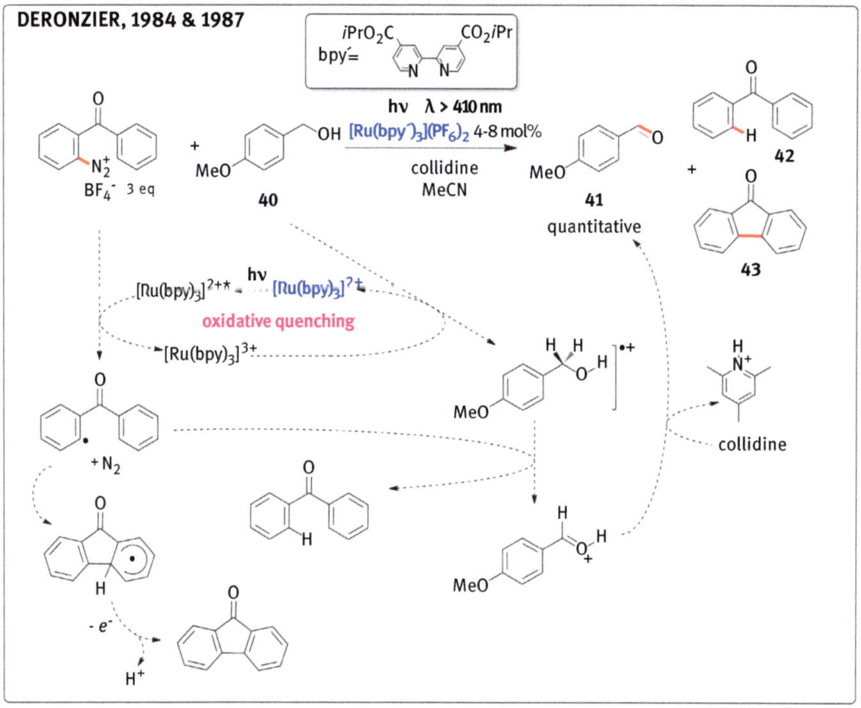

Figure 7.15: Photooxidation of carbinols to aldehydes via visible light photoredox catalysis with aryldiazonium salt as a sacrificial oxidant. [32, 33]

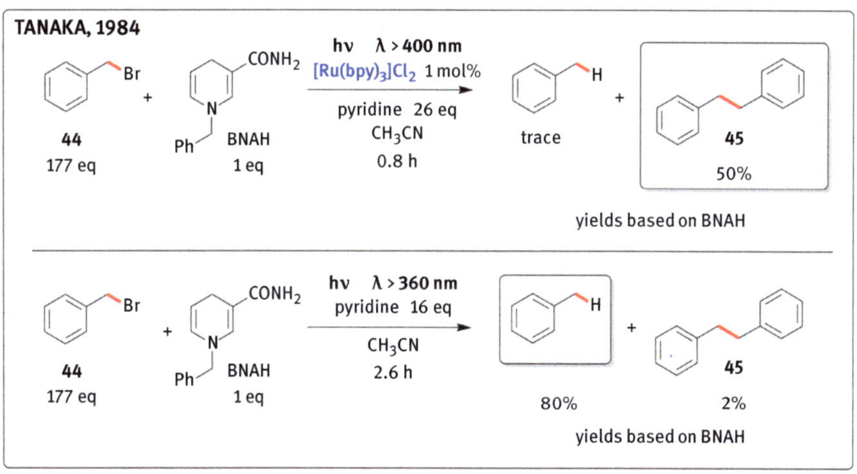

Figure 7.16: [Ru(bpy)3]2+-photosensitized reaction of 1-benzyl-1,4-dihydronicotinamide with benzyl bromide and nonphotosensitized process. [34]

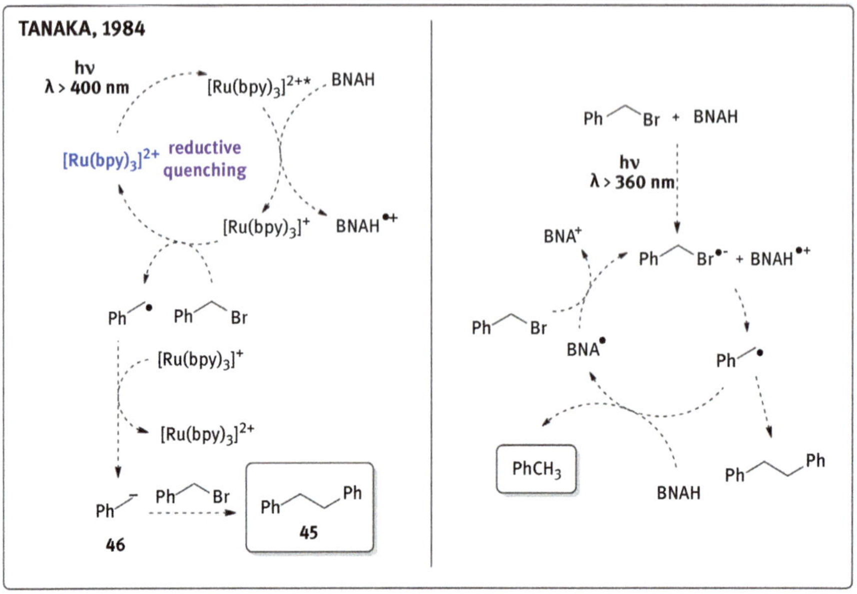

Figure 7.17: Comparison of photoinduced electron-transfer pathways in reaction of benzylbromide and BNAH in the presence (left) and absence (right) of $[Ru(bpy)_3]^{2+}$ photocatalyst. [34]

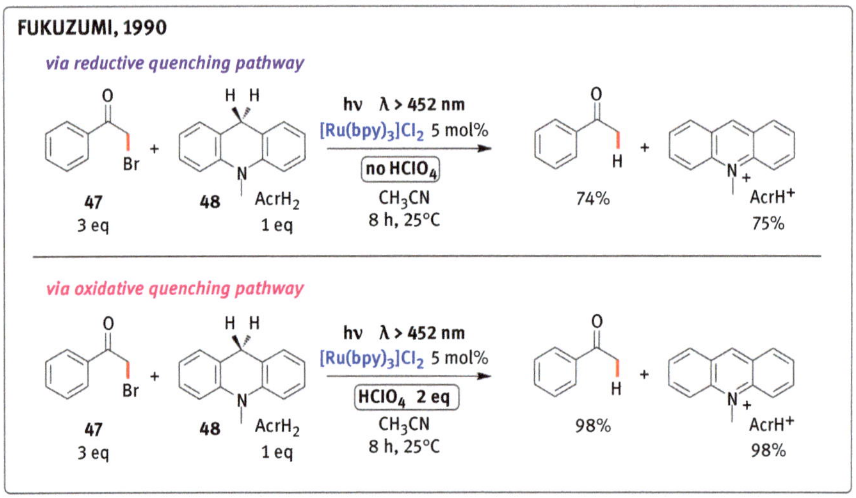

Figure 7.18: $[Ru(bpy)_3]^{2+}$-photosensitized reduction of phenacyl bromide by a NADH analogue in the absence and presence of acid additive. [36]

In 1994 Barton et al. reported additions of *Se*-phenyl p-tolueneselenosulfonate **55** to olefins promoted by $[Ru(bpy)_3]^{2+}$ and light (Figure 7.21). [41] Oxidative quenching pathway of excited $[Ru(bpy)_3]^{2+*}$ with *p*-$TolSO_2SePh$ (**55**) as the oxidant

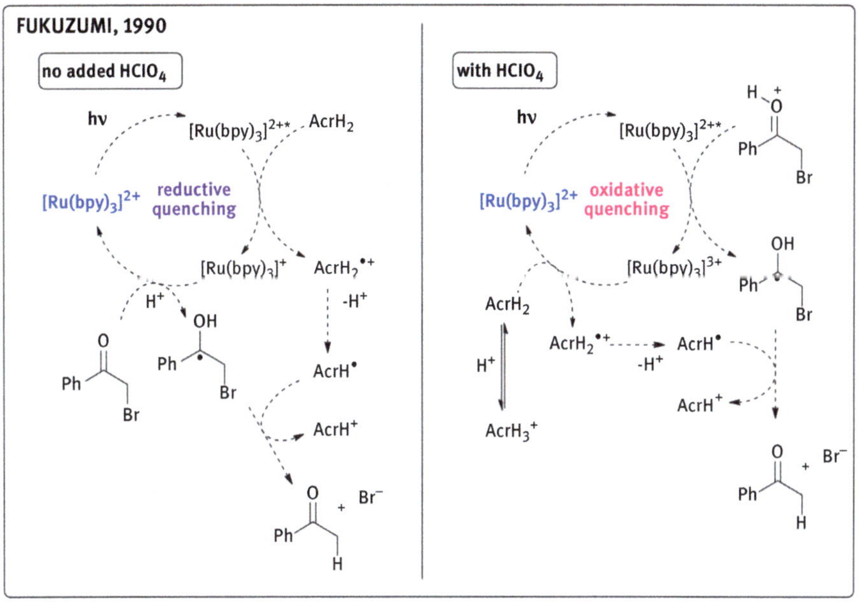

Figure 7.19: Electron-transfer pathways in reduction of phenacyl halides by dihydroacridine derivative **48** photocatalyzed by $[Ru(bpy)_3]^{2+}$ with and without perchloric acid. [36]

was proposed to initiate the radical chain reaction (Figure 7.21). Electron-rich olefins give products such as sulfone **57** in high yield. Interestingly, transformation **55**→**57** also worked with $Ru(CO)_3Cl_2$ (86 % yield of **57**) and $[Ru(bpy)_3]Cl_2$ catalysts. In this report, the authors predicted that $[Ru(bpy)_3]^{2+}$ in combination with light could be used as an initiator for various other radical reactions.

In 1993, Ohkubo et al. devised a photoredox catalytic process for oxidative homocoupling of 2-naphthol leading to nonracemic (*R*)-(+)-1,1′-bi-2-naphthol (BINOL). [42, 43] Visible light irradiation of the substrate in the presence of enantiopure Δ-$[Ru(menbpy)_3]^{2+}$ **60** afforded nonracemic BINOL with 16 % ee (Figure 7.22). In this transformation, which features oxidative quenching pathway of the excited ruthenium photocatalyst, $[Co(acac)_3]$ was used as a sacrificial oxidant. As the authors proposed, the enantioselectivity is presumably governed by the counterclockwise molecular helicity along the C_3 axis of the ruthenium complex during the oxidation of the radical intermediate **61**. Remarkably, the ruthenium photocatalytic system based on $[Ru(menbpy)_3]^{2+}$ is able to trigger a very efficient enantioselective photocatalytic reduction of $[Co(acac)_3]$ (Figure 7.23). [44] Δ enantiomer of the octahedral complex $[Co(acac)_3]$ reacts considerably faster as compared to the Λ enantiomer.

This transformation proceeds via oxidative quenching of the enantiopure ruthenium photocatalyst with racemic $[Co(acac)_3]$, which occurs with a significantly different rate for the Δ and Λ enantiomer in the mixture of ethanol–water 9 : 1 ($k_\Delta/k_\Lambda = 14.7$). With the rate being larger for consumption of Δ-$[Co(acac)_3]$, the reac-

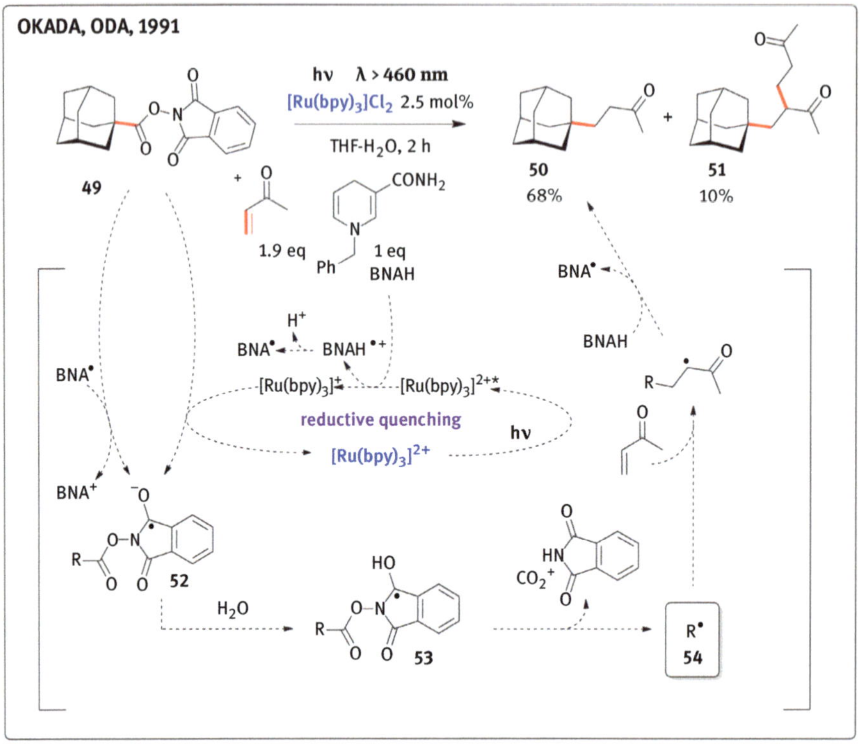

Figure 7.20: Decarboxylative Michael addition of N-(acyloxy)phthalimide **49** to electron-deficient olefin. [38]

tion mixture can be enriched in Λ-[Co(acac)₃] enantiomer (up to 94 % ee at 30 % conversion). Notably strong dependence on the composition of the solvent system has been observed. When the content of ethanol in the solvent system has been decreased to 8:1 relative to water, the ratio of the individual enantiomer consumption rates decreased considerably to $k_\Delta/k_\Lambda = 8.7$.

In a related work, the Ohkubo group described enantioselective photoredox synthesis of [Co(acac)$_3$] from racemic [Co(acac)$_2$(H$_2$O)$_2$] and acetylacetone using various nonracemic ruthenium photocatalysts in ethanol–water solvent system in the presence of molecular oxygen. [45] The best catalyst was Δ-[Ru(menbpy)$_3$]$^{2+}$ with which they achieved 37 % ee in favor of Λ-[Co(acac)$_3$] enantiomer. The authors proposed a reaction pathway, where the excited state of the ruthenium catalyst is oxidatively quenched by the molecular oxygen to generate strongly oxidizing Δ-[Ru(menbpy)$_3$]$^{3+}$. This in turn triggers the enantioselective key step that involves nonequal rates of formation of the individual enantiomers of [Co(acac)$_3$] favoring Λ-[Co(acac)$_3$] enantiomer. Although the previous examples by Ohkubo are more from the realm of inorganic chemistry, they might be inspiring to organic chemists interested in enantioselective electron-transfer reactions. [46]

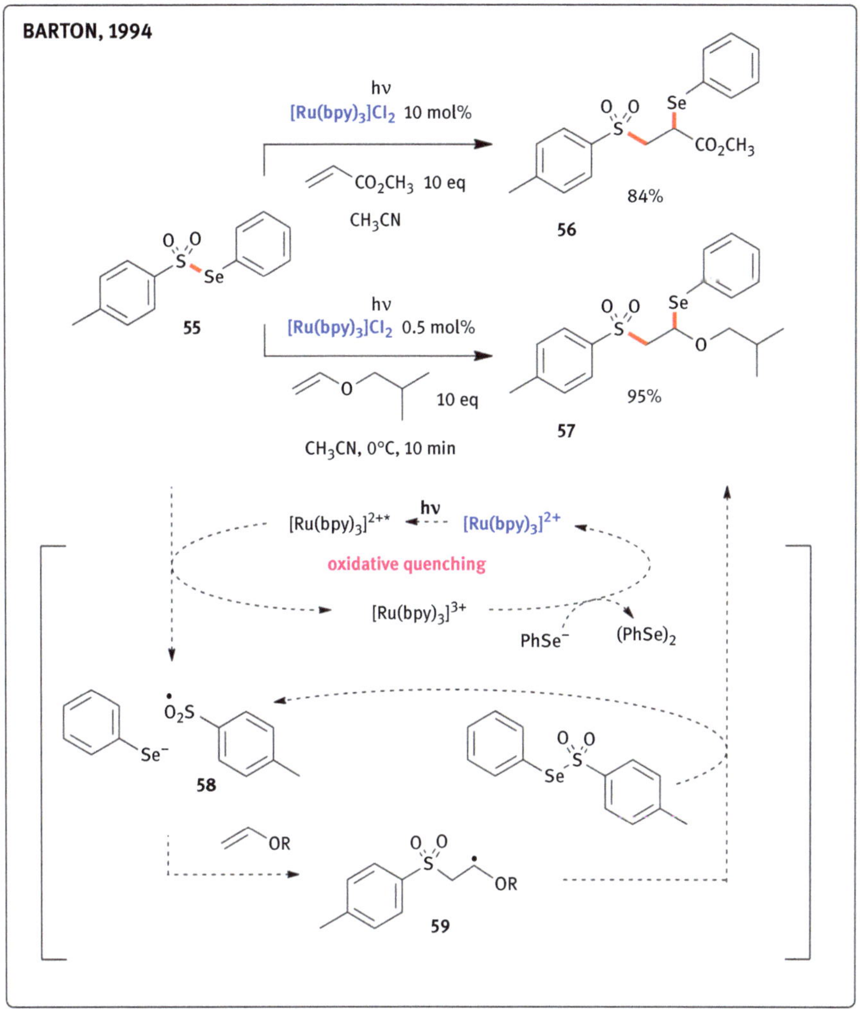

Figure 7.21: $[Ru(bpy)_3]^{2+}$ photocatalyzed addition of *Se*-phenyl p-tolueneselenosulfonate to electron rich olefins. [41]

In a seminal paper from 1999, the Kodadek group introduced $[Ru(bpy)_3]^{2+}$ photocatalyst as a new tool for chemical biology. Specifically, the oxidative quenching pathway of $[Ru(bpy)_3]^{2+*}$ has been utilized to affect a very rapid and efficient protein cross-linking (Figure 7.24). [47a] Ammonium persulfate was used as an oxidative quencher to generate $[Ru(bpy)_3]^{3+}$ as a potent one-electron oxidant. It has been proposed that this strongly oxidizing species, generated *in situ* using visible light excitation, can oxidize tyrosine moieties in proteins. The resultant radical was suggested to cross-link two associated proteins by at least two mechanisms depicted in Figure 7.24. In case another tyrosine residue is in proximity, aryl-aryl bond for-

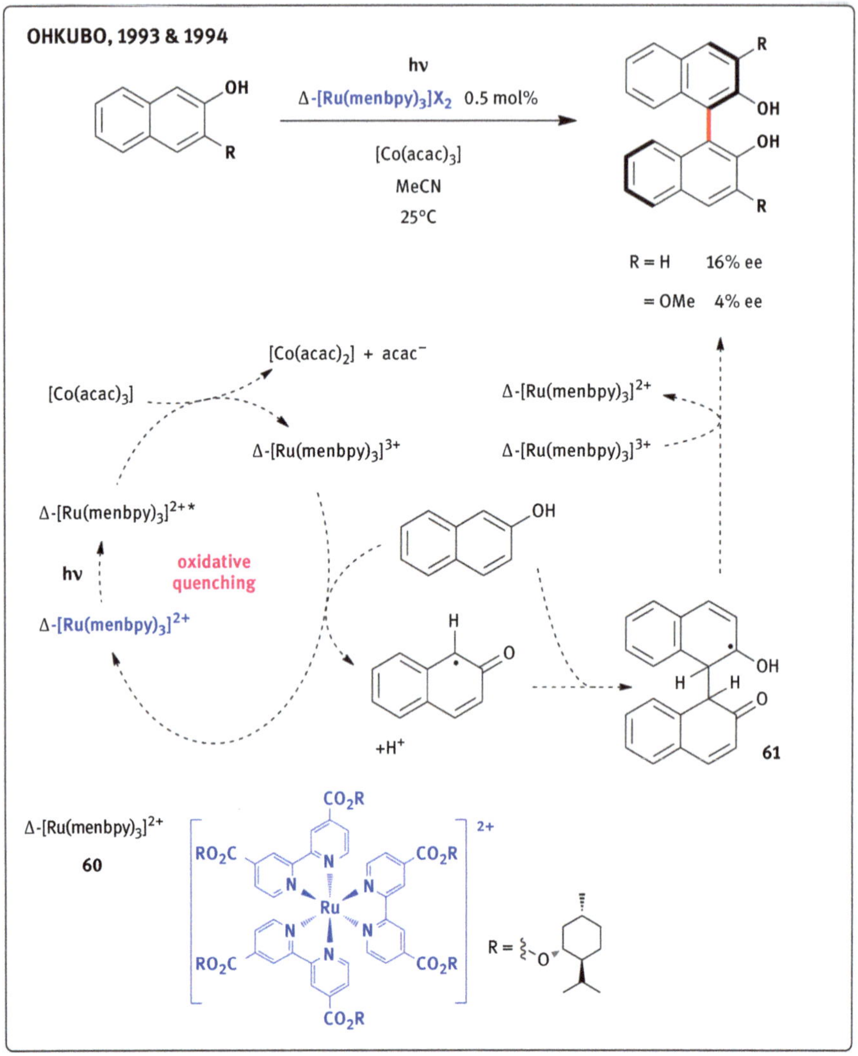

Figure 7.22: Enantioselective oxidative homocoupling of 2-naphthol by photoredox catalysis with Δ-[Ru(menbpy)$_3$]$^{2+}$. [42, 43]

mation can take place. Alternatively, a nucleophilic lysine or cysteine group can attack the phenol radical to afford a new heteroatom-arene bond. In both pathways, a hydrogen atom is lost to form stable products, and the sulfate radical produced during Ru(III) formation was proposed to play a key role in this step. Very high yields of cross-linked products were reported with irradiation times of less than 1 s. A later study on this photoinduced cross-linking of proteins revealed more details about the scope, limitations, and mechanistic aspects. [47b] This method was shown

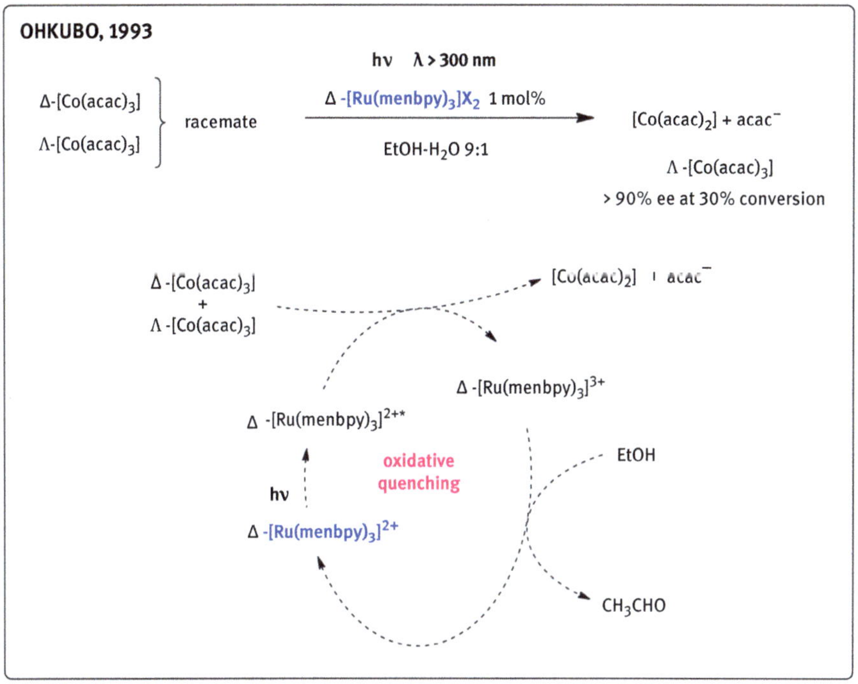

Figure 7.23: Enantioselective decomposition of $[Co(acac)_3]$ by photocatalysis with Δ-$[Ru(menbpy)_3]^{2+}$. [44]

to be useful for quantitative characterization of protein-protein and protein-peptide complexes. [48]

Notably, the chemical principle of the Kodadek method of making dityrosine linkages is similar to that of the Ohkubo method for synthesis of BINOL from 2-naphthol (cf. Figure 7.22 and 7.24). Both methods utilize the oxidative quenching cycle of Ru^{2+*} complex leading to generation of reactive phenol radical from phenolic substrate.

In general, biological applications of photoredox complexes related to $[Ru(bpy)_3]^{2+}$ are in focus of truly fascinating research endeavors with a promising future. As the more extensive coverage of this territory is beyond the scope of this review, interested readers are referred to excellent reviews summarizing the status and challenges in the area. [49]

7.4 Conclusion

In 2008 and 2009, the seminal studies by MacMillan, Yoon, and Stephenson groups initiated the era of rapid developments in the field of visible light photoredox catalysis.

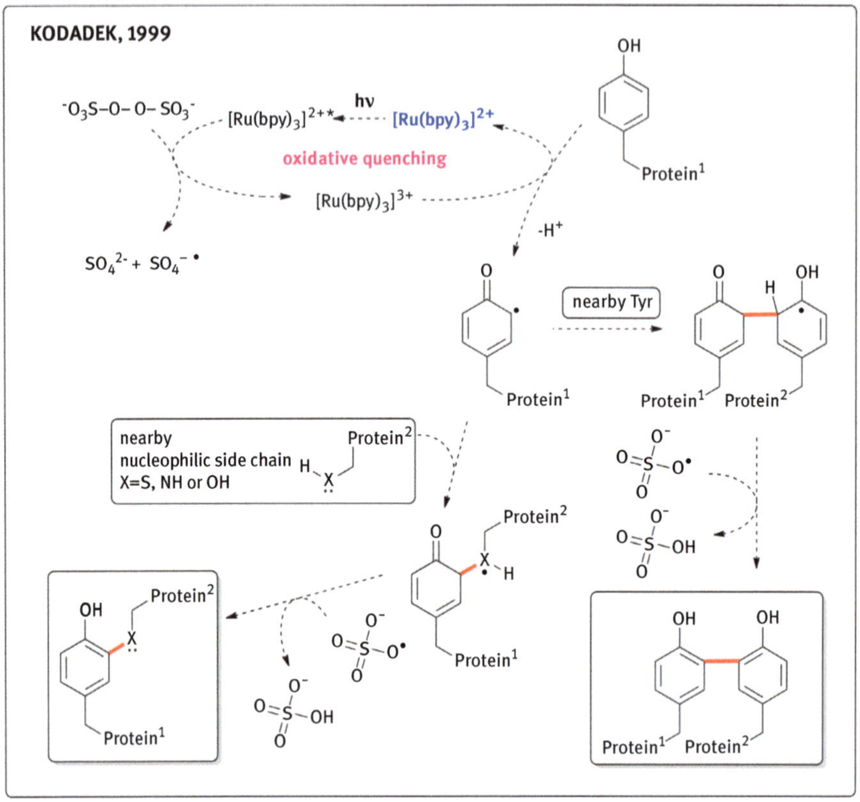

Figure 7.24: Phototriggered protein cross-linking using photoredox manifold of $[Ru(bpy)_3]^{2+}$ in conjunction with persulfate oxidative quencher. [47]

In 2010, the essence and promise of research on photoredox catalysis with visible light has been nicely summarized by Yoon:

"The versatility of $[Ru(bpy)_3]^{2+}$ arises from the ability to access either photooxidative or photoreductive reactivity by choosing the appropriate oxidative or reductive quencher, respectively (see Figure 7.3). In both regimes, the photophysical properties of $[Ru(bpy)_3]^{2+}$ enable a variety of inexpensive, readily available sources of visible light to be utilized, including sunlight. In addition, there exists a vast wealth of electrochemical literature that describes synthetically useful organic transformations initiated by one-electron redox processes. It is expected that photocatalytic systems exploiting the reactivity of $[Ru(bpy)_3]^{2+}$ should also be able to efficiently promote similar reactivity." [50]

It is clear that this newly defined research territory will have a profound impact on preparative organic chemistry in the future. This is documented by the fact that since 2011 the pace of research has markedly accelerated and many impressive contributions to the field of photoredox catalysis have been published. These reports focus on the introduction of novel photoredox catalysts and creative use of known catalysts in novel reaction types. [51–94] Innovative techniques such as reactions in flow systems have been devised. [95–98]

Before the fast acceleration in this fresh direction during the last four years, the limited number of scattered examples relevant to organic synthesis remained dormant in the literature. However, now, the early attempts published in the twentieth century might serve as an inspiration for the field, which is on the rise. Therefore, awareness of the pioneering efforts by the groups of Kellogg, Pac, Willner, Deronzier, Tanaka, Tomioka, Fukuzumi, Okada and Oda, Ohkubo, Barton, and Kodadek may help to nurture the current formative era and conceptual expansion of this branch of photocatalysis.

This work was supported by the Grant Agency of the Czech Republic (Grant No. P207/10/2391) and the Institute of Organic Chemistry and Biochemistry, Academy of Sciences of the Czech Republic, v.v.i. (RVO: 61388963).

7.5 Abbreviations

acac = acetylacetonato
BINOL = 1,1′-bi-2-naphthol
BNAH = 1-benzyl-1,4-dihydronicotinamide
bpy = 2,2′-bipyridine
dap = 2,9-bis(p-anisyl)-1,10-phenathroline
EDTA = ethylenediaminetetraacetic acid
menbpy = 4,4′-di(1*R*,2*S*,5*R*)-(−)-menthoxycarbonyl-2,2′-bipyridine
MLCT = metal-to-ligand charge-transfer
NADH = nicotinamide adenine dinucleotide
SET = single-electron transfer
TEOA = triethanolamine, tris(2-hydroxyethyl)amine
TPP = *meso*-tetraphenylporphine

7.6 References

[1] For excellent discussion on the beginnings of organic photochemistry, see: (a) Roth, H.D. (1989). The beginnings of organic photochemistry. Angew. Chem. Int. Ed Engl. 28, 1193–1207. For a visionary address on the future of photochemistry, see: (b) Ciamician, G. (1912). The photochemistry of the future. Science 36, 385–394.
[2] (a) Klán, P., and Wirz, J. (2009). Photochemistry of Organic Compounds: From Concepts to Practice (Wiley: Chichester); (b) Turro, N.J., Ramamurthy, V., and Scaiano, J.C. (2010). Modern Molecular Photochemistry of Organic Molecules (University Science Books).
[3] For selected reviews on photocatalytic formation of important bonds in organic chemistry, see: (a) Fagnoni, M., Dondi, D., Ravelli, D., and Albini, A. (2007). Photocatalysis for the formation of

the C–C bond. Chem. Rev. 107, 2725–2756; (b) Hoffmann, N. (2008). Photochemical reactions as key steps in organic synthesis. Chem. Rev. 108, 1052–1103; (c) Svoboda, J., and König, B. (2006). Templated photochemistry: Toward catalysts enhancing the efficiency and selectivity of photoreactions in homogeneous solutions. Chem. Rev. 106, 5413–5430; (d) Bach, T. (1998). Stereoselective intermolecular [2+2]-photocycloaddition reactions and their application in synthesis. Synthesis, 683–703. For photochemical synthesis of fine chemicals with sunlight, see: (e) Esser, P., Pohlmann, B., and Scharf, H.-D. (1994). The photochemical synthesis of fine chemicals with sunlight. Angew. Chem. Int. Ed. Engl. 33, 2009–2023.

[4] For reviews on photoredox catalysis with visible light, see: (a) Tucker, J.W., and Stephenson, C.R.J. (2012). J. Org. Chem. 77, 1617; (b) Teplý, F. (2011). Collect. Czech. Chem. Commun. 76, 859; (c) Narayanam, J.M.R., and Stephenson, C.R.J. (2011). Visible light photoredox catalysis: applications in organic synthesis. Chem. Soc. Rev. 40, 102–113; (d) Yoon, T.P., Ischay, M.A., and Du, J. (2010). Visible light photocatalysis as a greener approach to photochemical synthesis. Nat. Chem. 2, 527–532; (e) Zeitler, K. (2009). Photoredox catalysis with visible light. Angew. Chem. Int. Ed. 48, 9785–9789; (f) Renaud, P., and Leong, P. (2008). A light touch catalyzes asymmetric carbon-carbon bond formation. Science 322, 55–56; (g) Ravelli, D., and Fagnoni M. (2012). Dyes as visible light photoredox organocatalysts. ChemCatChem 4, 169–171; (h) Xuan, J., and Xiao, W.-J. (2012). Visible-light photoredox catalysis. Angew. Chem. Int. Ed. 51, 6828–6838; (i) Hoffmann, N. (2012). Homogeneous photocatalytic reactions with organometallic and coordination compounds-perspectives for sustainable chemistry. ChemSusChem 5, 352–371.

[5] Nicewicz, D.A., and MacMillan, D.W.C. (2008). Merging photoredox catalysis with organocatalysis: the direct asymmetric alkylation of aldehydes. Science 322, 77–80.

[6] Ischay, M.A., Anzovino, M.E., Du, J., and Yoon, T.P. (2008). Efficient visible light photocatalysis of [2+2] enone cycloadditions. J. Am. Chem. Soc. 130, 12886–12887.

[7] Narayanam, J.M.R., Tucker, J.W., and Stephenson, C.R.J. (2009). Electron-transfer photoredox catalysis: development of a tin-free reductive dehalogenation reaction. J. Am. Chem. Soc. 131, 8756–8757.

[8] Neumann, M., Füldner, S., König, B., and Zeitler. K. (2011). Metal-free, cooperative asymmetric organophotoredox catalysis with visible light. Angew. Chem. Int. Ed. 50, 951–954.

[9] Burstall F. H. (1936). Optical activity dependent on co-ordinated bivalent ruthenium. J. Chem. Soc., 173–175.

[10] Interestingly high configurational stability of $[Ru(bpy)_3]^{2+}$ was in stark contrast to analogous iron and nickel complexes studied previously by Blau and Werner: (a) Blau, F. (1898). Űber neue organische Metallverbindungen. Monatsh. Chem. 19, 647–689: (b) Werner, A. (1912). Űber Spiegelbild-Isomerie bei Eisenverbindungen. Chem. Ber. 45, 433–436. For configurationally stable octahedral iron complexes, see: (c) Monchaud, D., Jodry, J.J., Pomeranc, D., Heitz, V., Chambron, J.C., Sauvage, J.P., and Lacour, J. (2002). Ion-pair-mediated asymmetric synthesis of a configurationally stable mononuclear tris(diimine)-iron(II) complex. Angew. Chem. Int. Ed. 41, 2317–2319, and references therein.

[11] (a) Sabbatini, N., and Balzani, V. (1972). Photosensitized aquation of the hexacyanochromate(III) ion. Evidence against the doublet mechanism. J. Am. Chem. Soc. 94, 7587–7589; (b) Gafney, H.D., and Adamson, A.W. (1972). Excited state $Ru(bipyr)_3^{2+}$ as an electron-transfer reductant. J. Am. Chem. Soc. 94, 8238–8239; (c) Wrighton, M., and Markham, J. (1973). Quenching of the luminescent state of tris(2,2′-bipyridine)ruthenium(II) by electronic energy transfer. J. Phys. Chem. 77, 3042–3044; (d) Bock, C.R., Meyer, T.J., and Whitten, D.G. (1974). Electron transfer quenching of the luminescent excited state of tris(2,2′-bipyridine)ruthenium(II). Flash photolysis relaxation technique for measuring the rates of very rapid electron transfer reactions. J. Am. Chem. Soc. 96, 4710–4712; (e) Navon, G., and Sutin, N. (1974). Mechanism of the quenching of

the phosphorescence of tris(2,2′-bipyridine)ruthenium(II) by some cobalt(III) and ruthenium(III) complexes. Inorg. Chem. 13, 2159–2164.
[12] For reviews, see: (a) Campagna, S., Puntoriero, F., Nastasi, F., Bergamini, G., and Balzani, V. (2007). Photochemistry and photophysics of coordination compounds: ruthenium. Top. Curr. Chem. 280, 117–214; (b) Juris, A., Balzani, V., Barigelletti, F., Campagna, S., Belser, P., and von Zelewsky, A. (1988). Ru(II) polypyridine complexes: photophysics, photochemistry, electrochemistry, and chemiluminescence. Coord. Chem. Rev. 84, 85–277; (c) Kalyanasundaram, K. (1982). Photophysics, photochemistry and solar energy conversion with tris(bipyridyl)ruthenium(II) and its analogues. Coord. Chem. Rev. 46, 159–244; (d) Sutin, N. (1979). Light-induced electron transfer reactions. J. Photochem. 10, 19–40; (e) Whitten, D.G. (1980). Photoinduced electron transfer reactions of metal complexes in solution. Acc. Chem. Res. 13, 83–90.
[13] (a) Kalyanasundaram, K., Kiwi, J., and Grätzel, M. (1978). Hydrogen evolution from water by visible light, a homogeneous three component test system for redox catalysis. Helv. Chim. Acta 61, 2720–2730; (b) Grätzel, M., and Kiwi, J. (1979). Projection, size factors, and reaction dynamics of colloidal redox catalysts mediating light induced hydrogen evolution from water. J. Am. Chem. Soc. 101, 7214–7217.
[14] (a) Creutz, C., and Sutin, N. (1975). Reaction of tris(bipyridine)ruthenium(III) with hydroxide and its application in a solar energy storage system. Proc. Natl. Acad. Sci. USA 72, 2858–2862. For recent reviews on utilization of solar energy, see: (b) Lewis, N.S., and Nocera, D.G. (2006). Powering the planet: Chemical challenges in solar energy utilization. Proc. Natl Acad. Sci. USA 103, 15729–15735; (c) Balzani, V., Credi, A., and Venturi, M. (2008). Photochemical conversion of solar energy. ChemSusChem 1, 26–58. For excellent overviews on various relevant environmental issues, see: (d) Armaroli, N., and Balzani, V. (2011). The legacy of fossil fuels. Chem. Asian J. 6, 768–784; (e) Jenck, J.F., Agterberg, F., and Droescher, M.J. (2004). Products and processes for a sustainable chemical industry: a review of achievements and prospects. Green Chem. 6, 544–556.
[15] Ceroni, P., Bergamini, G., and Balzani, V. (2009). Old molecules, new concepts: $[Ru(bpy)_3]^{2+}$ as a molecular encoder–decoder. Angew. Chem. Int. Ed. 48, 8516–8518.
[16] Hedstrand, D.M., Kruizinga, W.H., and Kellogg, R.M. (1978). Light induced and dye accelerated reductions of phenacyl onium salts by 1,4-dihydropyridines. Tetrahedron Lett. 19, 1255–1258.
[17] van Bergen, T.J., Hedstrand, D.M., Kruizinga, W.H., and Kellogg, R.M. (1979). Chemistry of dihydropyridines. 9. Hydride transfer from 1,4-dihydropyridines to sp^3-hybridized carbon in sulfonium salts and activated halides. Studies with NAD(P)H models. J. Org. Chem. 44, 4953–4962.
[18] Mashraqui, S.H., and Kellogg, R.M. (1985). 3-Methyl-2,3-dihydrobenzothiazoles as reducing agent. Dye enhanced photoreactions. Tetrahedron Lett. 26, 1453–1456.
[19] Pac, C., Ihama, M., Yasuda, M., Miyauchi, Y., and Sakurai, H. (1981). $Ru(bpy)_3^{2+}$-mediated photoreduction of olefins with 1-benzyl-1,4-dihydronicotinamide: a mechanistic probe for electron-transfer reactions of NAD(P)H-model compounds. J. Am. Chem. Soc. 103, 6495–6497.
[20] Pac, C., Miyauchi, Y., Ishitani, O., Ihama, M., Yasuda, M., and Sakurai, H. (1984). Redox-photosensitized reactions. 11. $Ru(bpy)_3^{2+}$-photosensitized reactions of 1-benzyl-1,4-dihydronicotinamide with aryl-substituted enones, derivatives of methyl cinnamate, and substituted cinnamonitriles: electron-transfer mechanism and structure-reactivity relationships. J. Org. Chem. 49, 26–34.
[21] Ishitani, O., Pac, C., and Sakurai, H. (1983). Redox-photosensitized reactions. 10. Formation of a novel type of adduct between an NADH model and carbonyl compounds by photosensitization using $Ru(bpy)_3^{2+}$. J. Org. Chem. 48, 2941–2942.
[22] Ishitani, O., Yanagida, S., Takamuku, S., and Pac, C. (1987). Redox-photosensitized reactions. 13. $Ru(bpy)_3^{2+}$-photosensitized reactions of an NADH model, 1-benzyl-1,4-dihydronico-

tinamide, with aromatic carbonyl compounds and comparison with thermal reactions. J. Org. Chem. 52, 2790–2796.

[23] Maidan, R., Goren, Z., Becker, J.Y., and Willner, I. (1984). Application of multielectron charge relays in chemical and photochemical debromination processes. The role of induced disproportionation of N,N′-dioctyl-4,4′-bipyridinium radical cation in two-phase systems. J. Am. Chem. Soc. 106, 6217–6222.

[24] Goren, Z., and Willner, I. (1983). Photochemical and chemical reduction of vicinal dibromides via phase transfer of 4,4′-bipyridinium radical: the role of radical disproportionation. J. Am. Chem. Soc. 105, 7764–7765.

[25] Willner, I., Goren, Z., Mandler, D., Maidan, R., and Degani, Y. (1985). Transformation of single-electron transfer photoproducts into multielectron charge relays: the functions of water—oil two-phase systems and enzyme catalysis. J. Photochem. 28, 215–228.

[26] Maidan, R., and Willner, I. (1986). Photochemical and chemical enzyme catalyzed debromination of meso-1,2-dibromostilbene in multiphase systems. J. Am. Chem. Soc. 107, 1080–1082.

[27] For a related example, see: Mandler, D., and Willner, I. (1984). Solar light induced formation of chiral 2-butanol in an enzyme-catalyzed chemical system. J. Am. Chem. Soc. 106, 5352–5353.

[28] Willner, I., Tsfania, T., and Eichen, Y. (1990). Photocatalyzed and electrocatalyzed reduction of vicinal dibromides and activated ketones using ruthenium(I) tris(bipyridine) as electron-transfer mediator. J. Org. Chem. 55, 2656–2662.

[29] Tomioka, H., Ueda, K., Ohi, H., and Izawa, Y. (1986). Photochemical and chemical reduction of nitroalkenes using viologens as an electron phase-transfer catalyst. Chem. Lett. 1359–1362.

[30] Cano-Yelo, H., and Deronzier, A. (1984). Photocatalysis of the Pschorr reaction by tris-(2,2′-bipyridyl)ruthenium(II) in the phenanthrene series. J. Chem. Soc., Perkin Trans. 2, 1093–1098.

[31] Oxidative quenching of $[Ru(bpy)_3]^{2+*}$ with p-$RC_6H_4N_2^+$ has been also studied in the context of two-compartment photoelectrochemical cells, see: Cano-Yelo, H., and Deronzier, A. (1984). Photo-oxidation of tris(2,2′-bipyridine)ruthenium(II) by para-substituted benzene diazonium salts in acetonitrile. Two-compartment photoelectrochemical cell applications. J. Chem. Soc., Faraday Trans. 1 80, 3011–3019.

[32] Cano-Yelo, H., and Deronzier, A. (1984). Photo-oxidation of some carbinols by the Ru(II) polypyridyl complex-aryl diazonium salt system. Tetrahedron Lett. 25, 5517–5520.

[33] Cano-Yelo, H., and Deronzier, A. (1987). Photooxidation of alpha-phenylated alcohols by ruthenium polypyridyl complexes and aryl diazonium salts – comparison with electrochemical redox catalysis. New J. Chem. 11, 479–485.

[34] Hironaka, K., Fukuzumi, S., and Tanaka, T. (1984). Tris(bipyridyl)ruthenium(II)-photosensitized reaction of 1-benzyl-1,4-dihydronicotinamide with benzyl bromide. J. Chem. Soc., Perkin Trans. 2, 1705–1709.

[35] (a) Rakshys, Jr., J.W. (1971). Use of ^{19}F CIDNP in reaction mechanism studies. Evidence for the intermediacy of benzyl anion in the reaction of benzyl halides with sodium naphthalene. Tetrahedron Lett., 4745–4748; (b) Bank, S., and Bank, J.F. (1969). Reactions of aromatic radical anions V. Evidence for anion intermediacy in the reaction of alkyl chlorides, bromides and iodides with sodium naphthalene. Tetrahedron Lett., 4533–4536; (c) Kern, J.M., and Sauvage, J.P. (1987). Photoassisted C-C coupling via electron transfer to benzylic halides by a bis(di-imine) copper(I) complex. J. Chem. Soc., Chem. Commun., 546–548.

[36] Fukuzumi, S., Mochizuki, S., and Tanaka, T. (1990). Photocatalytic reduction of phenacyl halides by 9,10-dihydro-10-methylacridine: control between the reductive and oxidative quenching pathways of tris(bipyridine)ruthenium complex utilizing an acid catalysis. J. Phys. Chem. 94, 722–726.

[37] For the initial study on acid catalysis in photoinduced electron transfer from $[Ru(bpy)_3]^{2+}$ to substituted acetophenones, see: (a) Fukuzumi, S., Ishikawa, K., Hironaka, K., and Tanaka, T. (1987). Acid catalysis in thermal and photoinduced electron-transfer reactions. J. Chem. Soc., Perkin Trans. 2, 751–760. For a related study with ferrocene reductants in their ground state, see: (b) Fukuzumi, S., Mochizuki, S., and Tanaka T. (1989). Acid-catalyzed electron-transfer processes in reduction of alpha-haloketones by an NADH model compound and ferrocene derivatives. J. Am. Chem. Soc. 111, 1497–1499.
[38] Okada, K., Okamoto, K., Morita, N., Okubo, K., and Oda, M. (1991). Photosensitized decarboxylative Michael addition through N-(acyloxy)phthalimides via an electron-transfer mechanism. J. Am. Chem. Soc. 113, 9401–9402.
[39] Okada, K., Okubo, K., Morita, N., and Oda, M. (1992). Reductive decarboxylation of N-(acyloxy)phthalimides via redox-initiated radical chain mechanism. Tetrahedron Lett. 33, 7377–7380.
[40] Okada, K., Okubo, K., Morita, N., and Oda, M. (1993). Redox-mediated decarboxylative photophenylselenenylation of N-acyloxyphthalimides. Chem. Lett., 2021–2024.
[41] Barton, D.H.R., Csiba, M.A., and Jaszberenyi, J.C. (1994). $Ru(bpy)_3{}^{2+}$-mediated addition of *Se*-phenyl p-tolueneselenosulfonate to electron rich olefins. Tetrahedron Lett. 35, 2869–2872.
[42] Hamada, T., Ishida, H., Usui, S., Watanabe, Y., Tsumura, K., and Ohkubo, K. (1993). A novel photocatalytic asymmetric synthesis of (*R*)-(+)-1,1′-bi-2-naphthol derivatives by oxidative coupling of 3-substituted-2-naphthol with Δ-$[Ru(menbpy)_3]^{2+}$[menbpy = 4,4′-di(1*R*,2*S*,5*R*)-(−)-menthoxycarbonyl-2,2′-bipyridine], which posseses molecular helicity. J. Chem. Soc., Chem. Commun., 909–911.
[43] Hamada, T., Ishida, H., Usui, S., Tsumura, K., and Ohkubo, K. (1994). Enantioselective and photocatalytic oxidation of 1,1′-bi-2-naphthol with a chiral ruthenium complex which includes molecular helicity. J. Mol. Catal. 88, L1–L5.
[44] Ohkubo, K., Hamada, T., and Ishida, H. (1993). Novel enantioselective photocatalysis by chiral, helical ruthenium(II) complexes. J. Chem. Soc., Chem. Commun., 1423–1425.
[45] Ohkubo, K., Watanabe, M., Ohta, H., and Usui, S. (1996). Novel photocatalytic asymmetrical synthesis of Δ (or Λ)-$[Co(acac)_3]$ (acac, pentane-2,4-dione) from $[Co(acac)_2(H_2O)_2]$ and Hacac with helical ruthenium (II) complexes. J. Photochem. Photobiol. A 95, 231–233.
[46] (a) Rau, H. (1983). Asymmetric photochemistry in solution. Chem. Rev. 83, 535–547; (b) Inoue, Y. (1992). Asymmetric photochemical reactions in solution. Chem. Rev. 92, 741–770.
[47] (a) Fancy, D.A., and Kodadek, T. (1999). Chemistry for the analysis of protein–protein interactions: Rapid and efficient cross-linking triggered by long wavelength light. Proc. Natl. Acad. Sci. USA 96, 6020–6024; (b) Fancy, D.A., Denison, C., Kim, K., Xie, Y., Holdeman, T., Amini, F., and Kodadek, T. (2000). Scope, limitations and mechanistic aspects of the photo-induced cross-linking of proteins by water-soluble metal complexes. Chem.Biol. 7, 697–708.
[48] Amini, F., Denison, C., Lin, H.-J., Kuo, L., and Kodadek, T. (2003). Using oxidative crosslinking and proximity labeling to quantitatively characterize protein-protein and protein-peptide complexes. Chem. Biol. 10, 1115–1127.
[49] (a) Barton, J.K., Olmon, E.D., and Sontz, P.A. (2011). Metal complexes for DNA-mediated charge transport. Coord. Chem. Rev. 255, 619–634; (b) Herman, L., Ghosh, S., Defrancq, E., and Kirsch-De Mesmaeker, A. (2008). Ru(II) complexes and light: molecular tools for biomolecules. J. Phys. Org. Chem. 21, 670–681; (c) Szaciłowski, K., Macyk, W., Drzewiecka-Matuszek, A., Brindell, M., and Stochel, G. (2005). Bioinorganic photochemistry: frontiers and mechanisms. Chem. Rev. 105, 2647–2694.
[50] Ischay, M.A., Lu, X., and Yoon, T. P. (2010). [2+2] Cycloadditions by oxidative visible light photocatalysis. J. Am. Chem. Soc. 132, 8572–8574.

[51] Cherevatskaya, M., Neumann, M., Füldner, S., Harlander, C., Kümmel, S., Dankesreiter, S., Pfitzner, A., Zeitler, K., and König, B. (2012). Visible-light-promoted stereoselective alkylation by combining heterogeneous photocatalysis with organocatalysis. Angew. Chem. Int. Ed. 51, 4062–4066.

[52] Hari, D.P., Schroll, P., and König, B. (2012). Metal-free, visible-light-mediated direct C–H arylation of heteroarenes with aryl diazonium salts. J. Am. Chem. Soc. 134, 2958–2961.

[53] Pirtsch, M., Paria, S., Matsuno, T., Isobe, H., and Reiser, O. (2012). [$Cu(dap)_2Cl$] as an efficient visible-light-driven photoredox catalyst in carbon–carbon bond-forming reactions. Chem. Eur. J. 18, 7336–7340.

[54] Hernandez-Perez, A.C., Vlassova, A., and Collins, S. K. (2012). Toward a visible light mediated photocyclization: Cu-based sensitizers for the synthesis of [5]helicene. Org. Lett. 14, 2988–2991.

[55] Yoon, H.-S., Ho, X.-H., Jang, J., Lee, H.-J., Kim, S.-J., and Jang H.-Y. (2012). N719 dye-sensitized organophotocatalysis: enantioselective tandem michael addition/oxyamination of aldehydes. Org. Lett. 14, 3272–3275.

[56] DiRocco, D.A., and Rovis, T. (2012). Catalytic asymmetric α-acylation of tertiary amines mediated by a dual catalysis mode: N-heterocyclic carbene and photoredox catalysis. J. Am. Chem. Soc. 134, 8094–8097.

[57] Dadová, J., Svobodová, E., Sikorski, M., König, B., and Cibulka, R. (2012). Photooxidation of sulfides to sulfoxides mediated by tetra-O-acetylriboflavin and visible light. ChemCatChem 4, 620–623.

[58] Mitkina, T., Stanglmair, C., Setzer, W., Gruber, M., Kisch, H., and König, B. (2012). Visible light mediated homo- and heterocoupling of benzyl alcohols and benzyl amines on polycrystalline cadmium sulfide. Org. Biomol. Chem. 10, 3556–3561.

[59] Hari, D.P., and König, B. (2011). Eosin Y catalyzed visible light oxidative C–C and C–P bond formation. Org. Lett. 13, 3852–3855.

[60] Füldner, S., Mild, R., Siegmund, H.I., Schroeder, J.A., Gruber, M., and König, B. (2010). Green-light photocatalytic reduction using dye-sensitized TiO_2 and transition metal nanoparticles. Green Chem., 400–406.

[61] McNally, A., Prier, C.K., and MacMillan, D.W.C. (2011). Discovery of an α-amino C–H arylation reaction using the strategy of accelerated serendipity. Science 334, 1114–1117.

[62] Miyake, Y., Nakajima, K., and Nishibayashi, Y. (2012). Visible-light-mediated utilization of α-aminoalkyl radicals: addition to electron-deficient alkenes using photoredox catalysts. J. Am. Chem. Soc. 134, 3338–3341.

[63] Rueping, M., Leonori, D., and Poisson, T. (2011). Visible light mediated azomethine ylide formation-photoredox catalyzed [3+2] cycloadditions. Chem. Commun. 47, 9615–9617.

[64] Rueping, M., Zhu, S.-Q., and Koenigs, R.M. (2011). Visible-light photoredox catalyzed oxidative Strecker reaction. Chem. Commun. 47, 12709–12711.

[65] Schroll, P., Hari, D.P., and König, B. (2012). Photocatalytic arylation of alkenes, alkynes and enones with diazonium salts. ChemistryOpen 1, 130–133.

[66] Maji, T., Karmakar, A., and Reiser, O. (2011). Visible-light photoredox catalysis: dehalogenation of vicinal dibromo-, α-halo-, and α,α-dibromocarbonyl compounds. J. Org. Chem. 76, 736–739.

[67] Zou, Y.-Q., Chen, J.-R., Liu, X.-P., Lu, L.-Q., Davis, R. L., Jørgensen, K. A., and Xiao, W.-J. (2012). Highly efficient aerobic oxidative hydroxylation of arylboronic acids: photoredox catalysis using visible light. Angew. Chem. Int. Ed. 51, 784–788.

[68] Xuan, J., Feng, Z.-J., Duan, S.-W., and Xiao, W.-J. (2012). Room temperature synthesis of isoquino[2,1-a][3,1]oxazine and isoquino[2,1-a]pyrimidine derivatives via visible light photoredox catalysis. RSC Adv. 2, 4065–4068.

[69] Zou, Y.-Q., Duan, S.-W., Meng, X.-G., Hu, X.-Q., Gao, S., Chen, J.-R., and Xiao W.-J. (2012). Visible light induced intermolecular [2+2]-cycloaddition reactions of 3-ylideneoxindoles through energy transfer pathway. Tetrahedron, http://dx.doi.org/10.1016/j.tet.2012.06.011.

[70] Su, Y.-J., Zhang, L.-R., and Jiao, N. (2011). Utilization of natural sunlight and air in the aerobic oxidation of benzyl halides. Org. Lett. 13, 2168–2171.

[71] Cheng, Y., Yang, J., Qu, Y., and Li, P. (2012). Aerobic visible-light photoredox radical C-H functionalization: Catalytic synthesis of 2-substituted benzothiazoles. Org. Lett. 14, 98–101.

[72] Tucker, J.W., and Stephenson, C.R.J. (2011). Tandem visible light-mediated radical cyclization-divinylcyclopropane rearrangement to tricyclic pyrrolidinones. Org. Lett. 13, 5468–5471.

[73] Dai, C., Meschini, F., Narayanam, J.M.R., and Stephenson, C.R.J. (2012). Friedel–Crafts amidoalkylation via thermolysis and oxidative photocatalysis. J. Org. Chem. 77, 4425–4431.

[74] Freeman, D.B., Furst, L., Condie, A.G., and Stephenson, C.R.J. (2012). Functionally diverse nucleophilic trapping of iminium intermediates generated utilizing visible light photoredox catalysis. Org. Lett. 14, 94–97.

[75] Wallentin, C.-J., Nguyen, J.D., Finkbeiner P., and Stephenson, C.R.J. (2012). Visible light-mediated atom transfer radical addition via oxidative and reductive quenching of photocatalysts. J. Am. Chem. Soc. 134, 8875–8884.

[76] Zhu, M.-Z., and Zheng, N. (2011). Synthesis, 2223–2236.

[77] Zlotorzynska, M., and Sammis, G. M. (2011). Photoinduced electron-transfer-promoted redox fragmentation of N-alkoxyphthalimides. Org. Lett. 13, 6264–6267

[78] Courant, T., and Masson, G. (2012). Photoredox-initiated α-alkylation of imines through a three-component radical/cationic reaction. Chem. Eur. J. 18, 423–427.

[79] Rueping, M., Zoller, J., Fabry, D.C., Poscharny, K., Koenigs, R.M., Weirich, T.E., and Mayer, J. (2012). Light-mediated heterogeneous cross dehydrogenative coupling reactions: metal oxides as efficient, recyclable, photoredox catalysts in C-C bond-forming reactions. Chem. Eur. J. 18, 3478–3481.

[80] Rueping, M., Koenigs, R.M., Poscharny, K., Fabry, D.C., Leonori, D., and Vila, C. (2012). Dual catalysis: combination of photocatalytic aerobic oxidation and metal catalyzed alkynylation reactions-C-C bond formation using visible light. Chem. Eur. J. 18, 5170–5174.

[81] Parrish, J.D., Ischay, M.A., Lu, Z., Guo, S., Peters, N.R., and Yoon, T.P. (2012). Endoperoxide synthesis by photocatalytic aerobic [2+2+2] cycloadditions. Org. Lett. 14, 1640–1643.

[82] Lin, S., Padilla, C.E., Ischay, M.A., and Yoon, T.P. (2012). Visible light photocatalysis of intramolecular radical cation Diels–Alder cycloadditions. Tetrahedron Lett. 53, 3073–3076.

[83] Tyson, E.L., Farney, E.P., and Yoon, T.P. (2012). Photocatalytic [2+2] cycloadditions of enones with cleavable redox auxiliaries. Org. Lett. 14, 1110–1113.

[84] Guillo, P., Hamelin, O., Batat, P., Jonusauskas, G., McClenaghan, N.D., and Ménage, S. (2012). Photocatalyzed sulfide oxygenation with water as the unique oxygen atom source. Inorg. Chem. 51, 2222–2230.

[85] Zhao, G., Yang, C., Guo, L., Sun, H., Chen, C., and Xia, W. (2012). Visible light-induced oxidative coupling reaction: easy access to Mannich-type products. Chem. Commun. 48, 2337–2339.

[86] Ju, X., Liang, Y., Jia, P., Li, W., and Yu, W. (2012). Synthesis of oxindoles via visible light photoredox catalysis. Org. Biomol. Chem. 10, 3556–3561.

[87] Maity, S., Zhu, M.-Z., Shinabery, R.S., and Zheng, N. (2012). Intermolecular [3+2] cycloaddition of cyclopropylamines with olefins by visible-light photocatalysis. Angew. Chem. Int. Ed. 51, 222–226.

[88] Kohls, P., Yadav, D., Pandey, G., and Reiser, O. (2012). Visible light photoredox catalysis: generation and addition of N-aryltetrahydroisoquinoline derived α-amine radicals to Michael acceptors. Org. Lett. 14, 672–675.

[89] Yasu, Y., Koike, T., and Akita, M. (2012). Sunlight-driven synthesis of γ-diketones via oxidative coupling of enamines with silyl enol ethers catalyzed by $[Ru(bpy)_3]^{2+}$. Chem. Commun. 48, 5355–5357.
[90] Fu, W., Guo, W., Zou, G., and Xu, C. (2012). Selective trifluoromethylation and alkynylation of tetrahydroisoquinolines using visible light irradiation by Rose Bengal. J. Fluor. Chem. 140, 88–94.
[91] Ananthakrishnan, R., Gazi, S. (2012). $[Ru(bpy)_3]^{2+}$ aided photocatalytic synthesis of 2-arylpyridines via Hantzsch reaction under visible irradiation and oxygen atmosphere. Catal. Sci. Technol. 2, 1463–1471.
[92] Miyake, Y., Ashida, Y., Nakajima, K., and Nishibayashi, Y. (2012). Visible-light-mediated addition of α-aminoalkyl radicals generated from α-silylamines to α,β-unsaturated carbonyl compounds. Chem. Commun. 48, 6966–6968.
[93] Ischay, M.A., Ament, M.S., and Yoon, T.P. (2012). Crossed intermolecular [2+2] cycloaddition of styrenes by visible light photocatalysis. Chem. Sci., DOI: 10.1039/C2SC20658G.
[94] Cai, S., Zhao, X., Wang, X., Liu, Q., Li, Z., and Wang, D.Z. (2012). Visible-light-promoted C–C bond cleavage: photocatalytic generation of iminium ions and amino radicals. Angew. Chem. Int. Ed., DOI: 10.1002/anie.201202880.
[95] Neumann, M., and Zeitler, K. (2012). Application of microflow conditions to visible light photoredox catalysis. Org. Lett. 14, 2658–2661.
[96] Bou-Hamdan, F.R., and Seeberger, P.H. (2012). Visible-light-mediated photochemistry: Accelerating $Ru(bpy)_3^{2+}$-catalyzed reactions in continuous flow. Chem. Sci. 3, 1612–1616.
[97] Tucker, J.W., Zhang, Y., Jamison, T.F., and Stephenson, C.R.J. (2012). Visible-light photoredox catalysis in flow. Angew. Chem. Int. Ed. 51, 4144–4147.
[98] Andrews, R.S., Becker, J.J., and Gagné, M.R. (2012). A photoflow reactor for the continuous photoredox-mediated synthesis of C-glycoamino acids and C-glycolipids. Angew. Chem. Int. Ed. 51, 4140–4143.

Oliver Reiser, Georgiy Kachkovskyi, Viktor Kais, Paul Kohls, Suva Paria, Michael Pirtsch, Daniel Rackl, Hana Seo

8 Homogeneous visible light-mediated transition metal photoredox catalysis other than ruthenium and iridium

8.1 Introduction

The ability to change the oxidation state is an essential property of transition metal complexes for their application in photoredox chemistry, and in particular in visible light-mediated redox photocatalysis. In classical photochemistry activation of reactants is usually performed via direct irradiation with light of suitable wavelength that can be absorbed by the substrates. Since many organic reagents only absorb light in the UV region, its higher energy in comparison with visible light often results in the formation of byproducts and in a loss of selectivity. Shifting the irradiation into the visible range might be possible by applying sensitizers or photoredoxcatalysts, in the latter case invoking a mechanistic pathway that involves single-electron transfer as key reaction steps. An effective photocatalyst should possess a high extinction coefficient, a long triplet lifetime (10^{-8}–10^{-6} s) together with a high triplet yield, a suitable redox potential for a given reaction and photostability. [1] These requirements are met well in ruthenium and the iridium complexes, which have established themselves as the photoredoxcatalysts of choice over the last 30 years. In contrast, the application of other *d*-metal complexes, and especially of those in the third row of the periodic table such as copper or iron being more abundant is still very limited. [2] Here we would like to summarize the so far rare examples of visible light-mediated homogeneous redox photocatalysis with complexes based on transition metals excluding ruthenium and iridium. [1, 3, 4] We also do not discuss singlet oxygen generation in this chapter.

8.2 Copper in visible light catalysis

The potential of copper for photoredox catalysis was recognized already in 1980 when Mitani et al. reported the copper(I) chloride-catalyzed atom transfer radical addition reaction (ATRA) of activated organochlorides and organobromides such as dichloromethane, chloroform or bromobenzene to various olefins. [5, 6] Due to the lack of visible light absorption of CuCl this reaction must be performed by using a low pressure mercury lamp emitting light at 254 nm (Figure 8.1).

In a pioneering report, Sauvage et al. demonstrated in 1987 light-driven ($\lambda = 350$ nm) transformations of 4-nitrobenzyl bromide (**2**), mediated by $Cu(dap)_2Cl$ (**1**,

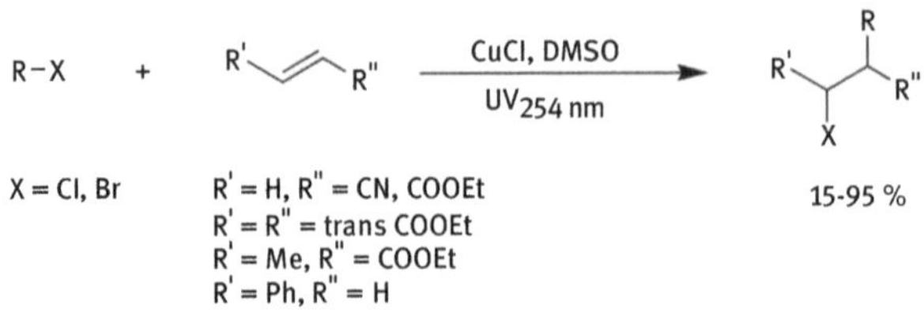

Figure 8.1: UV light-driven ATRA reaction.

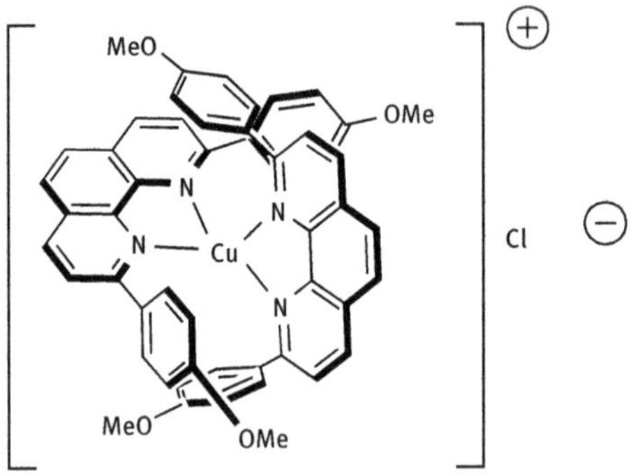

1

Figure 8.2: Structure of [Cu(dap)₂Cl] (1).

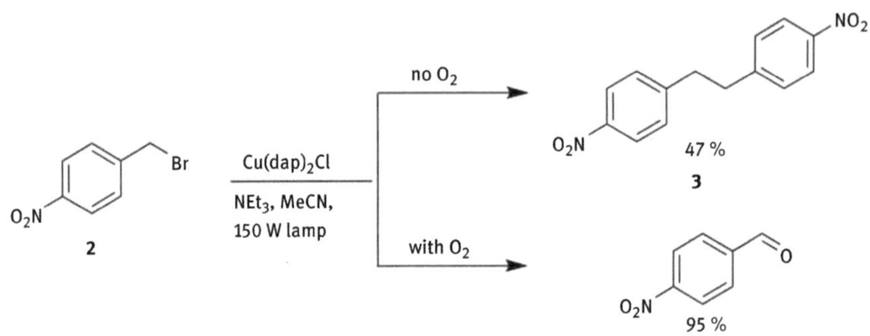

Figure 8.3: First visible light-driven Cu-catalyzed reactions.

dap = 2,9-bis(para-anisyl)-1,10-phenenthroline) as photoredoxcatalyst (Figure 8.2). [7] By carrying out the transformation either under oxygen-free or oxygen-containing conditions, photodimerization to 1,2-bis(4-nitrophenyl)ethane (**3**) or oxidation to nitrobenzaldehyde (**4**) was observed (Figure 8.3).

Assuming the initial step of both reactions to be a one-electron transfer from the catalyst to the organohalide, **1** shows strong analogies to the oxidative quenching cycles of related ruthenium and iridium complexes.

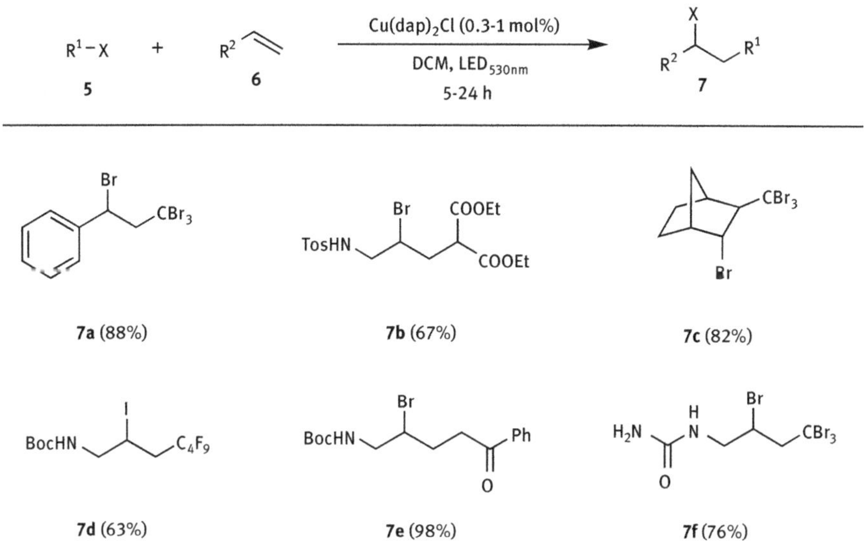

Figure 8.4: General scheme for $Cu(dap)_2Cl$ catalyzed ATRA reaction and selected products.

Furthermore the physical properties of $Cu(dap)_2Cl$ are reasonably close to ruthenium- and iridium-based photoredox catalysts: The lifetime of $Cu(dap)_2Cl$ (270 ns) is significantly shorter than $Ir[dF(CF_3)ppy_2(dtbbpy)]PF_6$ (2.3µs) [8] and $Ru(bpy)_3Cl_2$ (890 ns), but the reduction potential of excited $Cu(dap)_2Cl$ is even slightly stronger (−1.43 V) than both excited $Ru(bpy)_3^+$ (−1.33 V) and excited $Ir[dF(CF_3)ppy_2(dtbbpy)]PF_6$ (−1.21 V), making it attractive for photoredox-initiated reductive processes.

In 2012, Reiser et al. were able to demonstrate the utility of $Cu(dap)_2Cl$ for ATRA (Figure 8.4, 8.5) and allylation reactions (Figure 8.6). [9] Employing 0.3–1 mol% of $Cu(dap)_2Cl$ organobromides **5** such as tetrabromomethane, diethylbromomalonate or α-bromoacetophenone as well as perfluorinated organoiodides like *n*-(perfluoroiodo)butane could be successfully coupled with various olefins **6** upon irradiation at 530 nm to give addition products **7**.

It is proposed that initially exited $Cu(dap)_2Cl$ transfers an electron onto the organohalide **5**, which results in the formation of a carbon-centered radical **8** and a halide ion (Figure 8.5). The radical **8** next adds under thermodynamic and steric control to the olefin **6**, followed by the oxidation of the radical **9** to the cation **10**, thus regenerating the catalyst. This cation **10** is then trapped by the initially formed halide ion to give the final product **7**.

Moreover, $Cu(dap)_2Cl$ could efficiently catalyze the visible light-mediated allylation of activated α-halo and α,α-dihalo carbonyl compounds using allyltributyltin. [9] Using 1 mol% of $Cu(dap)_2Cl$ catalyst loading (Figure 8.6), the allylation of differ-

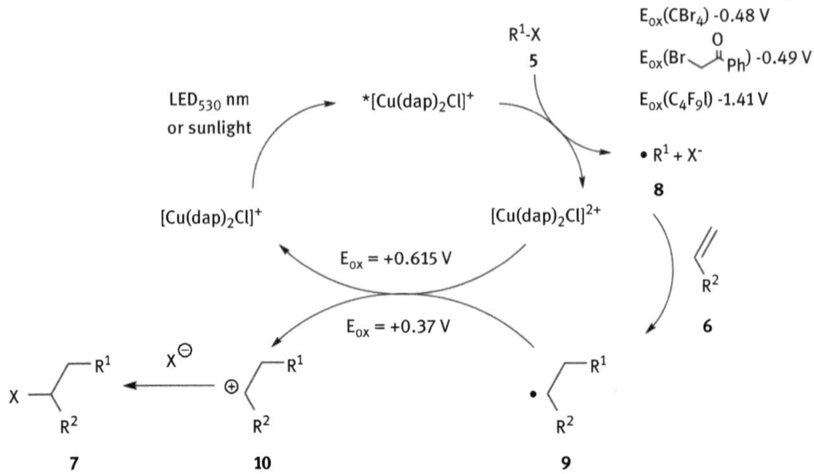

Figure 8.5: Suggested mechanism of the ATRA reaction.

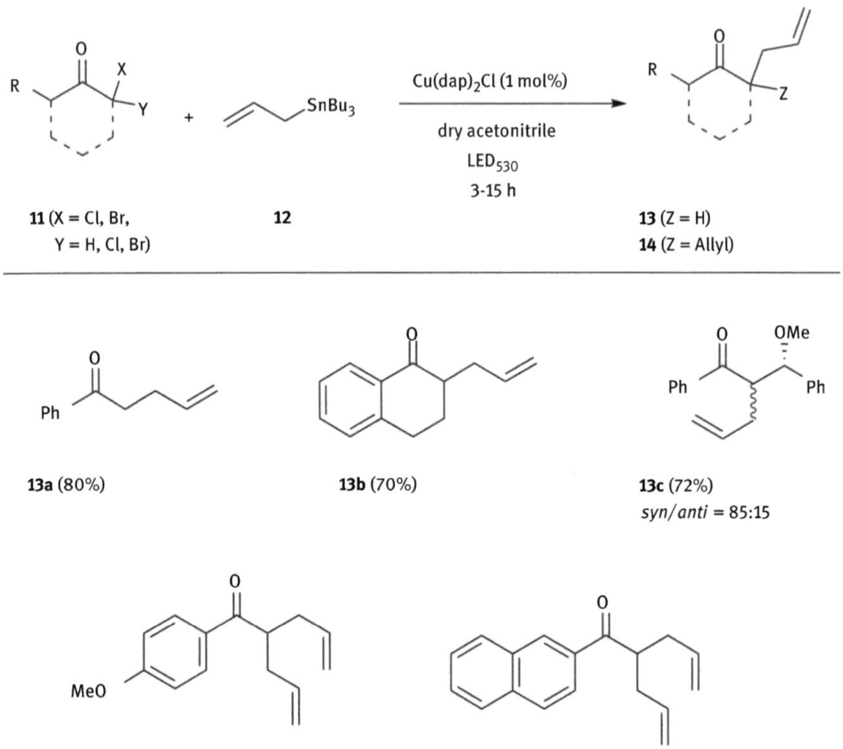

Figure 8.6: Cu(dap)₂Cl photoredox catalyzed allylation with **12** and selected products.

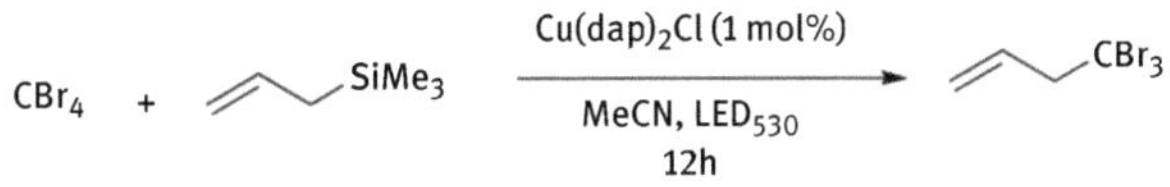

15 **16** **17**(60%)

Figure 8.7: Cu(dap)₂Cl photoredox catalyzed allylation with **16**.

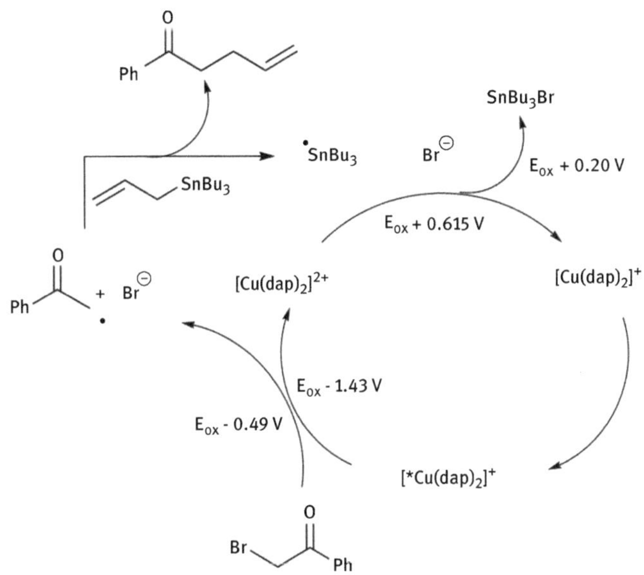

Figure 8.8: Suggested mechanism of the allylation.

ent α-halocarbonyl compounds and diallylation of α,α-dibromo or α,α-dichloro ketones was achieved at room temperature.

Allylation with environmentally more benign allyltrimethylsilane (**16**) was also possible, however, thus far only when CBr_4 (**15**) was used as halogen compound (Figure 8.7).

Again it seems plausible that $Cu(dap)_2Cl$ acts as an electron shuttle between the organohalide and the allylmetal reagents as exemplified between the reaction of α-bromoacetophenon and allyltributyltin (Figure 8.8).

8.3 Rhenium and platinum in visible light catalysis

Rhenium, the heaviest d^7 element, has been employed in a variety of photochemical reactions using UV light. [10] Generally, octahedral complexes such as **18–20**

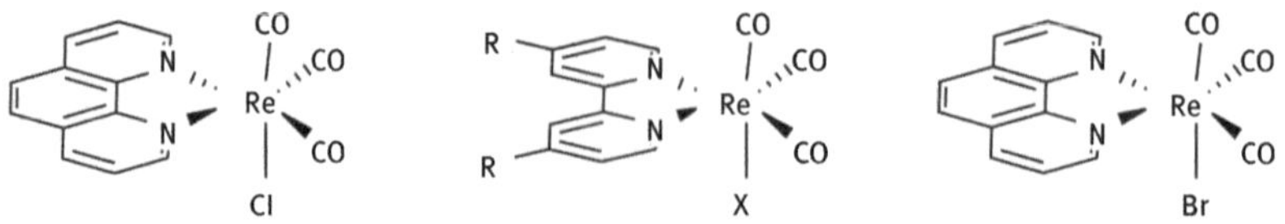

18 19 20

Figure 8.9: Rhenium based visible light photocatalysts (R = H, Me; X = Br, Cl, NCS).

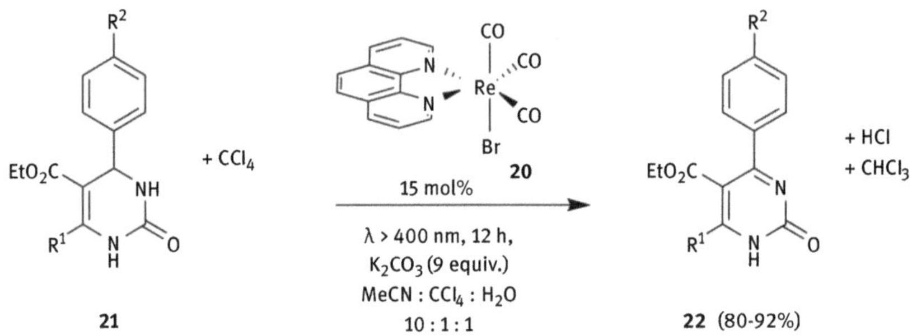

21 22 (80-92%)

Figure 8.10: Photooxidation of dihydropyrimidinones (21) to pyrimidinones (22) (R^1 = H, Me, R^2 = H, Me, OMe, Cl, Br).

(Figure 8.9) with chelating nitrogen ligands, e. g. bipyridine (bpy) or phenanthroline (phen), and carbon monoxide have been investigated. Such complexes have been used for the reduction of carbon dioxide to carbon monoxide under UV irradiation. [11]

Lehn et al. disclosed that CO_2 can be reduced under visible light irradiation by $Re(bpy)(CO)_3Cl$ (**18**) to CO in the presence of Et_4NCl and a mixture of DMF:TEA (5:1) as solvent. [12] In a more recent example, Rieger et al. employed complex **19** for the same purpose. [13]

Wu et al. demonstrated the oxidation of dihydropyrimidones (DHPM, **21**) to pyrimidones **22** (Figure 8.10). [14] $Re(bpy)(CO)_3Br$ (**20**, 15 mol%) proved to be a suitable photoredox catalyst when irradiated at wavelengths $\lambda > 400$ nm to yield **22** in good to excellent yields. In the key step of the reaction it is assumed that an electron mediated by the excited rhenium complex is transferred from DHPM **21** to the solvent CCl_4, forming a trichloromethyl radical and chloride. After subsequent deprotonation, the $CCl_3\bullet$ radical abstracts a hydrogen atom to form product **22**.

A conceptually interesting approach aiming at the utilization of solar energy for hydrogen production was reported by the same group. Platinum(II) terpyridyl acetylide complexes such as **25** excited by visible light ($\lambda > 450$ nm) were shown to oxidize dihydrohetarenes such as Hantzsch esters **23** [15] (Figure 8.11) or 2,5-dihydropyrroles

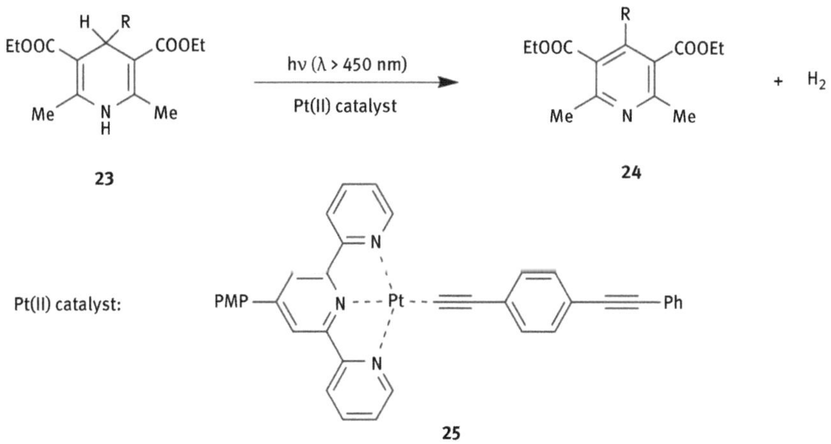

Figure 8.11: Pt(II) complex **25** catalyzed Hantzsch esters oxidation and hydrogen production.

[16] via two successive hydrogen atom abstractions almost quantitatively yielding molecular hydrogen. The catalyst shows a high turnover number (> 1000) and recyclability (up to 5 times) without loss in activity. [16]

8.4 Iron in visible light catalysis

8.4.1 Photocatalytic oxidation of hydrocarbons

Visible light irradiation ($\lambda > 425$ nm) of pyridine solutions containing the so-called Pacman catalyst **26** allows the oxidation of organic substrates, e. g. fluorene (**29**), diphenylmethane (**31**) toluene (**33**) or cumol (**35**) (Figure 8.12). The thermally inert Fe-O bonds of diiron(III)-μ-oxobisporphyrin **26** is cleaved by a photon to generate **27** containing a PFe^{II} and a ferryl $PFe^{IV} = O$ unit (P = porphyrin), the latter being crucial for the subsequent oxygenation of the hydrocarbon. In this course, the ferryl $PFe^{IV} = O$ pair will be reduced to iron(II) porphyrin **28** (Figure 8.12), from which the (DPDF)Fe_2O complex **26** is regenerated in the presence of O_2 and re-enters into the photocatalytic cycle. [17, 18, 19, 20]

8.4.2 Photocatalytic oxidative decarboxylation

MacFaul et al. and Taylor et al. have shown that solutions of carboxylic acids such as **38** and iron(III)porphyrin **40** (Figure 8.13) can be photolyzed with visible light ($\lambda > 390$ nm) to initiate decarboxylation. Iron(II)porphyrin and a carboxyl radical is

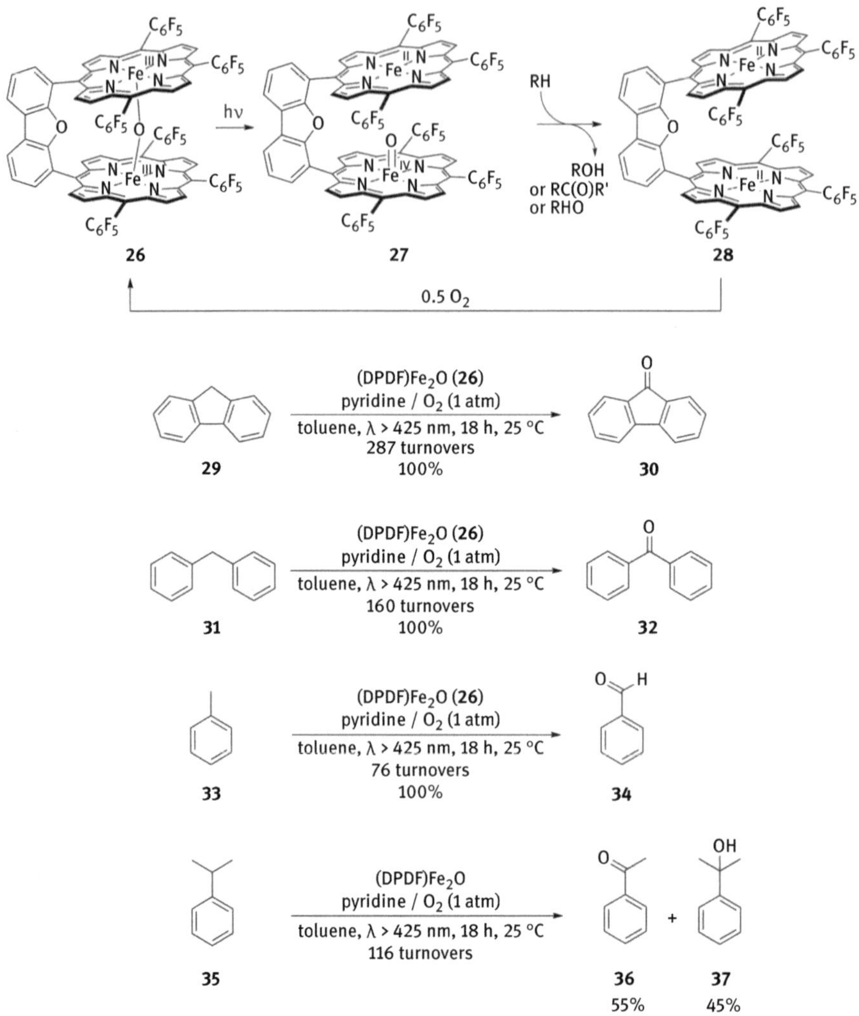

Figure 8.12: Photocatalytic oxidation of hydrocarbons and regeneration of (DPDF)Fe_2O (**26**) by O_2.

generated followed by oxidative decarboxylation, giving rise to **39**, whereas the iron(II)porphyrin rapidly oxidizes back to the iron(III) species **40** in the presence of dioxygen. [21, 22]

8.4.3 Oxidative degradation

Oxidative degradation procedures for organic compounds in aqueous media are of crucial importance for chemical waste water treatments. If catalytic or photochem-

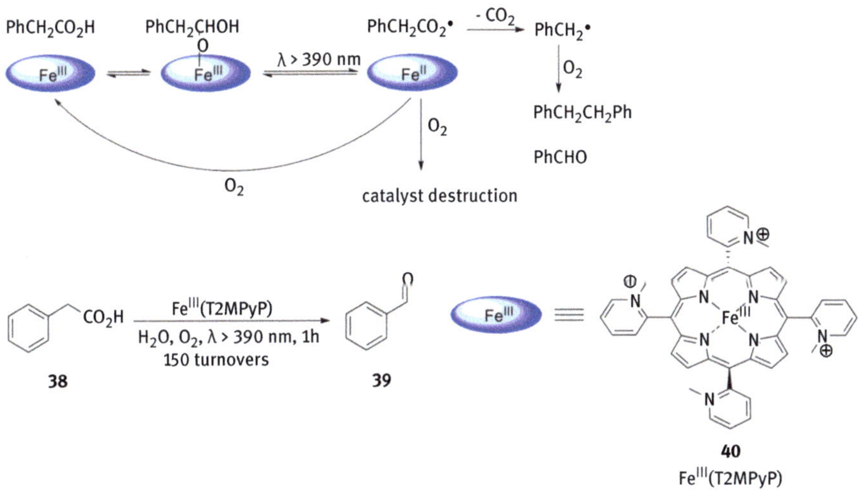

Figure 8.13: Photocatalytic decarboxylation of phenylacetic acid (**38**) to benzaldehyde (**39**) with iron(III)-5,10,15,20-tetra(2-*N*-methylpyridyl)porphyrin (**40**) as photocatalyst.

ical methods are applied, they are referred to as *advanced oxidation processes*. [23] These processes usually rely on the formation of hydroxyl radicals and the subsequent reaction of the organic substrate with such highly oxidizing species in order to achieve full mineralization. The (photo-)Fenton reaction uses iron salts and hydrogen peroxide to degrade a large variety of organic compounds, albeit the large amount of ferrous salts, a strong pH dependency and the over-stoichiometric employment of hydrogen peroxide limits the potential applications. [25] To lower the loading of ferrous salts, alter the light absorption characteristics of iron salts, and broaden the feasible pH range, while at the same time maintaining catalytic activity in water, organic ligands mainly based on porphyrin derivatives were mainly employed. Also Iron(III) tetrasulfophthalocyanine ([Fe^{III}(PcS)], **41**) was successfully used in the photodegradation of Rhodamine B under visible light illumination (Figure 8.14). [24]

Further improvement is necessary with respect to the consumption of hydrogen peroxide in these processes. H_2O_2 is considered an environmental friendly oxidant since its sole byproduct is water, nevertheless an even more environmentally benign oxidant is O_2. Zhao and coworkers demonstrated the use of iron(II) 4,4′-dicarboxy-2,2′-bipyridine ([$Fe^{III}(dcbpy)_3$]) as a catalyst that can activate considerable amounts of molecular oxygen to completely mineralize organic molecules under visible light illumination. With 2,4-dichlorophenol (**42**) as substrate more than 90 % degradation was achieved after an irradiation time of 8 hours using O_2 as the main oxidant (Figure 8.15). The mechanism for the molecular oxygen activation remains to be determined. [25]

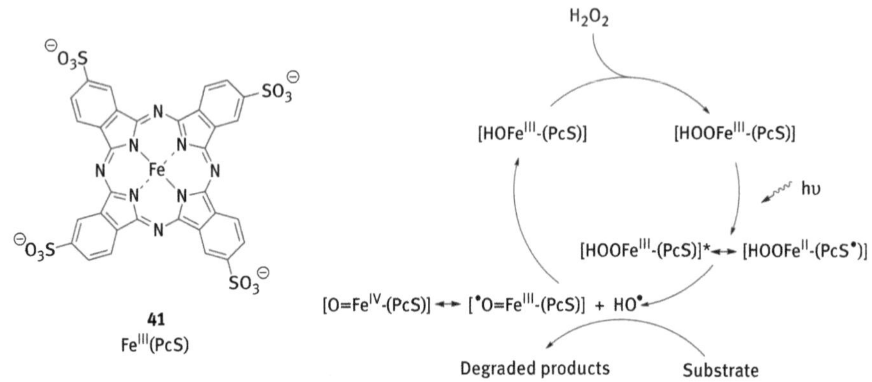

Figure 8.14: Structure of iron(III) tetrasulfophthalocyanine ($[Fe^{III}(PcS)]$) and proposed mechanism for the photochemical degradation process.

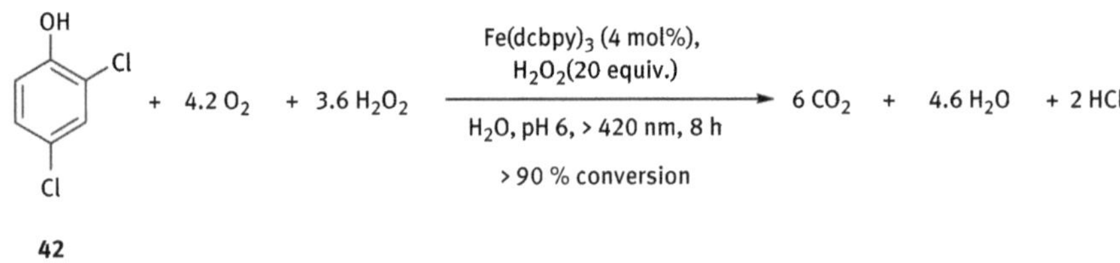

Figure 8.15: Degradation of 2,4-dichlorophenol (**42**) with $Fe(II)(dcbpy)_3$ as photocatalyst and O_2 as main oxidant.

8.4.4 Isomerization

It has been shown that iron carbonyl complexes $Fe(PF_3)_x(CO)_{5-x}$ ($x = 0 \ldots 4$) can catalyze the isomerization of alkenes under visible light photocatalysis. [26, 27] 1-Penten (**43**) can be completely isomerized to the thermodynamical equilibrium composition of *trans*-2-penten (**44**), *cis*-2-penten (**45**), and 1-penten (**43**) (79:18:2) under mild conditions upon irradiation with light at wavelengths between 400 and 450 nm (Figure 8.16). The quantum yields reaches values close to 500, therefore the catalytically active species is generated through the irradiation with visible light

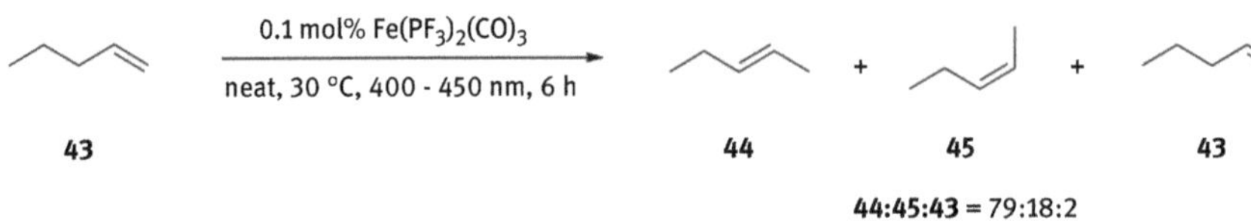

Figure 8.16: Isomerization of 1-penten catalyzed by photochemically activated $Fe(PF_3)_2(CO)_3$.

and the isomerization reaction itself must be thermal in nature. The catalytic intermediate is proposed to be a carbonyl-bridged diiron species.

8.5 Conclusion

Although the most broadly applied transition metal complexes for photoredox reactions are still based on ruthenium and iridium, a number of applications demonstrate the utility of rhenium and platinum and especially of iron and copper, being attractive with respect to abundance and low toxicity, as visible light photoredoxcatalysts as well.

8.6 References

[1] Tucker, J. W. and Stephenson, C. R. J. (2012). Shining Light on Photoredox Catalysis: Theory and Synthetic Applications. J. Org. Chem. USA 77, 1617–1622.

[2] Balzani, V., Bergamini, G., Campagna, S. and Puntoriero, F. (2007). Photochemistry and Photophysics of Coordination Compounds: Overview and General Concepts. Top. Curr. Chem. GER 280, 1–36.

[3] Narayanam, J. M. R. and Stephenson, C. R. J. (2011). Visible light photoredox catalysis: applications in organic synthesis. *Chem. Soc. Rev.* UK 40, 102–113.

[4] Zeitler, K. (2009). Photoredox Catalysis with Visible Light. Angew. Chem. Int. Ed. GER 48, 9785–9789.

[5] Koyama, K., Nakayama, M. and Mitani, M. (1980). The cuprous chloride catalyzed addition of halogen compounds to olefins under photo–irradiation. Tetrahedron Lett. UK 46, 4457–4460.

[6] Koyama, K., Kato, I. and Mitani, M. (1983). Photoaddition of alkyl halides to olefins catalyzed by copper(I) complexes. J. Am. Chem. Soc. USA 105, 6719–6721.

[7] Kern, J. and Sauvage, J. P. (1987). Photoassisted C-C coupling via electron transfer to benzylic halides by a bis(di-imine) copper(I) complex. J. Chem. Soc., Chem. Commun. UK, 546–548.

[8] Flamigni, L., Barbieri, A., Sabatini, C., Ventura, B. and Barigellett, F. (2007). Photochemistry and Photophysics of Coordination Compounds: Iridium. Top. Curr. Chem. GER 281, 143–203.

[9] Pirtsch, M., Paria, S., Matsuno, T., Isobe, H. and Reiser, O. (2012). $[Cu(dap)_2Cl]$ as an efficient visible-light-driven photoredox catalyst in carbon–carbon bond-forming reactions. GER 18, 7336-7340.

[10] Takeda, H., Koike, K., Morimoto, T., Inumaru, H., and Ishitani, O. (2011) Photochemistry and photocatalysis of rhenium(I) diimine complexes. Adv. Inorg. Chem. UK 63, 137–186.

[11] Kumar, A., Sun, S-S., and Lees, A. (2010) Photophysics and photochemistry of organometallic rheniumdiimine complexes. Top. Organomet. Chem. GER 29, 1–35.

[12] Hawecker, J., Lehn, J.-M., and Ziessel R. (1986) Photochemical and Electrochemical Reduction of Carbon Dioxide to Carbon Monoxide Mediated by (2,2'-Bipyridine)tricarbonylchlororhenium(I) and Related Complexes as Homogeneous Catalysts. Helv. Chim. Acta GER 69, 1990–2012.

[13] Bruckmeier, C., Lehenmeier, M. W., Reithmeier, R,. Rieger, B., Herranz, J., and Kavakli C. (2012) Binuclear rhenium(I) complexes for the photocatalytic reduction of CO_2. Dalton Trans. UK 41, 5026–5037.

[14] Liu, Q.; Li, Y.-N.; Zhang, H.-H., Chen, B., Tung, C.-H., and Wu L.-Z. (2011) Photochemical Preparation of Pyrimidin-2(1 H)-ones by Rhenium(I) Complexes with Visible Light. J. Org. Chem. USA 76, 1444–1447.
[15] Zhang, D., Wu, L.-Z., Zhou, L., Han, X., Yang, Q. Z., Zhang, L. P., and Tung, C. H. (2004) Photocatalytic Hydrogen Production from Hantzsch 1,4-Dihydropyridines by Platinum(II) Terpyridyl Complexes in Homogeneous Solution. J. Am. Chem. Soc. USA 126, 3440–3441.
[16] Wang, D.-H., Peng, M.-L., Han, Y., Chen, B., Tung, C.-H., and Wu, L.-Z. (2009). Facile Preparation of 3,4-Diarylpyrroles and Hydrogen by a Platinum(II) Terpyridyl Complex. Inorg. Chem. USA 48, 9995–9997.
[17] Latos-Grazynski, L., Cheng, R. J., La Mar G. N., and Balch A. L. (1982). Oxygenation patterns for substituted meso-tetraphenylporphyrin complexes of iron(II). Spectroscopic detection of dioxygencomplexes in the absence of amines. J. Am. Chem. Soc. USA 104, 5992.
[18] Balch, A. L. (1992). The reactivity of spectroscopically detected peroxycomplexes of iron porphyrins. Inorg. Chim. Acta USA, 198–200, 297.
[19] Balch, A. L., Chan, Y. W., Cheng, R. J., La Mar, G. N., Latos-Grazynski, L., and Renner, M. W. (1984). Oxygenation patterns of iron(II) porphyrins. Peroxo and ferryl ($F^{IV}O$) intermediates detected by ^1H nuclear magnetic resonance spectroscopy during the oxygenation of (tetramesitylpophyrin)iron(II). J. Am. Chem. Soc. USA 106, 7779.
[20] Rosenthal, J., Luckett, T. D, Hodgkiss, J. M., and Nocera, D. G. (2006). Photocatalytic oxidation of hydrocarbons by a bis-iron(III)-μ-oxo Pacman porphyrin using O_2 and visible light. J. Am. Chem. Soc. USA 128, 6546–6547.
[21] Gilbert, B. C., Hodges, G. R., Lindsay Smith, J. R., MacFaul, P., and Taylor, P. (1997). The photoreactions of the carboxylate complexes of 5,0,15,20-tetra(2-N-methylpyridyl)porphyrin. J. Mol. Catal. A: Chem. UK 117, 249–257.
[22] Gilbert, B. C., Lindsay Smith, J. R., MacFaul, P., Taylor P. (1996). Reactions of Carboxyl Radicals generated by the photocleavage of complexes of iron(III)tetra((2-N-methylpyridyl)porphyrin with unsaturated and aromatic carboxylic acids in aqueous solution. J. Chem. Soc., Perkin Trans. 2. UK 511–518.
[23] Legrini, O., Oliveros, E., and Braun, A. M. (1993). Photochemical Processes for Water Treatment. Chem. Rev. USA 93, 671–698.
[24] Tao, X., Ma, W., Zhang, T., and Zhao, J. (2002). A Novel Approach for the Oxidative Degradation of Organic Pollutants in Aqueous Solutions Mediated by Iron Tetrasulfophthalocyanine under Visible Light Radiation. Chem. Eur. J. GER 8, 1321–1326.
[25] Chen, X., Ma, W., Li, J., Wang, Z., Chen, C., Ji, H., and Zhao, J. (2011). Photocatalytic Oxidation of Organic Pollutants Catalyzed by an Iron Complex at Biocompatible pH Values: Using O_2 as Main Oxidant in a Fenton-like Reaction. J. Phys. Chem. USA 115, 4089–4095.
[26] Schroeder, M. A., and Wrighton, M. S. (1975). Pentacarbonyliron(0) Photocatalyzed Hydrogenation and Isomerization of Olefins, J. Am. Chem. Soc. USA 98, 551–558.
[27] Swartz, G. L., and Clark, R. J. (1980). Iron Carbonyl-Trifluorophosphine Compounds as Photocatalytic Precursors in Isomerization Studies. Inorg. Chem. 19, 3191–3195.

Kirsten Zeitler, Matthias Neumann

9 Synergistic Visible Light Photoredox Catalysis

9.1 Introduction

With respect to efficient and sustainable methods catalytic processes nowadays constitute a key technology to access the majority of synthetically derived chemicals on an industrial scale. [1] Apart from heterogeneous approaches, homogenous catalysis has become well established not only for the synthesis of fine chemicals. In the context of traditional catalysis the lowering of the energetic barrier for a transformation of interest relies on the activation of a single substrate with a specific catalyst and the subsequent reaction with an unactivated second substrate. This strategy has already given rise to a countless number of new chemical reactions. However, this approach is limited to the interaction with "reactive" substrates, which do not require any additional activation. Therefore, several strategies have been developed to overcome this challenging limitation either by multistep processes like domino [2], tandem [3] or cascade [4] reactions or by applying single-step procedures based on multiple activation of two or more substrates. Multiple activation approaches generally rely on a narrowing of the substrates' HOMO–LUMO gap (ΔE) by lowering the LUMO (lowest unoccupied molecular orbital) of the electrophile and simultaneously raising the nucleophile's HOMO (highest occupied molecular orbital) (Figure 9.1).

Multiple catalysis can be conducted in various modes and hence a differentiation into the following three distinct classes, depending on the nature of the catalyst(s) and their interplay in product formation, might be useful: cooperative catalysis, synergistic catalysis or reactions with multifunctional catalysts (Figure 9.2). [6] Cooperative catalysis and multifunctional catalysts involve a *single catalytic cycle* in which all catalytic sites, whether they are combined on a single catalyst or not,

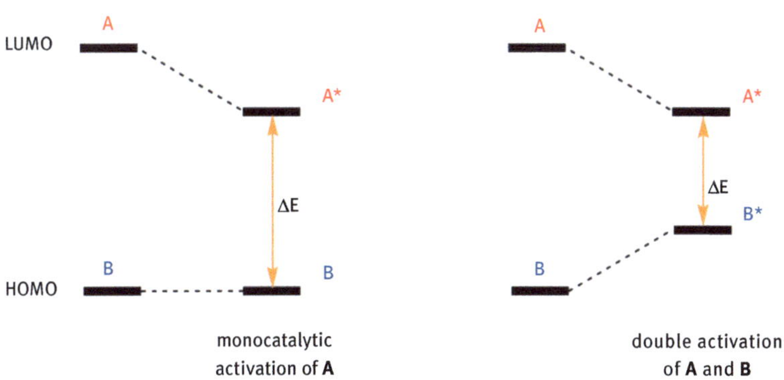

Figure 9.1: Comparison of catalytic activation modes (adapted from Ref. [5]).

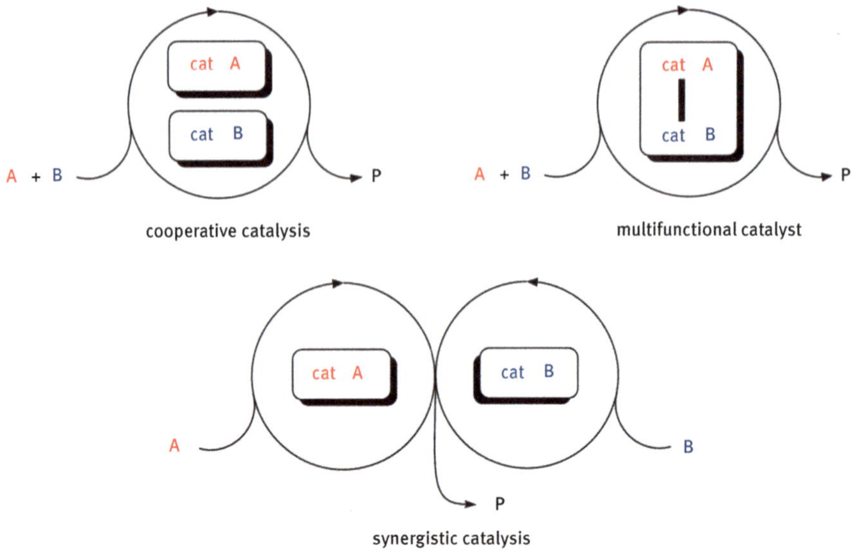

Figure 9.2: Multicatalytic activation modes.

enable product formation in a cooperative manner. Unlike, in synergistic (also termed: dual) catalysis both electrophile and nucleophile are activated simultaneously by two distinct catalysts in different but directly *coupled catalytic cycles.* [5] Here, one major challenge is the substoichiometric formation of two catalytic intermediates that will possess a sufficient lifetime for converging and bond formation along a desired pathway preferential to possible side reactions. Moreover, appropriate catalysts should not suffer from self-quenching, but should offer substrate selectivity and ideally regio- and stereocontrol during the reaction.

Visible light photoredox catalysis has experienced a renaissance in 2008 stimulated by the pioneering work of MacMillan [7] and has quickly turned into a vibrant area of research. [8]

In general terms, an excited photoredox catalyst can either be quenched reductively or by oxidation via single-electron transfer (SET) in the presence of a (sacrificial) substrate. The resulting intermediate catalyst species typically possesses high oxidative or reductive power depending on the nature of the prior quenching process and is – ideally – capable of performing a further SET to a different substrate or will undergo regeneration with a sacrificial electron acceptor or donor, respectively (Figure 9.3). Upon stepwise oxidation and reduction the photocatalyst mediates electron transfer reactions between the substrates and (if needed, i. e. not all electron transfers are "productive") sacrificial agents within a photoredox catalytic cycle. [9]

Up to now the total number of synergistic catalytic reactions merging visible light photoredox activation with other catalytic activation modes is rather small, but

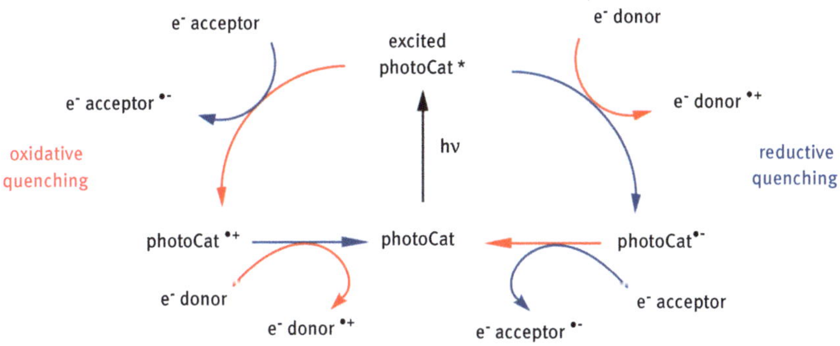

Figure 9.3: Quenching modes in photoredox catalysis.

this novel mechanistic approach has already proven to be highly powerful for the development of new chemical transformations. The following chapter will provide a short overview of selected, recent examples organized according to the type of involved photocatalytic-derived intermediates.

9.2 Stabilized iminium ions

Being easily oxidized and rather inexpensive, tertiary amines are widely used, attractive substrates in photoredox chemistry. Metal-mediated oxidative CH-functionalization [10] and sequential transformations, such as the Cu-catalyzed oxidation of tertiary amines to the corresponding iminium intermediates and their subsequent C-C [11] or C-P [12] coupling reactions, etc. are well established, but generally require the use of stoichiometric amounts of oxidants or an oxygen atmosphere. Building on these seminal results Stephenson et al. were able to develop photocatalytic conditions for the iminium generation and to transfer them to a first photooxidative aza-Henry reaction without the need of additional terminal oxidants. [13] Here an excited photocatalyst is reductively quenched by an *N*-aryl-1,2,3,4-tetrahydro*iso*quinoline (THIQ) derivative **1** resulting in the formation of the amine's corresponding radical cation **2** and the reduced photocatalyst species that is capable of reducing present atmospheric oxygen (and/or nitro derivatives) returning to its ground state. The hereby formed superoxide radical can subsequently oxidize and deprotonate the amino radical cation **2** to the highly electrophilic iminium ion **3** [14] which then can be trapped by nitroalkanes affording the C-C-coupled products **4** in very good yields of isolated products.

This proposed mechanism of the photooxidative aza-Henry reaction (see Figure 9.4) could later be partly confirmed by EPR detection of the intermediary superoxide radical by Wu and coworkers in the context of an eosin-catalyzed reaction. [15] Several variations of this reaction have been developed in the last few years. Besides

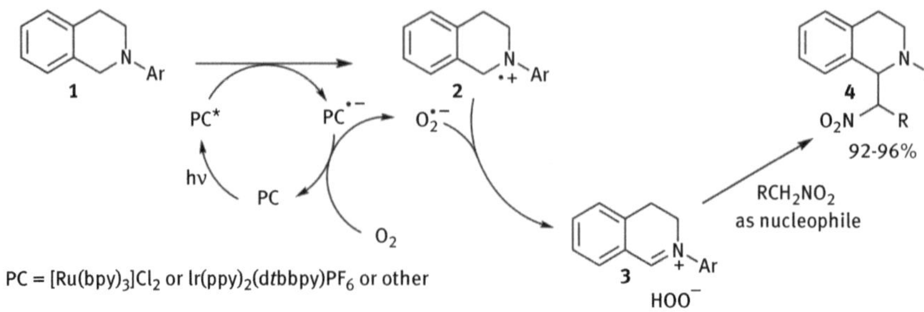

Figure 9.4: Photooxidative aza-Henry reaction.

the use of diverse (pro-)nucleophiles such as cyanide, [16] TMS-trifluoromethane, [16b] dialkyl malonates, [15–16] phosphonates [16a, 17] or alcohols and amines [18] photocatalytically generated iminium ions have also served as precursors for 1,3-dipoles and their subsequent cycloaddition reactions [19]; moreover the application of different photocatalysts, such as organic dyes, [15–16, 20] organic [21] and inorganic semiconductors [22] or MOFs [23] has been reported. Apart from these reactions, which often require excess nucleophiles, synergistic approaches implementing catalytically generated nucleophiles have also attracted increasing attention.

9.2.1 Secondary amine-catalyzed Mannich reactions

Rueping et al. initially showed the productive combination of the photooxidative generation of THIQ-derived iminium ions with organocatalytic enamine catalysis. [24] In this Mannich-type reaction (see Figure 9.5) the authors could only obtain good yields if the iminium ion had access to a sufficient concentration of the nucleophile to avoid undesired side reactions that would lower the product yield. The authors tested a number of light sources of different intensity and wavelength in order to find the best matching combination and could show that here rather mod-

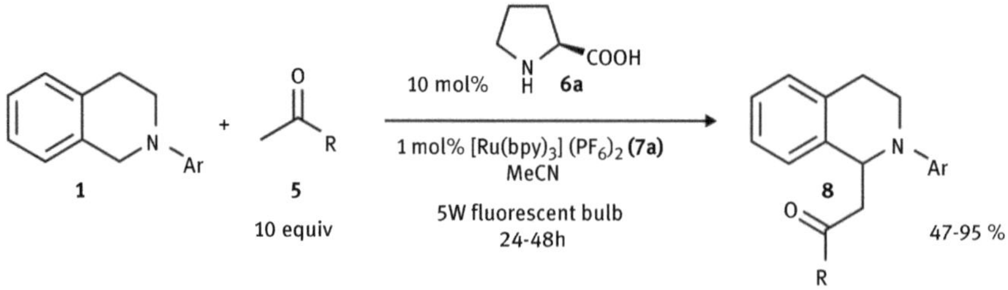

Figure 9.5: Organo-catalyzed photooxidative Mannich reaction.

Table 9.1: Enantioselectivity of different secondary amine organocatalysts 6 in oxidative Mannich reactions. (a) $VO(acac)_2$ was used as catalyst with ^tBuOOH as terminal oxidant instead of photooxidative conditions.

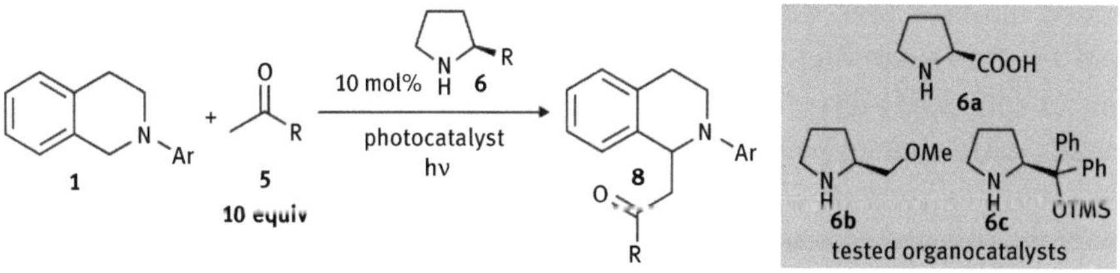

Entry	Organocatalyst	Aryl	Ketone	ee [%]	Photocat	Ref.
1	6a	Ph (**1a**)	Acetone (**5a**)	8	$[Ru(bpy)_3](PF_6)_2$	[24]
2	6a	*p*-(MeO)Ph (**1b**)	Acetone (**5a**)	7	Rose Bengal	[20]
3	6b	*p*-(MeO)Ph (**1b**)	Acetone (**5a**)	15	Rose Bengal	[20]
4[a]	6a	Ph (**1a**)	Acetone (**5a**)	7	–	[25]
5[a]	6c	Ph (**1a**)	Acetone (**5a**)	17	–	[25]

estly intense light from a 5 W fluorescent bulb is superior to, for instance, high power LEDs which are often favored in visible light photoredox catalysis. The low light intensity and hence potentially a slower turn over within the photocatalytic cycle assures a low iminium ion concentration compared to intermediary enamine allowing for pseudo first-order reaction conditions.

Although the authors used L-proline as a chiral organocatalyst they did not observe any significant enantioselectivity. This is in good accordance with earlier results of Klussmann et al. who noted slow racemization of the products **8** under their vanadium promoted oxidative conditions and who neither could achieve any useful enantioselectivity. [25] Similarly, other groups who transferred this synergistic Mannich approach to organic dye [20] or inorganic semiconductor photocatalysis [22a, 22c] could only detect virtually racemic products or only very little enantioenrichment (Table 9.1).

In contrast to the original Rueping protocol in the latter transformations higher light intensity proved to be beneficial as both systems are – by nature – less permeable for light either due to the very high molar extinction coefficients of the organic dyes or the reaction mixture's heterogeneity and the resulting light scattering in the case of solid semiconductors.

The scope of this transformation [20, 22, 24] is limited to simple ketones including cycloalkanones where unsymmetric ketones tend to regioselectively form the C–C bond from the less hindered (kinetic) enamine. Reactions proceed very successfully if the ketone is used as a solvent, but also can be conducted in acetonitrile in the presence of 10 equivalents of the respective ketone.

9.2.2 Coinage metal-catalyzed alkynylation reactions

To further extend the concept of synergistic catalysis combining photocatalytically generated iminium ions with other activation modes the Rueping group recently showed the first successful merger of photooxidative amine activation with transition metal-catalyzed alkynylations (Figure 9.6). [26]

Again, an elaborate balance had to be found to obtain sufficient yields of the desired coupling product **12** and to avoid detrimental side reactions. As both the nucleophilic intermediate **11** and the transition metal catalyst **9** are sensitive to reductive and/or oxidative conditions the appropriate tuning of the photocatalytic cycle proved to be essential.

Slowing down photocatalysis by the application of less intense light sources turned out to be beneficial for clean conversion (Table 9.2, entries 3+4) as described above (Section 9.1.2.1). Having established best conditions, the authors were able to show the scope of this reaction with a variety of terminal aryl and alkyl alkynes.

A large series of tested alkynes and THIQ derivatives afforded the cross-dehydrogenative coupling (CDC) products in moderate to very good yields, except for *p*-(*t*-butyl)phenyl alkyne **10b**. Here, only variation of the metal catalyst to a silver salt allowed the alkyne **10b** to be coupled to THIQ **1a** (see Figure 9.7), illustrating the modular character of this synergistic catalysis in which each catalytic cycle can be adapted to the substrates' demands.

A related, sequential multicatalytic alkynylation approach was reported by the Stephenson group [27] (see also Figure 9.8). In a first step a (bench stable) iminium bromide salt is generated using bromotrichloromethane as terminal oxidant and

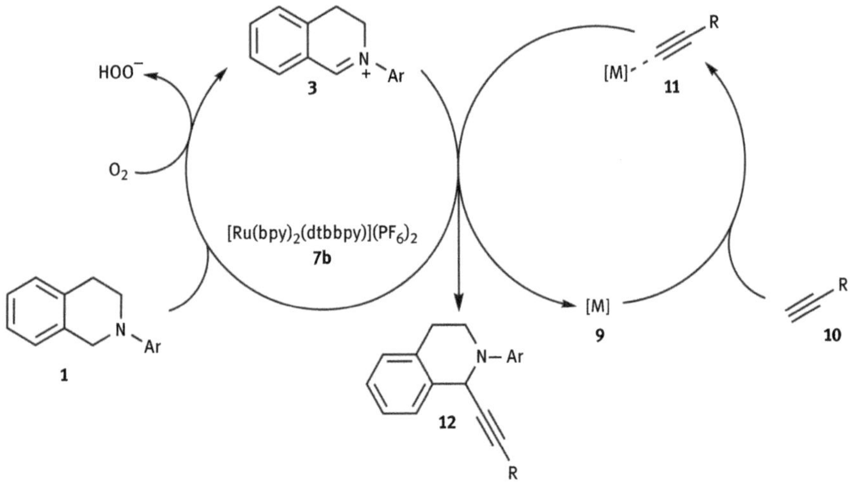

Figure 9.6: Photooxidative copper-catalyzed alkynylation.

counter anion source (Br⁻). After complete conversion of the THIQ **1a** to the corresponding iminium salt **3a** excess base, copper(I) bromide and excess terminal alkyne are added to provide the product in moderate to good yields in a subsequent "dark reaction."

Table 9.2: Survey of best reaction conditions (CSA: camphor sulfonic acid; OBz: benzoate; O_2CCF_3: trifluoroacetate).

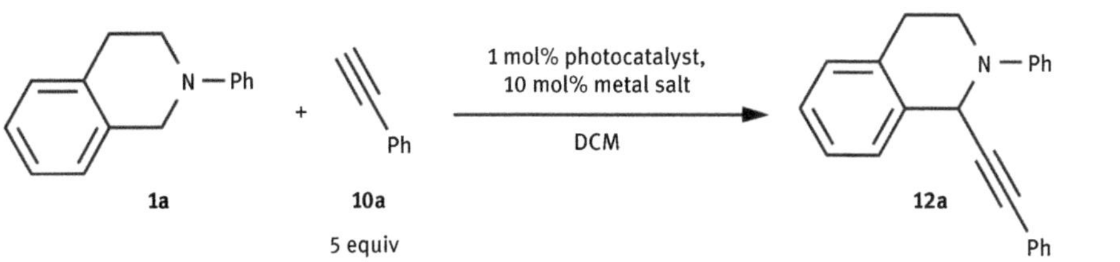

Entry	Photocatalyst	Light source	Metal salt	Time [h]	Yield [%]
1	7a	Blue LEDs	AgO_2CCF_3	24	–
2	7a	Blue LEDs	AgOBz	24	–
3	7a	Blue LEDs	AgOBz + CSA	72	32
4	7a	5 W lamp	AgOBz + CSA	24	50
5	7b	5 W lamp	AgOBz + CSA	16	74
6	7b	5 W lamp	$Cu(OTf)_2$	16	63
7	7b	5 W lamp	$(MeCN)_4CuPF_6$	16	88

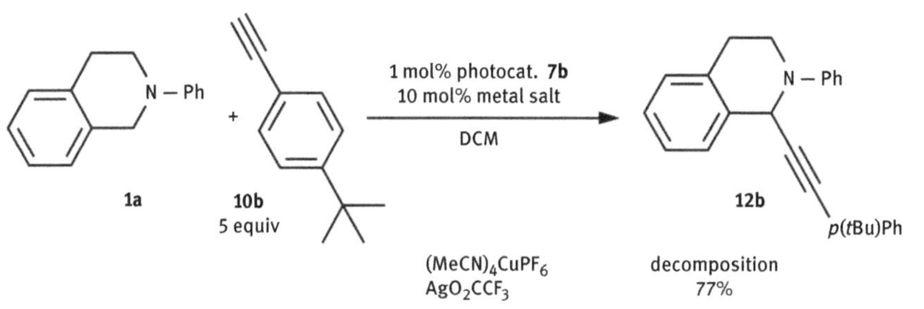

Figure 9.7: Limitation of scope and change of metal catalyst to adjust the dual catalytic cycles.

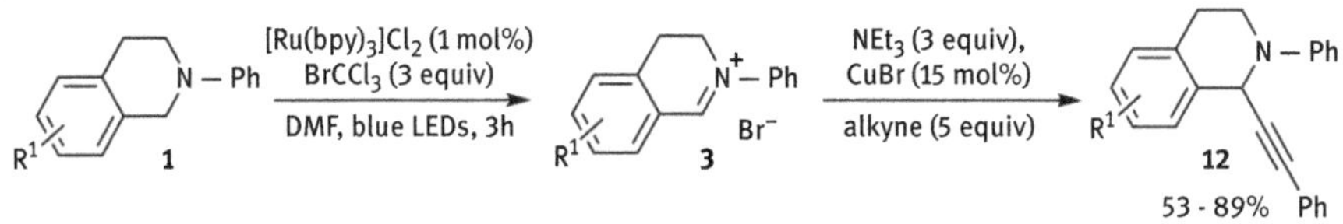

Figure 9.8: Sequential multicatalytic alkynylation.

9.2.3 NHC-catalyzed acylations

The concept of "umpolung" [28] (i. e. polarity reversal) allows for various nontraditional C-X and C-C bond formations. In this context the use of aldehydes as so-called "acylanion equivalents" for common C-C coupling reactions (e. g. benzoin or Stetter-type reactions, [29] etc.) is well established. For catalytic umpolung transformations *N*-heterocyclic carbenes (NHCs) are most frequently applied as catalysts to generate the Breslow intermediate (enaminol), which then serves as a nucleophile. [30] Apart from a number of NHC domino transformations [31] as well as cooperative combinations with Lewis acid [32] or with enamine and iminium catalysis [33], a first successful fusion of photocatalytic iminium ion generation with NHC-catalyzed umpolung resulting in an acylation of THIQs has been published very recently by the Rovis group. [34]

Unlike all aforementioned iminium ion generating methods using $[Ru(bpy)_3]^{2+}$ as photocatalyst, the Rovis group suggested an oxidatively quenched photoredox cycle mediated by *meta*-dinitrobenzene (*m*-DNB, **13**) as oxidizing agent, which either can be recycled by atmospheric oxygen or act as terminal oxidant under anaerobic conditions (Figure 9.9). The resulting Ru(III) species is a very strong oxidant ($E_{1/2} = 1.29$ V vs. SCE) capable of iminium ion formation from THIQ **1a,** which can then be attacked by Breslow intermediate **16** generated *in situ* from the chiral NHC catalyst **14** and the aldehyde **15.** Having established the best catalyst and best conditions the authors could apply this transformation for the synthesis of a variety of enantio-enriched acylation products from THIQs and different aldehydes (Figure 9.10).

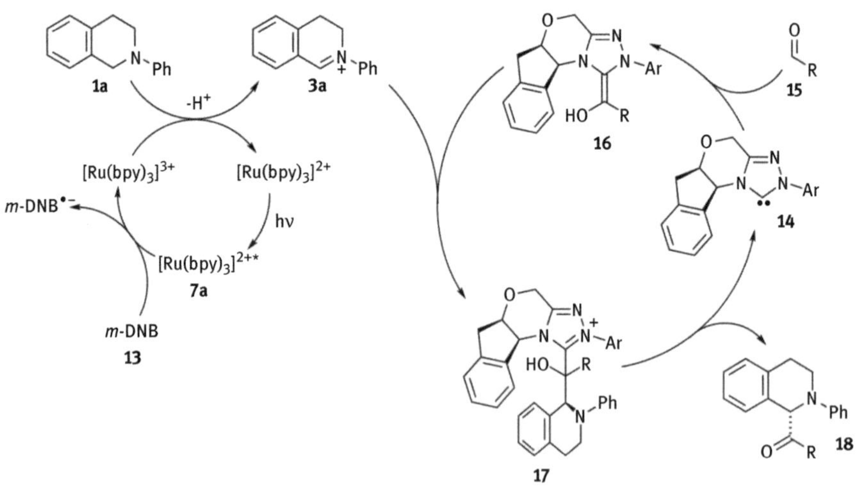

Figure 9.9: Mechanistic proposal for α-acylation of THIQs (*m*-DNB: *m*-dinitrobenzene).

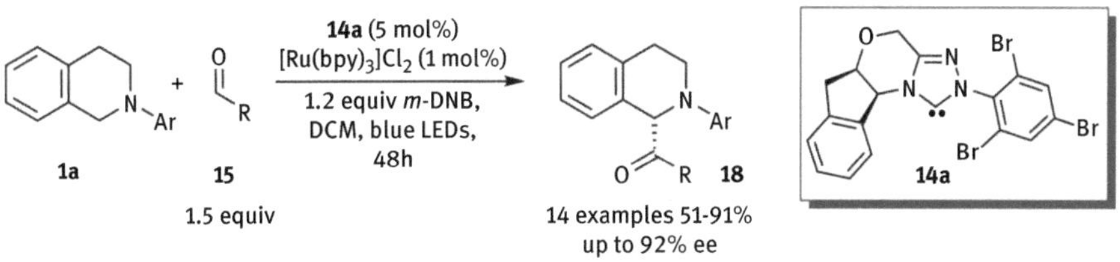

Figure 9.10: Synergistic NHC-catalyzed asymmetric α-acylation of THIQs.

9.3 Electrophilic carbon-centered radicals

A few seminal examples from the 1980s have demonstrated the possibility of using visible light photoredox chemistry to cleave *activated* carbon-halogen bonds [35] as well as related pseudo halogen-type bonds such as carbon-diazo bonds. [36] If these are adjacent to either carbonyl groups or aromatic systems, which are capable of accepting one electron from any common photocatalyst, cleavage can occur. The resulting, carbon centered radicals are trapped either by hydride donors or in intramolecular cyclization reactions. For at least two decades this approach was left unexploited; only a key publication of MacMillan and coworkers [7] inspired a number of different groups to enter the field of visible light photoredox catalysis. At least in the beginning their work mainly focused on follow-up chemistry of α-carbonyl halide-derived electrophilic radicals for C-C bond formations in intra- and intermolecular reactions. [8–9, 9c–e]

9.3.1 Secondary amine-catalyzed α-alkylation of aldehydes

Catalytic asymmetric functionalization of the α-position of aldehydes by alkylation has been a very challenging task for a long time, especially in the context of organocatalysis. [37] In 2007 the MacMillan group was able to present a first asymmetric α-allylation of aldehydes, introducing the concept of single occupied molecular orbital (SOMO) catalysis. [38] In this approach the intermediary enamines are oxidized by a stoichiometric amount of oxidant (e. g. cer ammonium nitrate (CAN)) to an electrophilic amine radical cation, which can be trapped by nucleophilic SOMOphiles, such as allylsilanes. The radical intermediates are in turn further oxidized to yield a stabilized carbocation, which then can undergo follow-up chemistry to afford the corresponding products. Later, the same group transferred this strategy to visible light photoredox catalysis. However, unlike to oxidative SOMO catalysis, as referred to above, in this synergistic photoredox catalysis the reactivities of the corresponding intermediates are switched (no formal "umpolung" of the nucleophilic enamine by oxidation as described for the SOMO catalysis).

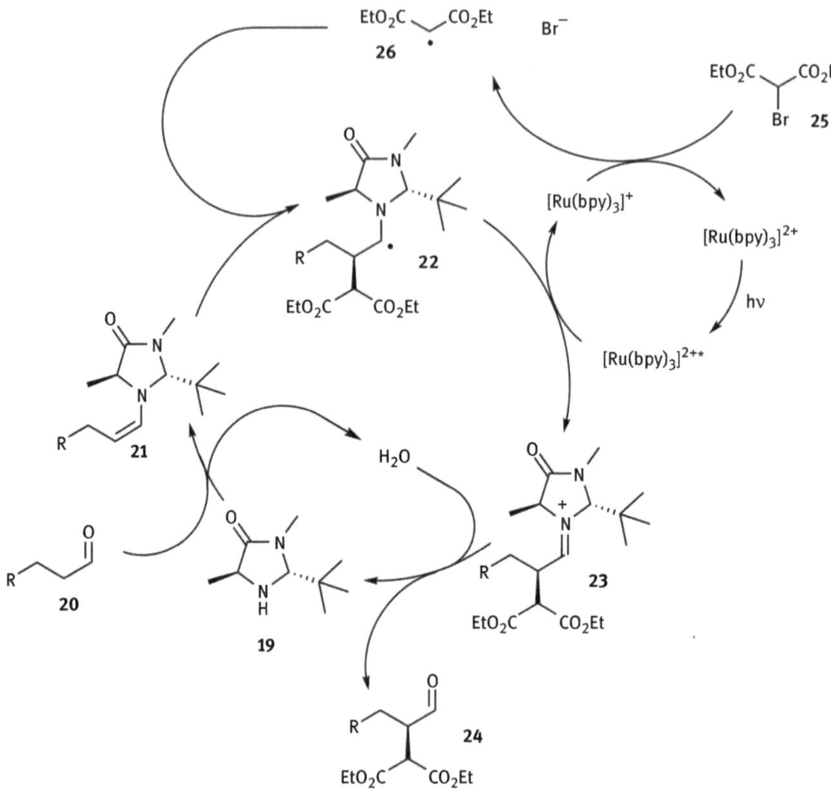

Figure 9.11: Asymmetric photoredox α-alkylation of aldehydes with bromodiethylmalonate.

In this secondary amine catalysis (see Figure 9.11) the intermediary enamine **21** is attacked by the electrophilic radical **26** stemming from a photoreductive cleavage of the activated bromoalkyl derivative **25** (*vide supra*). The newly formed C-C bond results in α-aminoradical **22,** which is then oxidized by the (excited) photocatalyst to an iminum ion **23** that readily releases the alkylated product **24** upon regeneration of the imidazolidinone catalyst **19.** As this synergistic catalysis does not require any additional auxiliary for electron shuffling or a sacrificial agent this approach was highly innovative at that stage.

With this powerful concept in hands the MacMillan group expanded the scope of electrophiles from the alkylations with α-carbonyl alkyl halogenides to trifluoromethylations [39] and benzylations [40] of a broad range of (enolizable) aldehydes (Table 9.3).

It should be mentioned that in the case of the α-benzylation reactions a "reversed" photocycle based on *fac*-Ir(ppy)$_3$ as catalyst is proposed. This Ir complex, which is typically used as a green triplet emitter in OLED applications [41] possesses a high reduction potential ($E_{1/2}$ Ir^{3+*} = −1.73 V vs. SCE in CH_3CN) in its excited state. [42]

Table 9.3: Representative examples for synergistic, asymmetric photoredox catalytic α-functionalizations of aldehydes (Conditions A: 19a×HOTf, $[Ru(bpy)_3]Cl_2$, DMF, RT; B: 19a×TFA, $Ir(dtbbpy)(ppy)_2PF_6$, DMF, −20 °C; C: 19b×HOTf, fac-$Ir(ppy)_3$, DMSO, RT; ppy: phenypyridine; HOTf: triflic acid; TFA: trifluoroacetic acid).

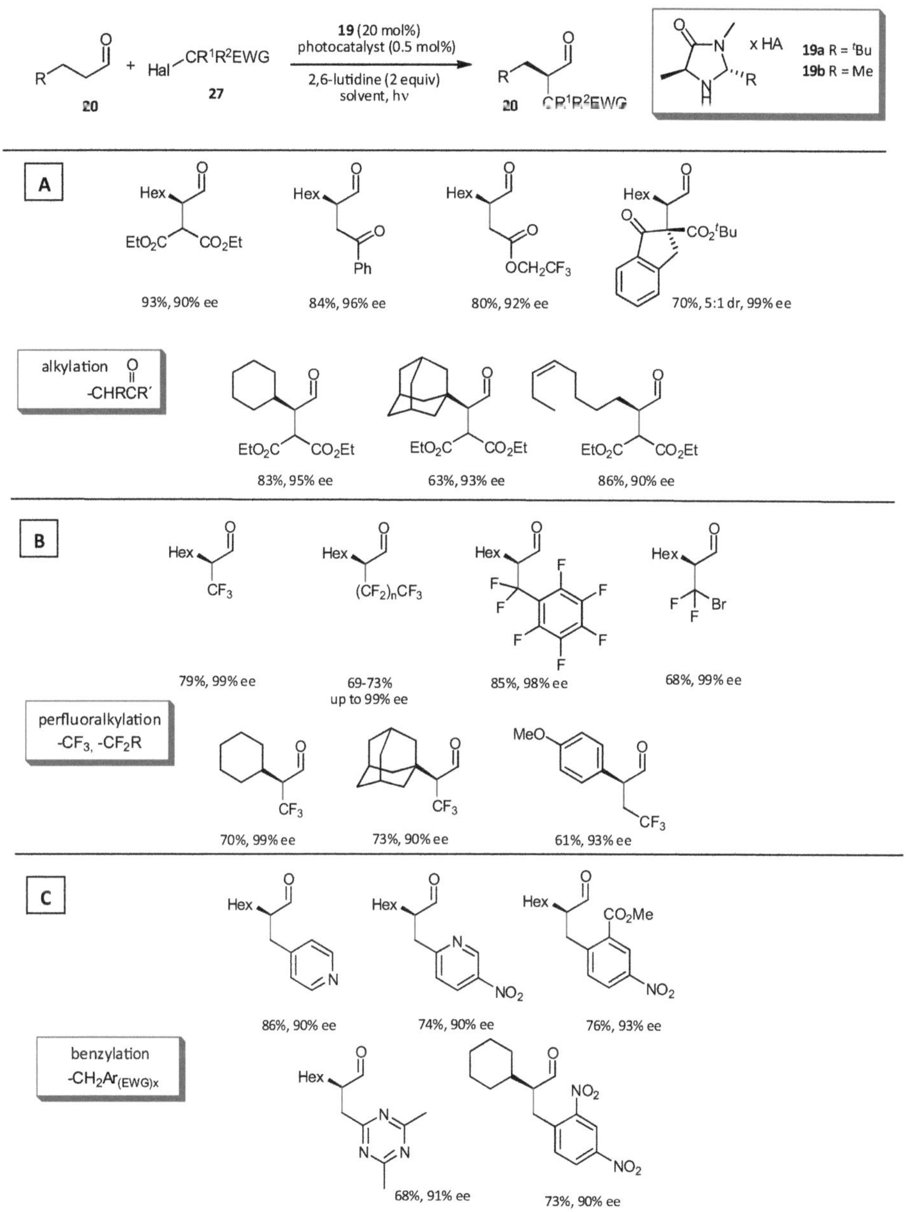

Hence, in contrast to the prior described mechanism, here via oxidative quenching the excited photocatalyst generates the carbon-centered radical which subsequently is attacked by the enamine. The oxidized photocatalyst is reduced to regenerate the *fac*-Ir(ppy)₃ by the intermediary α-amine radical **22** closing the photoredox cycle.

Besides transition metal-based photocatalysts, which might prove problematic in their future availability and due to possible metal traces within the products, simple, inexpensive organic dyes can also be used in photoredox catalysis as demonstrated by Zeitler et al. [43] Through the development of a fast screening reaction the group was able to identify the xanthene dye family as most promising candidates with eosin Y being the most potent. Besides photoredox reactions with (pro)nucleophiles they could also show the first example of an enantioselective, metal-free, synergistic photoredox catalysis (Figure 9.12) based on the asymmetric α-alkylation reactions described above.

In a subsequent study the same group has demonstrated the beneficial effects of combining visible light photoredox chemistry with microflow conditions. [44] Zeitler and coworkers were able to show a drastic rate acceleration in reactions with excess

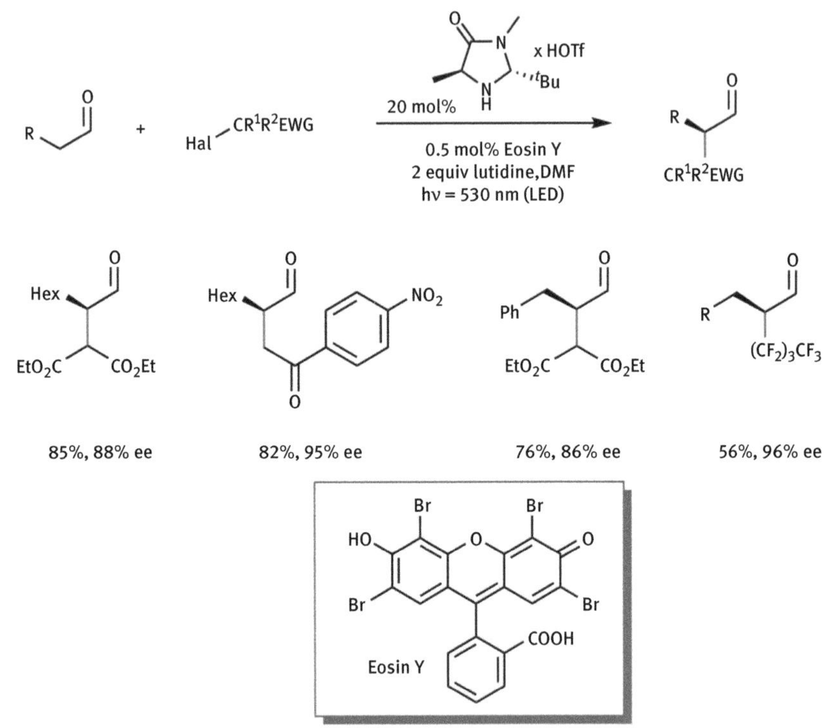

Figure 9.12: Metal-free asymmetric α-alkylations using eosin Y as photocatalyst.

Table 9.4: Comparison of the performance of different reactors.

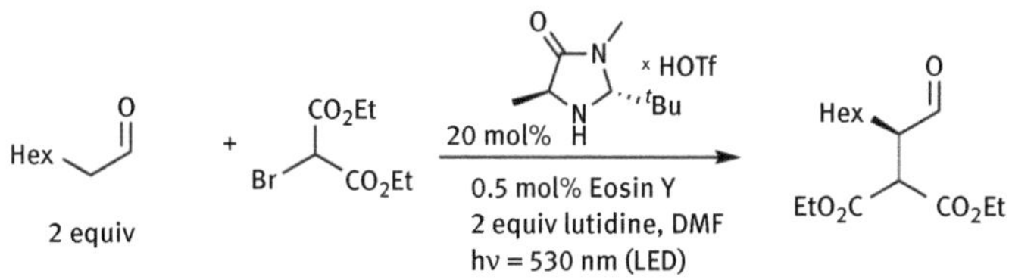

Entry	Reactor type	Yield [%]	ee [%]	Productivity [mmol/h]	Relative factor
1	Batch	85	88	0.018	1
2	Microreactor	86	87	0.037	2
3	Tube reactor	92	82	1.92	107

nucleophiles in aza-Henry reactions (cf. Section 9.1.2) if performed in a commercially available microstructured flow reactor (100 µl internal volume).

Furthermore, it was shown that the productivity (mmol/h) of the synergistic photoredox α-alkylations can be increased by at least two orders of magnitude if conducted in a simple, self-made FEP (fluorinated ethylene propylene copolymer) tubing reactor Remarkably, similar conditions as in the batch approach could be used without observing a significant loss in enantioselectivity [44] (Table 9.4).

9.3.2 Palladium-catalyzed C-H arylation

Transition metal-catalyzed C-H activation including arylations is of major interest in current organic synthesis. But besides recent advancements in functional group tolerance and substrate scope typical protocols still require elevated temperatures (>80°C). [45] Finding room temperature protocols for C-H arylations is hence very desirable. Recently, the Sanford group reported on a combination of palladium-catalyzed C-H activation (see Figure 9.13) and photoreductive phenyl radical generation resulting in formation of biaryl products. [46] The successful merger of the two known concepts (visible light photoredox catalytic access to phenyl radicals from diazonium salts [36] and ligand-directed C-H arylation [47]) by Sandford et al. now offers an easy-to-perform room temperature protocol.

Oxidative addition of a photogenerated phenyl radical **30** to palladacycle **32** can form a palladium(III) complex **33**, which is further oxidized to a palladium(IV) species **34** capable of reductive elimination to afford the desired arylated product **35** (Figure 9.14).

Using their previously developed best conditions a number of different directing groups were examined to further extend the scope of this transformation (see Figure 9.15). In addition to the original 2-pyridyl substrates also amides **36**, pyrazoles **37**, pyrimidines **38** and oxime ethers **39** could be employed to provide the corresponding arylation products in moderate to good yields.

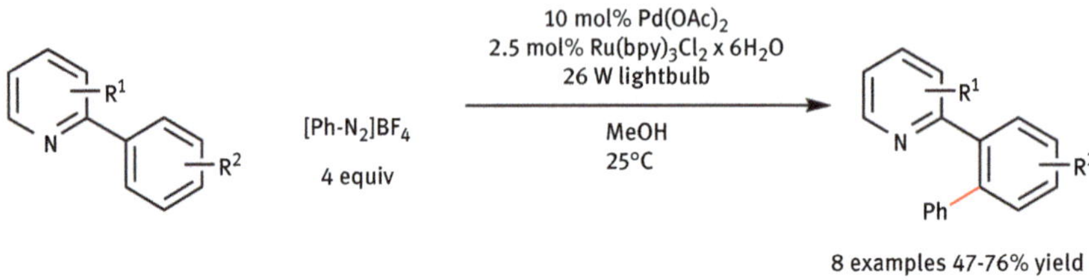

Figure 9.13: Palladium-catalyzed arylation of phenylpyridine derivatives.

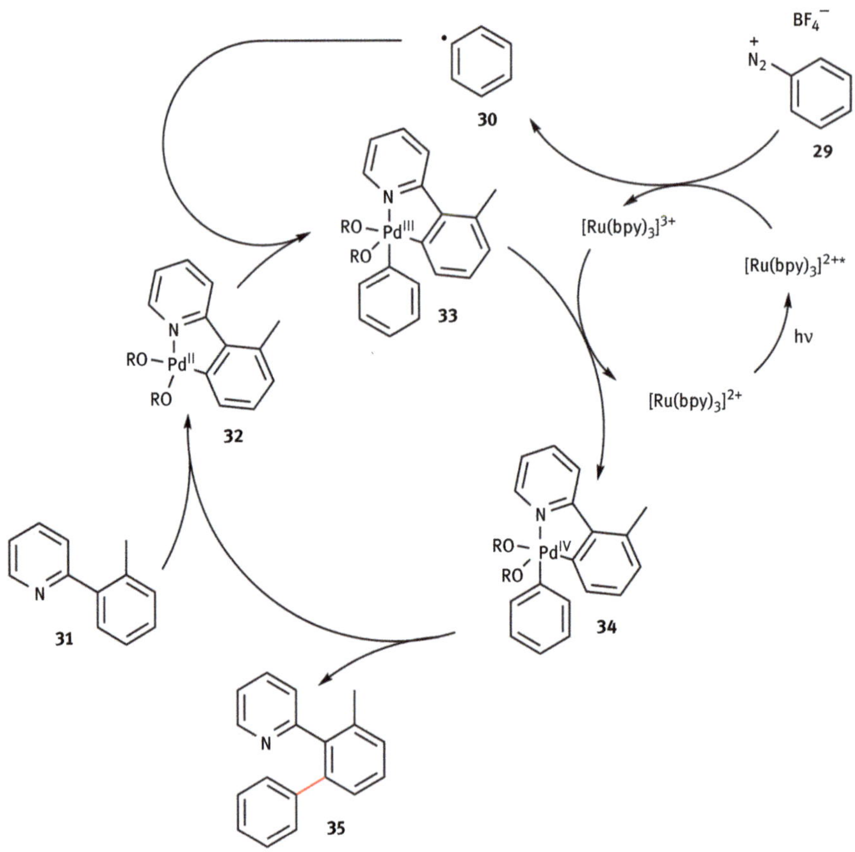

Figure 9.14: Mechanistic proposal for the combination of Pd-catalyzed C-H arylation and photocatalysis.

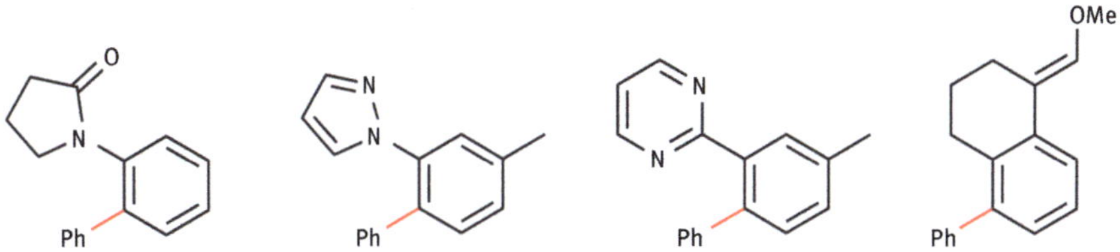

36 72% 37 53% 38 44% 39 49%

Figure 9.15: Extended substrate scope of C H arylation reactions.

9.3.3 Copper-catalyzed trifluoromethylation of aryl boronic acids

Fluorinated organic compounds are of considerable interest in the field of agrochemicals and especially pharmaceuticals. Finding new protocols for the introduction of fluorine and especially trifluoromethyl groups is hence a highly desirable goal in organic synthesis. Several protocols for direct trifluoromethylation of aryl compounds by Pd- and Cu-catalysis have been developed, but up to now they often suffer from several limitations such as expensive trifluoromethyl sources, or the need of elevated temperatures and strong acids or bases. An approach for the Cu-catalyzed cross-coupling of inexpensive CF_3I with commercially available aryl boronic acids at 60 °C was recently reported by Sanford et al. [48] (see also Figure 9.16).

The authors suppose a copper(III) species **42** as the key intermediate of this synergistic catalysis, which after reductive elimination forms the product **41** and regenerates the catalytic copper(I) species (Figure 9.17). It remains unclear whether the oxidative addition of trifluoromethyl radical to copper or the transmetalation of boron to copper resulting in intermediate **42** occurs first in the formation of intermediate **42**.

To investigate the full scope of this transformation several aryl boronic acids were tested examining electronic effects of substituents and functional group tolerance (see Figure 9.18).

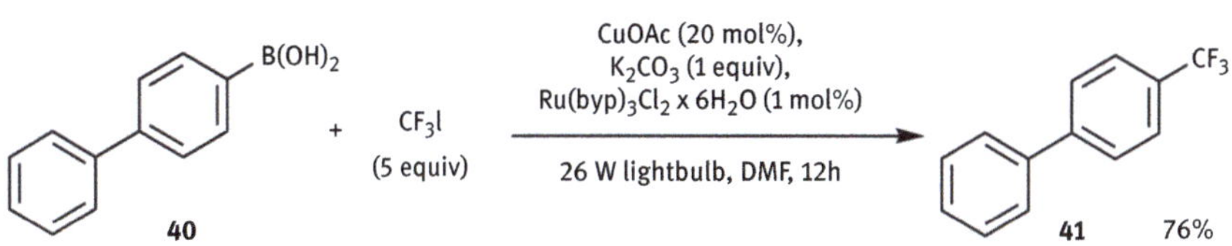

40 41 76%

Figure 9.16: Cu-catalyzed cross-coupling of aryl boronic acids and iodotrifluoromethane merged with photocatalysis.

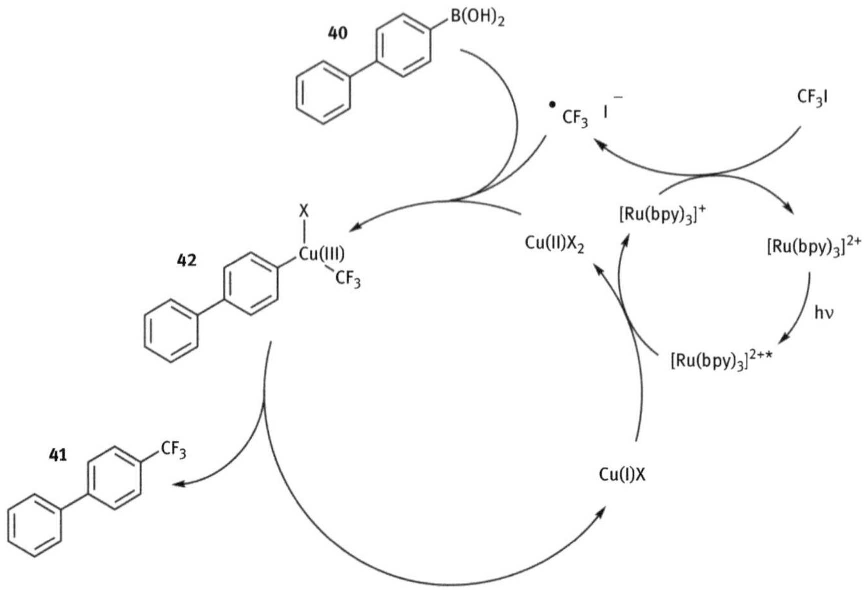

Figure 9.17: Mechanistic proposal for Cu-catalyzed trifluoromethylation.

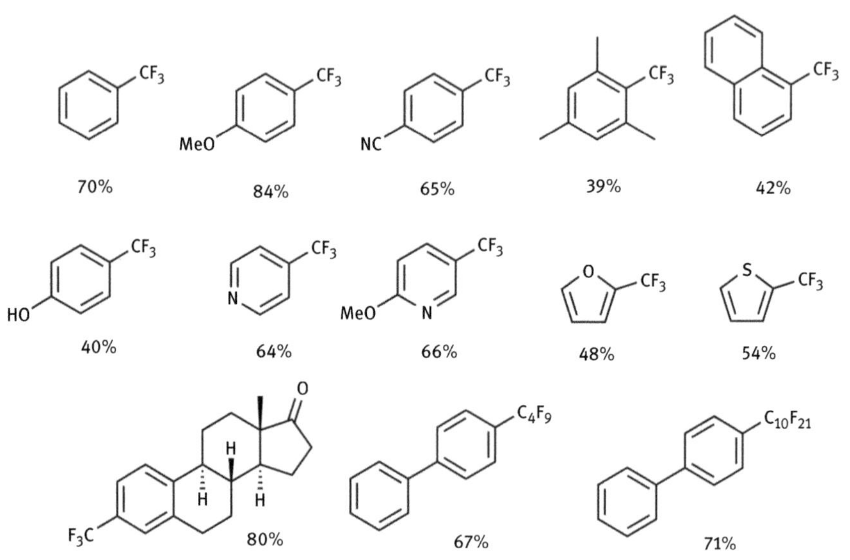

Figure 9.18: Representative examples for trifluoromethylations and perfluoroalkylations.

9.4 Conclusion

Although synergistic catalysis combining visible light photoredox catalysis with other catalytic activation modes may be perceived as considerably challenging the examples presented here should illustrate the benefits of this novel approach. The powerful strategy will certainly continue to grow allowing for new transformations not accessible by single catalyst methods so far.

9.5 References

[1] (a) Röper, M. (2006). Chem. unserer Zeit, 40, 126; (b) Schüth, F. (2006). Chem. unserer Zeit, 40, 92.
[2] Tietze, L. F., Brasche, G., Gericke, K. M. (2006). in Domino Reactions in Organic Synthesis, Wiley-VCH Verlag GmbH & Co. KGaA.
[3] Fogg, D. E., dos Santos, E. N. (2004). Coord. Chem. Rev., 248, 2365.
[4] Walji, A. M., MacMillan, D. W. C. (2007). Synlett, 1477.
[5] Allen, A. E., MacMillan, D. W. C. (2012). Chem. Sci, 3, 633.
[6] Wende, R. C., Schreiner, P. (2012). Green Chem., 14, 1821.
[7] Nicewicz, D. A., MacMillan, D. W. C. (2008). Science, 322, 77.
[8] Teplý, F. (2011). Collect. Czech. Chem. Commun., 76, 859.
[9] (a) Narayanam, J. M. R., Stephenson, C. R. J. (2011). Chem. Soc. Rev., 40, 102; (b) Xuan, J., Xiao, W.-J. (2012). Angew. Chem. Int. Ed., 51, 6828; (c) Tucker, J. W., Stephenson, C. R. J. (2012). J. Org. Chem, 77, 1617; (d) Yoon, T. P., Ischay, M. A., Du, J. (2010). Nat Chem, 2, 527; (e) Zeitler, K. (2009). Angew. Chem. Int. Ed., 48, 9785.
[10] (a) Yoo, W.-J., Li, C.-J. (2010). Top. Curr. Chem., 292, 281; (b) Li, C.-J. (2008). Acc. Chem. Res., 42, 335.
[11] Li, Z., Bohle, D. S., Li, C.-J. (2006). Proc. Nat. Acad. Sci. USA, 103, 8928.
[12] Basle, O., Li, C.-J. (2009). Chem. Commun., 4124.
[13] Condie, A. G., González-Gómez, J. C., Stephenson, C. R. J. (2010). J. Am. Chem. Soc., 132, 1464.
[14] Mayr, H., Kempf, B., Ofial, A. R. (2002). Acc. Chem. Res., 36, 66.
[15] Liu, Q., Li, Y.-N., Zhang, H.-H., Chen, B., Tung, C.-H., Wu, L.-Z. (2012). Chem. – Eur. J., 18, 620.
[16] (a) Hari, D. P., König, B. (2011). Org. Lett., 13, 3852; (b) Pan, Y., Wang, S., Kee, C. W., Dubuisson, E., Yang, Y., Loh, K. P., Tan, C.-H. (2011). Green Chem., 13, 3341; (c) Rueping, M., Zhu, S., Koenigs, R. M. (2011). Chem. Commun., 47, 12709.
[17] Rueping, M., Zhu, S., Koenigs, R. M. (2011). Chem. Commun., 47, 8679.
[18] (a) Xuan, J., Feng, Z.-J., Duan, S.-W., Xiao, W.-J. (2012). RSC Advances, 2, 4065; (b) Xuan, J., Cheng, Y., An, J., Lu, L.-Q., Zhang, X.-X., Xiao, W.-J. (2011). Chem. Commun., 47, 8337.
[19] (a) Rueping, M., Leonori, D., Poisson, T. (2011). Chem. Commun., 47, 9615; (b) Zou, Y.-Q., Lu, L.-Q., Fu, L., Chang, N.-J., Rong, J., Chen, J.-R., Xiao, W.-J. (2011). Angew. Chem. Int. Ed., 50, 7171.
[20] Pan, Y., Kee, C. W., Chen, L., Tan, C.-H. (2011). Green Chem., 13, 2682.
[21] Möhlmann, L., Baar, M., Rieß, J., Antonietti, M., Wang, X., Blechert, S. (2012). Adv. Synth. Catal., 354, 1909.
[22] (a) Cherevatskaya, M., Neumann, M., Füldner, S., Harlander, C., Kümmel, S., Dankesreiter, S., Pfitzner, A., Zeitler, K., König, B. (2012). Angew. Chem. Int. Ed., 51, 4062; (b) Mitkina, T., Stanglmair, C., Setzer, W., Gruber, M., Kisch, H., König, B. (2012). Org. Biomol. Chem., 10,

3556; (c) Rueping, M., Zoller, J., Fabry, D. C., Poscharny, K., Koenigs, R. M., Weirich, T. E., Mayer, J. (2012). Chem.-Eur. J., 3478.

[23] (a) Wang, C., Xie, Z., deKrafft, K. E., Lin, W. (2011). J. Am. Chem. Soc., 133, 13445; (b) Xie, Z., Wang, C., deKrafft, K. E., Lin, W. (2011). J. Am. Chem. Soc., 133, 2056.

[24] Rueping, M., Vila, C., Koenigs, R. M., Poscharny, K., Fabry, D. C. (2011). Chem. Commun., 47, 2360.

[25] Sud, A., Sureshkumar, D., Klussmann, M. (2009). Chem. Commun., 3169.

[26] Rueping, M., Koenigs, R. M., Poscharny, K., Fabry, D. C., Leonori, D., Vila, C. (2012). Chem. – Eur. J., 18, 5170.

[27] Freeman, D. B., Furst, L., Condie, A. G., Stephenson, C. R. J. (2011). Org. Lett., 14, 94.

[28] (a) Bugaut, X., Glorius, F. (2012). Chem. Soc. Rev., 41, 3511; (b) Seebach, D. (1979). Angew. Chem. Int. Ed., 18, 239.

[29] DiRocco, D. A., Rovis, T. (2011). Science of Synthesis, 835.

[30] (a) Zeitler, K. (2005). Angew. Chem. Int. Ed., 44, 7506; (b) Enders, D., Niemeier, O., Henseler, A. (2007). Chem. Rev., 107, 5606; (c) Biju, A. T., Kuhl, N., Glorius, F. (2011). Acc. Chem. Res., 44, 1182.

[31] Grossmann, A., Enders, D. (2012). Angew. Chem. Int. Ed., 51, 314.

[32] Cohen, D. T., Scheidt, K. A. (2012). Chem. Sci, 3, 53.

[33] (a) Ozboya, K. E., Rovis, T. (2011). Chem. Sci, 2, 1835; (b) Lathrop, S. P., Rovis, T. (2009). J. Am. Chem. Soc., 131, 13628.

[34] DiRocco, D. A., Rovis, T. (2012). J. Am. Chem. Soc., 134, 8094.

[35] (a) Fukuzumi, S., Mochizuki, S., Tanaka, T. (1990). J. Phys. Chem., 94, 722; (b) Goren, Z., Willner, I. (1983). J. Am. Chem. Soc., 105, 7764; (c) Hironaka, K., Fukuzumi, S., Tanaka, T. (1984). J. Chem. Soc., Perkin Trans. 2, 1705; (d) Maidan, R., Goren, Z., Becker, J. Y., Willner, I. (1984). J. Am. Chem. Soc., 106, 6217; (e) Mashraqui, S. H., Kellogg, R. M. (1985). Tetrahedron Lett., 26, 1453; (f) Willner, I., Tsfania, T., Eichen, Y. (1990). J. Org. Chem, 55, 2656.

[36] (a) Cano-Yelo, H., Deronzier, A., J. Chem. Soc. (1984). Perkin Trans. 2, 1093; (b) Cano-Yelo, H., Deronzier, A. (1984). Tetrahedron Lett., 25, 5517; (c) Cano-Yelo, H., Deronzier, A. (1987). New J. Chem., 11, 479.

[37] Alba, A.-N., Viciano, M., Rios, R. (2009). ChemCatChem, 1, 437.

[38] Beeson, T. D., Mastracchio, A., Hong, J.-B., Ashton, K., MacMillan, D. W. C. (2007). Science, 316, 582.

[39] Nagib, D. A., Scott, M. E., MacMillan, D. W. C. (2009). J. Am. Chem. Soc., 131, 10875.

[40] Shih, H.-W., Vander Wal, M. N., Grange, R. L., MacMillan, D. W. C. (2010). J. Am. Chem. Soc., 132, 13600.

[41] Rausch, A., Homeier, H., Yersin, H. (2010). Top. in Organomet. Chem., 29, 193.

[42] Flamigni, L., Barbieri, A., Sabatini, C., Ventura, B., Barigelletti, F. (2007). Top. Curr. Chem., 281, 143.

[43] Neumann, M., Füldner, S., König, B., Zeitler, K. (2011). Angew. Chem. Int. Ed., 50, 951.

[44] Neumann, M., Zeitler, K. (2012). Org. Lett., 14, 2658.

[45] (a) Chiusoli, G. P., Catellani, M., Costa, M., Motti, E., Della Ca', N., Maestri, G. (2010). Coord. Chem. Rev., 254, 456; (b) Daugulis, O. (2010). Top. Curr. Chem., 292, 57; (c) Lyons, T. W., Sanford, M. S. (2010). Chem. Rev., 110, 1147.

[46] Kalyani, D., McMurtrey, K. B., Neufeldt, S. R., Sanford, M. S. (2011). J. Am. Chem. Soc., 133, 18566.

[47] Yu, W.-Y., Sit, W. N., Zhou, Z., Chan, A. S. C. (2009). Org. Lett., 11, 3174.

[48] Ye, Y., Sanford, M. S. (2012). J. Am. Chem. Soc., 134, 9034.

Magnus Rüping, Rene M. Königs, Iuliana Atodiresei

10 Photoredox catalyzed α-functionalization of amines – Visible light mediated carbon-carbon and carbon-hetero bond forming reactions

10.1 Introduction

The direct functionalization of unreactive C–H bonds is one of the major challenges for synthetic organic chemists. It allows the sustainable formation of new carbon-carbon or carbon-heteroatom bonds without prefunctionalization of substrates and, thus, without the generation of undesirable amounts of waste.

In particular, cross-dehydrogenative coupling (CDC) allows an operationally simple and environmentally benign approach to the direct functionalization of C–H bonds in the presence of an oxidant, ideally molecular oxygen, with water being the sole by-product. [1]

Typically, various transition metal salts are applied to CDC reactions, in which a C–H bond, adjacent to a tertiary nitrogen atom, can be functionalized through oxidation by the transition metal catalyst to an intermediate iminium ion, which is then intercepted by various C–H nucleophiles under the formation of a new C–C bond (Figure 10.1).

Cross-dehydrogenative coupling, [1] pioneered by C. J. Li using copper catalysis in the early 2000s, [2] is an ideal blueprint for photoredox catalysis, [3] as photoredox catalysts allow the oxidation of tertiary amines to the corresponding iminium ions, [4, 5] which are the crucial intermediates in CDC reactions. Stephenson and co-workers demonstrated the applicability of CDC in photoredox catalysis in 2010 in their publication on the visible light mediated aza-Henry reaction. [6]

So far, different homogeneous and heterogeneous photoredox catalysts have been identified to be applicable to visible light mediated CDC reactions. The most commonly used are Ru(II) poly-bipyridyl and Ir(III) poly-pyridiyl complexes [7] (Figure 10.2) which were shown to be highly efficient catalysts for the oxidative functionalization of C–H bonds adjacent to nitrogen atoms. The substitution pattern of the bipyridiyl and aryl-pyridyl ligands has shown to be of critical importance for this

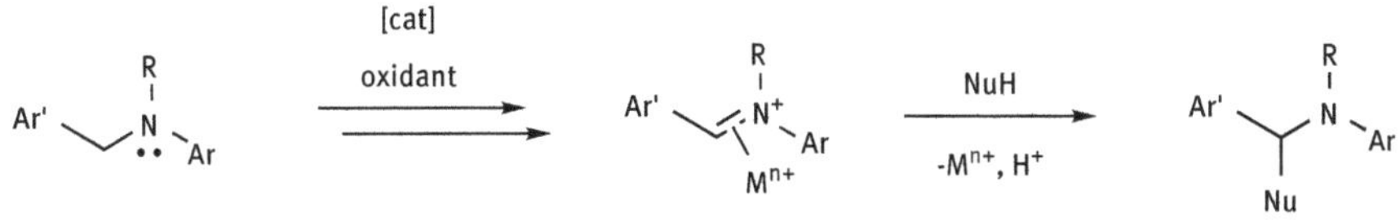

Figure 10.1: CDC reaction of tertiary amines with nucleophiles.

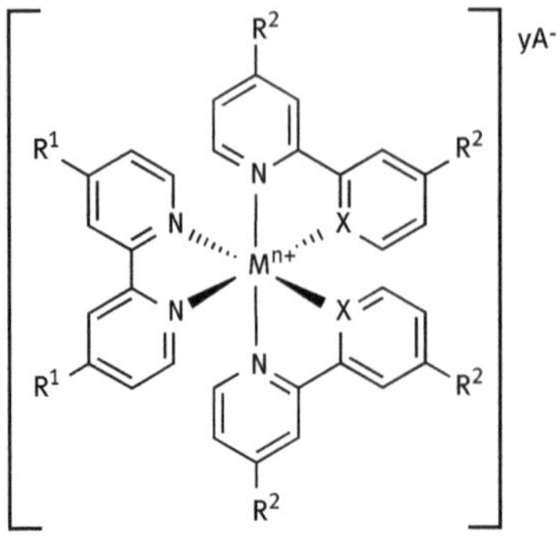

X = N: $M^{n+} = Ru^{2+}, Ir^{3+}$, y = n

X = C: $M^{n+} = Ir^{3+}$, y = n-2

Figure 10.2: Ru and Ir complexes applied in photoredox catalysis.

transformation and the formation of the intermediate iminium ion can be adapted to the subsequent trapping reaction resulting in less side product formation.

In addition to homogeneous transition metal complexes, organic dyes, such as Eosin Y and Rose Bengal (Figure 10.3), proved to be efficient catalysts for various visible light mediated cross-dehydrogenative transformations. [8] Organic dyes are of particular importance, as they are inexpensive (Eosin Y is about 50 times less expensive per gram than $Ru(bpy)_3Cl_2$; bpy = 2,2′-bipyridine) and commercially available in decent quantities. This is of particular importance for the synthetic applicability of photoredox catalysts, as they typically also possess relatively high molecular weights. For instance, the typically applied $[Ru(dtb\text{-}bpy)_2bpy](PF_6)_2$ (dtb-bpy = 4,4′-di-*tert*-butyl-2,2′-bipyridine) has a molecular weight of 1202 g/mol. Even though catalyst loadings of typically 1 mol% can be realized, this results in a significant amount of expensive catalyst, while the Eosin Y (648 g/mol) is only about a half of the molecular weight of a typical Ru(II) based catalyst. Thus, the development of organic dye catalyzed cross-dehydrogenative transformations is of considerable interest.

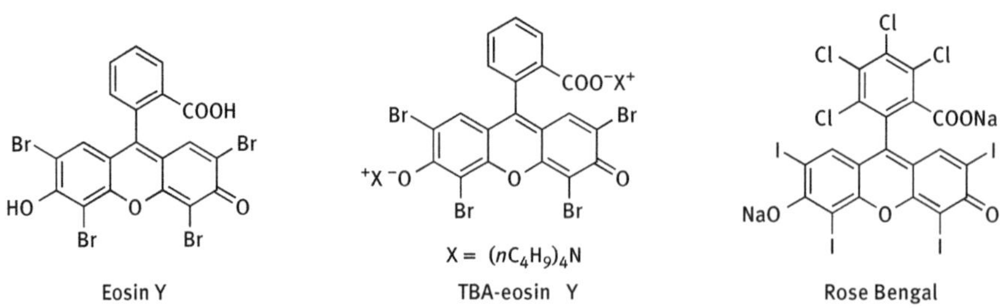

Figure 10.3: Organic dyes applied as photoredox catalysts.

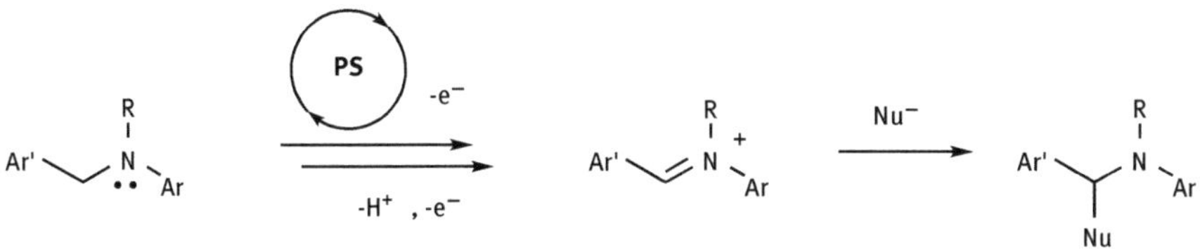

Figure 10.4: Photoredox catalyzed functionalization of tertiary amines.

Apart from homogeneous catalysts for visible light mediated cross-dehydrogenative coupling reactions, solid materials attracted the interest of organic chemists. Typically, semi-conducting materials are applied, [9] such as titanium dioxide, other transition metal oxides, cadmium, and zinc sulfide but also carbon nitride.

Before going into a detailed discussion on visible light photoredox catalyzed cross-dehydrogenative coupling reactions, the putative reaction mechanism will be briefly discussed. In a nutshell, the tertiary amine is first oxidized by the photoredox catalyst (photosensitizer, PS) and forms, after subsequent proton and electron transfer, the intermediate iminium ion which can then be trapped by a series of different nucleophiles (Figure 10.4).

The photochemical oxidation of tertiary amines dates back to the 1970s, when Whitten and co-workers examined the photophysical properties of Ru(II) polybipyridyl complexes and the quenching of the excited species by either reductive quenching with tertiary amines [10] or oxidative quenching with viologens (e. g. paraquat). [11] The putative reaction mechanism of photoredox catalyzed CDC transformations is depicted in Figure 10.5.

Irradiation with visible light generates the excited state of a photosensitizer PS*, which can then undergo reductive quenching through single electron transfer from a tertiary amine **A** [12] to the photoexcited state. The reduced photosensitzer PS can then be reoxidized by molecular oxygen to the ground state PS of the photocatalyst

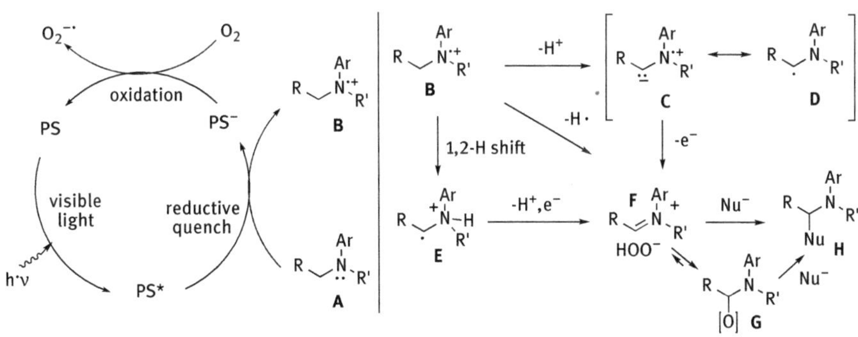

Figure 10.5: Photoredox oxidative functionalization of tertiary amines.

under formation of a superoxide anion. More importantly, the tertiary amine **A** is oxidized to the amine radical **B** in this process. The amine radical **B** can now undergo different pathways that finally result in the formation of the iminium ion **F**. Mechanistically three different pathways are possible. First, a hydrogen atom abstraction, e.g. by the superoxide anion, from the neighbouring carbon centre is possible, which would result in the direct formation of the iminium ion **F**. Second, after deprotonation an intermediate zwitterion **C** is formed, which can be also represented by a carbon centred radical **D**, which, after a second single electron transfer, can undergo oxidation to the iminium ion **F**. Last, a 1,2-H shift would result in the formation of the carbon centred radical **E** bearing an adjacent quaternary ammonium ion. This carbon centred radical can now undergo deprotonation and single electron oxidation, which ultimately also results in the formation of the iminium ion **F**, which can react to the hemiaminal **G**. In the final reaction step iminium ion **F** is intercepted by a range of different nucleophiles to provide the cross-dehydrogenative coupling products **H**.

The intermediate iminium ion represents a highly attractive species for further transformations and, not surprisingly, different groups focused on performing a second catalytic transformation, in which the nucleophile needs activation by a metal/organocatalyst prior to the addition to the iminium ion (Figure 10.6). Nicewicz and MacMillan performed seminal work on the application of combined photoredox catalysis and organocatalysis in 2008. [13] Yet, in the context of CDC, Rueping and co-workers were able to demonstrate for the first time a dual catalytic concept for the visible light mediated Mannich reaction [14] and alkynylations. [15] To perform such dual catalytic transformations, the adaptation of both catalytic cycles is crucial as decomposition/inactivation of the second catalyst by the photoredox catalyst through redox chemistry is possible.

Cross-dehydrogenative couplings have found widespread application in copper, iron, ruthenium and other transition metal catalyzed transformations. [1] Typically, they are performed at elevated reaction temperature and stoichiometric amounts of

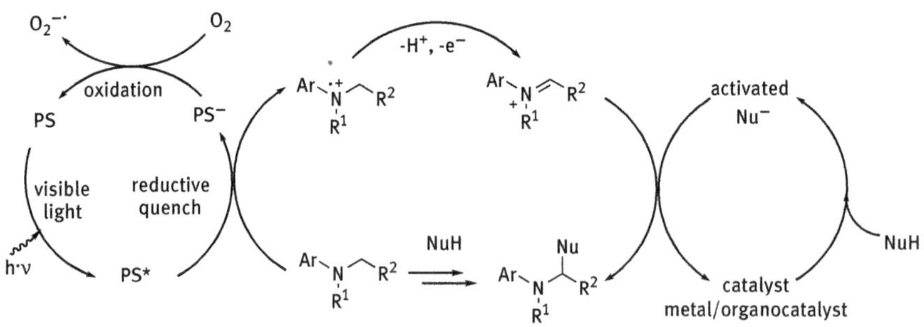

Figure 10.6: Oxidative functionalization of tertiary amines – dual catalytic systems.

oxidants, e. g. *t*BuOOH or TEMPO are applied, which both clearly outline shortcomings of these methods. In the following section, we will concentrate on the development of visible light mediated photoredox catalyzed cross-dehydrogenative coupling reactions.

10.2 Aza-Henry Reaction

In 2010, Stephenson and co-workers reported a visible light mediated photoredox catalyzed aza-Henry reaction (Figure 10.7). [6] They were the first to demonstrate that $[Ir(ppy)_2(dtbbpy)]PF_6$ (ppy = 2-phenyl-pyridine, dtbbpy = 4,4′-di-*tert*-butyl-2,2′-bipyridine) can be used as a photoredox catalyst for the efficient aerobic oxidation of *N*-aryl-tetrahydroisoquinoline derivatives **1** followed by a subsequent aza-Henry reaction yielding the desired reaction products **3** in very high chemical yield. Remarkably, the reaction can be performed at room temperature using a standard household fluorescent bulb as the source of energy with only 1 mol% of catalyst loading and oxygen from air as the terminal oxidant. This seminal publication set the basis for the future developments in this field.

The Tan, König, and Wu groups subsequently examined this transformation using organic dyes as photosensitizers. [16–18] Driven by the search for metal-free and a more economical way for photoredox catalysis Eosin Y, or Tetrabutylammonium (TBA) Eosin Y, respectively, and Rose Bengal were identified as promising candidates. Again, oxygen or air can be used as the terminal oxidant. Green LEDs were used as the source of light in the case of Rose Bengal and Eosin Y while Wu and co-

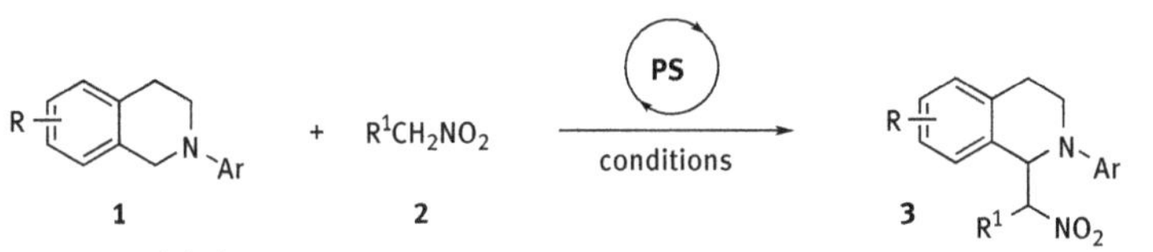

Homogeneous catalysis

Stephenson	König	Tan	Wu
$[Ir(ppy)_2(dtbbpy)]PF_6$ (1 mol %) visible light, rt, air 8 examples (90-96%)	Eosin Y (2 mol %) Green LEDs, rt, air 6 examples (66-80%)	Rose Bengal (5 mol%) Green LEDs, rt, air 5 examples (85-95%)	TBA-eosin Y (2 mol%) hν (>450 nm), rt, O_2 15 examples (46-85%)

Heterogeneous catalysis

Lin	Rueping	König	Blechert
visible light, rt, air and Ir-PCP (1 mol %) 8 examples (75-98%) or Ru-PCP (2 mol%) 8 examples (84-99%)	TiO_2 (1 equiv.) EtOH, 11W lamp, rt, air 9 examples, (60-96%)	CdS (1 equiv.) hν (440 nm), rt 3 examples (85-97%)	mpg-C_3N_4 (15 mg/0.25mmol THIQ) MeCN, 60W bulb, rt, O_2 12 examples, (81-92%)

Figure 10.7: Visible light-mediated aza-Henry reaction.

workers applied a 500 W high-pressure mercury lamp in the TBA-Eosin Y catalyzed oxidative aza-Henry reaction.

Apart from typical homogeneous photosensitizers, the groups of Lin, Rueping, König, and Blechert and Wang developed protocols to perform visible light mediated cross-dehydrogenative couplings in the presence of heterogeneous photocatalysts. [19–22]

The first to report on heterogeneous catalysts for the photochemical oxidative aza-Henry reaction were Lin and co-workers. [19] They developed porous-cross linked polymers bearing a bipyridine motif for binding of Ru(II) or Ir(III) based photoredox catalyst. The group of Blechert, Wang, and co-workers applied mesoporous-graphitic carbon nitride (mpg-C_3N_4) [22] while the groups of König [21] and Rueping [20] applied semi-conducting materials such as CdS or ZnO and TiO_2, in this visible light mediated transformation. Additionally, Rueping and co-workers demonstrated the recycling potential of the heterogeneous catalyst, which could be reused for at least five consecutive times. The advantage of these protocols is clearly the possibility of operationally facile catalyst recycling/reuse after simple filtration or centrifugation. Moreover, titanium dioxide or cadmium sulfide represent two of the economically most attractive catalysts, having low price and low molecular weight at the same time (TiO_2: 80 g/mol); the latter resulting in reasonable amounts of catalyst needed per reaction, although up to 1 equiv of catalyst is used.

The seminal publication by Stephenson and co-workers relies on the reductive quenching of the excited state of the photocatalyst. [6] The reduced photoredox catalyst is then reoxidized in the presence of oxygen to regenerate the ground state. In 2012, Stephenson and co-workers reported on the application of an alternative route for the formation of the iminium ion, which is crucial for CDC. [23] Bromotrichloromethane acts as an stoechiometric oxidant, which finally leads to the formation of the iminium ion **4** (Figure 10.8). After subsequent addition of the nucleophile, the desired aza-Henry product **3** can be obtained in a dark reaction with excellent chemical yield without the need for further irradiation. Advantageous to this one-pot protocol is the controlled formation of the intermediate iminium ion **4**. Moreover, the reaction product cannot be oxidized by the photoredox catalyst, which would lead to undesired side products. However, this protocol needs strict exclusion of oxygen and the reaction solution needs to be treated by three consecutive freeze-pump-thaw cycles to remove traces of oxygen, which is a limitation of this protocol.

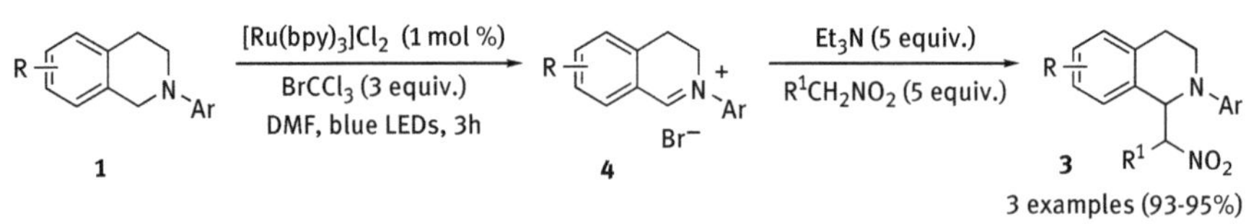

Figure 10.8: Two-step one-pot procedure for the aza-Henry reaction.

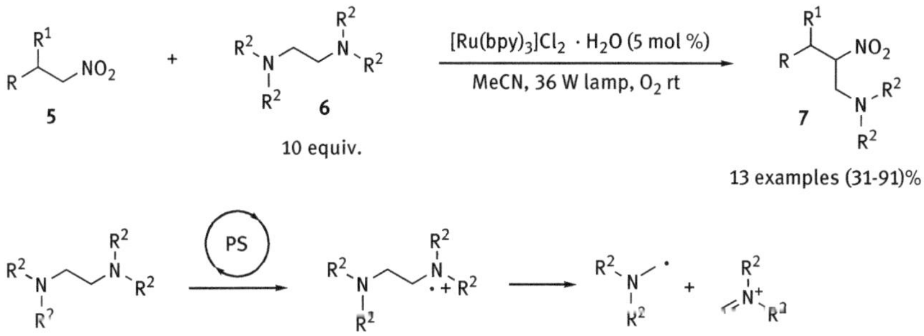

Figure 10.9: Photoredox catalyzed aza-Henry reaction with vicinal diamines.

Besides the oxidative CDC of *N*-aryl-tetrahydroisoquinoline derivatives, several groups concentrated on the oxidative aza-Henry reaction of aliphatic acyclic tertiary amines. In their publication from 2010, Stephenson and co-workers reported on the oxidative aza-Henry reaction of *N*-phenyl-pyrrolidine and nitromethane, [6] yet the desired CDC product could only be isolated with 27 % yield; clearly showing up the limitations of the photochemical CDC. However, this clearly set up the stage for further improvement of this transformation, which further flourished by the application of organic dyes as shown by the Tan [16] and König [17] groups. Tan and co-workers were able to obtain the CDC product of *N*,*N*-dimethyl-aniline and nitromethane with a moderate 48 % chemical yield.

In 2012 Wang, Li, and co-workers advanced the application of aliphatic amines (Figure 10.9). [24] 1,2-Bis-(sec-amino)-ethane undergoes, in the presence of visible light, an oxidative C–C bond cleavage resulting in a highly reactive iminium ion, which can be trapped with a series of β-aryl-nitro-ethanes or long-chain aliphatic nitroalkanes.

Furthermore, they were able to demonstrate, that *O*-protected 2-dimethyl-amino-ethanol derivatives can be selectively functionalized at one of both methyl groups by means of photoredox catalysis.

As photoredox catalysis relies on the use of visible light, Lambert–Beer's law has direct implications on the catalytic efficiency during scale-up processes. The microreactor technology provides a well-established platform to overcome this problem. [25] The high surface-to-volume ratio ensures efficient irradiation of the reaction solution and would thus allow the gram-scale synthesis of precious building blocks by photoredox chemistry.

Not surprisingly, the groups of Stephenson, [26] Zeitler [27] and others [28] concentrated on the development of efficient continuous-flow visible light mediated photochemical transformations.

As already outlined before, Stephenson and co-workers had been able to successfully demonstrate the selective formation of the valuable iminium ion intermediate using bromo-trichloromethane as terminal oxidant. [23] They were able to demon-

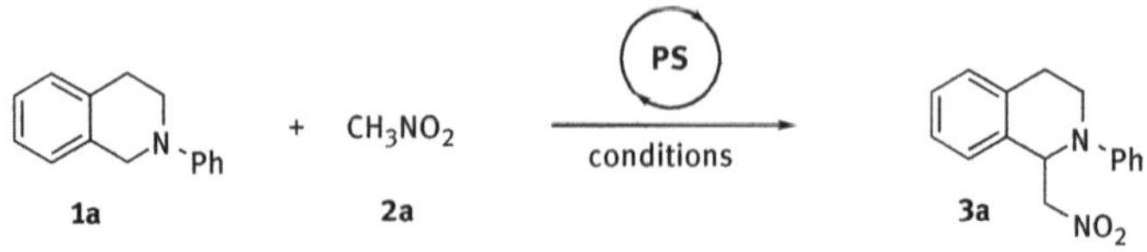

Batch[6]: **1a** (0.25 mmol), $[Ir(ppy)_2(dtbbpy)]PF_6$ (1 mol %), visible light, rt, air, 10h, 92%

Batch[22]: 1. **1a** (0.244 mmol), $[Ru(bpy)_3]Cl_2$ (1 mol %), $BrCCl_3$ (3 equiv.),DMF, blue LEDs, 3h
2. no light, Et_3N (5 equiv.), CH_3NO_2 (5 equiv.), 12h, 95%

Flow reactor[26]: 1. **1a** (4.7 mmol), $[Ru(bpy)_3]Cl_2$ (0.5 mol %), $BrCCl_3$ (3 equiv.), DMF, blue LEDs, t_R = 0.5min
2. no light, Et_3N (5.1 equiv.), CH_3NO_2 (5.1 equiv.), 1h, 89%

Flow reactor[27]: **1a** (0.98 mmol), $[Ir(ppy)_2(dtbbpy)]PF_6$ (1 mol %), visible light, rt, air, t_R = 30min, 93%

Figure 10.10: Comparison between photoredox catalyzed aza-Henry reaction in batch and flow modes.

strate that this protocol is not only applicable to batch reactions, but that it can also be run in a continuous-flow microreactor (Figure 10.10). [26] By generating the iminium ion in a microreactor, and adding this mixture to a flask containing the required nucleophile, the continuous-flow CDC reaction can be realized on a gram-scale. At the same time, Zeitler and co-workers reported on the application of $[Ir(ppy)_2bpy]PF_6$ in the continuous-flow photocatalyzed aza-Henry reaction using the standard reaction protocol, in which the reductive quenching of the photoexcited state by the tertiary amine is the key (Figure 10.10). [27] Comparable yields to the protocol of Stephenson and co-workers were obtained. Most interestingly, substrates, that proved poorly reactive for batch CDC reactions, could be efficiently subjected to CDC under continuous-flow conditions. In this manner CDC products of nitromethane with *N*-phenyl-pyrrolidin and *N,N*-dimethyl-*p*-methyl-anilin, respectively, could be isolated with good yields of 59 and 77 %, respectively. This nicely illustrates the potential of continuous-flow microreactors in photocatalyzed transformations.

10.3 Addition of malonates

After the work by Stephenson and co-workers, the area of visible light mediated photoredox catalyzed CDC reactions rapidly developed and a range of different nucleophiles have been investigated. The groups of König [17] and Wu [18] investigated the addition of malonates using organic dyes. Under otherwise the same reaction conditions as in the previously mentioned case, the desired CDC products **9** of *N*-aryl-tetrahydroisoquinolines **1** and malonates **8** could be obtained in moderate to very good yields using molecular oxygen as terminal oxidant (Figure 10.11). Blechert, Wang and co-workers applied mpg-C_3N_4 as heterogeneous recyclable catalyst. [22a] Remarkably, the reaction can be performed both neat and in solution. Stephenson and co-workers realized this transformation by preformation of the iminium ion with bromo-trichloromethane and subsequent dark reaction with malonates. [23]

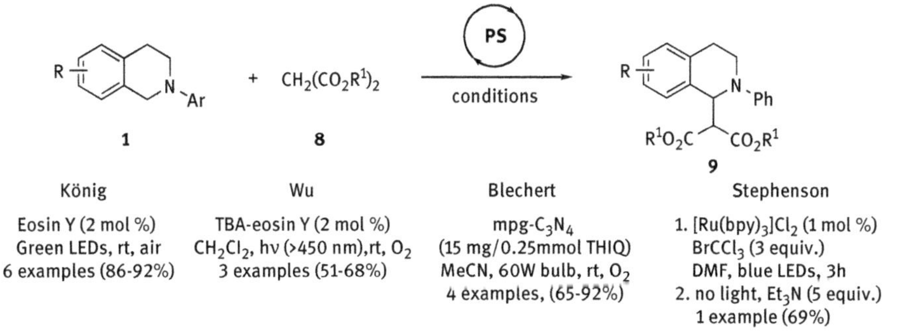

Figure 10.11: Addition of malonates to photogenerated iminium ions.

10.4 Mannich reaction

In early 2011 Rueping and co-workers reported in their seminal publication on the successful merger of organocatalysis and photoredox catalysis for the CDC reaction of *N*-aryl-tetrahydroisoquinolines **1** and linear aliphatic methyl-ketones **10** (Figure 10.12). [14] L-proline was found to be a good catalyst for an efficient ketone addition to intermediate iminium ions. As outlined in the introduction to this chapter, the adaptation of both catalytic cycles is crucial. This was achieved by adjusting the reaction rate of the photoredox catalytic cycle to the organocatalytic transformation by variation of the light source. Later, the groups of Wu [18] and Tan [16] subsequently investigated this transformation using organic dyes (Rose Bengal and

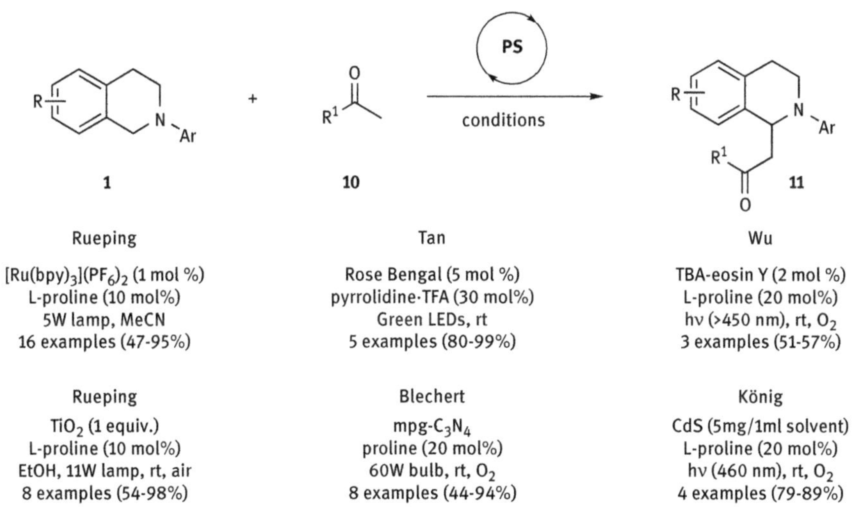

Figure 10.12: Dual catalytic systems for the Mannich reaction.

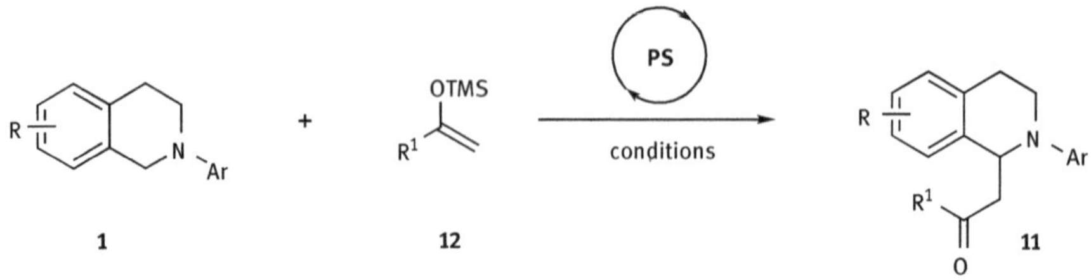

R^1 = alkyl: Stephenson

1. $[Ru(bpy)_3]Cl_2$ (1 mol %)
   $BrCCl_3$ (3 equiv.)
   DMF, blue LEDs, 3h
2. no light, Et_3N (5 equiv.)
   2 examples (59-65%)

R^1 = aryl: Xia

$[Ru(bpy)_3]Cl_2$ (5 mol %)
CH_3OH, blue LEDs, rt
15 examples (72-98%)

Figure 10.13: Photoredox catalyzed addition of silyl-enol-ethers to iminium ions.

TBA-eosin Y). In addition, the groups of Blechert and Wang [22] as well as König [29] and Rueping, [20] investigated the heterogeneous photoredox catalysis with mpg-carbon nitride, cadmium sulfide and titanium dioxide, respectively (Figure 10.12). In the case of Rose Bengal and CdS as photosensitizer, cyclohexanone can be used as nucleophile. A direct approach to the Mannich products can be alternatively achieved by the addition of silyl-enol-ethers **12**. In this context Stephenson demonstrated the addition of the silyl-enol-ether derived from acetone and 2-silyloxy-tetrahydrofurane (Figure 10.13). [23] Moreover, Xia and co-workers were able to show that silyl-enol-ethers derived from a range of different acetophenons can be subjected to visible-light mediated CDC transformations (Figure 10.13). [30]

10.5 Allylation

The controlled formation of the intermediate iminium ion can also be applied to subsequently perform efficient allylation reactions. Stephenson and co-workers reported on the addition of (2-methyl-allyl)-trimethylsilane to photochemically generated iminium ions. [23]

10.6 Cyanation of tertiary amines

The Strecker reaction is one of the best-studied transformations to yield α-amino acids. [31] Valuable α-amino nitriles can be obtained through the addition of cyanide to an iminium ion. Not surprisingly, different groups reported their work on the addition of cyanide to an iminium ion, generated through photoredox catalysis. Different sources of cyanide can be applied, e. g. KCN in the presence of AcOH, or

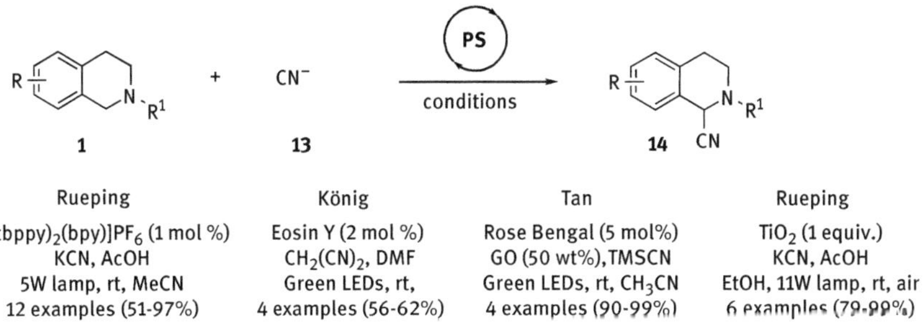

Figure 10.14: α-Cyanation of tetrahydroisoquinolines.

TMSCN and malodinitrile, which both liberate HCN in the presence of water (Figure 10.14). Rueping and co-workers reported their findings on the photoredox catalyzed Strecker reaction using homogeneous Ir(III) complexes as well as heterogeneous titanium dioxide as catalyst. [32] [20] Moreover, they were able to demonstrate, that under the cyanation conditions, *N,N*-dialkyl-anilines can be subjected to photoredox catalyzed CDC reactions. The groups of König [17] and Tan [33] reported on the application of organic dyes for this transformation.

10.7 Alkynylation

The intermediate iminium ion can alternatively be trapped in metal-catalyzed transformations. The groups of Rueping and Stephenson developed protocols to realize the combination of photoredox- and metal catalysis (Figure 10.15). Stephenson and co-workers relied on the initial formation of the iminium ion by employing bromotrichloromethane as the terminal oxidant. [23] In a subsequent dark reaction, the

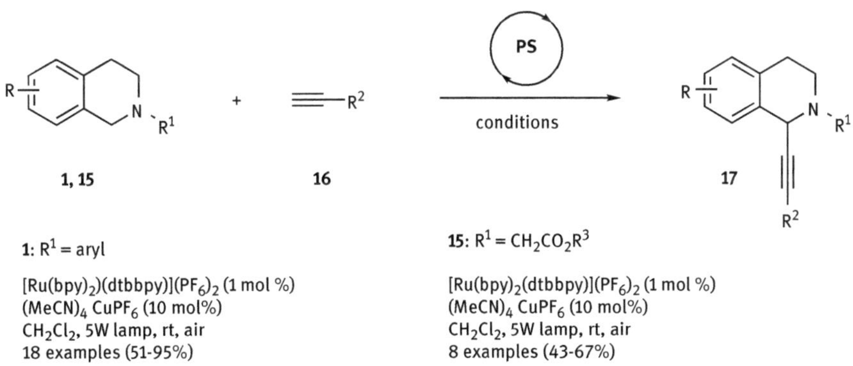

Figure 10.15: Dual catalysis – combining photoredox and metal catalysis for the alkynylation of THIQ.

copper catalyst and alkyne are added and the propargylic amines can be obtained in a one-pot procedure. Rueping and co-workers relied on a cascade reaction sequence. [15] The propargylic amines can be obtained via the reductive quenching pathway and, through careful optimization of reaction parameters, the interplay of both catalysts can be achieved. In addition, the authors were able to demonstrate that also *N*-glycinyl-ester derivates **15** can be subjected to the alkynylation reaction.

10.8 [3+2] cycloaddition reaction

The application of *N*-glycinyl esters or derivatives thereof, opens up the field of visible light mediated cross-dehydrogenative couplings toward annulation reactions. Oxidation of the tetrahydroisoquinoline core furnishes an iminium ion. In case of *N*-glycinyl-esters, the adjacent C–H bond can be easily deprotonated, resulting in the formation of an azomethine ylide, which is an ideal precursor for [3+2] cycloaddition reactions. [34] The groups of Rueping and Xiao reported on their findings on photochemically triggered [3+2] cycloaddition reactions, allowing the construction of tri- and tetracyclic systems (Figure 10.16). [35–36]

10.9 Acylation

In 2012, DiRocco and Rovis reported on an interesting application combining both photoredox and NHC catalysis (Figure 10.17). [37] Interestingly, the authors reported on the beneficial influence of *m*-dinitrobenzene, an oxidant that was found to be crucial for catalytic turnover, although the reaction was performed without exclusion of oxygen, which is typically sufficient for good turnover. Photochemical oxidation of *N*-aryl-tetrahydroisoquinolines **1** furnishes an iminium ion, which can be acylated with a range of different aliphatic (mostly linear) aldehydes **23** after the Umpolung reaction with the chiral NHC **25** to give products in moderate to high chemical yield and excellent enantioselectivity.

10.10 C-heteroatom (C–P, C–O, C–N) bond formation

The CDC protocols furthermore enable the efficient synthesis of C–P bonds (Figure 10.18). Rueping and co-workers demonstrated that photochemical oxidation of *N*-aryl-tetrahydroisoquinolines with $[Ir(ppy)_2bpy]PF_6$ furnishes iminium ions that can be readily reacted with a range of different phosphonates. [38] At the same time, König and co-workers reported on the application of Eosin Y in the oxidative phosphonylation reaction. [17] Furthermore, it was shown that ZnO can be used as an efficient heterogeneous photoredox catalyst. [20]

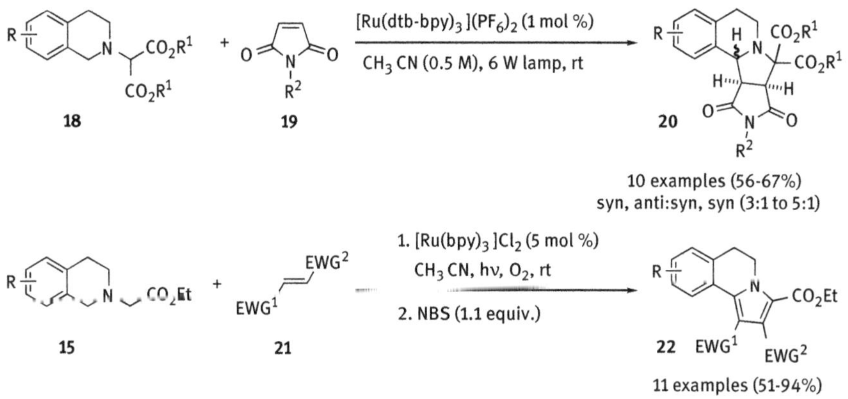

Figure 10.16: Photoredox catalyzed cycloaddition reactions.

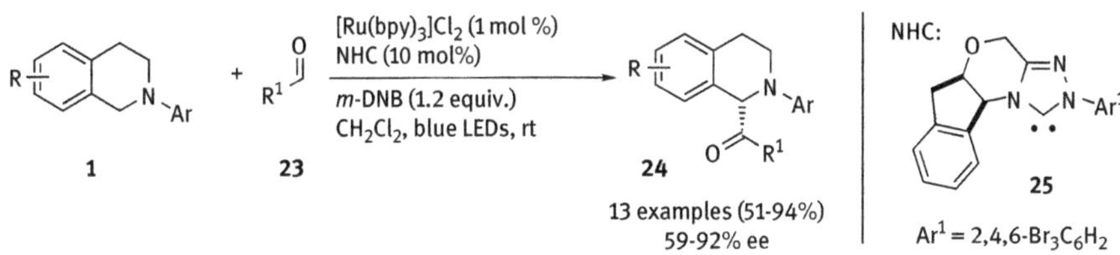

Figure 10.17: Dual catalytic system – asymmetric α-acylation of tertiary amines.

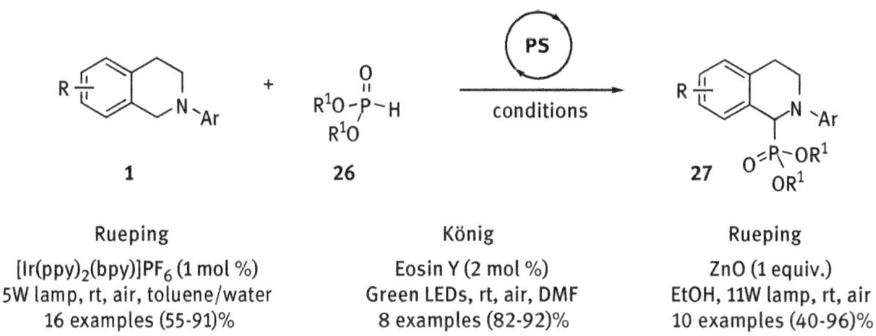

Figure 10.18: Visible light mediated oxidative phosphonylations of amines.

Xiao and co-workers recently demonstrated the efficient application of different oxygen and nitrogen-centred nucleophiles. Yet, protocols that have been developed so far only concentrate on intramolecular cyclization reactions. In a first report, Xiao and co-workers described an interesting functionalization of *N*-benzyl-*N'*-tosyl-ethylene-diamines (Figure 10.19). [39] Photochemical oxidation of the benzylamine furnishes the reactive intermediate iminium ion that is trapped in a diastereoselective cyclization reaction by the second nitrogen atom. This protocol allows the formation of valuable tetrahydroimidazoles **29**.

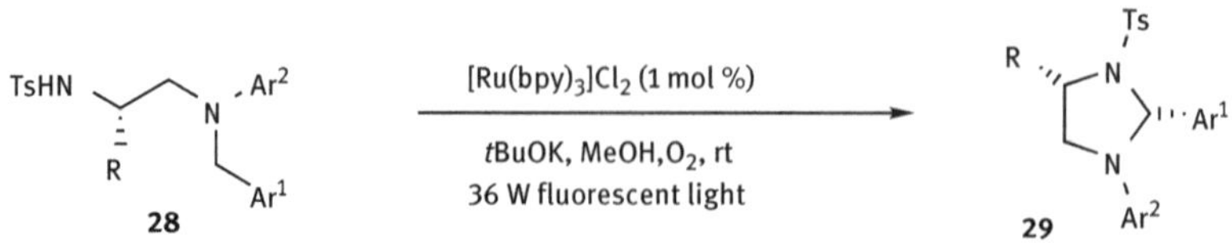

R = Bn: 9 examples (90-94%)
dr 8:1 to 19:1
$Ar^1 = Ar^2$ = Ph: 6 examples (45-94%)
dr 2:1 to 19:1

Figure 10.19: Photoredox catalyzed synthesis of tetrahydroimidazoles.

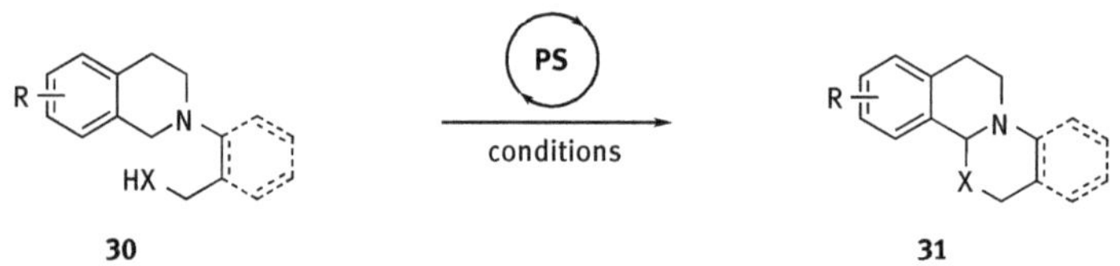

X = O	X = NTs	
$[Ir(dtb\text{-}bpy)(ppy)_2]PF_6$ (0.5 mol %)	$[Ir(dtb\text{-}bpy)(ppy)_2]PF_6$ (0.5 mol %)	Eosin Y Na salt (0.5 mol %)
MeOH, 36W fluorescent lamp, air, rt	$MeOH/CH_2Cl_2$, *t*BuOH (2 equiv.)	$MeOH/CH_2Cl_2$, *t*BuOH (2 equiv.)
9 examples (26-72%)	36W fluorescent lamp,air,rt	36W fluorescent lamp, air, rt
	4 examples (70-86%)	1 example (85%)

Figure 10.20: Visible light photoredox catalyzed intramolecular additions.

Later, Xiao and co-workers reported on the cyclization reaction of tetrahydroisoquinoline derivatives **30** bearing a proximal alcohol or sulfonated amine functional group, which can intramolecularly add to the photochemically generated iminium ions (Figure 10.20). [40] Furthermore, the authors were able to demonstrate that this transformation can be conducted under metal free conditions by using Eosin Y as well as on gram-scale with an iridium complex as photosensitizer.

10.11 Conclusion

In summary, visible light photoredox catalyzed cross-dehydrogenative coupling reactions have been extensively investigated over the past years, showing significant improvements in both the application of visible light in organic synthesis and the development of sustainable methods for the coupling of non-prefunctionalized building blocks. In this chapter a range of different methods for the functionalization of tertiary amines have been presented; yet, this strategy will continue growing to further exploit the benefits of CDC coupling reactions under photochemical reaction conditions.

10.12 References

[1] Reviews on CDC-reactions: a) Li, Z., Bohle, D. S., Li, C.-J. (2006). Proc. Natl. Acad. Sci. U.S.A., 103, 8928; b) Yoo, W.-J., Li, C.-J. (2010). Top. Curr. Chem., 292, 281; c) So, M.-H., Liu, Y., Ho, C.-M., Che, C.-M. (2009). Chem. Asian J., 4, 1551; d) Li, C.-J. (2009). Acc. Chem. Res., 42, 335; e) Yeung, C. S., Dong, V. M. (2011). Chem. Rev., 111, 1215; f) Sun, C.-L., Li, B.-J., Shi, Z.-J. (2011). Chem. Rev., 111, 1293; g) Liu, C., Zhang, H., Shi, W., Lei, A. (2011). Chem. Rev., 111, 1780.

[2] For Li's pioneering work, see: a) Li, Z., Li, C.-J. (2004). Org. Lett., 6, 4997; b) Li, Z., Li, C.-J. (2004). J. Am. Chem. Soc., 126, 11810; c) Li, Z., Li, C.-J. (2005). J. Am. Chem. Soc., 127, 3672; d) Li, Z., Li, C.-J. (2005). Eur. J. Org. Chem., 3173. e) Li, Z., Li, C.-J. (2005). J. Am. Chem. Soc., 127, 6968; f) Li, Z., Li, C.-J. (2006). J. Am. Chem. Soc., 128, 56.

[3] General reviews on photoredox catalyzed reactions: a) Zeitler, K. (2009). Angew. Chem. Int. Ed., 48, 9785; b) Yoon, T. P., Ischay, M. A., Du, J. N. (2010). Nat. Chem., 2, 527; c) Narayanam, J. M. R., Stephenson, C. R. J. (2011). Chem. Soc. Rev., 40, 102; d) Telpy, F. (2011). Collect. Czech. Chem. Commun., 76, 859; e) Xuan, J., Xiao, W.-J. (2012). Angew. Chem. Int. Ed., 51, 6828.

[4] Reviews on the photoredox-functionalization of amines: a) Maity, S., Theng, N. (2012). Synlett, 23, 1851; b) Shi, L., Xia, W. (2012). Chem. Soc. Rev., 41, 7687.

[5] For selected pioneering work in the field of photocatalyzed generation of iminium ions, see: a) Pandey, G., Kumaraswamy, G. (1988). Tetrahedron Lett., 29, 4153; b) Pandey, G., Kumaraswamy, G., Yella Reddy, P. (1992). Tetrahedron, 8, 8295; c) Pandey, G., Sudha Rani, K., Lakshamaiah, G. (1992). Tetrahedron Lett., 33, 5107; d) Pandey, G., Gadre, S. R. (2003). Arkivoc, 45.

[6] Condie, A. G., Gonzalez-Gomez, J. C., Stephenson, C. R. J. (2010). J. Am. Chem. Soc., 132, 1464.

[7] a) Balzani, V., Bergamini, G., Campagna, S., Puntoriero, F. (2007). in Top. Curr. Chem., 280, 1; b) Campagna, S., Puntoriero, F., Nastasi, F., Bergamini, G., Balzani, V. (2007). in Top. Curr. Chem., 280, 117; c) Flamigni, L., Barbieri, A., Sabatini, C., Ventura, B., Barigelletti, F. (2007). Top. Curr. Chem., 281, 143; d) Tucker, J. W., Stephenson, C. R. J. (2012). J. Org. Chem., 77, 1617.

[8] Highlight: a) Ravelli, D., Fagnoni, M. (2012). ChemCatChem, 4, 169; Review on photoorganocatalysis: b) Ravelli, D., Fagnoni, M., Albini, A. (2013). Chem. Soc. Rev., 42, 97.

[9] a) Fox, M. A., Dulay, M. T. (1993). Chem. Rev., 93, 341; b) Serpone, N., Emeline, A. V. (2005). Res. Chem. Intermed., 31, 391.

[10] a) DeLaive, P. J., Lee, J. T., Sprintschnik, H. W., Abruna, H., Meyer, T. J., Whitten, D. G. (1977). J. Am. Chem. Soc., 99, 7094; b) Delaive, P. J., Sullivan, B. P., Meyer, T. J., Whitten, D. G. (1979). J. Am. Chem. Soc., 101, 4007; c) Delaive, P. J., Foreman, T. K., Giannotti, C., Whitten, D. G. (1980). J. Am. Chem. Soc., 102, 5627.

[11] Bock, C. R., Meyer, T. J., Whitten, D. G. (1974). J. Am. Chem. Soc.,, 96, 4710.

[12] For general reviews on the photoinduced electron transfer reactions of amines, see: a) Pandey, G. (1992). Synlett, 546; b) Yoon, U. C., Mariano, P. S., Givens, R. S., Atwater, B. W. (1994). in Advances in Electron Transfer Chemistry, Vol. 4 (ed. P. S. Mariano), JAI-Press: Greenwich, CT, 117–205; c) Kumar, J. S. D., Das, S. (1997). Res. Chem. Intermed., 23, 755; d) Das, S., Suresh, V. (2001). in Electron Transfer in Chemistry Vol. 2, (ed. V. Balzani), Wiley-VCH New York, 379–456.

[13] Nicewicz, D. A., MacMillan, D. W. (2008). Science, 322, 77.

[14] Rueping, M., Vila, C., Koenigs, R. M., Poscharny, K., Fabry, D. C. (2011). Chem. Commun., 47, 2360.

[15] Rueping, M., Koenigs, R. M., Poscharny, K., Fabry, D. C., Leonori, D., Vila, C. (2012). Chem.-Eur. J., 18, 5170.

[16] Pan, Y., Kee, C. W., Chen, L., Tan, C.-H. (2011). Green Chem., 13, 2682.

[17] Hari, D. P., König, B. (2011). Org. Lett., 13, 3852.

[18] Liu, Q., Li, Y.-N., Zhang, H.-H., Chen, B., Tung, C.-H., Wu, L.-Z. (2012). Chem.-Eur. J., 18, 620.

[19] Xie, Z., Wang, C., deKrafft, K. E., Lin, W. (2011). J. Am. Chem. Soc., 133, 2056.

[20] Rueping, M., Zoller, J., Fabry, D. C., Poscharny, K., Koenigs, R. M., Weirich, T. E., Mayer, J. (2012). Chem.-Eur. J., 18, 3478.
[21] Mitkina, T., Stanglmair, C., Setzer, W., Gruber, M., Kisch, H., König, B. (2012). Org. Biomol. Chem., 10, 3556.
[22] a) Möhlmann, L., Baar, M., Rieß, J., Antonietti, M., Wang, X., Blechert, S. (2012). Adv. Synth. Catal., 354, 1909; b) Wang, X., Blechert, S., Antonietti, M. (2012). ACS Catalysis, 2, 1596.
[23] Freeman, D. B., Furst, L., Condie, A. G., Stephenson, C. R. J. (2012). Org. Lett., 14, 94.
[24] Cai, S., Zhao, X., Wang, X., Liu, Q., Li, Z., Wang, D. Z. (2012). Angew. Chem. Int. Ed., 51, 8050.
[25] a) Ehrfeld, W., Hessel, V., Löwe, H. (2000). Microreactors: new technology for modern chemistry, 1st edn, Wiley-VCH, Weinheim; New York,; b) Wirth, T. (2008). Microreactors in organic synthesis and catalysis, Wiley-VCH, Weinheim.
[26] Tucker, J. W., Zhang, Y., Jamison, T. F., Stephenson, C. R. J. (2012). Angew. Chem. Int. Ed., 51, 4144.
[27] Neumann, M., Zeitler, C. (2012). Org. Lett., 14, 2658.
[28] a) Bou-Hamdan, F. R., Seeberger, P. H. (2012). Chem. Sci., 3, 1612; b) Andrews, R. S., Becker, J. J., Gagné, M. R., Angew. Chem. (2012). Int. Ed., 51, 4140; c) Aida, S., Terao, K., Nishiyama, Y., Kakiuchi, K., Oelgemöller, M. (2012). Tetrahedron Lett., 53, 5578.
[29] Cherevatskaya, M., Neumann, M., Füldner, S., Harlander, C., Kümmel, S., Dankesreiter, S., Pfitzner, A., Zeitler, K., König, B. (2012). Angew. Chem. Int. Ed., 51, 4062.
[30] Zhao, G., Yang, C., Guo, L., Sun, H., Chen, C., Xia, W. (2012). Chem. Commun., 48, 2337.
[31] For the first synthesis of α-aminonitriles through Strecker reaction: a) Strecker, A. (1850). Liebigs. Ann. Chem., 75, 27; For the application of α-aminonitriles in synthesis: b) Enders, D., Shilvock, J. P. (2000). Chem. Soc. Rev., 29, 359.
[32] Rueping, M., Zhu, S., Koenigs, R. M. (2011). Chem. Commun., 47, 12709.
[33] Pan, Y., Wang, S., Kee, C. W., Dubuisson, E., Yang, Y., Loh, K. P., Tan, C.-H. (2011). Green Chem., 13, 3341.
[34] For a recent review on [3+2] dipolar cycloadditions, see: Kissane, M., Maguire, A. R. (2010). Chem. Soc. Rev., 39, 845.
[35] Rueping, M., Leonori, D., Poisson, T. (2011). Chem. Commun., 47, 9615.
[36] Zou, Y.-Q., Lu, L.-Q., Fu, L., Chang, N.-J., Rong, J., Chen, J.-R., Xiao, W.-J. (2011). Angew. Chem. Int. Ed., 50, 7171.
[37] DiRocco, D. A., Rovis, T. (2012). J. Am. Chem. Soc., 134, 8094.
[38] Rueping, M., Zhu, S., Koenigs, R. M. (2011). Chem. Commun., 47, 8679.
[39] Xuan, J., Cheng, Y., An, J., Lu, L.-Q., Zhang, X.-X., Xiao, W.-J. (2011). Chem. Commun., 47, 8337.
[40] Xuan, J., Feng, Z.-J., Duan, S.-W., Xiao, W.-J. (2012). RSC Advances, 2, 4065.

Sven Rau, Michael G. Pfeffer, Robert Staehle

11 Metal complexes for photohydrogenation and hydrogen evolution

The conversion and storage of solar energy is currently one of the most important challenges. [1] The photocatalytic water splitting yielding hydrogen and oxygen exhibits one possibility to reach this task. Even though solar water splitting is accessible by electrolysis when powered by photovoltaic cells, the method is expensive and still of limited efficiency. [2, 3] An alternative approach is the formation of hydrogen and oxygen from water using a photoredox catalyst, which allows the direct transformation and capture of light energy into redox energy. Such a catalyst can consist of two coupled redox catalysts, whereby the driving force for the water splitting is provided by solar energy as shown in Figure 11.1.

This concept is well known from natural photosynthesis. [4, 5] The processes of light absorption, charge separation and electron transfer can be described in the Z-scheme, which is shown in Figure 11.2.

In this scheme two photocenters (P680, photo system II and P700, photo system I) are connected either to the oxygen evolving complex (OEC, oxidation center) or the ferredoxin-NADP reductase (FNR, reduction center). The high efficiency in this system arises from the perfect interplay between one photocenter with one redoxcatalyst, respectively, and the optimized electron transfer chain between them. Based on the analysis, one possible improvement of the artificial water-splitting could be the separation into two half-reactions, the water reduction and the water oxidation. [6] Moreover, this leads to a reduction of side reactions and complexity.

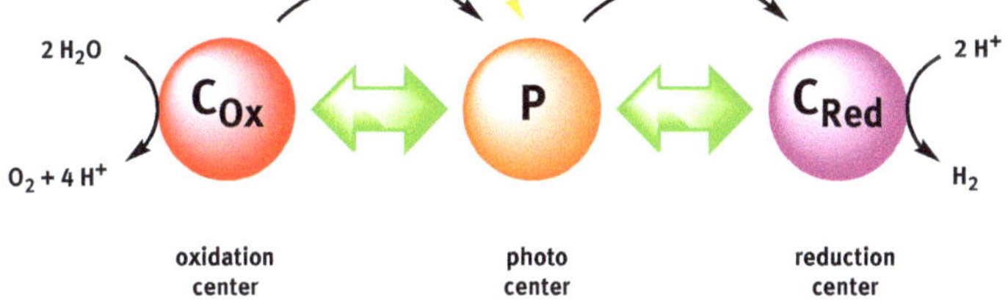

Figure 11.1: Photoredox catalyst for artificial water splitting: C_{Ox} = oxidation catalyst; P = photocenter, and C_{Red} = reduction catalyst.

Consequently, this will simplify the characterization and understanding, which in turn can help to optimize the interplay within the system. Thus, each subsystem consists of a chromophore, an electron transfer mediator, a redox catalyst, and an electron donor or acceptor, respectively (Figure 11.3). However, this simplification will only lead to overall water splitting if the oxidized donor, D^+, can play the role of the acceptor, A, see Figure 11.3.

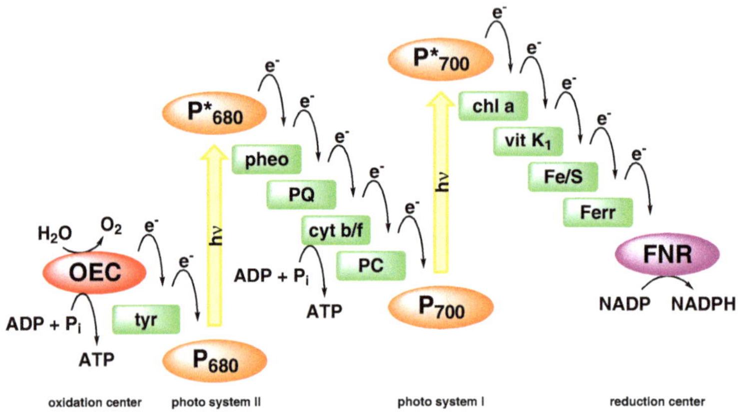

Figure 11.2: Z-scheme of photosynthesis (OEC = oxygen-evolving complex; tyr = tyrosine; pheo = pheophytin; P680 = photo center P680; PQ = plastoquinone; cyt b/f = cytochrome b/f; PC = plastocyanin; P700 = photo center P700; chl a = chlorophyll a; vit K1 = vitamin K1; Fe/S = Fe/S cluster; Ferr = ferredoxin; FNR = ferredoxin-NADP reductase). [7]

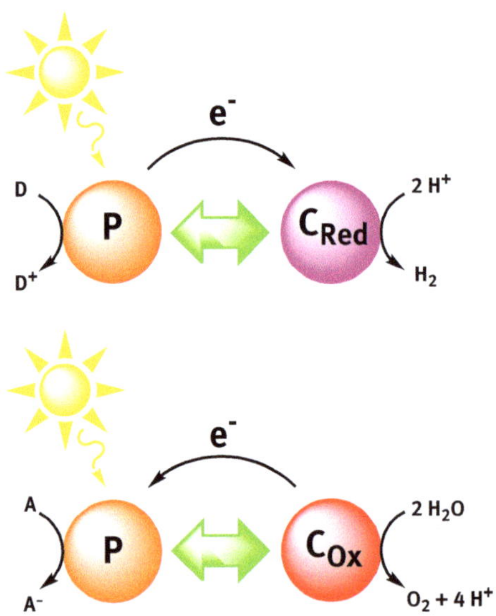

Figure 11.3: Half reactions: water reduction (top) and water oxidation (bottom).

First, intermolecular systems using ruthenium polypyridyl complexes as chromophore and visible light for water reduction and oxidation were developed since the late 1970s. [3, 8, 9, 10] A first example for the overall artificial water splitting in a two electrode system with UV light has been known since the early 1970s, the Fujishima–Honda cell. [11] Since then, the two electrode system has been constantly improved, especially in terms of the effective use of sunlight. [12, 13, 14] A special challenge here is the effective connection between the essential functions, such as light absorption, electron transfer, catalytic turnover and the available building blocks. These building blocks can be at the molecular level or a material and are available for the chromophore, catalyst and the electrode surface. [15, 16, 17, 18, 19, 20, 21, 22] To make better predictions of an effective interplay of the species, which are involved in the catalysis, it is advisable to have a closer look at the individual building blocks. Since the development of the first intermolecular systems in the late 70s, constant progress has been made.

11.1 Analysis of construction components of artificial photocatalytic systems

11.1.1 Chromophore

In photocatalytic systems, the chromophore is an essential requisite for the reaction. Therefore, it has to accomplish important functions:
- The possibility to generate a highly reactive and sufficiently long lived excited state by light absorption.
- The capability to be incorporated in directional transfer of electrons by a quenching mechanism
- The regeneration of the ground state via subsequent additional redox reaction.

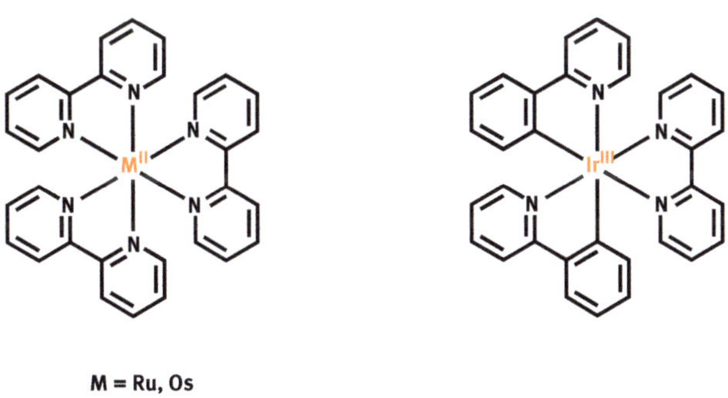

Figure 11.4: Generic structure of transition metal complexes used as chromophores.

Indeed, many chromophores for intermolecular systems for photocatalytic water reduction were tested, including acridine dyes, metalloporphyrins, metallophthalocyanines, and transition metal complexes in particular Ir, Os or Ru (see Figure 11.4). [3, 23, 24]

Mainly ruthenium polypyridyl complexes turned out to be suitable, because of their outstanding photo- and electrochemical properties as well as their sufficient synthetic availability. [25, 26, 27, 28] Due to the large variety of chromophores, we will confine ourselves in this chapter to the ruthenium polypyridyl type chromophores.

11.1.2 Electron relay

A direct water reduction by ruthenium polypyridine complexes is not found, but the excited chromophore can be used to initiate an electron transfer process. To trap the photoexcited electrons, redox mediators are needed. They are responsible for a fast and irreversible charge separation via quenching mechanisms and the transportation of the electrons between the photocenter and the redox catalysts. Commonly used mediators are bipyridinium ions (methylviologens like diquat and paraquat), phenanthrolinium ions, metal ions such as Eu^{3+}, V^{3+} or Cr^{3+}, metal complexes of Rh or Co, and proteins like cytochrome c3. [3]

11.1.3 Redox equivalents

Additionally, electron donors or acceptors are needed to regenerate the chromophore. In principal they may also be viewed as quenchers of the excited state to increase the efficiency of the electron transfer. Therefore, they also have the purpose to hinder a possible charge recombination. Commonly used electron donors are tertiary amines like EDTA, triethylamine (TEA) or triethanolamine (TEOA), reduced metal complexes like Eu^{2+}- salts or other reducing agents like ascorbate, H_2S or NADPH. Peroxo species like $K_2S_2O_8$ or oxidizing complexes such as $[Ce^{IV}(NO_3)_6](NH_4)_2$ or $[Co^{III}(NH_3)_5Cl]Cl_2$ are most frequently used as electron acceptors. The common feature among them is the instability of the reduced state, which upon fast irreversible decomposition inhibits back electron transfers.

11.1.4 Reduction catalysts

The reduction of water yielding molecular hydrogen is a two electron process. For an efficient catalysis, an overpotential has to be applied to overcome the barriers in

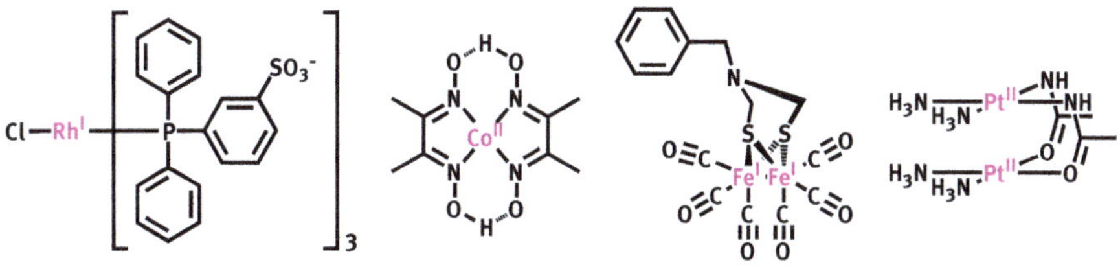

a b c d

Figure 11.5: Selected homogeneous reduction catalysts: a) $[Rh^{I}Cl(dpm)_3]^{3-}$ (dqm = diphenylphosphinobenzene-m-sulphonate) [38, 39], b) $[Co^{II}(dmgH)_2]$ (dmgH = dimethylglyoximato) [40], c) $[\{(\mu\text{-}SCH_2)_2N(CH_2C_6H_5)\}Fe_2(CO)_6]$ [41], d) $[Pt^{II}_2(NH_3)_4(\mu\text{-acetamidato})_2]$ [42, 43, 44].

this multi-electron reaction. Various heterogeneous materials were tested with respect to their catalytic activity. It has been shown that mainly late transition metals and their oxides are effective, either as bulk materials, in colloidal form or deposited on electrodes. [3, 29, 30] In nature the water reduction is performed by hydrogenases, transition metal dependent enzymes. [31, 32, 33] With the intention to mimic these enzymes, the common $[Fe_2S_2]$ motif of iron-only compounds is the basis for the development of many bioinspired systems. [34, 35, 36] Other homogenous systems are commonly transition metal complexes. [37, 38, 39, 40, 41, 42] Selected Examples are shown in Figure 11.5.

11.1.5 Intramolecular hydrogen evolving photocatalysts

The functionality of the above discussed systems is determined by several hard to control factors such as the collision probability between its components. This is obviously a limitation of the effectiveness of the photocatalytic process. In addition, nature's examples work with supramolecularly, spatially well-defined catalytic components thus largely avoiding any collision probability dependence. It is possible to combine the above mentioned building blocks. With the intention to improve the photocatalytic activity, particular attention has to be paid to an effective charge separation and a subsequent directed electron transport. It is a feasible approach to link the individual molecular building blocks, thus generating a supramolecular photocatalyst. First efforts to develop a complete supramolecular photocatalyst which can mimic one of the two sub-reactions of water splitting were known since 2001. [4] Since then various systems for the intramolecular water reduction have been developed (Figure 11.6).

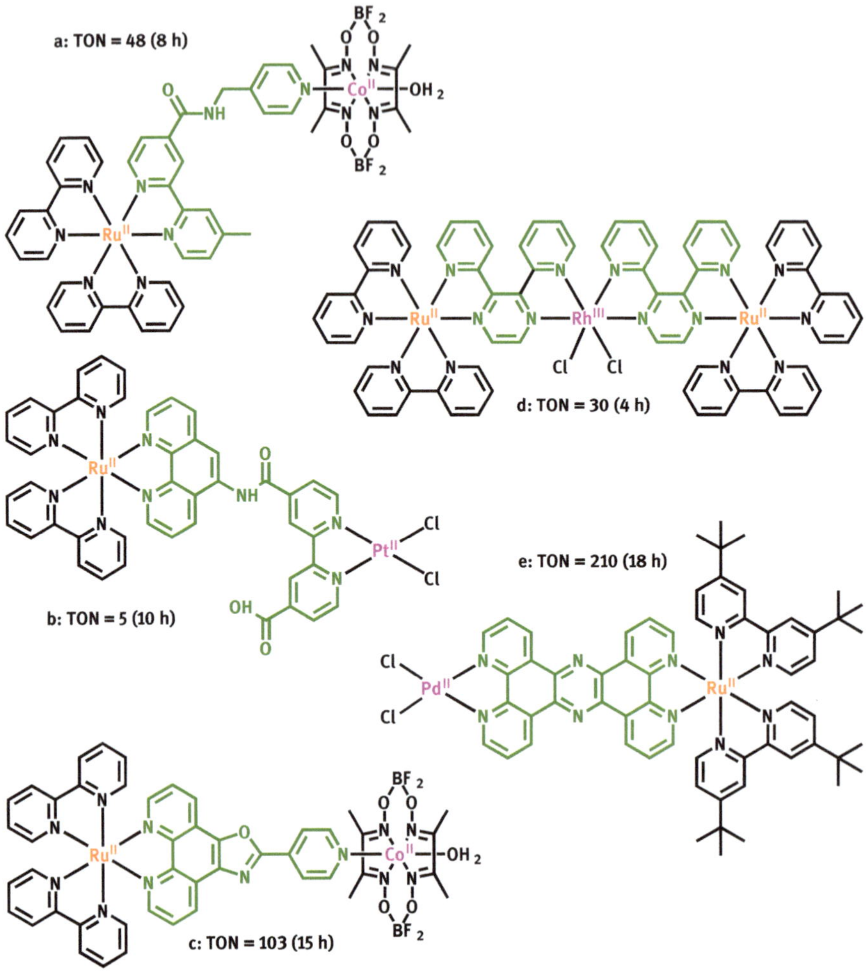

Figure 11.6: Representative systems for the intramolecular hydrogen evolution (TON = turnover number; irradiation time in brackets): a) Sun [45], b) Sakai [46], c) Artero [47, 48], d) Brewer [24, 49], e) Rau [50, 51].

11.1.6 Oxidation catalysts

Besides the water reduction, the water oxidation is needed to accomplish the overall water splitting. For an effective catalysis, it is necessary to reduce the overpotential which is required to overcome the barriers in this multi-electron reaction. [2] In nature this process is catalyzed by the oxygen evolving complex, consisting of a μ-oxo bridged Mn_4Ca cluster (Figure 11.7d). [52, 53] Artificial approaches include not only metal oxides [31] or a cobalt phosphate [54] functionalized electrode surface, but also homogeneous transition metal complexes. [3, 53, 55, 56, 57, 58, 59] Selected samples are shown in Figure 11.7a–c.

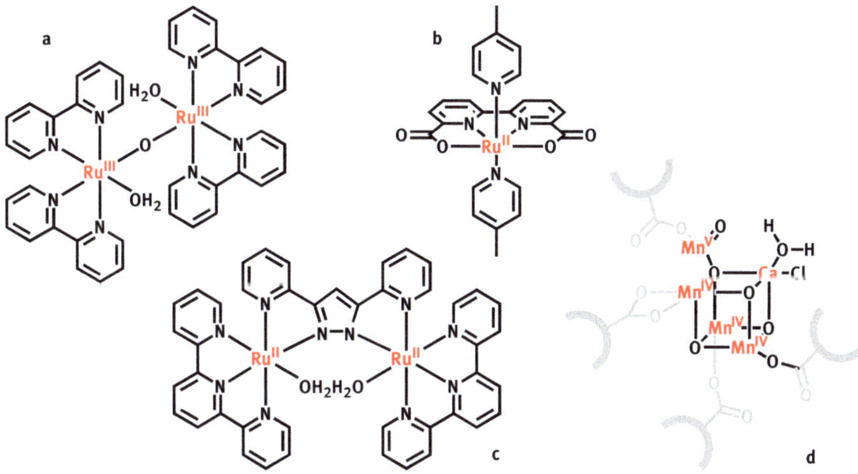

Figure 11.7: Selected transition metal complexes for the water oxidation: a) $[(\mu\text{–O})\{Ru^{III}(H_2O)(bpy)_2\}_2]^{4+}$ (bpy = 2,2′-bipyridine) [57, 58], b) $[Ru^{II}(bda)(pic)_2]^{2+}$ (H_2bda = 2,2-bipyridine-6,6-dicarboxylic acid; pic = 4-picoline), c) $[(\mu\text{–bpp})\{Ru^{II}(H_2O)(tpy)\}_2]^{3+}$ (Hbpp = bis(2-pyridyl)-3,5-pyrazolate; tpy = 2,6-bis(2-pyridyl)pyridine) [59], d) active center of the OEC [53]. Compounds a and b work as chemical water oxidation catalysts whereas c represents a photochemically activated water oxidation catalyst.

11.1.7 Intramolecular oxidation catalysts

A possibility for improving the electron transfer between the chromophore and the catalyst is to bring them into spatial proximity. There are approaches which either stabilize the catalyst by a ligand of the chromophore [12] or immobilize the catalyst next to the chromophore with a membrane. [13, 60] Recently the first intramolecular water oxidation catalyst working in its molecular form was developed (Figure 11.8). [20] Here, the delicate balance between favorable photochemical electron transfer and the charge recombination is even more difficult to find than for the more numerous hydrogen evolving photocatalysts.

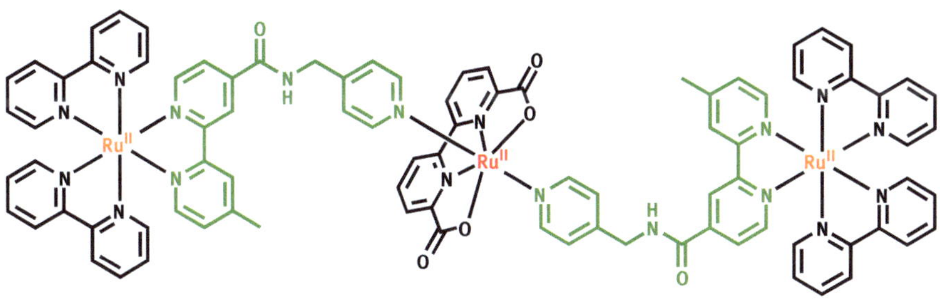

Figure 11.8: Intramolecular catalyst for the photoinduced water oxidation. [20]

11.1.8 Comparison of inter- and intramolecular photocatalysis

As shown, for the artificial water reduction and oxidation, both inter- and intramolecular systems are known. Excellent reviews have covered the field of intermolecular photocatalysis in great detail. [62] The much younger developments within the field of intramolecular photocatalysis are not yet comprehensively covered. It is very apparent that the linkage between the chromophore and the catalyst through a bridging ligand has implications for the catalysis itself and the applicability in devices, e. g. photochemical water splitting cells. [12, 15, 16, 17, 18, 19] The environment of the catalytic center itself will be determined by the chemical structure of the bridging ligand (ligand tuning of catalytic metal centers). For an effective electron transfer between the chromophore and the catalyst, a spatial proximity between both is necessary for both the inter- and the intramolecular catalysis systems. Consequently, the connection of chromophore and catalyst reduces the dependence on collision processes and is associated therewith the easier insertion in devices. [12, 15] Furthermore, the spectroscopic data of the intramolecular systems give information of the interplay and their influence on each other. However, the construction of these intramolecular systems is often attended by a difficult multistep synthesis and a difficult characterization due to aggregation effects. [61] Furthermore, these huge molecules show an instability with respect to various influences, whereas a sustentative repair mechanism is missing. [62, 63, 64, 65] Moreover, a task of the bridging ligand in intramolecular systems is the directional multielectron transfer either towards (oxidation) or away from (reduction) the chromophore (see Figure 11.9). [66, 67, 68] These directional multielectron processes, are on the one hand poorly understood, i. e. we have only for the first electron transfer step a relatively firm

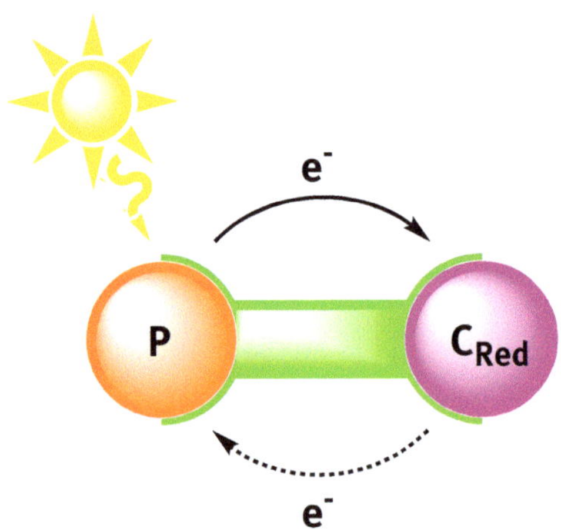

Figure 11.9: Photoinduced electron transfer in intramolecular photocatalysts and potentially harmful back electron transfer.

theoretical understanding and spectroscopic tools to characterize it. Furthermore, they require a sufficient stability of the charge separated intermediates in order to allow additional electron transfer steps to occur. This stability is influenced by the nature of the bridging ligand, the stability of different redox states of the catalytic center and several other factors.

It is very tempting to analyze the bridging ligand. [69] One classification often applied is the Robin and Day grouping as it defines the extent of electronic communication between metal centers in the ground state. [70] A possible answer may emerge from a detailed analysis of already known bridging ligands. Therefore it is useful to have a look at the communication between catalyst and chromophore inside intramolecular catalysts in the ground and the excited state (Figure 11.9).

For example, the Brewer as well as the Rau catalyst shows a strong emission quenching compared to complexes without their reduction catalyst. [50, 67–71, 72] However, only the Brewer system shows an interaction between the chromophore and the catalyst in the ground state. According to literature analysis, there are metal complexes with the utilized 2,3-dpp (dpp = 2,3-bis-2-pyridylpyrazine) like-ligand system for the Robin and Day class II. In contrast, the ruthenium-palladium complex $[(tbbpy)_2Ru(tpphz)PdCl_2]^{2+}$ (tbbpy = 4,4′-di-tert-butyl-2,2′-bipyridine; tpphz = tetrapyrido[3,2-a:2′,3′-c:3″,2″-h:2‴,3‴-j]phenazine) utilizes the tpphz ligand. There is no communication between the metal centers bound to this ligand. [73] Utilization of a 2,2′-bipyrimidine bridging ligand which allows strong communication between metal centers for ruthenium-palladium complexes results in an inactive system. The tolerance of ligand mediated communication for ruthenium-rhodium systems and thus facilitated back electron transfer may be explained with the possibility of stabilization of the catalytic metal in its reduced Rh(II)-state. It is therefore apparent that a more detailed analysis is necessary. [74, 75]

An indication that this analysis will yield very important knowledge is the fact that one and the same bridging ligand can be used either for the water reduction or oxidation (see Figure 11.6 and 11.8). [20, 45]

11.2 Intramolecular photocatalysts for hydrogen production and hydrogenation

According to the above mentioned reasons, a homogeneous and monomolecular system which is capable of transforming the energy of sunlight into chemical energy (hydrogen) at room temperature has been of great scientific interest for many years. [45, 46, 47, 48, 49, 50, 51] One of the first examples in this regard is the well-studied photocatalyst $[(tbbpy)_2Ru(tpphz)PdCl_2](PF_6)_2$ (**Ru(tpphz)Pd**, Figure 11.10). [50]

This supramolecular catalyst, containing a bis(bipyridine)ruthenium(II) moiety as chromophor, which is linked by tpphz as an electron relaying bridge to a catalytic palladium(II) centre, combines all single components that are required for light-

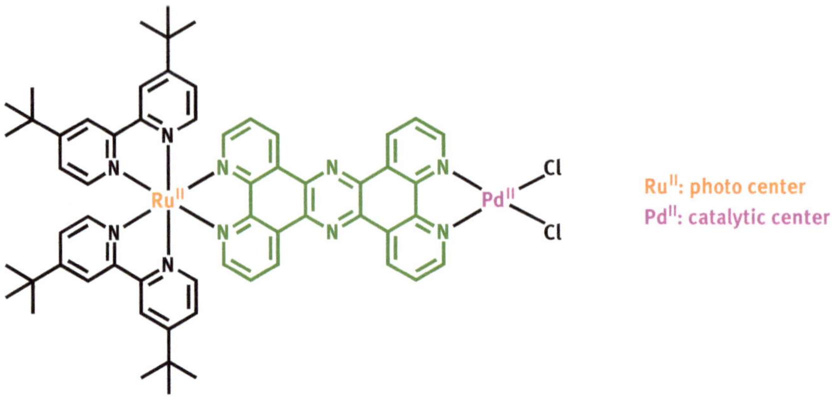

Figure 11.10: Monomolecular photocatalyst **Ru(tpphz)Pd**.

driven hydrogen generation. The following subchapters give an insight into its application as photocatalyst for hydrogen production, photohydrogenation, photophysics and derivatisation.

11.2.1 Hydrogen production

The general idea is to develop a photocatalyst that facilitates an intramolecular photo-induced 2-electron transfer from the photocentre to the catalytic centre at which a subsequent reduction of protons to molecular hydrogen occurs (Figure 11.11).

The **Ru(tpphz)Pd** photocatalyst effectively generates hydrogen under visible light irradiation at room temperature with up to 238 turnovers at the maximum. [76]

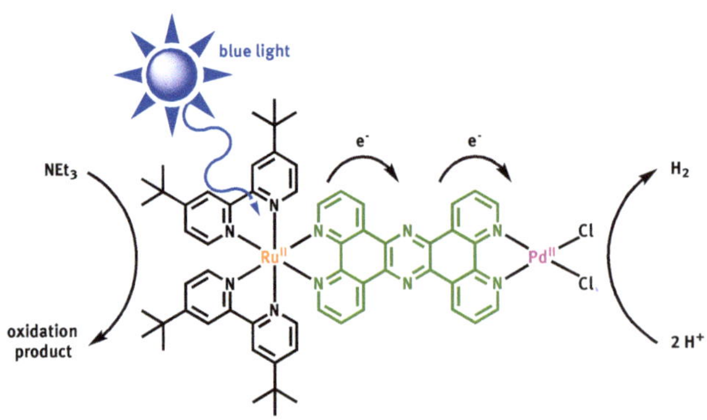

Figure 11.11: Basic mechanism of photocatalytic hydrogen production with **Ru(tpphz)Pd** as catalyst.

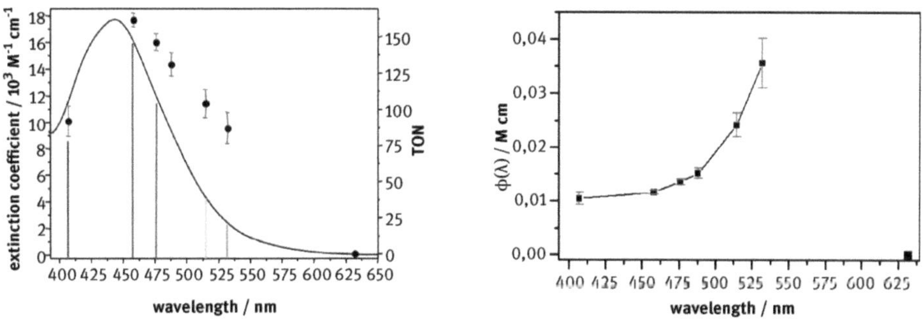

Figure 11.12: Left: absorption spectrum of **Ru(tpphz)Pd** in acetonitrile (—) and TON spectrum (●); Right: Efficiency spectrum ϕ(λ) of the photon-to-hydrogen conversion of the catalytic process calculated as the ratio of the TON values and the extinction coefficient.

Thereby the catalytic cycle starts with an excitation of the ruthenium(II) photocentre within the MLCT (metal to ligand charge transfer) absorption band that promotes the system into a ^{1}MLCT state (^{1}MLCT = singlet metal to ligand charge transfer), which can either be located on the terminal tbbpy ligands or on the phenazine unit of the tpphz bridging ligand. After an ISC (inter system crossing) to a ^{3}MLCT state (^{3}MLCT = triplet metal to ligand charge transfer), the electron located on the phenanthroline unit is transferred further to the palladium(II) centre, leading to a subsequent chloride dissociation [67, 50] (detailed photophysics are described in Section 11.2.3). Simultaneously, a sacrificial electron donor (e. g. triethylamine) is required to reduce the temporarily formed ruthenium(III). After another electron transferred from the ruthenium to the palladium centre, protons can be reduced to molecular hydrogen by taking up electrons from the catalytic centre.

In general, the strategy for the design of novel molecular photocatalysts aims at high catalytic efficiency which could only be established by identifying a structure-activity correlation. In this context the localisation of the initially populated excited state plays an important part. [50, 66, 77]

Figure 11.12 shows the dependency of the catalytic activity (TON = turnover number) on the irradiation wavelengths together with the MLCT absorption spectrum. Obviously, the extinction coefficient tails off much more steeply in the range of 450 to 550 nm than the TON (λ) value (Figure 11.12, left). Thus, the catalytic activity of **Ru(tpphz)Pd** does not strictly follow the absorption spectrum, what becomes even clearer by converting the TON values into an efficiency spectrum ϕ(λ) of the catalysis, which was calculated from the ratio of the TON values and the extinction coefficient. It turns out that ϕ(λ) increases with increasing wavelength in the range between 400 and 530 nm. According to this, the absorption of photons with lower energy leads to a higher catalytic activity than the absorption of high energy photons. The explanation for this behavior is the wavelength dependency of the localization of the initially and predominantly populated MLCT state.

Resonance Raman spectroscopy (RR), supported by TDDFT-calculations, revealed that when the excitation wavelength is shifted to lower energy, the electron density of the excited state shifts from the terminal tbbpy ligands to the tpphz bridging ligand. [66] A comparison of the electron density distribution of the initially photoexcited state and the efficiency spectrum clearly shows, that the catalytic efficiency correlates strongly with the electron density located on the tpphz bridging ligand. This means a more active photocatalyst should generate a high portion of a bridging ligand-localized excited state at all incident wavelengths, especially with polychromatic sunlight considered as a future energy source. This might be realized by exchanging the terminal tbbpy-ligands by ligands which do not contribute to the localization of the first excited state, but still stabilize the ruthenium(II) centre. Bibenzimidazoles could serve as an appropriate class of ligands in this context. [78]

With regard to future design and tuning of novel hydrogen evolving photocatalysts, beside the initial excitation step, also the electron-storage capabilities of the bridging ligand [68], rapid electron and energy transport within the supramolecular system, [27, 79, 80] and the stability of the photocatalyst must be taken into account.

11.2.2 Photohydrogenation

A "hydrogen free" photocatalytic hydrogenation of olefins and acetylenes can serve as an interesting alternative to common homogeneous efforts [81, 82, 83, 84], which usually require a hydrogen atmosphere or at least the external addition of molecular hydrogen. In 1995 Willner et al. published an eosin-modified Co(II)-reconstituted myoglobin (Eo^{2-}-Mb-Co(II)), which is capable of catalytically generating hydrogen from an aqueous solution under visible light irradiation ($\lambda > 475$ nm) in the presence of a sacrificial electron donor. [85] As soon as acetylene is added to the catalytic mixture, hydrogen evolution is eliminated and the acetylene is hydrogenated to ethylene. Moreover, the selective hydrogenation of acetylene dicarboxylic acid by Eo^{2-}-Mb-Co(II) to maleic acid (> 92 %) indicates an addition of the acetylene substrate to the catalytic Co(II) centre where the orientation of the substrate is induced by the protein to selectively form the *cis*-product. However, detailed mechanistic studies seem to be difficult due to the structural complexity of the protein surrounding.

By analogy, the **Ru(tpphz)Pd** photocatalyst can serve as hydrogenation catalyst under similar conditions as used for hydrogen production. By addition of tolane to the reaction mixture with triethylamine (TEA) as sacrificial electron donor, the evolution of hydrogen is replaced by a highly selective reduction of tolane to cis-stilbene (Figure 11.13). [50] This opens, on the one hand, a novel synthetic route for a selective hydrogenation of alkynes and delivers, on the other hand, important structural information on the $PdCl_2$-centre during catalysis. In this context the formation of Pd(0) colloids has to be taken into account, since it is known from high-temperature Heck reactions that Pd(II) catalysts, irrespective of the nature of their precursor,

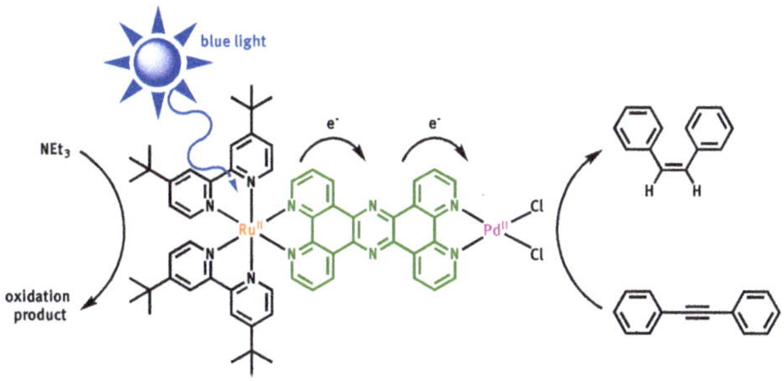

Figure 11.13: Photohydrogenation of tolane to cis-stilbene.

are rapidly reduced to Pd(0), which accumulates as metal colloids. [86] This can also be the case for light induced catalysis using **Ru(tpphz)Pd**-type photocatalysts, meaning that the *in situ* generated Pd(0) detaches from the bridging ligand and forms colloidal Pd(0). [62] Against this background, the observed high selectivity towards the hydrogenation of tolane to *cis*-stilbene takes on significance as it strongly suggest an intramolecular pathway with **Ru(tpphz)Pd** functioning as a monomolecular assembly. Since heterogeneous hydrogenation by means of Pd(0) colloids working as catalytic active species usually lack a high selectivity and furthermore homogeneous Pd(0) catalysts bearing a bidentate nitrogen ligand are known to highly selective *cis*-hydrogenate alkynes, a colloidal Pd(0) species as active species is rather unlikely. [87] Moreover, an amine effect as observed for the Lindlar catalyst can be excluded. [88, 89] When reacting the **Ru(tpphz)Pd** catalyst with tolane under absence of TEA as electron donor and addition of hydrogen, a selective reduction to *cis*-stilbene still occurs, and this again confirms the N,N′-chelated palladium as a catalytic centre.

11.2.3 Photophysics

In order to directly tune and increase the activity of the **Ru(tpphz)Pd** photocatalyst and in order to create novel photocatalysts for light-driven hydrogen production, a detailed elucidation of the electron transfer cascade from the ruthenium photocentre to the catalytic palladium, is required.

Since the multitude of all electron transfers, taking place during catalysis consists of several single electron transfer processes which occur in different time frames, analysis methods have to be applied which cover a time range from 20 fs up to several nanoseconds. For this concern, the combination of resonance Raman (RR) and femtosecond time-resolved pump-probe experiments is an appropriate technique to observe the excited state dynamics. [90]

As illustrated in Figure 11.14, both the mononuclear complex **Ru(tpphz)** (**Ru(tpphz)** = $[(tbbpy)_2Ru(tpphz)](PF_6)_2$) and the dinuclear complex **Ru(tpphz)Pd**

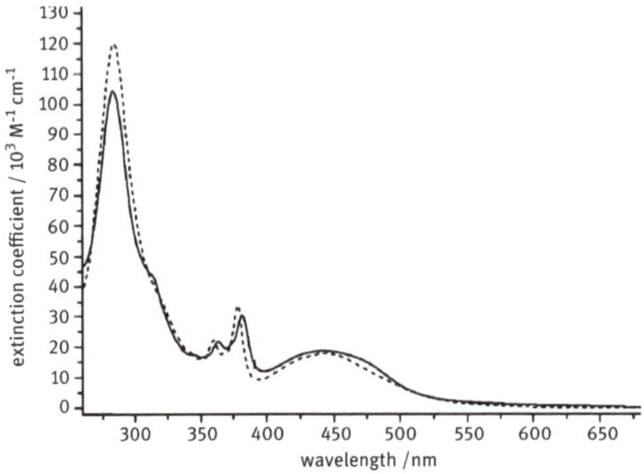

Figure 11.14: Steady-state absorption spectra of **Ru(tpphz)** (solid, **Ru(tpphz)** = [$(tbbpy)_2$Ru(tpphz)]$(PF_6)_2$) and **Ru(tpphz)Pd** (dashed) in acetonitrile.

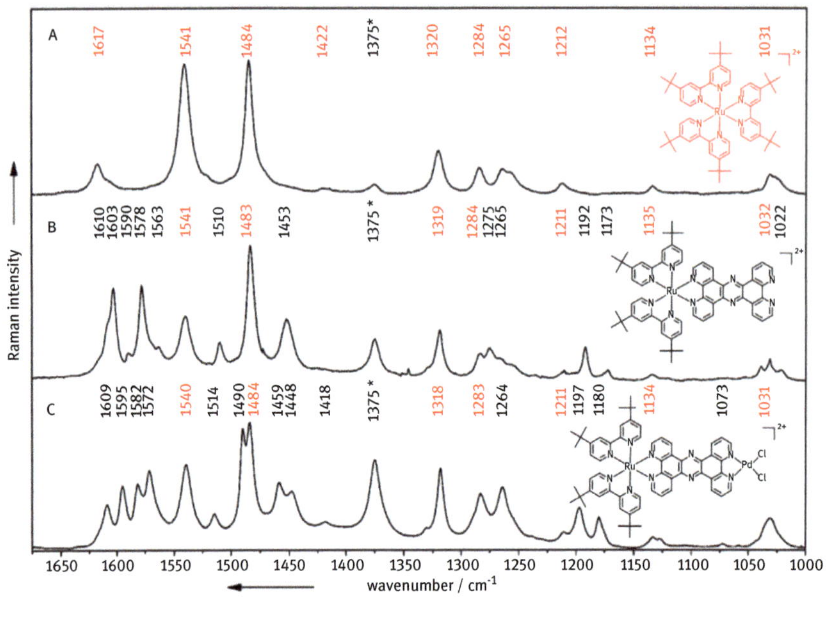

Figure 11.15: RR spectra of [$Ru(tbbpy)_3$]$(PF_6)_2$ (A), **Ru(tpphz)** (B) and **Ru(tpphz)Pd** (C) in acetonitrile.

show absorption characteristics, which are typical of the class of ruthenium polypyridine compounds. [67] In general, **Ru(tpphz)** and **Ru(tpphz)Pd** reveal qualitatively similar absorption spectra with four maxima between 250 and 700 nm: The maximum at around 280 nm can be assigned to the ligand-centred π–π^* transition of tbbpy, the two maxima between at 350 and 390 nm belong to π–π^* transitions of tpphz, and the energetically lowest lying broad transition at about 450 nm is assigned to MLCT transitions between the ruthenium centre and the coordinated ligands.

As already mentioned in Section 11.2., the localization of the initial MLCT state has a significant influence on the efficiency of the photocatalytic hydrogen production. To investigate the sub-20 fs dynamics of the population of the ^{1}MLCT, RR was applied. Figure 11.15 shows the RR spectra of the homoleptic reference complex $[Ru(tbbpy)_3]^{2+}$, **Ru(tpphz)** and the dinuclear photocatalyst **Ru(tpphz)Pd**. All spectra were recorded in acetonitrile with an excitation wavelength of 458 nm to excite within the MLCT absorption band. A detailed spectral analysis of the RR spectra, which are summarized in Table 11.1, shows that the bands of the reference compound $[Ru(tbbpy)_3]^{2+}$, which mainly contains vibrational modes that are associated with the tbbpy ligands, also show up in the RR spectra of **Ru(tpphz)** and **Ru(tpphz)Pd**. In addition, the spectra of both **Ru(tpphz)** and **Ru(tpphz)Pd** contain bands which belong to vibrational modes located on the tpphz ligand [68]; that clearly shows that the initial light absorption of **Ru(tpphz)** and **Ru(tpphz)Pd** produces an ensemble of electronically excited states, which are spread over the terminal tbbpy ligands as well as over the bridging tpphz ligand.

Table 11.1: RR bands [cm^{-1}] of $[Ru(tbbpy)_3]^{2+}$, **Ru(tpphz)** and **Ru(tpphz)Pd**; *Solvent bands.

$[Ru(tbbpy)_3]^{2+}$	Ru(tpphz)	Ru(tpphz)Pd
–	1022	–
1031	1032	1031
–	1039	–
1134	1135	1134
–	1173	1180
–	1192	1197
1212	1211	1211
1265	–	1264
–	1275	–
1284	1284	1283
1320	1319	1318
1375*	1375*	1375*
1422	–	1418
–	–	1448
–	1453	1459
1484	1483	1484
–	–	1490
–	1510	1514
1541	1541	1540
–	1563	1572
–	1578	1582
–	1590	1595
–	1603	–
–	1610	1609
1617	–	–

A comparison of the mononuclear complex **Ru(tpphz)** and the dinuclear complex **Ru(tpphz)Pd** reveals some differences in the RR spectra, for example, for the bands at 1418, 1448 and 1275 cm^{-1}, which are associated with modes of the tpphz ligand. These changes are probably due to some long-range effect of Pd, which alters the initial sub-20 fs excited-state dynamics.

Beside the population of the first excited state, the dynamics of the subsequent electron transfer from the ruthenium centre to the catalytic palladium centre were investigated by means of femtosecond time-resolved differential absorption spectroscopy. By comparing the obtained data with the absorption spectrum of the reduced phenazine moiety, which contributes one part of the tpphz ligand [91] and data on related ruthenium complexes, [79, 92, 93] a detailed elucidation of the individual light-triggered electron transfer steps in **Ru(tpphz)** and **Ru(tpphz)Pd** was accomplished. The electron transfer dynamics are schematically illustrated in Figure 11.16.

As shown above, for both compounds **Ru(tpphz)** and **Ru(tpphz)Pd,** the initially populated excited state is a mixture of tbbpy- and tpphz-centred MLCT states. Subsequently, a relaxation into a MLCT state, located on the phenanthroline sphere of tpphz ligand occurs, which goes along with a so-called inter ligand hopping transferring electron density from the terminal tbbpy ligands to the phenanthroline part of the tpphz bridging ligand. Formation of this phenanthroline-centred ^{3}MLCT state from the initially populated mixture occurs with a time-constant of 1.2 ps for **Ru(tpphz)** and 0.8 ps for **Ru(tpphz)Pd**. Both decay-associated spectra [67] of **Ru(tpphz)** and **Ru(tpphz)Pd** show similarities for the fastest component and for the subsequent component (240 ps for **Ru(tpphz)** and 5 ps for **Ru(tpphz)Pd**), which is assigned to the population of an excited state centred on the phenazine moiety of the bridging ligand. However, the coordination of the $PdCl_2$ unit results in a 40 nm red shift of the decay-associated spectrum [67] of the corresponding process, which is assigned to decay of the equilibrated tpphz-centred ^{3}MLCT state into an intra-ligand charge transfer (ILCT) localized on the phenazine unit. Beside this spectral difference, the time-constant from the ILCT changes drastically with the introduction of the Pd(II) centre, by means of a 50-fold increase of the ILCT rate. This is very likely due to the Pd(II) centre influencing the electronic structure of the bridging ligand to cause a significantly higher driving force for the formation of the phenanthroline-centred ILCT state. In contrast to **Ru(tpphz)**, this state further decays in **Ru(tpphz)Pd**, by means of an ligand to metal charge transfer (LMCT) from the phenazine unit of the bridging ligand to the Pd(II) centre to populate a long-lived state, which decay is not observed within the time range accessible. The final reduction of the Pd(II) centre is expected to go along with the dissociation of a Cl^-. This is supported by the finding that the catalytic activity of **Ru(tpphz)Pd** is completely inhibited by external addition of Cl^- during catalysis and by quantum chemical calculations. [50]

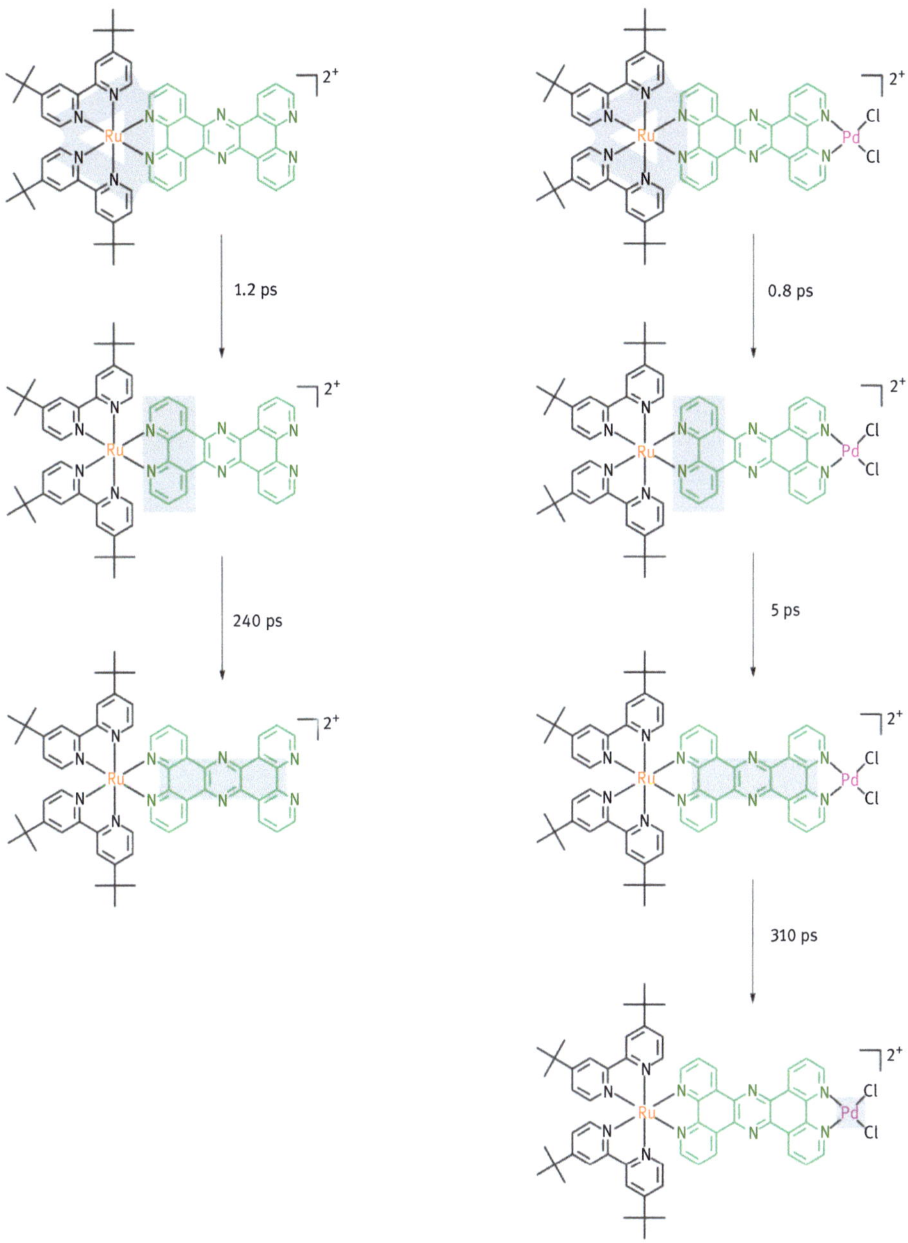

Figure 11.16: Electron transfer dynamics of **Ru(tpphz)** and **Ru(tpphz)Pd**. Localization of the photoexcited electron is illustrated by shaded areas.

11.2.4 Ru(tpphz)Pd-type catalysts as photochemical molecular devices (PMD)

In order to develop a highly active photocatalyst for light-driven hydrogen evolution, detailed information on the structure-activity relationship is required. Therefore a specific tuning of **Ru(tpphz)Pd** was performed by modifying either the bridging ligand or the catalytic center. The – in this respect – synthesized derivates of **Ru(tpphz)Pd** and the corresponding maximal TONs of each photoctalyst are shown in Figure 11.17.

Transient absorption experiments analogous to **Ru(tpphz)Pd** (Section 11.2.3) showed that a bromination of the ruthenium facing phenanthroline moiety of the bridging ligand results in a deceleration of the electron transfer from the photocenter towards the catalytic center. [76] This deceleration is probably causing the observed reduced activity of **Ru(Br_2tpphz)Pd** (Figure 11.17, max. TON = 69, Br_2tpphz = 2,7-dibromotetrapyrido[3,2-a:2′3′-c:3‴,2″-h:2‴,3‴-j]phenazine).

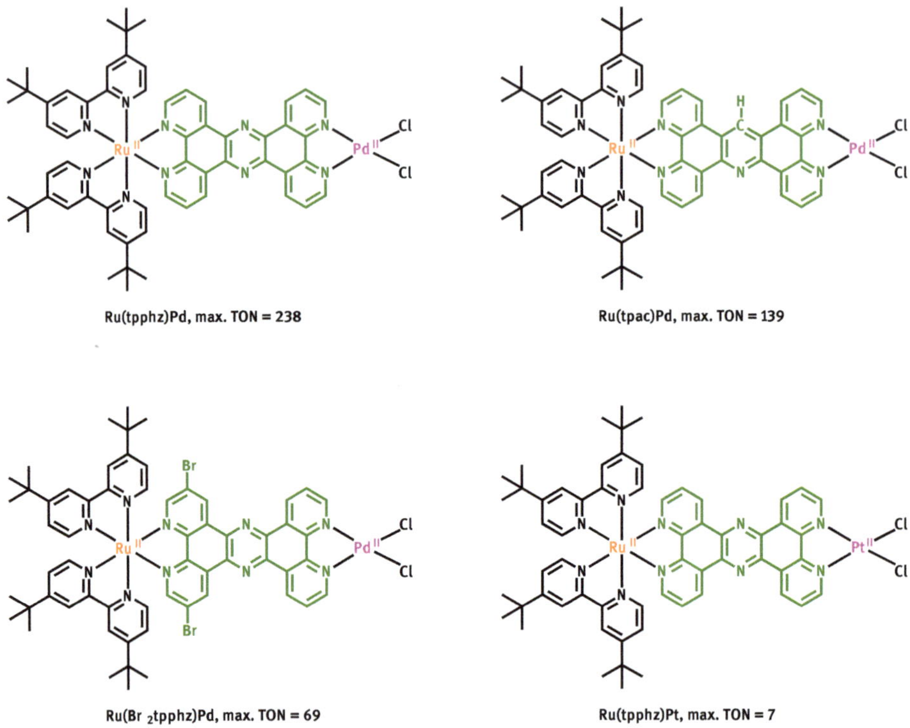

Figure 11.17: Derivates of **Ru(tpphz)Pd** and the corresponding maximal TONs, with tpac = tetrapyrido[3,2-a:2′3′-c:3‴,2″-h:2‴,3‴-j]acridine and Br_2tpphz = 2,7-dibromotetrapyrido[3,2-a:2′3′-c:3‴,2″-h:2‴,3‴-j]phenazine.

The same effect was observed for the substitution of one nitrogen of the phenazine unit of the bridging ligand, leading to **Ru(tpac)Pd** (Figure 11.17, max. TON = 139, tpac = tetrapyrido[3,2-a:2′3′-c:3‴,2″-h:2‴,3‴-j]acridine), which also resulted in a slower LMCT and so a decreased catalytic activity compared to the tpphz containg counterpart. [51]

Beside the modification of the bridging ligand which mainly affects the intramolecular electron transfer characteristics, substitution of the catalytic metal center where the formation of molecular hydrogen is expected to occur was performed. Therefore, the $PdCl_2$ unit was exchanged by $PtCl_2$, which offers the possibility to influence the catalytic reaction directly. Indeed, the novel catalyst **Ru(tpphz)Pt** (Figure 11.17, max. TON = 7) turned out to be less active with respect to light-driven hydrogen evolution, but is accessible for further analysis methods like ^{195}Pt-NMR or XAS (X-ray absorption spectroscopy) which allow insights into the actual mechanism taking place at the catalytic center. [94] In this concern, a high stability of **Ru(tpphz)Pt** under catalytic conditions was proven, which is one of the key features with regard to the development of a structure-activity relationship.

As shown in Figure 11.18, a photochemical molecular device (PMD) was designed, [95] which allows to tune either the bridging ligand, its structure mainly determines the electron transfer from the photo- to the catalytic center, or the terminal metal center to directly affect the catalytic reaction. Moreover, as already mentioned in Section 11.2.1, also the photocenter can be tuned by modifying or exchanging the terminal ligands. [96] Despite modifying single components of the supramolecular assembly, the general function as a hydrogen evolving system is maintained, and gives the chance to establish a consistent structure-activity relationship and to modularly design a durable and highly active PMD.

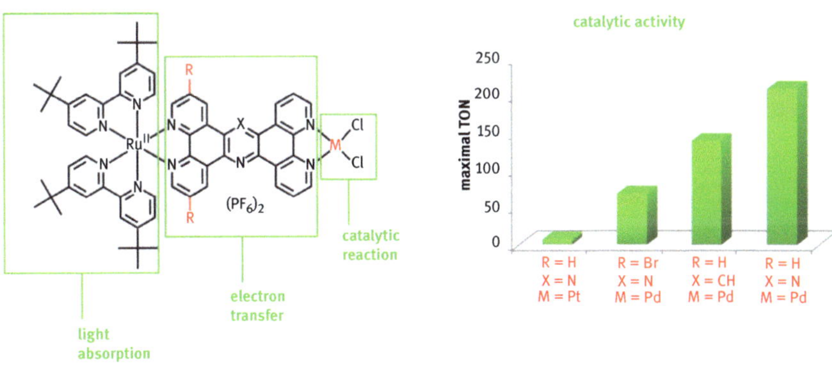

Figure 11.18: Ru(tpphz)M-type (M = Pd or Pt) catalyst as tunable photochemical molecular device (PMD).

11.3 Conclusion

The rapid development of intramolecular photocatalysis provides a growing number of structural examples of active catalysts. In some instances this is already coupled with an increased understanding of the structure activity relationships based on detailed photophysical investigations. Within this context, the analysis of the nature of the photocatalytically active species has resulted in the identification of PMD-like properties. If this analysis can be generalized, an increase of photocatalytic activity of the PMD as a whole can be accomplished by tuning the independent building blocks. This is exemplified for the above discussed tpphz like structures, keeping the photocentre constant and varying the other components. However, it is still not possible to forecast the activity of the catalysts.

Based on the example of biological photosynthesis a general direction of further development of photocatalysis with PMD's has to focus on coupling the two half reactions. This task is accomplished in nature via the redoxchain as shown in Figure 11.2. A very intriguing concept is the utilization of photoelectrochemical cells using ideas from dye sensitized solar cells, Figure 11.19. [97]

Here the activity of the immobilized PMD will depend on several factors already known from optimizations on the molecular level and on the electron transfer probabilities between the excited photocenter and the electrode. The chemical stability of the immobilization will be of crucial importance for the overall process as well. First examples of photocatalysis within similar cells have already been shown by the research groups of Sun and Mallouk. [12, 13, 98] The further development of this field will ultimately depend on an interdisciplinary research effort to fully understand the interplay between light driven excitation, electron transfer and catalytic activation.

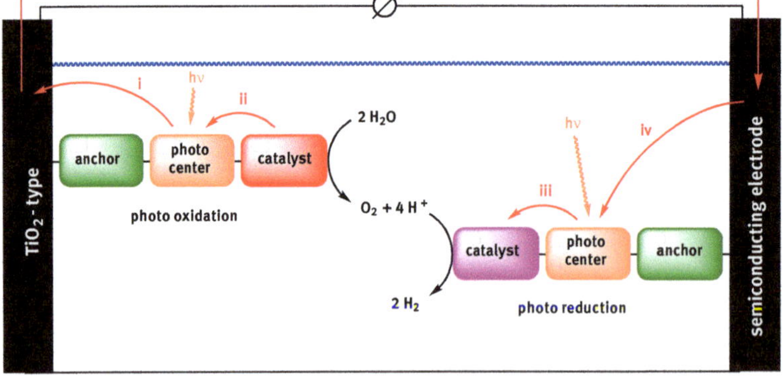

Figure 11.19: Principal layout of a photoelectrochemical cell based on immobilized PMD's, two photoinduced electron transfer reactions with inverse direction may allow overall watersplitting.

11.4 References

[1] Lewis, N.S., and Nocera, D.G. (2006). Powering the planet: Chemical challenges in solar energy utilization. PNAS 103, 15729–15735.
[2] Dau, H., Limberg, C., Reier, T., Risch, M., Roggan, S., and Strasser, P. (2010). The mechanism of water oxidation: From electrolysis via homogeneous to biological catalysis. ChemCatChem 2, 724–761.
[3] Amouyal, E. (1995). Photochemical production of hydrogen and oxygen from water: A review and state of the art. Sol. Energy Mater. Sol. Cells 38, 249–276.
[4] Sun, L., Hammarström, L., Åkermark, B., and Styring, S. (2001). Towards artificial photosynthesis: Ruthenium–manganese chemistry for energy production. Chem. Soc. Rev. 30, 36–49.
[5] Renger, G. (2010). The light reactions of photosynthesis. Curr. Sci. 98, 1305–1319.
[6] Bolton, J.R. (1978). Solar fuels. Science 202, 705–711.
[7] Hohmann-Marriott, M.F., and Blankenship, R.E. (2011). Evolution of photosynthesis. Annu. Rev. Plant Biol. 62, 515–548.
[8] Lehn, J., and Sauvage, J. (1977). Chemical storage of light energy. Catalytic generation of hydrogen by visible light or sunlight. Irradiation of neutral aqueous solutions. Nouv. J. Chim 1, 449–451.
[9] Kalyanasundaram, K., Kiwi, J., and Grätzel, M. (1978). Hydrogen evolution from water by visible light, a homogeneous three component test system for redox catalysis. Helv. Chim. Acta 61, 2720–2730.
[10] Kirch, M., Lehn, J.-M., and Sauvage, J.-P. (1979). Hydrogen generation by visible light irradiation of aqueous solutions of metal complexes. An approach to the photochemical conversion and storage of solar energy. Helv. Chim. Acta 62, 1345–1384.
[11] Fujishima, A., and Honda, K. (1972). Electrochemical photolysis of water at a semiconductor electrode. Nature 238, 37–38.
[12] Youngblood, W.J., Lee, S.-H.A., Kobayashi, Y., Hernandez-Pagan, E. a, Hoertz, P.G., Moore, T.A., Moore, A.L., Gust, D., and Mallouk, T.E. (2009). Photoassisted overall water splitting in a visible light-absorbing dye-sensitized photoelectrochemical cell. J. Am. Chem. Soc. 131, 926–927.
[13] Li, L., Duan, L., Xu, Y., Gorlov, M., Hagfeldt, A., and Sun, L. (2010). A photoelectrochemical device for visible light driven water splitting by a molecular ruthenium catalyst assembled on dye-sensitized nanostructured TiO_2. Chem. Commun. 46, 7307–7309.
[14] Reece, S.Y., Hamel, J.A., Sung, K., Jarvi, T.D., Esswein, A.J., Pijpers, J.J.H., and Nocera, D.G. (2011). Wireless solar water splitting using silicon-based semiconductors and earth-abundant catalysts. Science 334, 645–648.
[15] Li, F., Zhang, B., Li, X., Jiang, Y., Chen, L., Li, Y., and Sun, L. (2011). Highly efficient oxidation of water by a molecular catalyst immobilized on carbon nanotubes. Angew. Chem. Int. Ed. 50, 12276–12279.
[16] Chen, Z., Concepcion, J.J., Jurss, J.W., and Meyer, T.J. (2009). Single-site, catalytic water oxidation on oxide surfaces. J. Am. Chem. Soc. 131, 15580–15581.
[17] O'Regan, B., and Grätzel, M. (1991). A low-cost, high-efficiency solar cell based on dye-sensitized colloidal TiO_2 films. Nature 353, 737–740.
[18] Argazzi, R., Murakami Iha, N.Y., Zabri, H., Odobel, F., and Bignozzi, C.A. (2004). Design of molecular dyes for application in photoelectrochemical and electrochromic devices based on nanocrystalline metal oxide semiconductors. Coord. Chem. Rev. 248, 1299–1316.
[19] Polo, A.S., Itokazu, M.K., and Murakami Iha, N.Y. (2004). Metal complex sensitizers in dye-sensitized solar cells. Coord. Chem. Rev. 248, 1343–1361.

[20] Li, F., Jiang, Y., Zhang, B., Huang, F., Gao, Y., and Sun, L. (2012). Towards a solar fuel device: Light-driven water oxidation catalyzed by a supramolecular assembly. Angew. Chem. Int. Ed. 51, 2417–2420.
[21] Duan, L., Tong, L., Xu, Y., and Sun, L. (2011). Visible light-driven water oxidation – from molecular catalysts to photoelectrochemical cells. Energy Environ. Sci. 4, 3296–3313.
[22] Pellegrin, Y., Le Pleux, L., Blart, E., Renaud, A., Chavillon, B., Szuwarski, N., Boujtita, M., Cario, L., Jobic, S., Jacquemin, D., and Odobel, F. (2011). Ruthenium polypyridine complexes as sensitizers in NiO based p-type dye-sensitized solar cells: Effects of the anchoring groups. J. Photochem. Photobiol., A 219, 235–242.
[23] Losse, S., Vos, J.G., and Rau, S. (2010). Catalytic hydrogen production at cobalt centres. Coord. Chem. Rev. 254, 2492–2504.
[24] Inagaki, A., and Akita, M. (2010). Visible-light promoted bimetallic catalysis. Coord. Chem. Rev. 254, 1220–1239.
[25] Juris, A., Balzani, V., Barigelletti, F., Campagna, S., Belser, P., and von Zelewsky, A. (1988). Ru(II) polypyridine complexes: Photophysics, photochemistry, eletrochemistry, and chemiluminescence. Coord. Chem. Rev. 84, 85–277.
[26] Abrahamsson, M., Jäger, M., Osterman, T., Eriksson, L., Persson, P., Becker, H.-C., Johansson, O., and Hammarström, L. (2006). A 3.0 μs room temperature excited state lifetime of a bistridentate Ru^{II}-polypyridine complex for rod-like molecular arrays. J. Am. Chem. Soc. 128, 12616–12617.
[27] Balzani, V., Juris, A., Venturi, M., Campagna, S., and Serroni, S. (1996). Luminescent and redox-active polynuclear transition metal complexes. Chem. Rev. 96, 759–833.
[28] Damrauer, N.H. (1997). Femtosecond dynamics of excited-state evolution in [Ru(bpy)3]2+. Science 275, 54–57.
[29] Maeda, K., and Domen, K. (2010). Photocatalytic water splitting: Recent progress and future challenges. J. Phys. Chem. Lett. 1, 2655–2661.
[30] Bolton, J.R. (1996). Solar photoproduction of hydrogen: A review. Sol. Energy 57, 37–50.
[31] Gordon, J.C., and Kubas, G.J. (2010). Perspectives on how nature employs the principles of organometallic chemistry in dihydrogen activation in hydrogenases. Organometallics 29, 4682–4701.
[32] Song, L.-C., Luo, X., Wang, Y.-Z., Gai, B., and Hu, Q.-M. (2009). Synthesis, characterization and electrochemical behavior of some N-heterocyclic carbene-containing active site models of [FeFe]-hydrogenases. J. Organomet. Chem. 694, 103–112.
[33] Thauer, R.K., Kaster, A.-K., Goenrich, M., Schick, M., Hiromoto, T., and Shima, S. (2010). Hydrogenases from methanogenic archaea, nickel, a novel cofactor, and H_2 storage. Annu. Rev. Biochem. 79, 507–536.
[34] Thomas, C.M., Liu, T., Hall, M.B., and Darensbourg, M.Y. (2008). Series of mixed valent Fe(II)Fe(I) complexes that model the H_{ox} state of [FeFe]hydrogenase: Redox properties, density-functional theory investigation, and reactivities with extrinsic CO. Inorg. Chem. 47, 7009–7024.
[35] Zhang, P., Wang, M., Na, Y., Li, X., Jiang, Y., and Sun, L. (2010). Homogeneous photocatalytic production of hydrogen from water by a bioinspired [Fe_2S_2] catalyst with high turnover numbers. Dalton Trans. 39, 1204–1206.
[36] Wang, M., Chen, L., Li, X., and Sun, L. (2011). Approaches to efficient molecular catalyst systems for photochemical H2 production using [FeFe]-hydrogenase active site mimics. Dalton Trans. 40, 12793–12800.
[37] Gärtner, F., Sundararaju, B., Surkus, A.-E., Boddien, A., Loges, B., Junge, H., Dixneuf, P.H., and Beller, M. (2009). Light-driven hydrogen generation: Efficient iron-based water reduction catalysts. Angew. Chem. Int. Ed. 48, 9962–9965.

[38] Oishi, S. (1987). A water-soluble Wilkinson's complex as homogeneous catalyst for the photochemical reduction of water. J. Mol. Catal. 39, 225–232.
[39] Bauer, R., and Werner, H.A.F. (1994). Investigations on a homogeneous Wilkinson's catalyst for the water photolysis. Int. J. Hydrogen Energy 19, 497–499.
[40] Hawecker, J., Lehn, J., and Ziessel, R. (1983). Efficient homogeneous photochemical hydrogen generation and water reduction mediated by cobaloxime or macrocyclic cobalt complexes. Nouv. J. Chim. 7, 271–277.
[41] Na, Y., Wang, M., Pan, J., Zhang, P., Åkermark, B., and Sun, L. (2008). Visible light-driven electron transfer and hydrogen generation catalyzed by bioinspired [2Fe2S] complexes. Inorg. Chem. 47, 2805–2810.
[42] Ozawa, H., and Sakai, K. (2011). Photo-hydrogen-evolving molecular devices driving visible-light-induced water reduction into molecular hydrogen: structure-activity relationship and reaction mechanism. Chem. Commun. 47, 2227–2242.
[43] Yamauchi, K., Masaoka, S., and Sakai, K. (2009). Evidence for Pt(II)-based molecular catalysis in the thermal reduction of water into molecular hydrogen. J. Am. Chem. Soc. 131, 8404–8406.
[44] Erxleben, A., Mutikainen, I., and Lippert, B. (1994). Conversion of acetonitrile into acetamide in the co-ordination spheres of cis- and trans-MII(amine)2 (M = Pt or Pd). Solution and crystal structural studies. J. Chem. Soc., Dalton Trans. 3667–3675.
[45] Li, C., Wang, M., Pan, J., Zhang, P., Zhang, R., and Sun, L. (2009). Photochemical hydrogen production catalyzed by polypyridyl ruthenium–cobaloxime heterobinuclear complexes with different bridges. J. Organomet. Chem. 694, 2814–2819.
[46] Ozawa, H., Haga, M., and Sakai, K. (2006). A photo-hydrogen-evolving molecular device driving visible-light-induced EDTA-reduction of water into molecular hydrogen. J. Am. Chem. Soc. 128, 4926–4927.
[47] Fihri, A., Artero, V., Pereira, A., and Fontecave, M. (2008). Efficient H2-producing photocatalytic systems based on cyclometalated iridium- and tricarbonylrhenium-diimine photosensitizers and cobaloxime catalysts. Dalton Trans. 5567–5569.
[48] Fihri, A., Artero, V., Razavet, M., Baffert, C., Leibl, W., and Fontecave, M. (2008). Cobaloxime-based photocatalytic devices for hydrogen production. Angew. Chem. Int. Ed. 47, 564–567.
[49] Elvington, M., Brown, J., Arachchige, S.M., and Brewer, K.J. (2007). Photocatalytic hydrogen production from water employing a Ru, Rh, Ru molecular device for photoinitiated electron collection. J. Am. Chem. Soc. 129, 10644–10645.
[50] Rau, S., Schäfer, B., Gleich, D., Anders, E., Rudolph, M., Friedrich, M., Görls, H., Henry, W., and Vos, J.G. (2006). A supramolecular photocatalyst for the production of hydrogen and the selective hydrogenation of tolane. Angew. Chem. Int. Ed. 45, 6215–6218.
[51] Karnahl, M., Kuhnt, C., Heinemann, F.W., Schmitt, M., Rau, S., Popp, J., and Dietzek, B. (2012). Synthesis and photophysics of a novel photocatalyst for hydrogen production based on a tetrapyridoacridine bridging ligand. Chem. Phys. 393, 65–73.
[52] Loll, B., Kern, J., Saenger, W., Zouni, A., and Biesiadka, J. (2005). Towards complete cofactor arrangement in the 3.0 Å resolution structure of photosystem II. Nature 438, 1040–1044.
[53] Guskov, A., Kern, J., Gabdulkhakov, A., Broser, M., Zouni, A., and Saenger, W. (2009). Cyanobacterial photosystem II at 2.9-Å resolution and the role of quinones, lipids, channels and chloride. Nat. Struct. Mol. Biol. 16, 334–342.
[54] Kanan, M.W., and Nocera, D.G. (2008). In situ formation of an oxygen-evolving catalyst in neutral water containing phosphate and Co2+. Science 321, 1072–1075.
[55] Duan, L., Xu, Y., Zhang, P., Wang, M., and Sun, L. (2010). Visible light-driven water oxidation by a molecular ruthenium catalyst in homogeneous system. Inorg. Chem. 49, 209–215.

[56] Muckerman, J.T., Polyansky, D.E., Wada, T., Tanaka, K., and Fujita, E. (2008). Water oxidation by a ruthenium complex with noninnocent quinone ligands: Possible formation of an O–O bond at a low oxidation state of the metal. Inorg. Chem. 47, 1787–1802.
[57] Liu, F., Concepcion, J.J., Jurss, J.W., Cardolaccia, T., Templeton, J.L., and Meyer, T.J. (2008). Mechanisms of water oxidation from the blue dimer to photosystem II. Inorg. Chem. 47, 1727–1752.
[58] Gersten, S.W., Samuels, G.J., and Meyer, T.J. (1982). Catalytic oxidation of water by an oxo-bridged ruthenium dimer. J. Am. Chem. Soc. 104, 4029–4030.
[59] Sens, C., Romero, I., Rodríguez, M., Llobet, A., Parella, T., and Benet-Buchholz, J. (2004). A new Ru complex capable of catalytically oxidizing water to molecular dioxygen. J. Am. Chem. Soc. 126, 7798–7799.
[60] Brimblecombe, R., Koo, A., Dismukes, G.C., Swiegers, G.F., and Spiccia, L. (2010). Solar driven water oxidation by a bioinspired manganese molecular catalyst. J. Am. Chem. Soc. 132, 2892–2894.
[61] Hong, Y., Lam, J.W.Y., and Tang, B.Z. (2009). Aggregation-induced emission: Phenomenon, mechanism and applications. Chem. Commun. 4332–4353.
[62] Lei, P., Hedlund, M., Lomoth, R., Rensmo, H., Johansson, O., and Hammarström, L. (2008). The role of colloid formation in the photoinduced H2 production with a RuII-PdII supramolecular complex: A study by GC, XPS, and TEM. J. Am. Chem. Soc. 130, 26–27.
[63] Rau, S., Schwalbe, M., Losse, S., Görls, H., McAlister, C., MacDonnell, F.M., and Vos, J.G. (2008). Photoinduced ligand transformation in a ruthenium polypyridophenazine complex. Eur. J. Inorg. Chem. 2008, 1031–1034.
[64] Lutterman, D.A., Surendranath, Y., and Nocera, D.G. (2009). A self-healing oxygen-evolving catalyst. J. Am. Chem. Soc. 131, 3838–3839.
[65] Du, P., Schneider, J., Li, F., Zhao, W., Patel, U., Castellano, F.N., and Eisenberg, R. (2008). Bi- and terpyridyl platinum(II) chloro complexes: Molecular catalysts for the photogeneration of hydrogen from water or simply precursors for colloidal platinum? J. Am. Chem. Soc. 130, 5056–5058.
[66] Tschierlei, S., Karnahl, M., Presselt, M., Dietzek, B., Guthmuller, J., González, L., Schmitt, M., Rau, S., and Popp, J. (2010). Photochemical fate: The first step determines efficiency of H_2 formation with a supramolecular photocatalyst. Angew. Chem. Int. Ed. 49, 3981–3984.
[67] Tschierlei, S., Presselt, M., Kuhnt, C., Yartsev, A., Pascher, T., Sundström, V., Karnahl, M., Schwalbe, M., Schäfer, B., Rau, S., Schmitt, M., Dietzek, B., and Popp, J. (2009). Photophysics of an intramolecular hydrogen-evolving Ru-Pd photocatalyst. Chem. Eur. J. 15, 7678–7688.
[68] onance raman studies of photochemical molecular devices for multielectron storage. J. Raman Spectrosc. 39, 557–559.
[69] Balzani, V., Juris, A., Venturi, M., Campagna, S., and Serroni, S. (1996). Luminescent and redox-active polynuclear transition metal complexes. Chem. Rev. 96, 759–833.
[70] Shriver, D. F., Atkins, P. W., and Langford, C. H. (1995). Inorganic Chemistry, second edition (Oxford University Press).
[71] Elvington, M., and Brewer, K.J. (2006). Photoinitiated electron collection at a metal in a rhodium-centered mixed-metal supramolecular complex. Inorg. Chem. 45, 5242–5244.
[72] White, T. A., Arachchige, S.M., Sedai, B., and Brewer, K.J. (2010). Emission spectroscopy as a probe into photoinduced intramolecular electron transfer in polyazine bridged Ru(II),Rh(III) supramolecular complexes. Materials 3, 4328–4354.
[73] Bolger, J., Gourdon, A., Ishow, E., and Launay, J.-P. (1996). Mononuclear and binuclear tetrapyrido[3,2-a:2′,3′-c:3″,2″-h:2‴-3‴-j]phenazine (tpphz) ruthenium and osmium complexes. Inorg. Chem. 35, 2937–2944.
[74] Brown, G.M., Chan, S.F., Creutz, C., Schwarz, H.A., and Sutin, N. (1979). Mechanism of the formation of dihydrogen from the photoinduced reactions of tris(bipyridine)ruthenium(II) with tris(bipyridine)rhodium(III). J. Am. Chem. Soc. 101, 7638–7640.

[75] Creutz, C., Keller, A.D., Sutin, N., and Zipp, A.P. (1982). Poly(pyridine)ruthenium(II)-photoinduced redox reactions of bipyridinium cations, poly(pyridine)rhodium complexes, and osmium ammines. J. Am. Chem. Soc. 104, 3618–3627.
[76] Karnahl, M., Kuhnt, C., Ma, F., Yartsev, A., Schmitt, M., Dietzek, B., Rau, S., and Popp, J. (2011). Tuning of photocatalytic hydrogen production and photoinduced intramolecular electron transfer rates by regioselective bridging ligand substitution. ChemPhysChem 12, 2101–2109.
[77] Konduri, R., Ye, H., MacDonnell, F.M., Serroni, S., Campagna, S., and Rajeshwar, K. (2002). Ruthenium photocatalysts capable of reversibly storing up to four electrons in a single acceptor ligand: A step closer to artificial photosynthesis. Angew. Chem. Int. Ed. 41, 3185–3187.
[78] Dietzek, B., Kiefer, W., Blumhoff, J., Böttcher, L., Rau, S., Walther, D., Uhlemann, U., Schmitt, M., and Popp, J. (2006). Ultrafast excited-state excitation dynamics in a quasi-two-dimensional light-harvesting antenna based on ruthenium(II) and palladium(II) chromophores. Chem. Eur. J. 12, 5105–5115.
[79] Chiorboli, C., Rodgers, M.A.J., and Scandola, F. (2003). Ultrafast processes in bimetallic dyads with extended aromatic bridges. Energy and electron transfer pathways in tetrapyridophenazine-bridged complexes. J. Am. Chem. Soc. 125, 483–491.
[80] Sauvage, J.P., Collin, J.-P., Chambron, J.-C., Guillerez, S., Coudret, C., Balzani, V., Barigelletti, F., De Cola, L., and Flamigni, L. (1994). Ruthenium(II) and osmium(II) bis(terpyridine) complexes in covalently-linked multicomponent systems: Synthesis, electrochemical behavior, absorption spectra, and photochemical and photophysical properties. Chem. Rev. 94, 993–1019.
[81] Chow, Y.L., Li, H., and Yang, M.-S. (1988). Light-promoted catalytic hydrogenation of olefins with nickel complexes: The formation of nickel hydrides from the reaction of Ni(I) complexes with hydrogen. Can. J. Chem. 66, 2920–2927.
[82] Samuel, E. (1980). Photocatalytic homogeneous hydrogenation of olefins with alkyltitanocene compounds. J. Organomet. Chem. 198, C65–C68.
[83] Strohmeier, W., and Steigerwald, H. (1977). Photoinduzierte, homogene und selektive Hydrierung von 1,3-Cyclohexadien in Substanz mit IrClCO(PPh3)2 bei Umsatzzahlen UZ bis 100 000. J. Organomet. Chem. 125, C37–C39.
[84] Bogdan, P.L., Sullivan, P.J., Donovan, T.A., and Atwood, J.D. (1984). Photocatalysis of hydrogenation and isomerization of alkenes by cis-$HMn(CO)_4PPh_3$. J. Organomet. Chem. 269, C51–C54.
[85] Willner, I., Zahavy, E., and Heleg-Shabtai, V. (1995). Eosin-modified reconstituted Co(II) protoporphyrin IX myoglobin: A semisynthetic photoenzyme for H2 evolution and hydrogenation. J. Am. Chem. Soc. 117, 542–543.
[86] de Vries, J.G. (2006). A unifying mechanism for all high-temperature Heck reactions. The role of palladium colloids and anionic species. Dalton Trans. 421–429.
[87] van Laren, M.W., and Elsevier, C.J. (1999). Selective homogeneous palladium(0)-catalyzed hydrogenation of alkynes to (Z)-alkenes. Angew. Chem. Int. Ed. 38, 3715–3717.
[88] Lindlar, H. (1952). Ein neuer Katalysator für selektive Hydrierungen. Helv. Chim. Acta 57, 446–450.
[89] Lindlar, H., and Dubuis, R. (1966). Palladium catalyst for partial reduction of acetylenes. Org. Synth. 46, 89–91.
[90] Lochbrunner, S., Szeghalmi, A., Stock, K., and Schmitt, M. (2005). Ultrafast proton transfer of 1-hydroxy-2-acetonaphthone: Reaction path from resonance Raman and transient absorption studies. J. Chem. Phys. 122, 244315.
[91] Fujita, M., Ishida, A., Majima, T., and Takamuku, S. (1996). Lifetimes of radical anions of dicyanoanthracene, phenazine, and anthraquinone in the excited state from the selective electron-transfer quenching. J. Chem. Phys. 100, 5382–5387.

[92] Chiorboli, C., Bignozzi, C.A., Scandola, F., Ishow, E., Gourdon, A., and Launay, J. (1999). Photophysics of dinuclear Ru(II) and Os(II) complexes based on the tetrapyrido[3,2-a:2′,3′-c:3″,2″-h:2‴-3‴-j]phenazine (tpphz) bridging ligand. Inorg. Chem. 38, 2402–2410.
[93] Flamigni, L., Encinas, S., Barigelletti, F., MacDonnell, F.M., Kim, K.-J., Puntoriero, F., and Campagna, S. (2000). Excited-state interconversion between emissive MLCT levels in a dinuclear Ru(II) complex containing a bridging ligand with an extended Π system. Chem. Commun. 1185–1186.
[94] unpublished results.
[95] Balzani, V. (2003). Photochemical molecular devices. Photochem. Photobiol. Sci. 2, 459–476.
[96] Schwalbe, M., Karnahl, M., Tschierlei, S., Uhlemann, U., Schmitt, M., Dietzek, B., Popp, J., Groake, R., Vos, J.G., and Rau, S. (2010). The switch that wouldn't switch-unexpected luminescence from a ruthenium(II)-dppz-complex in water. Dalton Trans. 39, 2768–2771.
[97] Hagfeldt, A., and Grätzel, M. (2000). Molecular photovoltaics. Acc. Chem. Res. 33, 269–277.
[98] Kaschak, D.M., Johnson, S.A., Waraksa, C.C., Pogue, J., and Mallouk, T.E. (1999). Artificial photosynthesis in lamellar assemblies of metal poly(pyridyl) complexes and metalloporphyrins. Coord. Chem. Rev. 185–186, 403–416.

Arno Pfitzner, Stephan Dankesreiter, Anna Eisenhofer, Maria Cherevatskaya

12 Heterogeneous semiconductor photocatalysis

12.1 Inorganic semiconductors

12.1.1 General features of a photocatalyst

When talking about semiconductor photocatalysts which might be either inorganic or organic solids, several parameters of the material have to be considered. Not every solid compound is suitable for photocatalysis, despite appropriate physico-chemical parameters, e. g. color or large surface area. In this chapter a general overview on some key features of a semiconductor photocatalyst is given. The basic principles hold both for inorganic and organic semiconductors. However, since the literature is much richer and manifold in the case of inorganic semiconductor photocatalysts, we will mainly focus on these materials throughout this chapter.

12.1.1.1 Band structure and band gap

The first process of photocatalysis is the absorption of a photon by the photocatalyst. In the case of semiconductors, a closer look at the band structure of such a material is inevitable in order to understand the basic principle of this light absorption step. The band structure gives an idea of the electronic structure of a solid, it describes allowed and forbidden energetic states of the electrons, which are referred to as "bands" and "band gap", respectively. The energetically highest bands populated by electrons in the ground state are referred to as "valence band" (VB), the lowest unpopulated bands are referred to as "conduction band" (CB). [1, 2] To understand photochemical effects, one needs to take a more comprehensive look at the band structure.

The description of such electronic structures is a tough task in direct space. Because of Heisenberg's uncertainty principle, it is impossible neither to give the exact energy and position nor to give the momentum and position of an electron. However, it is possible to determine the momentum and the energy. With this assumption, the whole problem can be transferred to reciprocal space.

Every vector $\vec{r}$, which is composed of a multiple of the unit vectors a_1, a_2, and a_3, of the direct space, is transferred to a reciprocal space vector $\vec{k}$, which is composed of the reciprocal unit vectors b_1, b_2, and b_3, by the following equations: [3]

$$
\begin{aligned}
b_1 &= \frac{a_2 \times a_3}{a_1 \cdot (a_2 \times a_3)} \\
b_2 &= \frac{a_3 \times a_1}{a_1 \cdot (a_2 \times a_3)} \\
b_3 &= \frac{a_1 \times a_2}{a_1 \cdot (a_2 \times a_3)}
\end{aligned}
.
$$

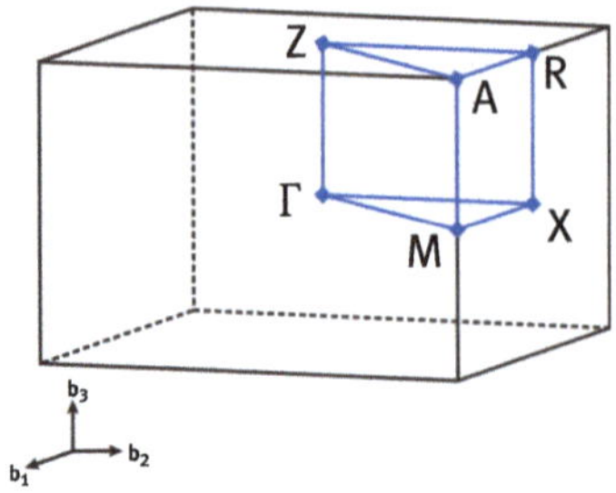

Figure 12.1: First Brillouin zone of a primitive tetragonal lattice in reciprocal space. [4] Critical points, which are areas of high symmetry [1], are marked. Γ indicates the center of the zone, M and R a center of an edge, A is a corner, and X and Z are centers of a plane.

With this vector set, the Brillouin zone can be defined, which describes bands and band gaps in reciprocal space as a function of the $\vec{k}$ vector. Figure 12.1 describes the Brillouin zone for a primitive tetragonal lattice. [4] Points of high symmetry, also referred as "critical points" [1], are indicated. **Γ** is the center of the zone, **R** and **M** mark the center of an edge, **A** gives the corner and **X** and **Z** the center of a plane of the Brillouin zone (Figure 12.1).

Using the Brillouin zone, the band structure of a material can be calculated as a function of the wave vector $\vec{k}$ – so to say the position in reciprocal lattice.

The band structure of a material also allows the calculation of the band gap, i. e. the difference between the highest populated energetic state and the lowest non-occupied energetic state of a solid. However, electrons can be excited across this band gap, e. g. by absorption of light. There are two different pathways of an electronic excitation from the VB to the CB. The band gap is called "direct band gap" when the $\vec{k}$ vector of the maximum of the VB is equal to the $\vec{k}$ vector of the minimum of the CB. This means an electron is changed in its energy but remains constant in momentum upon excitation. Such band gaps can be overcome by photon excitation with light and are observable by UV/Vis spectroscopy or fluorescence measurement, respectively. The second kind of band gap is referred to as "indirect band gaps". To generate an electron-hole pair in this case, the energy and momentum of an electron have to change during the excitation process, i. e. the $\vec{k}$ vector before and after excitation differs. To fulfill this condition, the electron has to interact with the photon and a lattice vibration, i. e. a phonon, to change its energy and momentum, respectively. [5] However, this process is comparably slow and more unlikely than other processes, which leads to the less efficient light absorption of semiconductors with indirect band gaps. [6]

The band structure of a solid compound is, of course, strongly influenced by both its chemical composition and crystal structure. As a consequence, a material which is composed of identical types and amounts of elements shows completely different band structures for different modifications. An example for of this fact is TiO_2. It exists in three different modifications, i. e., rutile, anatase, and brookite. They all have their composition in common and also the basic building units which

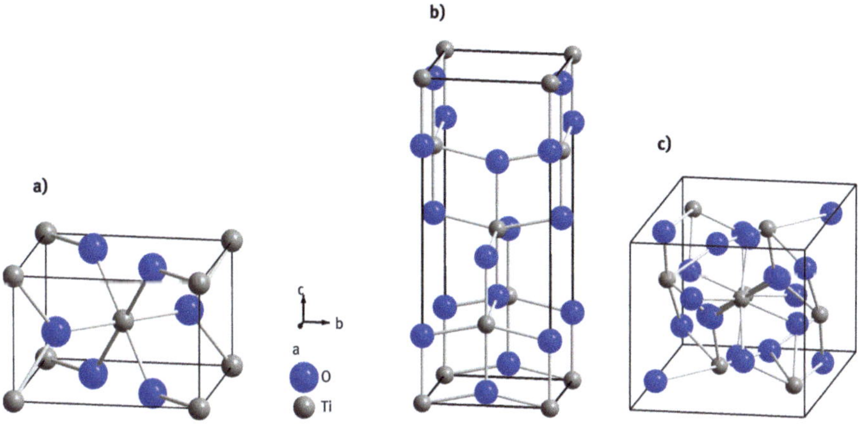

Figure 12.2: Unit cells of TiO_2 crystallizing in the rutile (a), anatase (b), and brookite (c) modification. Rutile and anatase crystallize with tetragonal symmetry but the physical properties of all modifications show significant differences.

are [TiO_6] octahedra. However, they differ in the way in how these octahedra are connected, see Figure 12.2.

Rutile crystallizes in the tetragonal crystal system with space group $P4_2$/mnm, Figure 12.2 a. [7] Anatase, by contrast, also forms tetragonal crystals, but in the space group $I4_1$/amd, Figure 12.2 b. [8] Brookite is orthorhombic, space group *P*bca, Figure 12.2 c. [9] The major difference between these three modifications can be rationalized by counting the number of shared edges of the basic [TiO_6] building units. This number is 2 for rutile, 4 for anatase, and 3 for brookite. Since the band structure is a measure for the chemical bonding it differs significantly for these different arrangements.

This difference in the crystal structure has great effects on the band structure and also on the band gap of TiO_2 (Figure 12.3). Thus, the band gap of rutile is about 3.03 eV [10] while the band gap of anatase is about 3.23 eV [10].

It becomes obvious that the chemical composition of a material and also its modification have a huge influence on its photocatalytic activity. This means, that only the correct chemical composition in combination with the desired modification make a compound a valuable photocatalytic active material.

To simplify the band gap considerations, the process of light absorption and electron excitation shall be simplified here to our needs. The term “band gap” will be used for the direct energy gap between VB and CB which can be observed by spectroscopic means.

As mentioned before, a photon of appropriate energy can excite an electron from the VB to the CB. The electron can be “lifted” to a state in the CB when the

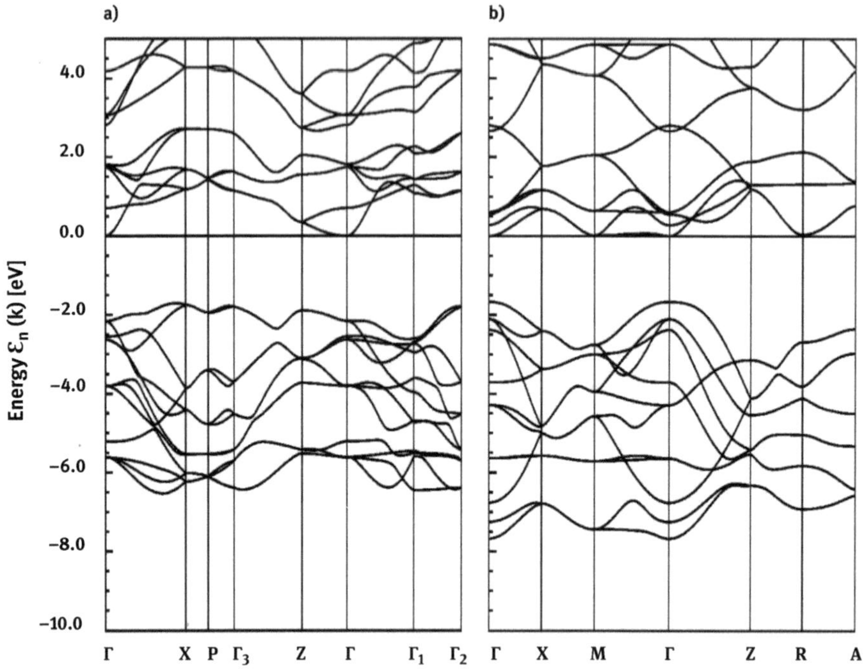

Figure 12.3: Calculated band structures of rutile (a) and anatase (b) TiO_2. [114] Per definition, the highest electronically occupied energy, the Fermi energy (E_F) is set equal zero. [115]

energy of the photon which depends on the wavelength λ of the photon beam is equal to or greater than the band gap energy ΔE_{BG}: [5]

$$h \cdot \upsilon = \frac{h \cdot c}{\lambda} \geq \Delta E_{BG} ,$$

with h is as the Planck constant, ν the frequency of the photon and c is the speed of light.

This process generates electron-hole pairs within the semiconductor which now behaves as a microelectrochemical cell. [2] Thus, the electron in the CB can be transferred to an electron accepting agent **A** forming the reduced species $\mathbf{A}^-$. The hole, on the other hand, can react as an oxidant and oxidize an electron donor compound **D** to $\mathbf{D}^+$.

These reactions of course can only take place when the compounds A and D have suitable HOMO and LUMO energies.

12.1.1.2 The Fermi level and charge separation

In order to estimate whether a desired reaction can be photocatalyzed with a certain semiconducting material, the energetic positions of the VB and CB have to be considered. The maximum energy of the VB is usually set equal to 0 in band gap calculations, because of difficulties in calculating an absolute potential. This maximum energy of the VB is usually referred to as the Fermi level $\boldsymbol{E_F}$. In addition, the temperature of the system in band gap calculations is set to $T = 0\,\text{K}$. At this temperature, energetic states below $\boldsymbol{E_F}$ are occupied by electrons and states above are empty. [1] However, at finite temperatures, few electrons are lifted to the CB. The probability of an electron, to get exited this way, is described by Fermi–Dirac statistics. In this statistics $\boldsymbol{E_F}$ describes a virtual electronic state of an electron, which is occupied with a probability of 0.5. Therefore, $\boldsymbol{E_F}$ is located at the middle of the band gap of a semiconductor (Figure 12.4). [11]

When the semiconductor is irradiated by light and electrons are excited to the CB this energy gives the available oxidation potential of a generated hole. The reduction potential of the electron, on the other hand, is determined by $\boldsymbol{E_{CB}}$ and the band gap:

$$E_{CB}[V] = E_{VB}[V] + \frac{E_{BG}[eV]}{e^0[e]},$$

with $\boldsymbol{E_{CB}}$ as the potential of the CB, $\boldsymbol{E_{VB}}$ the potential of the VB, e^0 the number of electrons, excited by a photon, which is equal to 1, and $\boldsymbol{E_{BG}}$ the band gap energy. [12] Therefore, $\boldsymbol{E_F}$ is most important when talking about photocatalytic processes.

Another effect associated with charge generation is charge separation. This aspect is fundamental for good photocatalytic performance since the material can act as a photocatalyst only when the electron-hole pair has a certain lifetime. After an electronic excitation to a state in the CB we can expect different ways for how the system loses energy. The first one is the desired reactive pathway, i. e. the photocatalytic reduction and oxidation processes, respectively. The second one, on the other hand, is the recombination of the excited electron and the hole. In order to

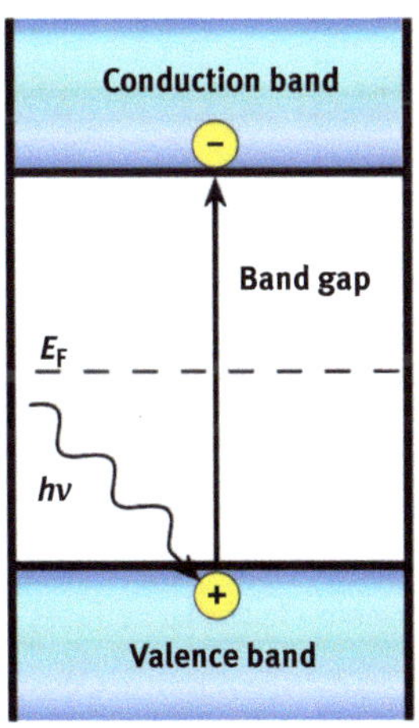

Figure 12.4: Simplified band structure of a solid and electron excitation. An electron of the VB is excited to the CB by absorption of a photon and an electron-hole pair is formed.

make the electron-hole pair, the so called exciton, needs to be available for a chemical reaction and a certain lifetime is necessary and also a spatial separation of both.

In general, the recombination time of electron and hole depends strongly on the nature of the material and is in the range of some ns to µs. In case of TiO_2 the recombination is a very fast process and takes only about 30 ps. [13] After this recombination the electron and hole are lost for a redox reaction. On the other hand, efficient charge separation and conduction leads to an increased photocatalytic performance. [14] It is well known that crystal defects and impurities act as charge trapping areas, where electron-hole recombination processes take place. [14, 15] As a consequence, photocatalysts with relatively poor crystallinity or a large amount of impurities favor the recombination of the electron-hole pair rather than the separation of light-generated electrons and holes. This leads to smaller quantum yields during photochemical reactions. However, sample crystallinity, crystal structure of the material, and particle size are parameters that strongly influence the recombination rate. [14]

In terms of photocatalytic activity, it is desirable to synthesize crystals without crystal defects and an infinite periodic structure. However, real crystals show more or less large amounts of crystal defects, which are supposed to be "inner surfaces", and impurities. Therefore, real crystals can be described as an intergrown aggregate of ideal crystals of various finite sizes. As a consequence, large crystals are supposed to show more structural defects than smaller ones (Figure 12.5). [16] A small crystal can be regarded as showing a smaller number of inner defects but of course the surface to volume ratio is increased in this case and the surface has also to be regarded as a defect. Smaller crystals exhibit a larger fraction of the highly crystalline areas leading to a lower tendency of charge recombination. If the particle size gets small enough, one particle consists of only one ideal crystal without defects. [14] Because of this, research into photocatalytically active nanoparticles becomes more and more important over the last years. Compared to bulk material, nanoparti-

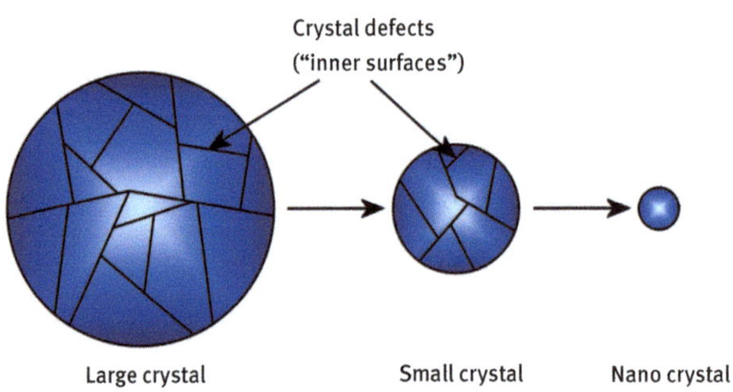

Figure 12.5: The amount of inner crystal defects is directly proportional to the crystal size. Large crystals can be described as composed of ideal crystal areas with finite dimensions. Small crystals are also composed of ideal crystals, but due to their smaller dimensions, only a few inner surfaces are present. Nanoparticles, by contrast, are composed of only one ideal crystal with finite dimensions and show excellent crystallinity.

cles show large surface area, a high degree of crystallinity, and short charge diffusion ways to the particle surface.

Another feature influencing the charge separation and conduction is the crystal structure of the photocatalyst. Materials consisting of layered structures show excellent charge separation and conduction and give excellent performance in photocatalytic processes. [17–19] It is proposed that the nature of the different layers of the compound can be closely related to the CB or VB, respectively. The two generated charges are well separated within the crystal. This results in a low tendency of recombination and drastically increases catalytic performance.

Nanoparticles show quantum size effects, which describe the increasing band gap and result in a blue shift of light absorption. [20, 21] The blue shift is exclusively due to particle size while other parameters like lattice constants remain constant. For a photocatalyst, this band gap shift can be beneficial, e. g. to drive the absorption into the energetically higher region of visible light, but it can also limit the applicability of a certain material when the absorption is driven to the UV range. [22]

12.1.2 How to tune a photocatalyst

12.1.2.1 Doping and Co-Catalysts

There are some semiconductors known that have a great oxidation and reduction power, so can be involved in many processes like air and water purification, solar cell applications, and photocatalytic water photolysis. The most famous ones are TiO_2, ZnO, and $SrTiO_3$. But great oxidation and reduction power means a wide band gap that consequently makes the semiconductor insensitive to visible light excitation and shifts its absorption to the UV region.

Nowadays interest in photocatalytic processes involving visible light sensitization of semiconductors grows rapidly. So, tuning the optical properties of well-known efficient semiconductors towards visible light sensitization is one promising direction.

One way of semiconductor modification is metal-ion and nonmetal doping. This method makes the semiconductor sensitive to visible light by decreasing its wide band-gap. One can find a lot of literature using doping as a modification method on TiO_2 as the most promising, abundant semiconductor with a 3.2 eV band-gap. A number of non-metals were examined for semiconductor doping, among them are H, C [23, 24], B, S, and N [25] atoms. As an example, good results in band-gap narrowing were achieved by N-doping, reported by Asahi et al. [25] and this area was developed successfully by the groups of H. Kisch [26], C. Burda [27, 28], C. S. Gopinath [29] and M. Anpo [30]. A variety of methods such as sputtering, ion implantation, chemical vapor deposition, sol-gel, oxidation of TiN, and decomposition of N-containing metal organic precursors are used for preparation and visible-light absorption occurs because of the mixing of nitrogen 2p states

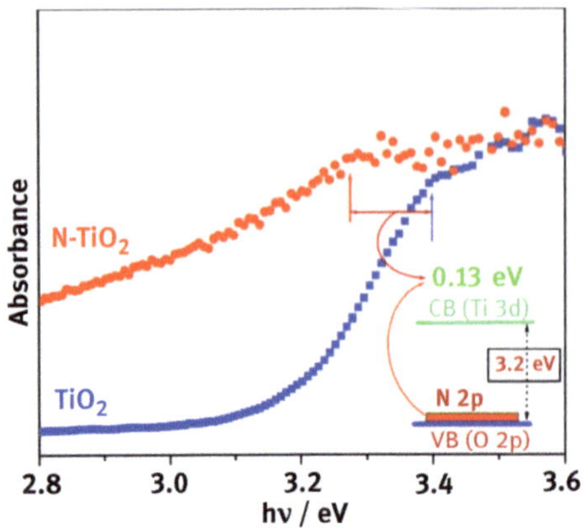

Figure 12.6: Band Gap modification on N-Doping in TiO_2. [29]

with O 2p states on the top of the valence band or a creation of N-induced mid-gap level (Figure 12.6).

The other possibility to tune the visible light sensitivity of wide band-gap semiconductors is doping with metal ions (Figure 12.7). [31] Among them are noble-metals, transition metals, and rare earth ions. Commonly, metal ion doping suppresses the recombination of the photogenerated electron-hole pairs and shifts the optical absorption edge. Still, not every metal enhances the visible-light sensitivity with this method, but doping of an already modified semiconductor increases the photocatalytic properties effectively. For example, Cu^{2+} co-doping of cerium (Ce) modified ZnO (0.1%Cu^{2+}-$Ce_{0.1}Zn_{0.9}O$) decomposes acetaldehyde into CO_2 effectively under visible-light irradiation. [32] Such dopants as Ag, Rb, Y, and La show similar UV-Vis absorbance spectra as undoped TiO_2. Modification of TiO_2 with rare earth metal ions [33] was done successfully with Sm^{3+} [34] and analyzed on methylene blue dye degradation experiments.

Chromium, molybdenum, [35] and vanadium [36, 37] cations were investigated in several groups because of the great influence on the photocatalytic activity and visible light absorbance of semiconductors as the consequence of changing the absorption edge of the semiconductor by the transition metals. Recent advantages in dye degradation were represented by T. Jafari [38] and computational methods by J. Ye. [39] For hydrogen evolution, the $SrTiO_3$ semiconductor was doped with Cr and Ta and showed enhanced activity under visible-light irradiation (Figure 12.8). [40] Also, modification of TiO_2 with Sb^{4+} cations was successful for dye modification with generated OH radicals upon visible-light irradiation. [41] Another efficient photocatalyst – ZnO – was successfully modified with Mn^{2+} and examined in dye degradation under visible light irradiation. [42]

The doping methods and semiconductor properties were investigated and described widely with TiO_2. [43]

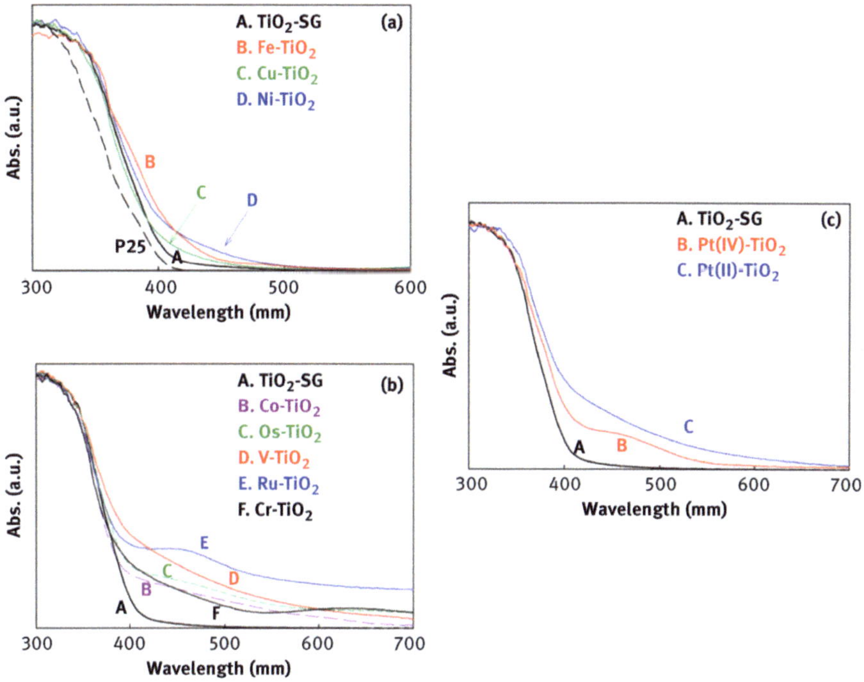

Figure 12.7: UV-Vis diffuse reflectance spectra (DRS) for various M-TiO_2 samples (0.3 at. % doping). [31]

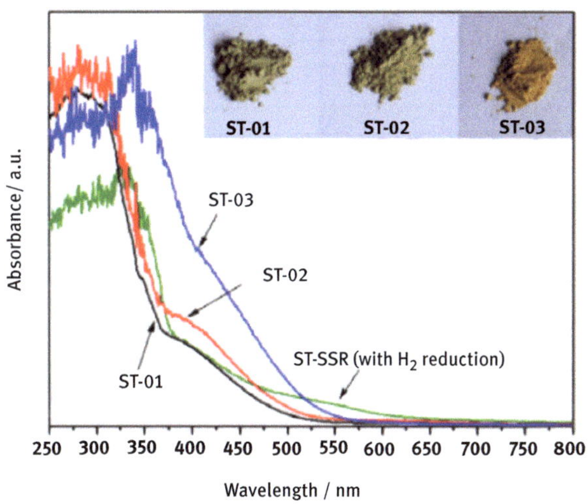

Figure 12.8: UV-Vis spectra of the final tubular (Cr, Ta) codoped $SrTiO_3$ (ST-01, ST-02, and ST-03) products and (Cr, Ta)-codoped $SrTiO_3$ as reference (ST-SSR). [40]

12.1.2.2 Particle size effect

As described previously in this chapter, there is a dependence between band gap and particle size of a semiconductor. In general, a decreasing particle size results in an increasing redox potential. However, enlargement makes a semiconductor less effective to sunlight activation. The recombination of electron-hole pairs is also dependent on the size and particle shape. Very fast recombination occurs at the bulk material core and reduces with decreasing particle size. The effect of electron-hole recombination grows again with further particle size decreasing because in this case it occurs on the semiconductor surface.

T. Ohno recently investigated the influence of pH and amounts of shape-control reagent – polyvinyl alcohol (PVA) added during the semiconductor preparation. [44] And among different particle sizes from 25 nm to 60 nm it was observed that the 40 nm size particle showed excellent activity on photocatalytic decomposition of acetaldehyde under UV irradiation.

Calcination temperature plays a crucial role in crystalline formation when preparing a semiconductor. Increased temperature and solid state synthesis are favorable for bulk material formation. But decreasing the temperature and performing the synthesis in solution (hydrothermal method), give a smaller particle size as was determined on $FeNbO_4$ semiconductor synthesis (Figure 12.9). [45] When controlling the temperature at 600 °C, the particle size was determined to be 10–20 nm and showed good photocatalytic activity for Rhodamin B dye degradation under visible light irradiation. In turn, the particle size influences all the optical and other properties of a semiconductor. The specific surface area and reduction power increase along with particle size decreasing, but absorption maximum shifts to UV region.

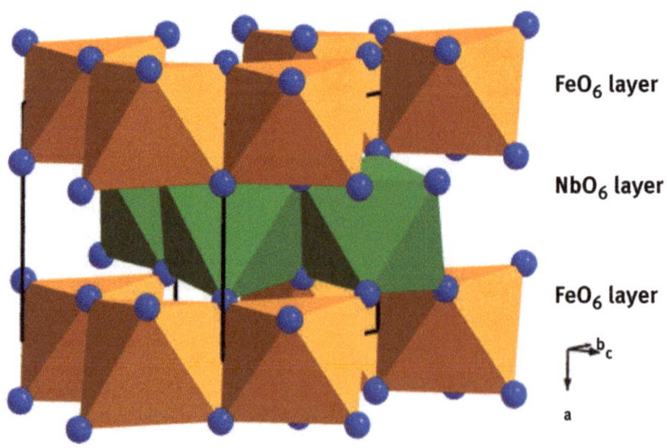

Figure 12.9: Structure of $FeNbO_4$ based upon the zigzag chain stacking sequence of FeO_6 and NbO_6 octahedra. [45]

12.1.3 Selected examples of photocatalysts and their application to organic synthesis

Research on new and efficient systems using the only sustainable energy source of our planet – sunlight – increased tremendously during the second half of the past century. This trend was further boosted by the decreasing availability of fossil fuels and an increasing demand of energy worldwide. The investigation of new and improved photocatalysts is nowadays one of the hot topics in modern science. In this chapter, a few selected examples of semiconductor photocatalysts and their uses in synthetic chemistry are presented in order to give an idea of the possibilities of such materials.

12.1.3.1 TiO_2 – an UV active photocatalyst

TiO_2 for photocatalytic usage has been intensively investigated during the past decades. The material is non-toxic, biologically and chemically inert. [22] The production and recyclability are well established. It shows high photocatalytic activity [46] which led to various applications during the past decades. Most of the studies were done with mixtures of rutile and anatase, which are commercially available (e. g. Degussa P-25; ratio rutile : anatase = 1 : 3 [47]) and show excellent physico-chemical properties, like specific surface area and purity grade. The first applications for TiO_2 were photochemical water splitting [48–51] and waste water purification. [22, 52, 53] Later on more complex reactions, e. g. the selective reduction or oxidation of organic compounds, were developed.

TiO_2 shows interesting properties for photocatalytic oxygenation of naphthalene and its derivates under UV-light irradiation. [54] The great advantage of this reaction is its high diversity of applications. Soana et al. investigated a photochemically catalyzed reaction of aromatic systems forming different products, depending on the reaction conditions (Figure 12.10).

It is noted, that the rate of oxidation of a substrate is highly influenced by the solvent. Excited TiO_2 reacts as a strong oxidant in water and oxidizes it to OH radicals, which react unselectively with the substrate. In organic solvents, by contrast, excited TiO_2 shows a smaller oxidation potential. Therefore, the solvent itself cannot be oxidized. The proposed mechanism is an electron transfer from the substrate to TiO_2 and from TiO_2 to O_2, respectively (Figure 12.10). [54]

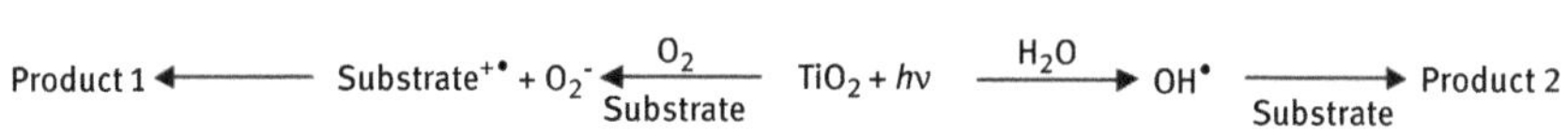

Figure 12.10: Possible pathways of the photocatalytically induced oxygenation. The high oxidation potential of TiO_2 in H_2O leads to OH radicals, which results in different products as compared to reactions under anhydrous conditions with a direct electron transfer mechanism. (Adapted from Ref. [54])

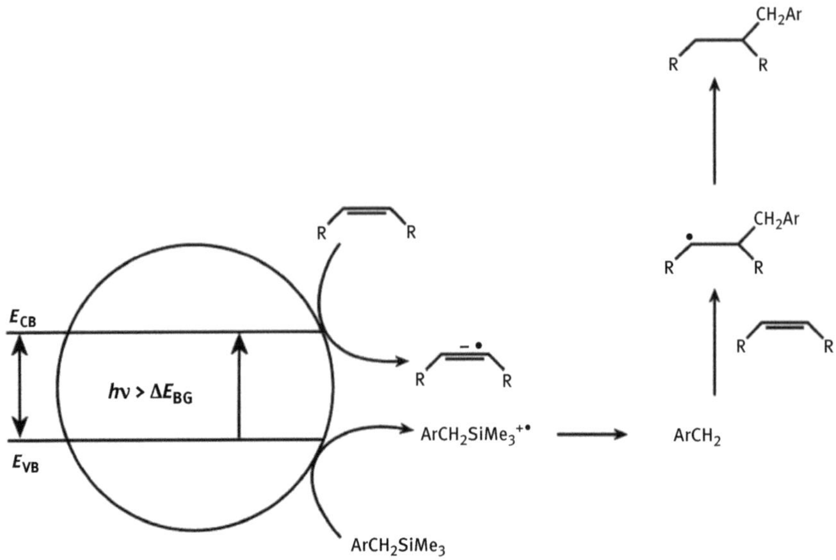

Figure 12.11: Schematic overview of the photocatalytic C–C bond formation. The potential of the VB (E_{VB}) of TiO_2 is much more positive than the potential of the electron donor ($ARCH_2SiMe_3$). The energy of the CB (E_{CB}) is higher in energy than the LUMO of the olefin (electron acceptor). When light with an energy higher than the band gap of TiO_2 is applied ($h\nu > \Delta E_{BG}$), the formation of the product is observable. (Adapted from Ref. [55])

One of the most interesting applications for TiO_2 is the photocatalytic C–C-bond formation with UV light. Albini and co-workers considered a radical alkylation mechanism of electron-withdrawing substituted olefins. [55]

Electrons in the VB of TiO_2 are excited with UV-light to the CB. The electron is transferred to the olefin, e. g. maleic acid, with a more positive potential than the CB of TiO_2. The electron donor, e. g. 4-methoxybenzyl(trimethyl) silane, on the other side, needs a less positive potential than the VB of TiO_2 (Figure 12.11). [55]

Another important application for TiO_2 as a photocatalyst is the photocatalytic reduction of CO_2 in aqueous solutions. [56–59] With this technique, the main contributor to the anthropogenic greenhouse effect can be transformed to chemicals like methanol and methane. The whole process is based on the reduction of CO_2 to its radical anion, but it is not fully understood yet. It is proposed that CO_2 and H_2O react with the photo-excited state of TiO_2. H_2O is oxidized to $H^•$- and $OH^•$-radicals, which react with the carbon species formed from CO_2. The products are CH_4 and CH_3OH. [60]

The photocatalytic decomposition of acetic acid for the production of hydrocarbons is another application for TiO_2. [61] Mozia and co-workers observed methane, ethane, propane, CO_2 and traces of butane and H_2 when irradiating acetic acid in an aqueous dispersion of TiO_2 with UV light ($\lambda_{max} = 365\,nm$). They designed a special

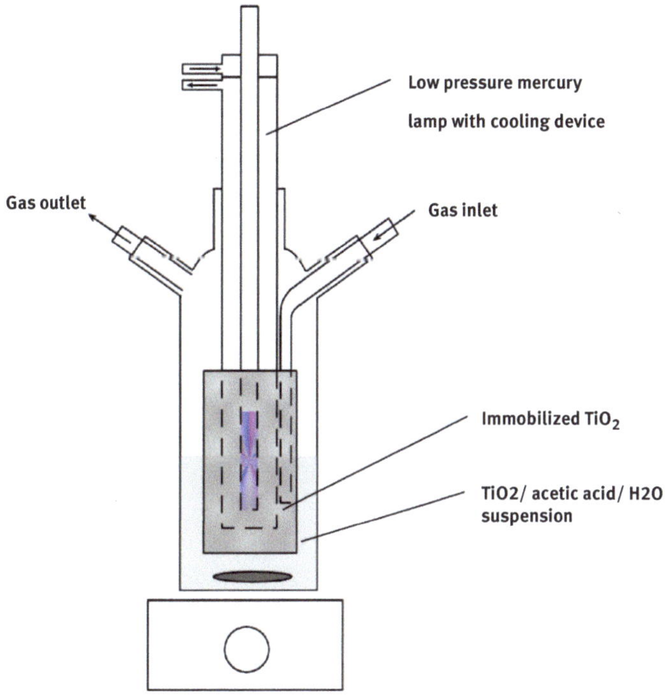

Figure 12.12: Reaction cell proposed by Mozia et al. for the photocatalytic degradation of acetic acid to form hydrocarbons, H_2 and CO_2. The combination of immobilized TiO_2 and TiO_2 in suspension gives good results of the photocatalytic process. (Adapted from Ref. [61])

reactor with reverse light irradiation and gas outlet for continuous product extraction (Figure 12.12). The yield of all products could be increased by immobilizing the semiconductor on glass fibers, but best results were obtained by a combination of immobilized and dispersed photocatalyst.

With their experiments, Mozia et al. showed a combination of photocatalytic degradation and assembly of organic compounds. [61] Since one of the products of this reaction is CO_2, a suitable combination of a photocatalytic CO_2 conversion to CH_4 can open a whole new source for organic chemicals in the future.

12.1.3.2 Selected examples of visible light active photocatalysts

TiO_2 shows excellent performance in photocatalysis. The advantages of this photocatalyst were already mentioned; however, the big disadvantage of TiO_2 is its wide band gap, which limits its use to the UV-range. Special light sources need to be installed and the reaction itself must be carried out in quartz glassware to guarantee the penetration of the reactor with UV-light. In addition, it is highly doubtful, whether reactions needing artificial UV light, are the key to establishing sustainable photochemistry.

The goal to find visible light active photocatalysts led to the development of a huge amount of materials with a band gap appropriate to the visible range of light during the past years. A broad spectrum of applications came up, like waste water treatment [62] and water splitting for hydrogen formation. [17, 18, 50, 63–65] Furthermore, some photocatalysts were developed for synthetic reactions. In principle, reactions which are catalyzed by TiO_2 can also be performed by other semiconductor photocatalysts. Nevertheless, some of these photocatalysts suffer from self-digesting and photo bleaching, are expensive to prepare, or are sensitive to common organic and inorganic solvents.

Therefore, nowadays each photocatalyzed reaction eventually needs a specifically designed photocatalyst. At first view, this seems to be disadvantageous. However, these reactions are of great potential in terms of selectivity for oxidation or reduction processes.

Recent experiments showed that immobilized Ir^{3+} complexes which are covalently attached on mesoporous SiO_2 can catalyze the selective oxidation of trans-stilbene into benzaldehyde and 1-naphthol into 1,4-naphthoquinone under visible light irradiation (Xe lamp with cut-off filter $\lambda > 420$ nm) in the presence of O_2 (Figure 12.13). [66]

Mori and co-workers found, that neither $O_2^{-\bullet}$ radicals nor $OH^\bullet$ radicals are the reactive species for the product formation. They proposed an energy transfer from the excited Ir^{3+}-complex to O_2. Oxygen is excited to the singlet state, which reacts with the substrates. By adding thiourea, a well-known $^1O_2$ scavenger, [67] the photo-oxidation process was significantly inhibited. They also proposed that the structure of the catalyst carrier plays the most important role for the catalytic reaction. The larger the specific surface area of SiO_2, the greater the catalytic performance of the Ir^{3+} catalyst. Mori et al. suggest SiO_2 providing a microenvironment that alters the photochemical properties of the metal complexes. This leads to an increased photocatalytic reactivity. [66]

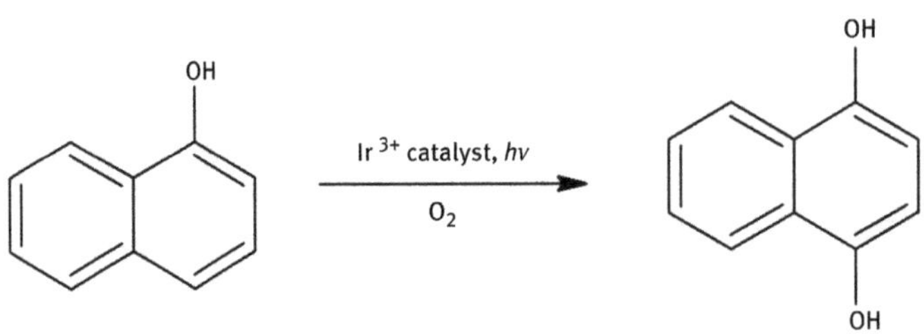

Figure 12.13: Oxidation of 1-naphthol into 1,4,-naphthoquinone using Ir^{3+} catalyst immobilized on silica. O_2 was used as oxidant.

Most interesting in terms of synthetic applications is the photocatalytic conversion of alcohols to their corresponding aldehydes. CdS particles are promising candidates for visible light active photocatalysts for this reaction. The band gap of this compound is about 2.49 eV, [68, 69] which corresponds to the blue-green part of visible light.

It was found that pure CdS particles are able to photocatalyze the oxidation of ethanol and isopropanol to their aldehydes. However, a crucial problem of these particles is the oxidation of the catalyst itself. In the presence of H_2O, which is needed to perform the oxidation reaction, S^{2-} of the photocatalyst is oxidized to SO_4^{2-} and the catalyst is destroyed. Another disadvantage of this reaction is a rather complicated reaction setup using alcoholic vapors instead of solutions. [68]

Furthermore, polycrystalline CdS was used for oxidative coupling of benzyl alcohol, its *p*-substituted derivates and benzyl amines. [70] The photocatalyst was dispersed in oxygen-free acetonitrile. Benzyl alcohol was dissolved in this mixture and the whole system was irradiated with blue light for 24 h. Mitkina et al. isolated benzaldehyde, hydrobenzoin, and benzoin. Photoconversion of the alcohol and product composition was influenced by variation of reaction temperature and concentration of initial benzyl alcohol. This photocatalytic system is also suitable for the conversion of benzyl amines. *N,N*-dimethylbenzyl amine was converted to 1,2-diphenyl-*N,N,N′,N′*-tetramethylethylendiamine and benzaldehyde. Mitkina and coworkers showed that this photocatalytic reaction is also applicable on derivatives of benzyl amine. They proposed CdS as an effective tool for photocatalytic C–C-coupling with visible light. [70]

One more photooxidative carbon-carbon bond formation with CdS was demonstrated in a Mannich type reaction of *N*-aryltetrahydroisoquinolines **1** with ketones **2a–c**, involving proline as organocatalysts. Upon irradiation with blue light of 460 nm the products **3a–d** can be obtained in good yields of 76–89 % (Table 12.1). [19] The reaction can successfully be performed in neat ketone and also with a significantly reduced amount of ketone (see Table 12.1, entry 1a–c) in CH_3CN as a solvent.

A different strategy was developed by Zhang and co-workers proposing the formation of a core-shell photocatalyst based on CdS. CdS nanoparticles with Pd cores were prepared via hydrothermal synthesis. The resulting Pd@CdS nanocomposite showed excellent results for the selective oxidation of e. g. benzyl alcohol, p-flurobenzyl alcohol and p-methoxy-benzyl alcohol. [69]

Yields, selectivity and conversion of all reactions could be increased tremendously by the use of the Pd@CdS core-shell photocatalyst as compared to pure CdS as photocatalyst. Thang and co-workers explain the improved catalytic performance as a coupling of the light absorption behavior of CdS with the charge trapping ability of Pd. According to this model, the photo generated electrons of CdS are transferred to Pd, which then acts as the reduction site. The holes, by contrast, are localized on CdS and oxidize the alcohol (Figure 12.14). This separation of charges and

Table 12.1: Photocatalytic Mannich reaction of *N*-aryltetrahydroisoquinolines 1 with ketones 2a–c and l-proline on CdS [a].

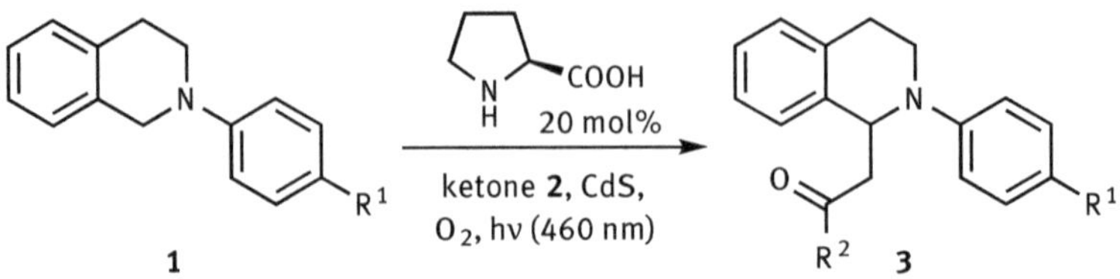

entry	R^1	ketone	product	reaction time [h]	yield [%][b]
1a	H	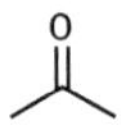 **15a**	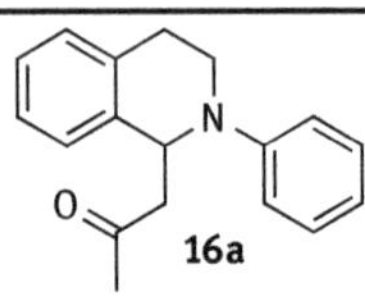 **16a**	24	86[c]
1b				24	90[d]
1c				24	100[e]
1d				24	87
2	OMe	 **15a**	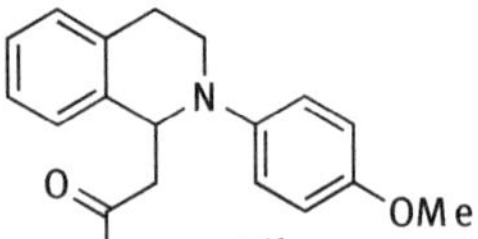 **16b**	18	89
3	H	 **15b**	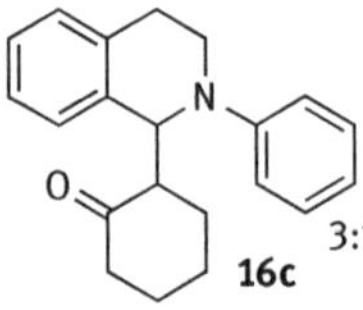 **16c** 3:1 dr.	24	79
4	H	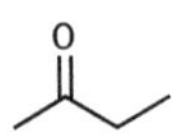 **15c**	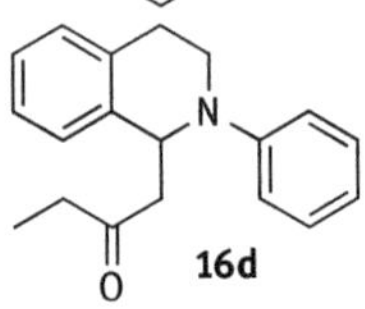 **16d**	15	76

[a] Unless otherwise noted all experiments were performed with amine (1 eq) and l-proline (0.2 eq) in a 5 mg/ml mixture of CdS in neat ketone (c_{amine} = 0.25 mol/l). Reactions were run in schlenk tubes with an attached oxygen balloon and irradiated with high power LEDs (460 nm) for the time indicated. [b] Given yields correspond to isolated product. [c] Reaction performed in CH_3CN with 2 equiv. of acetone; the conversion was determined by GC analysis. [d] Reaction performed in CH_3CN with 5 equiv. of acetone; the conversion was determined by GC analysis. [e] Reaction performed in CH_3CN with 10 equiv. of acetone; the conversion was determined by GC analysis.

reactive sites is common in heterogeneous catalysts and is most important for good catalytic performance. [14]

Layer structured substances offer charge separation ability by nature. Recent publications showed $PbBiO_2Br$, a semiconductor with a crystal structure related to the anti-$ThCr_2Si_2$ structure type (Figure 12.15), [71] to be a good substitute for TiO_2 as photocatalyst for several reactions. [19, 62, 72] The big advantages of $PbBiO_2Br$ are the excellent stability against organic and inorganic solvents, the easy and inexpensive synthesis and the material's band gap of 2.47 eV, which is in the range of visible light.

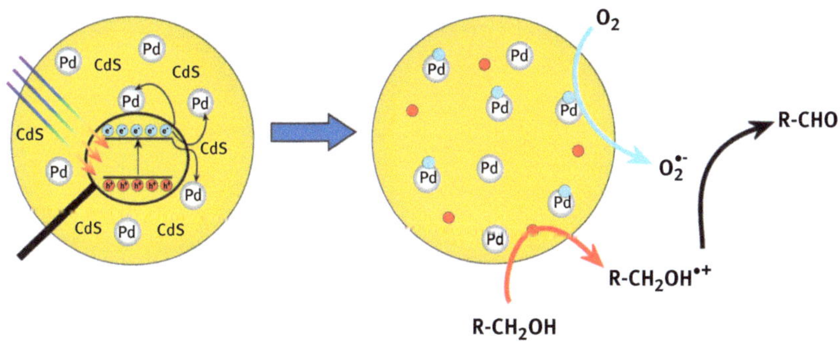

Figure 12.14: Proposed mechanism of the oxidation of alcohols to aldehydes in the presence of oxygen. The catalyst for this reaction is composed of CdS doped with Pd. Absorbed light energy generates electron-hole pairs in CdS. The electrons are transferred to Pd, which is the reactive site for the reduction of O_2, the holes are localized within CdS, where the oxidation of the alcohol takes place. (Adapted from Ref. [69])

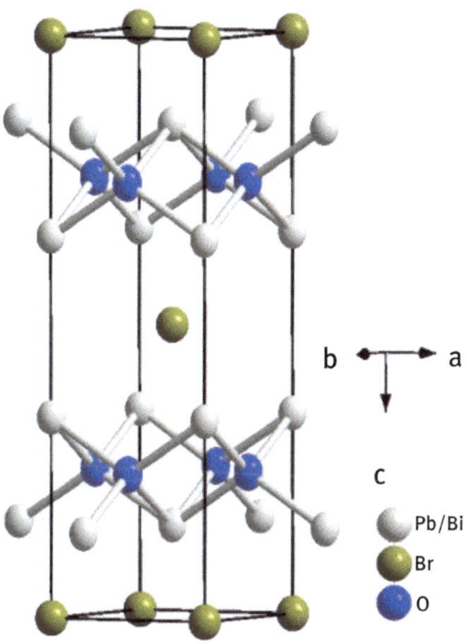

Figure 12.15: Unit cell of $PbBiO_2Br$. The structure consists of $[M_2O_2]^+$ layers ($M = Pb/Bi$) perpendicular to [001] which are separated by bromide ions. [71]

The material is quite useful as a photocatalyst for waste water purification [62] and also valuable synthetic reactions can be photocatalyzed with it. Thus, it is possible to perform the reductive formation of aniline from nitrobenzene. [72] Füldner et al. were able to reduce the nitro group of nitrobenzene and some of its derivates to the corresponding amines at almost full conversion. It was shown that $PbBiO_2Br$ produces hydrogen in the presence of triethanolamine as an electron donor. So a selective, clean, and complete reduction of the substrates to their corresponding

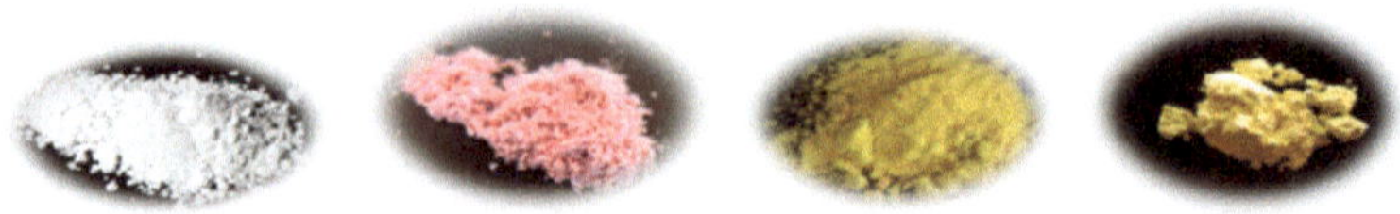

1 2 3 4

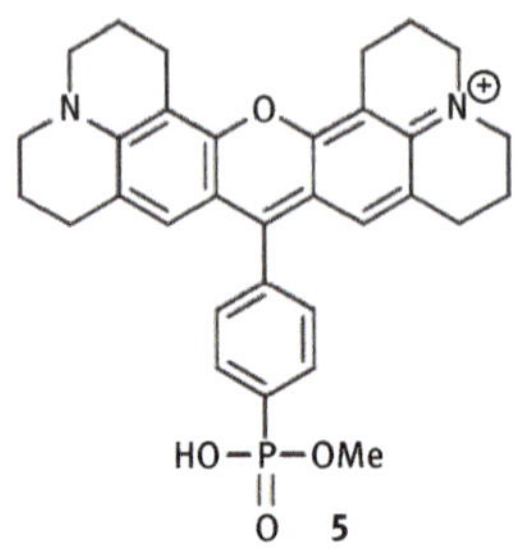

5

Figure 12.16: Semiconductors (1–4) used in C–C bond formation. [19] Phos–Texas–Red (5), compound for covalent surface immobilization on TiO_2.

anilines becomes possible under blue light irradiation. Furthermore, the photocatalyst is reusable for at least six catalytic cycles without any loss of catalytic performance by simple sonication of the particles before reuse. This photocatalyst was also successfully used with sunlight. [72]

One more example of heterogeneous photocatalysis in organic synthesis was reported by Cherevatskaya et al. [19] The work explores enantioselective C–C bond formations upon reduction of halogen precursors with photocatalysts excited by visible light (Figure 12.16). Several semiconductors were compared in this study and showed different activity. The above mentioned $PbBiO_2Br$ photocatalyst was prepared as nano and bulk crystals and used in the α-alkylation of aldehyde (**7**, Figure 12.17) in the presence of 20 mol% of secondary amine (**8**, Figure 12.17 as chiral catalyst [73]. After 20 hours of irradiation with blue light the product (**9**, Figure 12.17) was obtained in 69 % yield with 71 % *ee* at r.t. and 40 % yield with 84 % *ee* at −10 °C in case of bulk crystals (**3**, Figure 12.16) because its surface area is only 0.17 m2/g. The nanocrystalline $PbBiO_2Br$ (**4**, Figure 12.16) which was obtained with a larger surface area of 10.8 m2/g, gave an increased yield of 84 % (72%*ee*) under the same conditions. Lower temperature (−10°C) increases enantioselectivity to 83 % *ee* at, but with only 49 % yield. It is possible to reuse the nanocrystalline $PbBiO_2Br$ (**4**, Figure 12.16), but black organic surface deposits lead to significantly slower conversions. When the microreactor with 1 mm diameter tubes was used to increase the light intensity the irradiation times were reduced to 3 and 10 hours at r.t. and −10°C with product yields of 41 % and 69 %, respectively.

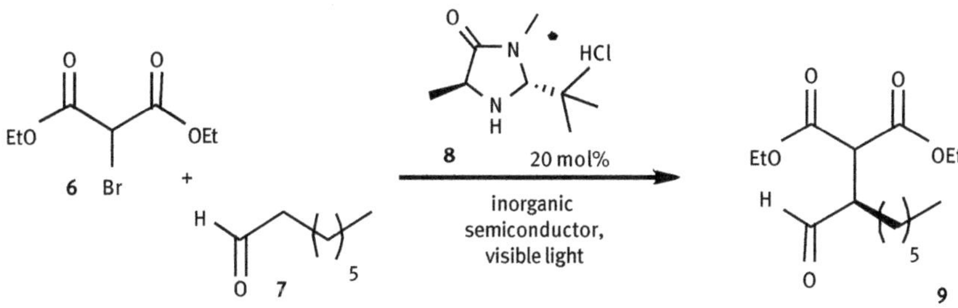

Figure 12.17: Enantioselective α-alkylation of an aldehyde catalyzed by heterogeneous semiconductor under irradiation with visible light. The chiral secondary amine provides the steric information to form the product. [19]

Moderate yield and good enantioselectivity of the product (**9**, Figure 12.17, 55 % yield, 71 % *ee*) was obtained when unmodified TiO_2 was used after 20 hours irradiation with 440 nm as it absorbs only a small fraction of visible light. The microreactor setup helps to increase light intensity and thus reduces the reaction time to three hours affording good yield of the product (**9**, Figure 12.17, 76 % yield, 74 % *ee*). Surface modified TiO_2 (**2**, Figure 12.16) allows the reaction to run with green light (530 nm) (**9**, Figure 12.17, 65 % yield, 81 % ee, −10 °C).

The nanocrystalline $PbBiO_2Br$ photocatalyst (**4**, Figure 12.16) in combination with homogeneous organic catalyst is also applicable to bromoacetophenone (**10a**) and 2,4-dinitro benzylbromide (**10b**) as different bromo-precursors (Figure 12.18). This represents the first application of heterogeneous photocatalysis for enantioselective transformations.

As described above, comparable reactions have been developed for TiO_2 before, but without stereo control of the product. [55, 74, 75]

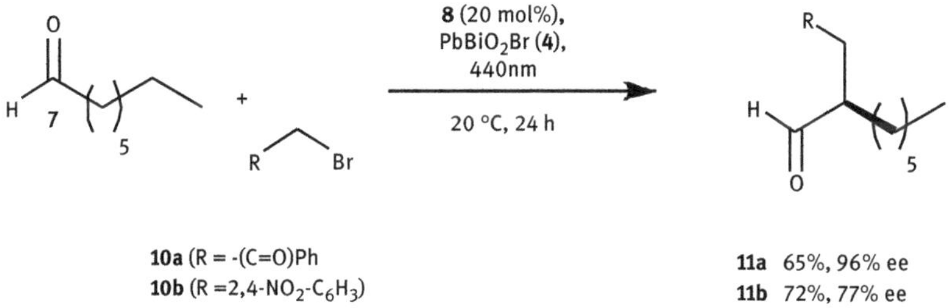

Figure 12.18: Alkylations using bromoacetophenone (**10a**) in CH_3CN or 2,4-dinitrobenzylbromide (**10b**) in DMSO, chiral amine 8, $PbBiO_2Br$ (4) and blue light.

12.2 Organic semiconductors

Organic semiconductors are materials with an extended, conjugated π-electron system. Depending on the molecular weight they can be classified into small molecules and polymers. Figure 12.19 depicts the chemical structure of different organic semiconductor materials. The combination of the unique electronic and optical properties with the ease of processing as well as the possibility to change the properties from isolator to metal by doping, enable broad applications in organic electronics, like organic field effect transistors, [76, 77] organic light emitting diodes, [78] and organic solar cells. [79] Optimization of device performance can be accomplished by alignment of the HOMO and LUMO energy levels and the band gap by structural modifica-

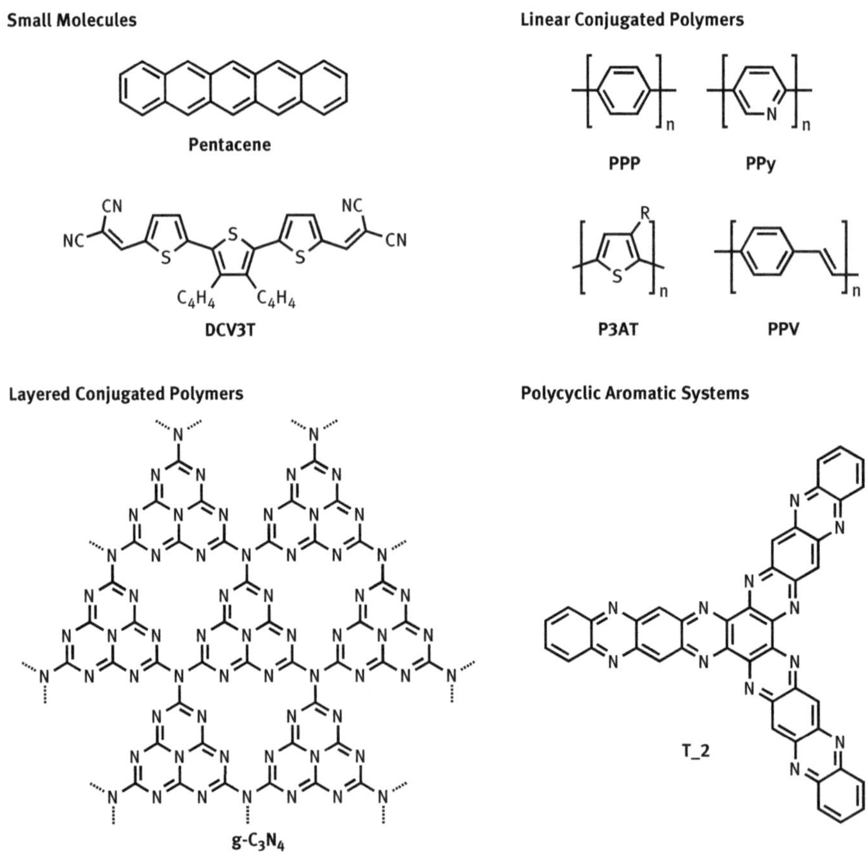

Figure 12.19: Chemical structures of conjugated small molecules and polymers. (DCV3T = dicyano-vinyl-terthiophene [91]; PPP = poly(*p*-phenylene); PPy = poly(pyridine-2,5-diyl); P3AT = poly(3-alkyl-thiophene-2,5-diyl); PPV = poly(*p*-phenylene-vinylene); g-C_3N_4 = polymeric graphitic carbon nitride; T_2 = tris(aza)pentacene [116]).

tion. [80, 81] The possibility to manipulate the optical and electrochemical properties by molecular design makes organic semiconductor materials appealing for application in photocatalysis. Tuning of the physical properties by structural engineering requires basic knowledge of the energy structure and charge transfer on the molecular level. A short overview is presented in the following chapters.

12.2.1 Basic properties of organic semiconductors

12.2.1.1 Band structure and band gap

The semiconductor properties arise from the alternating single-double-bond structure resulting in a delocalization of π-electrons along the σ-bond backbone. In ideal organic crystals interaction of the π- or π*-orbitals, respectively, within the backbone and between adjacent molecules, causes hybridization of the energy levels leading to a three-dimensional band structure. [81] Analogous to inorganic semiconductors, the highest occupied band (which originates from the HOMO of a single repeating unit) is defined as VB, while the lowest unoccupied band (emerging from the LUMO of a single monomer unit) is called the CB. The energy difference is called the band gap $\boldsymbol{E}_{BG}$. [81] The degree of delocalization depends on the nature of bonding as well as the crystallinity of the organic semiconductor materials. Organic solids are composed of individual molecules. Intermolecular bonding is caused by weak van-der-Waals interactions, hydrogen bonds and π-π-interaction typically producing rather disordered, polycrystalline or amorphous materials. [77] Compared to the strong covalently-bonded (or ionic) three-dimensional crystal structure in inorganic semiconductors, the electronic delocalization is rather low resulting in small bandwidths. [82] In principle, an enhanced overlap of π-orbitals implies a higher electron delocalization leading to a broadening of the bands and a narrowing of the band gap. Appropriate choice of conjugated backbone and substituents determine planarity and molecular organization thus crystallinity as well as the position of the energetic levels. [80]

12.2.1.2 Photoinduced electron transfer – Exciton generation and dissociation

The weak intermolecular interactions in organic solids have also implications for the electronic excited states. Absorption of a photon ($h\nu = \boldsymbol{E}_{BG}$) excites an electron from the VB to the CB generating a strongly bound electron-hole pair, a Frenkel exciton. In comparison to inorganic materials, the optical excitations are usually localized on one molecule due to the low extension of the electronic wavefunction. [82, 83] Coulomb forces between the electron and the hole result in binding energies typically between 0.5 to 1.0 eV, which correspond to an exciton diffusion length of few nanometers. Interactions of the electron-hole pair with electron acceptors or donors leads to exciton dissociation by electron transfer reactions (chemical doping). [82] The dissociation of the electron-hole pair is favorable if the difference of energy between the ionization potential of the excited donor and the electron affinity of the acceptor is larger than the exciton binding energy. Beside the desired reaction, the electron-hole pair can recombine in various loss mechanisms under dissipation of energy. [80] The purity of organic semiconductors compared to their inorganic counterparts is rather low. Conjugated semiconductors exhibit a high density of traps, which emerge from chemical and physical defects and can act as sites for exciton recombination or as doping agents.

12.2.2 Application of conjugated polymers in photocatalysis

Conjugated organic semiconductors allow the tuning of their physical properties by structural modification. Adjustment of the energy levels can be used to enhance the thermodynamic driving force of photocatalytic reactions. Control of the band gap energy enables absorption in the visible part of the solar radiation spectrum. However, in comparison to inorganic semiconductors, this class of materials is rarely investigated as photocatalysts. In this chapter selected examples of undoped, metal-free, polymeric organic semiconductor materials as photocatalysts are presented.

12.2.2.1 Linear conjugated polymers

The photocatalytic degradation of water pollutants by the inorganic semiconductor TiO_2 under UV irradiation was intensively studied due to its low toxicity, high stability, and excellent photocatalytic performance. [46] However, the lack of absorption in the visible part of the solar spectrum constitutes a major drawback. In order to

Table 12.2: Energy levels, band gaps and potentials of π-conjugated polymers.

π-conjugated polymer	E_{HOMO} [eV]	E_{LUMO} [eV]	E_{BG} [eV]	E_{VB} [V][a]	E_{CB} [V][a]	Ref.
Poly(3-octylthiophene-2,5-diyl)	5.2[c]	3.4[d]	1.9	1.2[b]		[119]
Poly(pyridine-2,5-diyl)	5.8[e]	3.4[d]	2.4	1.1[f]	−1.3[g]	[90]
Poly(*p*-phenylene)	6.1[e]	3.2[d]	2.9	1.4[f]	−1.5[g]	[90]
Polymeric graphitic carbon nitride			2.7	1.15	−1.55	[107]
Graphite oxide	5.9–6.4[h]	–	2.4–4.3[i]	–	–	[120]

[a] All values are reported in reference to the SCE electrode. If necessary the original values were converted according to reference [121]. E_{VB} determines the top of the VB. E_{CB} depicts the bottom of CB. [b] Oxidation onset measured by cyclic voltammetry. [c] Estimated from the oxidation onset using the following equation: $E_{HOMO} = E_{ox,onset} - E_{1/2}(Fc/Fc^+) - 4.8$ eV. [d] Calculated according to: $E_{LUMO} = E_{HOMO} - E_{BG}$; [e] Measured by atmospheric UV photoelectron analyser. [f] Calculated from $E_{VB} = E_{HOMO} - 4.7$; 4.7 eV was assumed as a scale factor relating SCE to vacuum. [g] Calculated from $E_{CB} = E_{VB} - E_{BG}$. [h] The top energy level of the VB to vacuum, was obtained by DFT calculations for 12.5 % and 25 % O coverage. [i] The bandgap energy depends on the oxidation level.

use solar energy more effectively, stable linear π-conjugated organic polymers with band gaps in the visible range were developed for the degradation of organic pollutants like textile dyes [84] and agrochemicals. [85–87] Yamamoto et al. investigated the photocatalytically active poly(3-octylthiophene-2,5-diyl) and poly(2,5-dihexoxy-*p*-phenylene) films for wastewater treatment. [85, 86] The degradation rate could be increased by introduction of molecular oxygen. Based on the detection of different active oxygen species and decomposed intermediates, a degradation mechanism involving hydroxyl radicals, generated from superoxide radical anions in several steps, was proposed. [86] Superoxide radical anions are formed by the initial electron transfer from the photoexcited polymer to molecular oxygen. Despite the formation of active oxygen species the π-conjugated polymer films were stable. [86]

Besides the photochemical activation of molecular oxygen, π-conjugated linear polymers were applied for the heterogeneous photocatalytic reduction of water [88–90], carbonyl compounds [89, 90], alkenes [89, 90] and CO_2 [92–94] in the presence of tertiary amines as sacrificial electron donors (Figure 12.20).

The approach to use organic semiconductors for selective photoreductions was developed and established by Yanagida and co-workers. They showed that the interfaces of linear poly(arylene) systems provide efficient charge separation and elec-

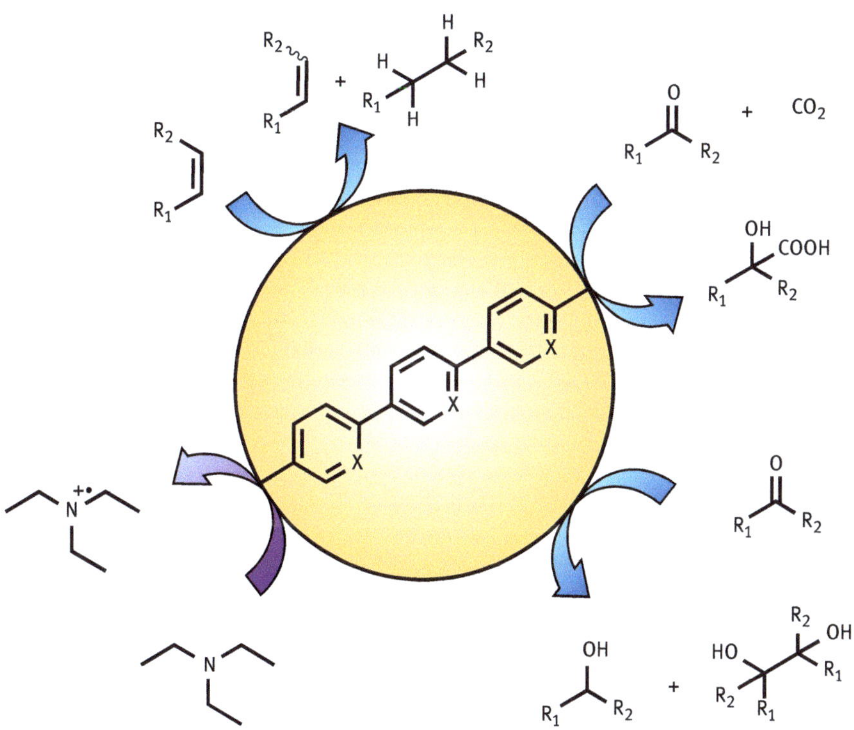

Figure 12.20: Application of poly(arylene) systems as photocatalyst in organic synthesis. [89, 90, 92]

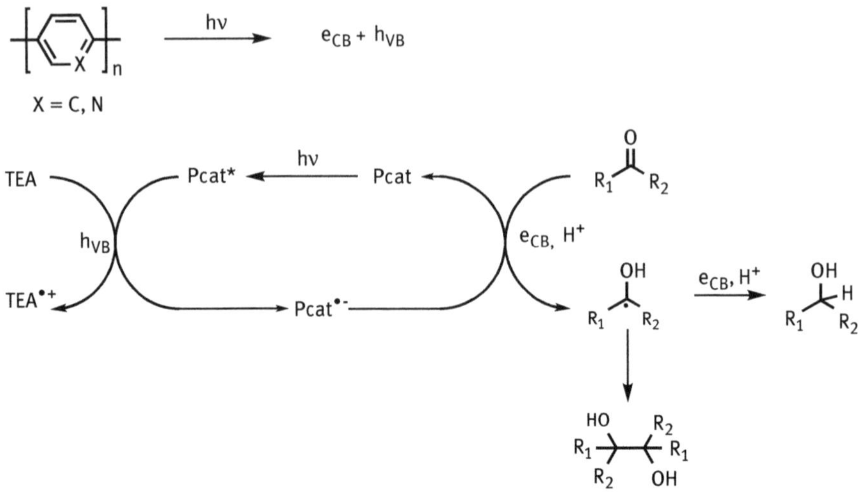

Figure 12.21: Photoreduction of carbonyl compounds using linear π-conjugated polymers; e_{CB} = electron in the CB of the semiconductor; h_{VB} = hole in the VB of the semiconductor; TEA = triethylamine. [89, 90]

tron transfer upon photoexcitation. The photoreduction of aromatic carbonyl compounds [89, 90] could be selectively controlled by the structure of the polymeric photocatalyst. Under visible light irradiation ($\lambda > 400$ nm), poly(pyridine-2,5-diyl) (PPy) exclusively afforded the corresponding alcohol in good yields. Poly(*p*-phenylene) (PPP) reduced the carbonyl compound to the alcohol and/or the one-electron-transfer coupling product – the pinacol – dependent on the reduction potential of the intermediary hydroxyl radical, the steric demand of the substituents and the reaction conditions (Figure 12.21). It is worth noting that the photoinduced reduction of benzaldehyde in methanol using PPP provided selectively the corresponding pinacol. [89, 90]

The preference in the product formation could be explained by mechanistic investigations. Photoexcited state quenching of PPP* with triethylamine provided a PPP-radical anion, which transfers electrons to the carbonyl compound leading to the observed reduction products, while oxidative quenching of the excited PPy* leads to sequential electron or hydride transfer. The lack of repulsion in PPy between the hydrogen atom at the 6 position and the nitrogen atom of the adjacent repeating unit facilitated the formation of coplanar chains enabling efficient electron delocalization of the radical anion $PPy^{\bullet -}$ and stabilization of hydrogenated coplanar quinoid-like PPy-intermediates (Figure 12.22). Similar to the NADH-reduction mechanism, the hydrogenated PPy could transfer hydrides to aromatic carbonyl compounds. This pathway is assumed to be decisive for the conversion of carbonyl compounds whose reduction potential is more negative than the CB potential of PPy (Table 12.2). [90] Rouch et al. investigated the influence of Brønsted and Lewis acids

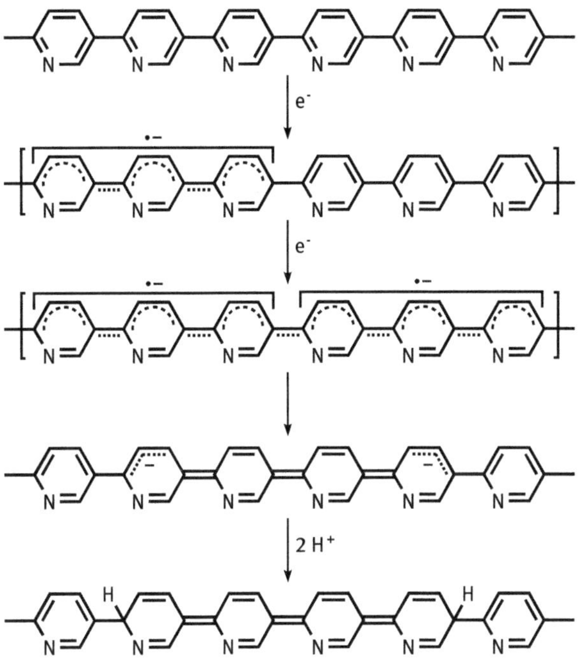

Figure 12.22: Hydride mechanism – Formation of hydrogenated PPy. [90]

on the reaction rate of the PPP-catalyzed photoreduction of benzaldehyde to the pinacol-coupling product. [95] The addition of Brønsted acids accelerated the reaction rate remarkably. Oxalic acid increased the relative rate by a factor of 39.5 in comparison to the photocatalytic system without acid. Additionally, an enhanced product formation was observed (90 % (oxalic acid) vs. 76 % (no acid)). The addition of a Lewis acid also results in an enhancement of the reaction rate, but the yield decreased significantly (rel. rate of 7.8, 53 % yield for aluminium chloride). [95]

12.2.2.2 Conjugated polymers with layered structure

Carbon nitride is one of the oldest synthetic polymers. Berzelius and Liebig already synthesized and described a polymeric derivative in 1834. [96] However, the first reported application as metal-free catalyst emerged 2006. [97, 98] Polymeric graphitic carbon nitride (g-C_3N_4) is the most stable allotrope of carbon nitride. The scaffold is constructed from N-bridged tri-*s*-triazine units forming a π-conjugated layer with remarkable thermal (up to 600 °C in air) and chemical stability. [117] Carbon nitride not only provides the requirements of a photocatalyst, but also features surface basicity due to terminating amino groups and the lone-pair on the heteroatoms of the poly(tri-*s*-triazine) framework (Figure 12.23). [99]

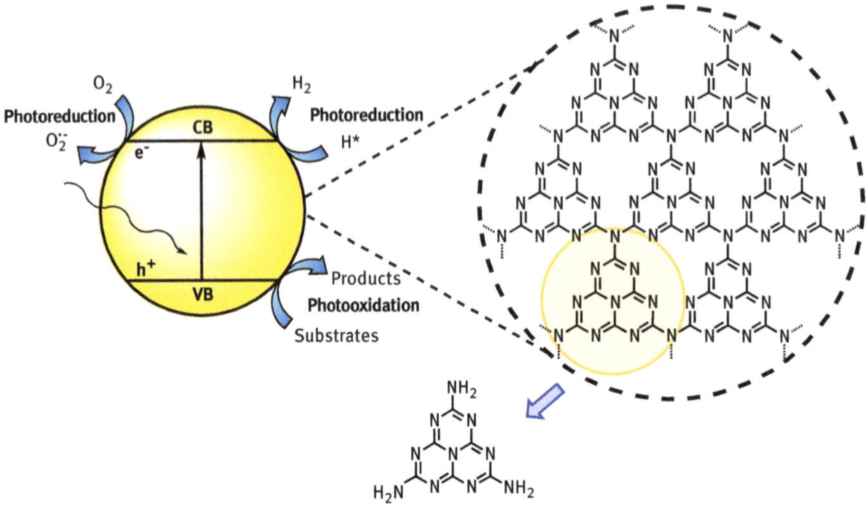

Figure 12.23: Possible photocatalytic strategies and chemical structure of g-C_3N_4 and the tri-*s*-triazine building block. [117]

Besides several applications in photochemical water splitting [100–102] as well as photodegradation of dyes and pollutants, [103–106] g-C_3N_4 was successfully applied to organic synthesis as visible light photocatalyst (Figure 12.23). Antonietti et al. investigated selective oxidation reactions of alcohols and amines by carbon nitride photocatalysis with visible light ($\lambda > 420$ nm) and molecular oxygen as terminal oxidant (Figure 12.24). [99, 107, 108] To improve the photocatalytic performance of the organic semiconductor, they enhanced the specific surface area from 8 m^2/g (bulk g-C_3N_4) to ca. 200 m^2/g (mpg-C_3N_4) by introduction of porosity using silica nanoparticles as templates. [99] Photoexcited mesoporous polymeric g-C_3N_4 (mpg-C_3N_4) exhibits a high oxidation potential of the CB (−1.55 V at pH 7 vs. SCE (Table 12.2)) capable of activating molecular oxygen by electron transfer. Subsequent photooxidation of the substrates by the hole in the VB (1.15 V at pH 7 vs. SCE (Table 12.2)) and deprotonation/dehydrogenation by the activated radical oxygen species afforded the corresponding aldehydes/ketones or imines, respectively. The high surface basicity of mpg-C_3N_4 facilitates deprotonation, additionally. [99] The oxidative dehydrogenation reaction of primary amines tolerated the introduction of electron-withdrawing and donating groups on the aromatic moiety. [107] The photogenerated primary imines underwent consecutive coupling reactions with unconsumed starting material. The catalyst could be recycled several times for the aerobic oxidative coupling of benzylamine into *N*-benzylidene benzylamine without loss of photocatalytic activity. The high reactivity of the generated primary imines was used in a one-pot synthesis photocatalytic cascade reaction providing benzoheteroazoles with high conversion. Benzimidazoles and benzothiazoles could be obtained with

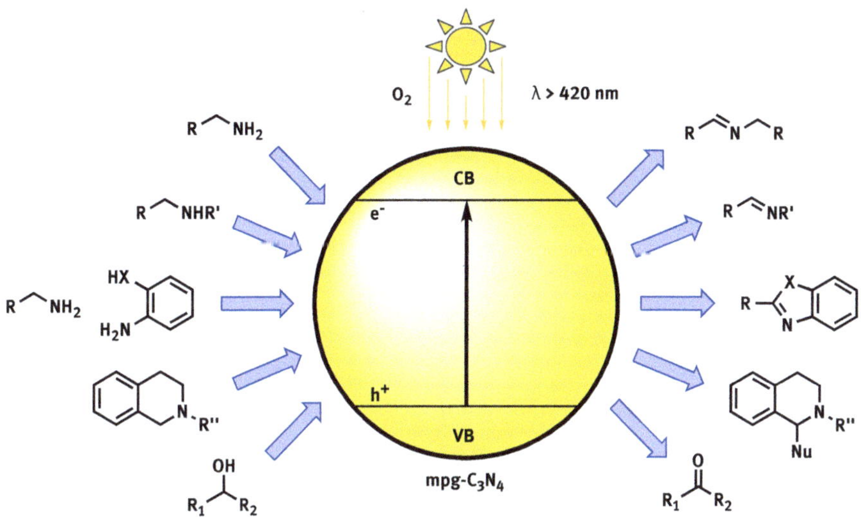

Figure 12.24: Carbon nitride photocatalyzed oxidation reactions. R = (hetero)aryl; R′ = alkyl,aryl; R″ = aryl; Nu = $-CH_2-NO_2$, $-CH_2(CO_2Me)_2$, $-CH_2COMe$; R_1 = alkyl, aryl; R_2 = H, alkyl. [99, 107, 108]

high selectivity (91–98 %). However, the one-pot synthesis of benzoxazoles proceeded with low to moderate selectivity (24–75 %). [107] Tertiary iminium ions were trapped with various nucleophiles in a Mannich-type reaction forming highly functionalized amines in good to excellent yield. [108]

Another strategy to improve the photocatalytic performance of carbon nitride is the formation of semiconductor composites with graphene-based materials. [109, 110] The enhancement of the photocatalytic efficiency is mainly ascribed to the facilitated electron-hole pair separation on the hybrid interface and the suppressed recombination of charge carriers. Carbon-semiconductor hybrid photocatalysts, especially inorganic semiconductors like TiO_2, have been investigated and applied to the photodegradation of organic environmental contaminants and for photoinduced hydrogen production. [111, 112]

Liao et al. recently reported the modification of g-C_3N_4 with an electron accepting graphene oxide layer to enhance the photocatalytic capability under visible light irradiation ($\lambda > 400$ nm). [109] Graphene oxide (GO) can be considered as the oxidative equivalent of graphene exhibiting several oxygen-containing functional groups, e. g. epoxides, alcohols, lactols, ketones, and carboxyl groups (Figure 12.25). The degree of oxidation determines the band gap and conductivity, fully oxidized GO is an electrical insulator, whereas partially oxidized GO shows semiconducting properties. [122] The kinetic constant for the degradation of rhodamine B by the GO/g-C_3N_4 hybrid photocatalyst exceeds the value for unmodified g-C_3N_4 by a factor of 3.80. The enhancement results from efficient transfer of the photogenerated electron from

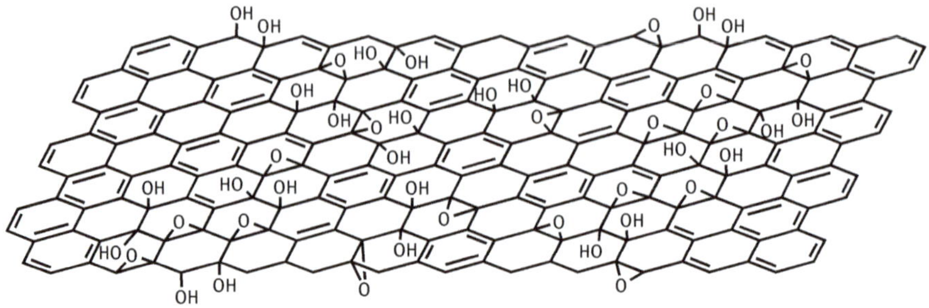

Figure 12.25: Simplified chemical structure of a single graphene oxide (GO)-layer according to the Lerf–Klinowski model. [118]

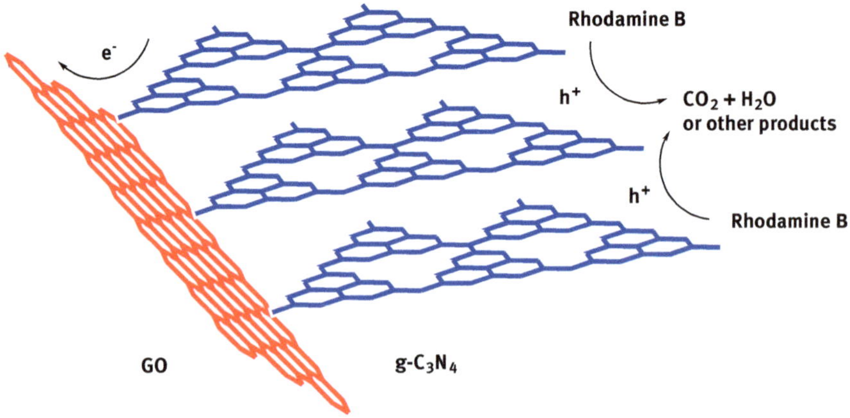

Figure 12.26: Proposed mechanism of the visible-light-induced oxidative degradation of rhodamine B on a GO/g-C_3N_4 interface. [109]

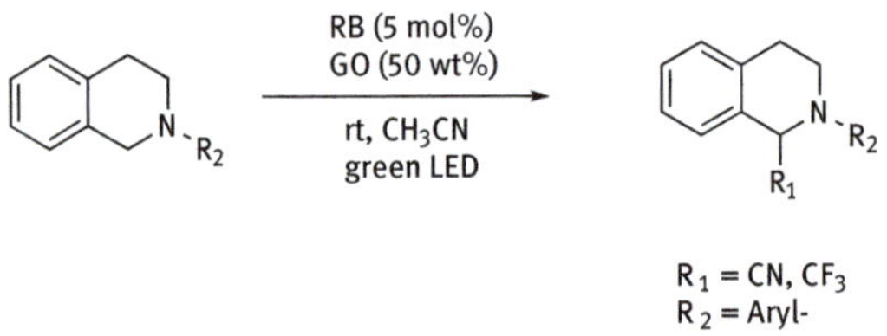

Figure 12.27: Visible light induced oxidative C-H functionalization of tertiary amines. [113]

the CB of g-C_3N_4 to GO. The hole generated in the VB of g-C_3N_4 oxidatively degrades rhodamine B (Figure 12.26). [109]

Tan and co-workers used GO as a co-catalyst in the photoinduced α-functionalization of tertiary amines by Rose Bengal (RB) using green light (Figure 12.27). [113] The addition of GO to the reaction system increased the reaction rate, whereas natural, hydrophobic graphite and activated carbon showed no effect. The α-cyanation of *N*-aryl-tetrahydroisoquinoline with TMSCN in the presence of the carbon catalyst GO increased the yield from 60 % to 97 %. Mechanistic investigations showed that GO exhibits no photocatalytic activity itself, could be recycled, and that the π-π interaction between GO and RB is low. Based on the preliminary results the catalytic effect could not be rationalized. [113]

12.3 References

[1] Ibach, H.; Lüth, H. (2010) The Electronic Bandstructure of Solids. In: Solid-State Physics – An Introduction to Principles of Materials Science, (Berlin: Springer-Verlag), pp 159–189.

[2] Rajh, T.; Poluejtov, O. G.; Thurnauer, M. C. (2003). Charge Seperation in Titanium Oxide Nanocrystalline Semiconductors Revealed by Magnetic Resonance. In: Chemical Physics of Nanostructured Semiconductors, Kokorin, A. I.; Bahnemann, D. W., Eds. (Utrecht: VSP), pp 1–34.

[3] Ibach, H.; Lüth, H. (2010) Diffraction from Periodic Structures. In: Solid-State Physics – An Introduction to Principles of Materials Science, (Berlin: Springer-Verlag), pp 51–82.

[4] Setyawan, W.; Curtarolo, S. (2010). High-throughput electronic band structure calculations: Challenges and tools. Comput. Mater. Sci., 2, 49, 299–312.

[5] Davies, J. H. (1998). The Physics of Low-Dimensional Semiconductors: An Introduction. (Cambridge: Cambridge University Press).

[6] Hybertsen, M. S. (1994). Absorption and emission of light in nanoscale silicon structures. Phys. Rev. Lett., 10, 72, 1514–1517.

[7] Rasmussen, S. E. (2003). Relative merits of reflection and transmission techniques in laboratory powder diffraction. Powder Diffr., 04, 18, 281–284.

[8] Horn, M.; Schwerdtfeger, C. F.; Meagher, E. P. (1972). Refinement of the structure of anatase at several temperatures. Z. Kristallogr., 3–4, 136, 273–281.

[9] Djerdj, I.; Tonejc, A. M. (2006). Structural investigations of nanocrystalline TiO_2 samples. J. Alloys Compd., 1–2, 413, 159–174.

[10] Goodenough, J. B.; Hammet, A.; Huber, G.; Hullinger, F.; Leiß, M.; Ramasesha, S. K.; Werheit, H. (1984). Group III: Crystal and Solid State Physics. (Berlin: Springer-Verlag).

[11] Bludau, W. (1995) Elektronische Eigenschaften undotierter und dotierter Halbleiter. In: Halbleiter – Optoelektronik: Die physikalischen Grundlagen der LED's, Diodenlaser und pn-Photodioden, (München: Carl Hanser Verlag), pp 22–40.

[12] Roy, A. M.; De, G. C.; Sasmal, N.; Bhattacharyya, S. S. (1995). Determination of the flatband potential of semiconductor particles in suspension by photovoltage measurement. Int. J. Hydrogen Energy, 8, 20, 627–630.

[13] Skinner, D. E.; Colombo, D. P.; Cavaleri, J. J.; Bowman, R. M. (1995). Femtosecond Investigation of Electron Trapping in Semiconductor Nanoclusters. J. Phys. Chem., 20, 99, 7853–7856.

[14] Kudo, A.; Miseki, Y. (2009). Heterogeneous photocatalyst materials for water splitting. Chem. Soc. Rev., 1, 38, 253–278.

[15] Coldren, L. A.; Corzine, S. W.; Mašanović, M. L. (2012) Gain and Current Relations. In: Diode Lasers and Photonic Integrated Circuits, (Hoboken, NJ, USA: John Wiley & Sons, Inc.), pp 157–246.

[16] Müller, U. (2007) Symmetry. In: Inorganic Structural Chemistry, (John Wiley & Sons, Ltd), pp 12–29.

[17] Kato, H.; Kobayashi, H.; Kudo, A. (2002). Role of Ag^+ in the Band Structures and Photocatalytic Properties of $AgMO_3$ (M: Ta and Nb) with the Perovskite Structure. J. Phys. Chem. B, 48, 106, 12441–12447.

[18] Zou, Z.; Ye, J.; Arakawa, H. (2000). Structural properties of $InNbO_4$ and $InTaO_4$: correlation with photocatalytic and photophysical properties. Chem. Phys. Lett., 3–4, 332, 271–277.
[19] Cherevatskaya, M.; Neumann, M.; Füldner, S.; Harlander, C.; Kümmel, S.; Dankesreiter, S.; Pfitzner, A.; Zeitler, K.; König, B. (2012). Visible-Light-Promoted Stereoselective Alkylation by Combining Heterogeneous Photocatalysis with Organocatalysis. Angew. Chem. Int. Ed., 17, 51, 4062–4066.
[20] Zhang, H.; Bayne, M.; Fernando, S.; Legg, B.; Zhu, M.; Penn, R. L.; Banfield, J. F. (2011). Size-Dependent Bandgap of Nanogoethite. J. Phys. Chem. C, 36, 115, 17704–17710.
[21] Gilbert, B.; Frandsen, C.; Maxey, E. R.; Sherman, D. M. (2009). Band-gap measurements of bulk and nanoscale hematite by soft x-ray spectroscopy. Phys. Rev. B, 3, 79, 035108.
[22] Hoffmann, M. R.; Martin, S. T.; Choi, W.; Bahnemann, D. W. (1995). Environmental Applications of Semiconductor Photocatalysis. Chem. Rev., 1, 95, 69–96.
[23] Chu, D.; Yuan, X.; Qin, G.; Xu, M.; Zheng, P.; Lu, J.; Zha, L. (2008). Efficient carbon-doped nanostructured TiO_2 (anatase) film for photoelectrochemical solar cells. J. Nanopart. Res., 2, 10, 357–363.
[24] Irie, H.; Watanabe, Y.; Hashimoto, K. (2003). Carbon-doped Anatase TiO_2 Powders as a Visible-light Sensitive Photocatalyst. Chem. Lett., 8, 32, 772–773.
[25] Asahi, R.; Morikawa, T.; Ohwaki, T.; Aoki, K.; Taga, Y. (2001). Visible-Light Photocatalysis in Nitrogen-Doped Titanium Oxides. Science, 5528, 293, 269–271.
[26] Sakthivel, S.; Janczarek, M.; Kisch, H. (2004). Visible Light Activity and Photoelectrochemical Properties of Nitrogen-Doped TiO_2. J. Phys. Chem. B, 50, 108, 19384–19387.
[27] Burda, C.; Lou, Y.; Chen, X.; Samia, A. C. S.; Stout, J.; Gole, J. L. (2003). Enhanced Nitrogen Doping in TiO_2 Nanoparticles. Nano Lett., 8, 3, 1049–1051.
[28] Liu, Y.; Chen, X.; Li, J.; Burda, C. (2005). Photocatalytic degradation of azo dyes by nitrogen-doped TiO_2 nanocatalysts. Chemosphere, 1, 61, 11–18.
[29] Sathish, M.; Viswanathan, B.; Viswanath, R. P.; Gopinath, C. S. (2005). Synthesis, Characterization, Electronic Structure, and Photocatalytic Activity of Nitrogen-Doped TiO_2 Nanocatalyst. Chem. Mater., 25, 17, 6349–6353.
[30] Kitano, M.; Funatsu, K.; Matsuoka, M.; Ueshima, M.; Anpo, M. (2006). Preparation of Nitrogen-Substituted TiO_2 Thin Film Photocatalysts by the Radio Frequency Magnetron Sputtering Deposition Method and Their Photocatalytic Reactivity under Visible Light Irradiation†. J. Phys. Chem. B, 50, 110, 25266–25272.
[31] Choi, J.; Park, H.; Hoffmann, M. R. (2009). Effects of Single Metal-Ion Doping on the Visible-Light Photoreactivity of TiO_2. J. Phys. Chem. C, 2, 114, 783–792.
[32] Anandan, S.; Miyauchi, M. (2011). Ce-doped ZnO ($Ce_xZn_{1-x}O$) becomes an efficient visible-light-sensitive photocatalyst by co-catalyst (Cu^{2+}) grafting. Phys. Chem. Chem. Phys., 33, 13, 14937–14945.
[33] Xu, A.-W.; Gao, Y.; Liu, H.-Q. (2002). The Preparation, Characterization, and their Photocatalytic Activities of Rare-Earth-Doped TiO_2 Nanoparticles. J. Catal., 2, 207, 151–157.
[34] Xiao, Q.; Si, Z.; Zhang, J.; Xiao, C.; Yu, Z.; Qiu, G. (2007). Effects of samarium dopant on photocatalytic activity of TiO_2 nanocrystallite for methylene blue degradation. J. Mater. Sci., 22, 42, 9194–9199.
[35] Wilke, K.; Breuer, H. D. (1999). The influence of transition metal doping on the physical and photocatalytic properties of titania. J. Photochem. Photobiol. A: Chem., 1, 121, 49–53.
[36] Wu, J. C. S.; Chen, C.-H. (2004). A visible-light response vanadium-doped titania nanocatalyst by sol-gel method. J. Photochem. Photobiol. A: Chem., 3, 163, 509–515.
[37] Gu, D.-E.; Yang, B.-C.; Hu, Y.-D. (2007). A Novel Method for Preparing V-doped Titanium Dioxide Thin Film Photocatalysts with High Photocatalytic Activity Under Visible Light Irradiation. Catal. Lett., 3, 118, 254–259.

[38] Khakpash, N.; Simchi, A.; Jafari, T. (2012). Adsorption and solar light activity of transition-metal doped TiO_2 nanoparticles as semiconductor photocatalyst. J. Mater. Sci. – Mater. Electron., 3, 23, 659–667.
[39] Reunchan, P.; Umezawa, N.; Ouyang, S.; Ye, J. (2012). Mechanism of photocatalytic activities in Cr-doped $SrTiO_3$ under visible-light irradiation: an insight from hybrid density-functional calculations. Phys. Chem. Chem. Phys., 6, 14, 1876–1880.
[40] Shi, J.; Shen, S.; Chen, Y.; Guo, L.; Mao, S. S. (2012). Visible light-driven photocatalysis of doped $SrTiO_3$ tubular structure. Opt. Express, S2, 20, A351–A359.
[41] Xu, W.; Jain, P. K.; Beberwyck, B. J.; Alivisatos, A. P. (2012). Probing Redox Photocatalysis of Trapped Electrons and Holes on Single Sb-doped Titania Nanorod Surfaces. J. Am. Chem. Soc., 9, 134, 3946–3949.
[42] Ullah, R.; Dutta, J. (2008). Photocatalytic degradation of organic dyes with manganese-doped ZnO nanoparticles. J. Hazard. Mater., 1–3, 156, 194–200.
[43] Nah, Y.-C.; Paramasivam, I.; Schmuki, P. (2010). Doped TiO_2 and TiO_2 Nanotubes: Synthesis and Applications. ChemPhysChem, 13, 11, 2698–2713.
[44] Murakami, N.; Kawakami, S.; Tsubota, T.; Ohno, T. (2012). Dependence of photocatalytic activity on particle size of a shape-controlled anatase titanium(IV) oxide nanocrystal. J. Mol. Catal. A: Chem., 0, 358, 106–111.
[45] Cho, I.-S.; Lee, S.; Noh, J. H.; Choi, G. K.; Jung, H. S.; Kim, D. W.; Hong, K. S. (2008). Visible-Light-Induced Photocatalytic Activity in $FeNbO_4$ Nanoparticles. J. Phys. Chem. C, 47, 112, 18393–18398.
[46] Linsebigler, A. L.; Lu, G.; Yates, J. T. (1995). Photocatalysis on TiO_2 Surfaces: Principles, Mechanisms, and Selected Results. Chem. Rev., 3, 95, 735–758.
[47] Ohno, T.; Sarukawa, K.; Tokieda, K.; Matsumura, M. (2001). Morphology of a TiO_2 Photocatalyst (Degussa, P-25) Consisting of Anatase and Rutile Crystalline Phases. J. Catal., 1, 203, 82–86.
[48] Karakitsou, K. E.; Verykios, X. E. (1993). Effects of altervalent cation doping of titania on its performance as a photocatalyst for water cleavage. J. Phys. Chem., 6, 97, 1184–1189.
[49] Duonghong, D.; Borgarello, E.; Graetzel, M. (1981). Dynamics of light-induced water cleavage in colloidal systems. J. Am. Chem. Soc., 16, 103, 4685–4690.
[50] Bard, A. J.; Fox, M. A. (1995). Artificial Photosynthesis: Solar Splitting of Water to Hydrogen and Oxygen. Acc. Chem. Res., 3, 28, 141–145.
[51] Fujishima, A.; Honda, K. (1972). Electrochemical Photolysis of Water at a Semiconductor Electrode. Nature, 5358, 238, 37–38.
[52] Wold, A. (1993). Photocatalytic properties of titanium dioxide (TiO_2). Chem. Mater., 3, 5, 280–283.
[53] Zhou, X.-T.; Ji, H.-B.; Huang, X.-J. (2012). Photocatalytic Degradation of Methyl Orange over Metalloporphyrins Supported on TiO_2 Degussa P25. Molecules, 2, 17, 1149–1158.
[54] Soana, F.; Sturini, M.; Cermenati, L.; Albini, A. (2000). Titanium dioxide photocatalyzed oxygenation of naphthalene and some of its derivatives. J. Chem. Soc., Perkin Trans. 2, 4, 699–704.
[55] Cermenati, L.; Richter, C.; Albini, A. (1998). Solar light induced carbon-carbon bond formation via TiO_2 photocatalysis. Chem. Commun., 7, 805–806.
[56] Wu, J. C. S.; Lin, H.-M. (2005). Photo reduction of CO_2 to methanol via TiO_2 photocatalyst. Int. J. Photoenergy, 3, 7, 115–119.
[57] Tan, S. S.; Zou, L.; Hu, E. (2006). Photocatalytic reduction of carbon dioxide into gaseous hydrocarbon using TiO_2 pellets. Catal. Today, 1–4, 115, 269–273.
[58] Kočí, K.; Obalová, L.; Matějová, L.; Plachá, D.; Lacný, Z.; Jirkovský, J.; Šolcová, O. (2009). Effect of TiO_2 particle size on the photocatalytic reduction of CO_2. Appl. Catal., B, 3–4, 89, 494–502.

[59] Dey, G. R. (2007). Chemical Reduction of CO_2 to Different Products during Photo Catalytic Reaction on TiO_2 under Diverse Conditions: an Overview. J. Nat. Gas Chem., 3, 16, 217–226.
[60] Yamashita, H.; Fujii, Y.; Ichihashi, Y.; Zhang, S. G.; Ikeue, K.; Park, D. R.; Koyano, K.; Tatsumi, T.; Anpo, M. (1998). Selective formation of CH_3OH in the photocatalytic reduction of CO_2 with H_2O on titanium oxides highly dispersed within zeolites and mesoporous molecular sieves. Catal. Today, 1–4, 45, 221–227.
[61] Mozia, S.; Heciak, A.; Darowna, D.; Morawski, A. W. (2012). A novel suspended/supported photoreactor design for photocatalytic decomposition of acetic acid with simultaneous production of useful hydrocarbons. J. Photochem. Photobiol. A: Chem., 0, 236, 48–53.
[62] Shan, Z.; Wang, W.; Lin, X.; Ding, H.; Huang, F. (2008). Photocatalytic degradation of organic dyes on visible-light responsive photocatalyst $PbBiO_2Br$. J. Solid State Chem., 6, 181, 1361–1366.
[63] Graetzel, M. (1981). Artificial photosynthesis: water cleavage into hydrogen and oxygen by visible light. Acc. Chem. Res., 12, 14, 376–384.
[64] Kalyanasundaram, K.; Borgarello, E.; Duonghong, D.; Grätzel, M. (1981). Cleavage of Water by Visible-Light Irradiation of Colloidal CdS Solutions; Inhibition of Photocorrosion by RuO_2. Angew. Chem. Int. Ed., 11, 20, 987–988.
[65] Zou, Z.; Ye, J.; Sayama, K.; Arakawa, H. (2001). Direct splitting of water under visible light irradiation with an oxide semiconductor photocatalyst. Nature, 6, 414, 625–627.
[66] Mori, K.; Tottori, M.; Watanabe, K.; Che, M.; Yamashita, H. (2011). Photoinduced Aerobic Oxidation Driven by Phosphorescence Ir(III) Complex Anchored to Mesoporous Silica. J. Phys. Chem. C, 43, 115, 21358–21362.
[67] Crank, G.; Mursyidi, A. (1992). Photochemistry of some organosulfur pesticides. J. Photochem. Photobiol. A: Chem., 3, 68, 289–297.
[68] Nasalevich, M. A.; Kozlova, E. A.; Lyubina, T. P.; Vorontsov, A. V. (2012). Photocatalytic oxidation of ethanol and isopropanol vapors on cadmium sulfide. J. Catal., 0, 287, 138–148.
[69] Zhang, N.; Liu, S.; Fu, X.; Xu, Y.-J. (2012). Fabrication of coenocytic Pd@CdS nanocomposite as a visible light photocatalyst for selective transformation under mild conditions. J. Mater. Chem., 11, 22, 5042–5052.
[70] Mitkina, T.; Stanglmair, C.; Setzer, W.; Gruber, M.; Kisch, H.; Konig, B. (2012). Visible light mediated homo- and heterocoupling of benzyl alcohols and benzyl amines on polycrystalline cadmium sulfide. Org. Biomol. Chem., 17, 10, 3556–3561.
[71] Pfitzner, A.; Pohla, P. (2009). Syntheses and Crystal Structures of $PbSbO_2Br$, $PbSbO_2I$, and $PbBiO_2Br$. Z. Anorg. Allg. Chem., 635, 1157–1159.
[72] Füldner, S.; Pohla, P.; Bartling, H.; Dankesreiter, S.; Stadler, R.; Gruber, M.; Pfitzner, A.; König, B. (2011). Selective photocatalytic reductions of nitrobenzene derivatives using $PbBiO_2X$ and blue light. Green Chem., 3, 13, 640–643.
[73] Beeson, T. D.; Mastracchio, A.; Hong, J.-B.; Ashton, K.; MacMillan, D. W. C. (2007). Enantioselective Organocatalysis Using SOMO Activation. Science, 5824, 316, 582–585.
[74[Idriss, H.; Libby, M.; Barteau, M. A. (1992). Carbon-carbon bond formation on metal oxides: From single crystals toward catalysis. Catal. Lett., 1, 15, 13–23.
[75] Fagnoni, M.; Dondi, D.; Ravelli, D.; Albini, A. (2007). Photocatalysis for the Formation of the C–C Bond. Chem. Rev., 6, 107, 2725–2756.
[76] Yamashita, Y. (2009). Organic semiconductors for organic field-effect transistors. Sci. Technol. Adv. Mater., 2, 10, 024313.
[77] Murphy, A. R.; Fréchet, J. M. J. (2007). Organic Semiconducting Oligomers for Use in Thin Film Transistors. Chem. Rev., 4, 107, 1066–1096.
[78] Hertel, D.; Müller, C. D.; Meerholz, K. (2005). Organische Leuchtdioden: Bilderzeugung. Chem. Unserer Zeit, 5, 39, 336–347.

[79] Günes, S.; Neugebauer, H.; Sariciftci, N. S. (2007). Conjugated Polymer-Based Organic Solar Cells. Chem. Rev., 4, 107, 1324–1338.
[80] Kroon, R.; Lenes, M.; Hummelen, J. C.; Blom, P. W. M.; de Boer, B. (2008). Small Bandgap Polymers for Organic Solar Cells (Polymer Material Development in the Last 5 Years). Polym. Rev., 3, 48, 531–582.
[81] van Mullekom, H. A. M.; Vekemans, J. A. J. M.; Havinga, E. E.; Meijer, E. W. (2001). Developments in the chemistry and band gap engineering of donor–acceptor substituted conjugated polymers. Mater. Sci. Eng., R, 1, 32, 1–40.
[82] Brütting, W. (2006). Introduction to the Physics of Organic Semiconductors. In: Physics of Organic Semiconductors, (Wiley-VCH Verlag GmbH & Co. KGaA), pp 1–14.
[83] Sworakowski, J.; Ulánski, J. (2003). 4 Electrical properties of organic materials. Annu. Rep. Prog. Chem., Sect. C.: Phys. Chem., 99, 87–125.
[84] Muktha, B.; Madras, G.; Guru Row, T. N.; Scherf, U.; Patil, S. (2007). Conjugated Polymers for Photocatalysis. J. Phys. Chem. B, 28, 111, 7994–7998.
[85] Hasegawa, K.; Wen, C.; Kotani, T.; Kanbara, T.; Kagaya, S.; Yamamoto, T. (1999). Photocatalyzed degradation of agrochemicals using poly(2,5-dihexoxy-*p*-phenylene) and poly(3-octylthiophene-2,5-diyl) films. J. Mater. Sci. Lett., 13, 18, 1091–1093.
[86] Wen, C.; Hasegawa, K.; Kanbara, T.; Kagaya, S.; Yamamoto, T. (2000). Visible light-induced catalytic degradation of iprobenfos fungicide by poly(3-octylthiophene-2,5-diyl) film. J. Photochem. Photobiol. A: Chem., 1–2, 133, 59–66.
[87] Wen, C.; Hasegawa, K.; Kanbara, T.; Kagaya, S.; Yamamoto, T. (2000). Photocatalytic properties of poly(3-octylthiophene-2,5-diyl) film blended with sensitizer for the degradation of iprobenfos fungicide. J. Photochem. Photobiol. A: Chem., 1, 137, 45–51.
[88] Yanagida, S.; Kabumoto, A.; Mizumoto, K.; Pac, C.; Yoshino, K. (1985). Poly(*p*-phenylene)-catalysed photoreduction of water to hydrogen. J. Chem. Soc., Chem. Commun., 8, 474–475.
[89] Shibata, T.; Kabumoto, A.; Shiragami, T.; Ishitani, O.; Pac, C.; Yanagida, S. (1990). Novel visible-light-driven photocatalyst. Poly(*p*-phenylene)-catalyzed photoreductions of water, carbonyl compounds, and olefins. J. Phys. Chem., 5, 94, 2068–2076.
[90] Matsuoka, S.; Kohzuki, T.; Kuwana, Y.; Nakamura, A.; Yanagida, S. (1992). Visible-light-induced photocatalysis of poly(pyridine-2,5-diyl). Photoreduction of water, carbonyl compounds and alkenes with triethylamine. J. Chem. Soc., Perkin Trans. 2, 4, 679–685.
[91] Uhrich, C.; Schueppel, R.; Petrich, A.; Pfeiffer, M.; Leo, K.; Brier, E.; Kilickiran, P.; Baeuerle, P. (2007). Organic thin-film photovoltaic cells based on oligothiophenes with reduced bandgap. Adv. Funct. Mater., 15, 17, 2991–2999.
[92] Wada, Y.; Ogata, T.; Hiranaga, K.; Yasuda, H.; Kitamura, T.; Murakoshi, K.; Yanagida, S. (1998). Visible light-induced photofixation of CO_2 into benzophenone: roles of poly(*p*-phenylene) as photocatalyst and two-electron mediator in the presence of quaternary onium salts. J. Chem. Soc., Perkin Trans. 2, 9, 1999–2004.
[93] Ogata, T.; Hiranaga, K.; Matsuoka, S.; Wada, Y.; Yanagida, S. (1993). Visible-light Induced Photocatalytic Fixation of CO_2 into Benzophenone Using Poly(*p*-phenylene) as a Photocatalyst. Chem. Lett., 6, 22, 983–984.
[94] Wada, Y.; Kitamura, T.; Yanagida, S. (2000). CO_2-fixation into organic carbonyl compounds in visible-light-induced photocatalysis of linear aromatic compounds. Res. Chem. Intermed., 2, 26, 153–159.
[95] Rouch, W. D.; Zhang, M.; McCulla, R. D. (2012). Conjugated polymers as photoredox catalysts: a new catalytic system using visible light to promote aryl aldehyde pinacol couplings. Tetrahedron Lett., 37, 53, 4942–4945.
[96] Liebig, J. V. (1834). Ann. Pharm., 10, 10.

[97] Goettmann, F.; Fischer, A.; Antonietti, M.; Thomas, A. (2006). Chemical Synthesis of Mesoporous Carbon Nitrides Using Hard Templates and Their Use as a Metal-Free Catalyst for Friedel–Crafts Reaction of Benzene. Angew. Chem. Int. Ed., 27, 45, 4467–4471.
[98] Goettmann, F.; Fischer, A.; Antonietti, M.; Thomas, A. (2006). Metal-free catalysis of sustainable Friedel-Crafts reactions: direct activation of benzene by carbon nitrides to avoid the use of metal chlorides and halogenated compounds. Chem. Commun., 43, 4530–4532.
[99] Su, F.; Mathew, S. C.; Lipner, G.; Fu, X.; Antonietti, M.; Blechert, S.; Wang, X. (2010). mpg-C_3N_4-Catalyzed Selective Oxidation of Alcohols Using O_2 and Visible Light. J. Am. Chem. Soc., 46, 132, 16299–16301.
[100] Wang, X.; Maeda, K.; Thomas, A.; Takanabe, K.; Xin, G.; Carlsson, J. M.; Domen, K.; Antonietti, M. (2009). A metal-free polymeric photocatalyst for hydrogen production from water under visible light. Nat. Mater., 1, 8, 76–80.
[101] Wang, X.; Maeda, K.; Chen, X.; Takanabe, K.; Domen, K.; Hou, Y.; Fu, X.; Antonietti, M. (2009). Polymer Semiconductors for Artificial Photosynthesis: Hydrogen Evolution by Mesoporous Graphitic Carbon Nitride with Visible Light. J. Am. Chem. Soc., 5, 131, 1680–1681.
[102] Maeda, K.; Wang, X.; Nishihara, Y.; Lu, D.; Antonietti, M.; Domen, K. (2009). Photocatalytic Activities of Graphitic Carbon Nitride Powder for Water Reduction and Oxidation under Visible Light. J. Phys. Chem. C, 12, 113, 4940–4947.
[103] Yan, S. C.; Li, Z. S.; Zou, Z. G. (2009). Photodegradation Performance of g-C_3N_4 Fabricated by Directly Heating Melamine. Langmuir, 17, 25, 10397–10401.
[104] Yan, S. C.; Li, Z. S.; Zou, Z. G. (2010). Photodegradation of Rhodamine B and Methyl Orange over Boron-Doped g-C_3N_4 under Visible Light Irradiation. Langmuir, 6, 26, 3894–3901.
[105] Dong, F.; Wu, L.; Sun, Y.; Fu, M.; Wu, Z.; Lee, S. C. (2011). Efficient synthesis of polymeric g-C_3N_4 layered materials as novel efficient visible light driven photocatalysts. J. Mater. Chem., 39, 21, 15171–15174.
[106] Cui, Y.; Ding, Z.; Liu, P.; Antonietti, M.; Fu, X.; Wang, X. (2012). Metal-free activation of H_2O_2 by g-C_3N_4 under visible light irradiation for the degradation of organic pollutants. Phys. Chem. Chem. Phys., 4, 14, 1455–1462.
[107] Su, F.; Mathew, S. C.; Möhlmann, L.; Antonietti, M.; Wang, X.; Blechert, S. (2011). Aerobic Oxidative Coupling of Amines by Carbon Nitride Photocatalysis with Visible Light. Angew. Chem. Int. Ed., 3, 50, 657–660.
[108] Möhlmann, L.; Baar, M.; Rieß, J.; Antonietti, M.; Wang, X.; Blechert, S. (2012). Carbon Nitride-Catalyzed Photoredox C–C Bond Formation with *N*-Aryltetrahydroisoquinolines. Adv. Synth. Catal., 10, 354, 1909–1913.
[109] Liao, G.; Chen, S.; Quan, X.; Yu, H.; Zhao, H. (2012). Graphene oxide modified g-C_3N_4 hybrid with enhanced photocatalytic capability under visible light irradiation. J. Mater. Chem., 6, 22, 2721–2726.
[110] Du, A.; Sanvito, S.; Li, Z.; Wang, D.; Jiao, Y.; Liao, T.; Sun, Q.; Ng, Y. H.; Zhu, Z.; Amal, R.; Smith, S. C. (2012). Hybrid Graphene and Graphitic Carbon Nitride Nanocomposite: Gap Opening, Electron–Hole Puddle, Interfacial Charge Transfer, and Enhanced Visible Light Response. J. Am. Chem. Soc., 9, 134, 4393–4397.
[111] An, X.; Yu, J. C. (2011). Graphene-based photocatalytic composites. RSC Adv., 8, 1, 1426–1434.
[112] Xiang, Q.; Yu, J.; Jaroniec, M. (2012). Graphene-based semiconductor photocatalysts. Chem. Soc. Rev., 2, 41, 782–796.
[113] Pan, Y.; Wang, S.; Kee, C. W.; Dubuisson, E.; Yang, Y.; Loh, K. P.; Tan, C.-H. (2011). Graphene oxide and Rose Bengal: oxidative C-H functionalisation of tertiary amines using visible light. Green Chem., 12, 13, 3341–3344.
[114] Mo, S.-D.; Ching, W. Y. (1995). Electronic and optical properties of three phases of titanium dioxide: Rutile, anatase, and brookite. Phys. Rev. B, 19, 51, 13023–13032.

[115] Pielnhofer, F., Regensburg, 2012.
[116] Tehfe, M.-A.; Lalevée, J.; Telitel, S.; Contal, E.; Dumur, F.; Gigmes, D.; Bertin, D.; Nechab, M.; Graff, B.; Morlet-Savary, F.; Fouassier, J.-P. (2012). Polyaromatic Structures as Organo-Photoinitiator Catalysts for Efficient Visible Light Induced Dual Radical/Cationic Photopolymerization and Interpenetrated Polymer Networks Synthesis. Macromolecules, 11, 45, 4454–4460.
[117] Wang, Y.; Wang, X.; Antonietti, M. (2012). Polymeric Graphitic Carbon Nitride as a Heterogeneous Organocatalyst: From Photochemistry to Multipurpose Catalysis to Sustainable Chemistry. Angew. Chem. Int. Ed., 1, 51, 68–89.
[118] He, H.; Klinowski, J.; Forster, M.; Lerf, A. (1998). A new structural model for graphite oxide. Chem. Phys. Lett., 1–2, 287, 53–56.
[119] Friedel, B.; McNeill, C. R.; Greenham, N. C. (2010). Influence of Alkyl Side-Chain Length on the Performance of Poly(3-alkylthiophene)/Polyfluorene All-Polymer Solar Cells. Chem. Mater., 11, 22, 3389–3398.
[120] Yeh, T.-F.; Syu, J.-M.; Cheng, C.; Chang, T.-H.; Teng, H. (2010). Graphite Oxide as a Photocatalyst for Hydrogen Production from Water. Adv. Funct. Mater., 14, 20, 2255–2262.
[121] Pavlishchuk, V. V.; Addison, A. W. (2000). Conversion constants for redox potentials measured versus different reference electrodes in acetonitrile solutions at 25 °C. Inorg. Chim. Acta, 1, 298, 97–102.
[122] Lahaye, R. J. W. E.; Jeong, H. K.; Park, C. Y.; Lee, Y. H. (2009). Density functional theory study of graphite oxide for different oxidation levels. Phys. Rev. B, 12, 79, 125435.

Carsten Streb, Katharina Kastner, Johannes Tucher

13 Polyoxometalates in photocatalysis

13.1 Introduction

Metal oxides are considered to be one of the technically most important and chemically most intriguing families of inorganic compounds. They feature a variety of catalytic and photocatalytic properties with applications in fundamental as well as in applied research. [1] Besides the classical solid-state metal oxides, there is the fascinating class of molecular metal oxide clusters, the so-called polyoxometalates (POMs) which can be considered molecular analogues of traditional solid state metal oxides. [2–4]

13.1.1 Polyoxometalates – Molecular metal oxide clusters

The POMs are molecular metal oxide cluster compounds which are typically formed in self-assembly reactions by oligo-condensation of small oxometalate precursors, e. g. VO_3^-, MoO_4^{2-} or WO_4^{2-}, often in the presence of templating anions such as SO_4^{2-}, PO_4^{3-} or SiO_4^{4-}, see Figure 13.1. During the cluster assembly, precise control of secondary reaction parameters such as solvent, solution pH, temperature and pressure, redox-agents or counterions allows the adjustment of the final cluster architecture. [2–4]

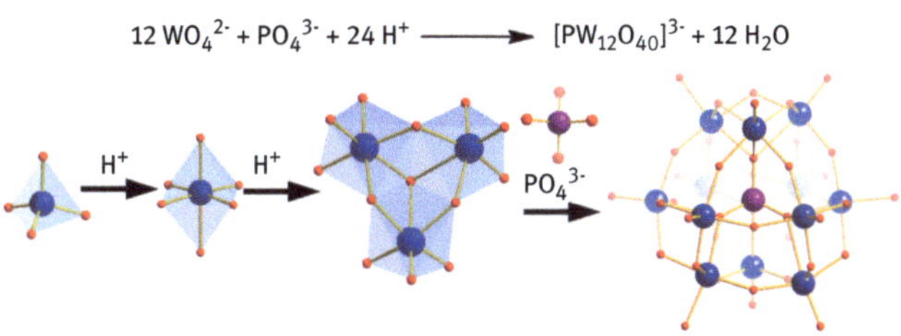

Figure 13.1: Formation of the archetypal Keggin anion $[PW_{12}O_{40}]^{3-}$ .Top: Scheme showing the oligo-condensation reactions of tungstate ions $[WO_4]^{2-}$ in acidic aqueous media. In the presence of phosphate templates, this reaction leads to the formation of the Keggin anion $[PW_{12}O_{40}]^{3-}$. Bottom: Formal self-assembly of the Keggin anion $[PW_{12}O_{40}]^{3-}$. Step 1: Proton-induced expansion of the coordination shell, resulting in the transformation of a tetrahedral $[WO_4]^{2-}$ to an octahedral $[H_xWO_6]^{n-}$ fragment. Step 2: Postulated self-aggregation into secondary building units (here, a so-called $[W_3O_{13}]$ triad, N.B.: these units are postulated intermediates and are typically not isolated). Step 3: Template-assisted cluster formation; the cluster structure is highly dependent on the template geometry. Colour scheme: W: blue, O: red, P: purple.

The most well-studied subclasses of POMs are based on group 5 and 6 metal oxoanions such as vanadates, molybdates, and tungstates. [5] The molybdate and tungstate chemistry is dominated by the formation of heteropolyoxometalates $[X_xM_yO_z]^{n-}$ (M = Mo, W), where heteroelements X are incorporated in the cluster shell, leading to the formation of highly stable and reactive molecular units, see Figure 13.1. Further, molybdate and tungstate solution chemistry is characterised by the aggregation of very large metal oxo clusters with up to 368 metal centres combined in one molecular unit. [6] In contrast, vanadate chemistry is more focussed on small to medium-sized clusters, typically featuring between 4 and 30 vanadium centres. [7] Vanadates feature a large structural variety mostly due to the wide variety of coordination geometries adopted by the vanadium metal centres which range from tetrahedral $[VO_4]$ to square pyramidal $[VO_5]$ and octahedral $[VO_6]$ units. [7] In contrast, the metal centres in Mo- and W-based clusters are often found in an octahedral $[MO_6]$ coordination environment, thereby somewhat restricting the accessible cluster shells.

13.1.2 Concepts in polyoxometalate photochemistry

It has been known for several decades that polyoxometalates feature a rich photochemical activity. [8] To understand why polyoxometalates have attracted as much interest as photoactive materials, [9] the electronic consequences of cluster irradiation need to be considered. As the clusters typically feature fully oxidized d^0 metal centres, light absorption is mainly controlled by O→M ligand-to-metal charge transfer (LMCT) bands in the region of $\lambda = 200–500$ nm. [10, 11] As a result of photon absorption, an electron is promoted from a doubly occupied bonding orbital (HOMO) to an empty, antibonding orbital (LUMO), resulting in the generation of an oxo-centered radical. [12] The photoexcited cluster species therefore is highly reactive and is a better oxidizing agent (higher electron affinity E_{ea}) and a better reducing agent (lower ionisation energy E_I) than the ground-state cluster species, see Figure 13.2. [11]

Besides the facile photoexcitation using near-visible or UV-light, POMs offer advantages as homogeneous or heterogeneous photocatalysts:

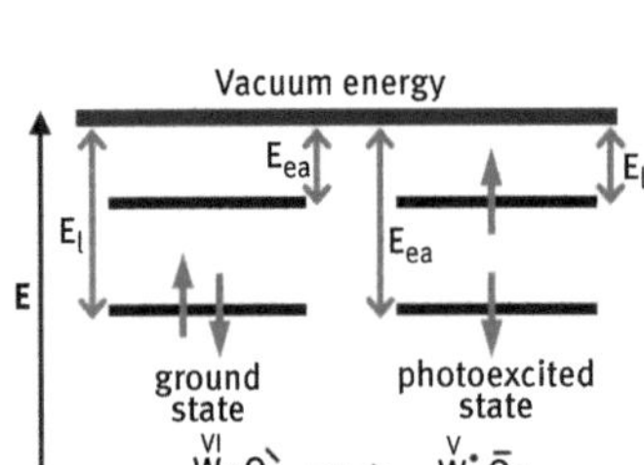

Figure 13.2: Simplified orbital diagram illustrating the ground state and excited state orbital occupation: photoexcitation results in the promotion of an electron from a bonding to an antibonding orbital. The resulting excited species features lower ionisation energy (E_I) and higher electron affinity (E_{ea}) than the ground state species and is, therefore, both a better oxidizing agent and a better reducing agent than the ground state species.

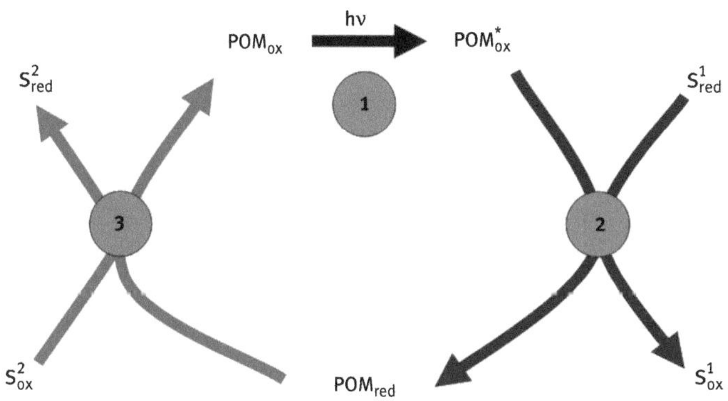

Figure 13.3: General scheme of a polyoxometalate-based photoredox-cycle: (1) Photoexcitation of the oxidised cluster, POM_{ox} by photon absorption; (2) Oxidation of Substrate (S_1) and reduction of the cluster giving POM_{red}; (3) Reduction of a second substrate (S_2) and re-oxidation of the cluster.

1. POMs show strong light absorption with high absorption coefficients $\varepsilon > 1 \times 10^4\, M^{-1}\, cm^{-1}$), although the absorption maxima are often found in the UV region. Tailing of these LMCT bands into the visible region can be used to tune light absorption, particularly for molybdate and vanadate clusters (vide infra).
2. POMs show interesting redox-activity and can undergo photoredox reactions to catalyze substrate oxidation or substrate reduction and can be used with a wide range of substrate molecules, see Figure 13.3.
3. The structural integrity of the cluster shell is often maintained during the photoredox processes, thereby allowing the application of the POM as a catalytic species.
4. Due to the large number of and type of metal centres, POMs can undergo multi-electron redox-events, making them interesting compounds for multi-step photoredox-systems.
5. Re-oxidation of the reduced species is often possible using molecular oxygen or hydrogen peroxide; without degradation of the cluster compounds even under harsh reaction conditions.

13.1.3 The basics of POM photochemistry

Although POM photoreactivity had been known previously, the systematic study of polyoxometalate photocatalysis became a mainstream topic only in the 1980s and was pioneered by the groups of Hill, Papaconstantinou, Yamase, and others. [8, 10, 11, 13] These initial studies were focussed on the photooxidation of organic substrates using three types of clusters, i. e. heteropolyoxometalates of the Keggin-type $[XM_{12}O_{40}]^{n-}$ and Dawson-type $[X_2M_{18}O_{62}]^{n-}$, respectively, (M = Mo, W; X = Si, P, etc.) as well as the decatungstate ion $[W_{10}O_{32}]^{4-}$. [14]

13.1.4 Traditional photooxidation of organic substrates

The polyoxometalate-catalyzed photooxidation of organic substrates follows a general photoredox-cycle as outlined in Figure 13.3 with S_1 = organic substrate (i. e. electron donor) and S_2 = electron acceptor, e.g. O_2, H_2O_2, H^+. [8, 10, 11, 15–18] In general, two possible oxidation mechanisms are discussed, depending on whether the reaction proceeds under strictly inert conditions, in the absence of water, or in the presence of water molecules. [19]

(I) In the presence of water, it is generally accepted that water molecules are pre-associated with the cluster shell via hydrogen-bonds. After photoexcitation of the cluster M–O bonds, this pre-association allows the facile hydrogen-atom abstraction by homolytic H–OH bond cleavage, resulting in the formation of a hydroxyl radical $OH^•$ and a protonated, reduced M–OH species. [10, 11, 18]

(II) In the absence of water, the substrate pre-associates with the cluster, and upon photoexcitation, hydrogen atom abstraction from the substrate is achieved directly by the cluster-based oxo-radical. [11, 20]

13.2 Recent developments in POM photochemistry

Over the last decade it has been accepted worldwide that a shift of energy sources to sustainable, carbon-neutral systems is required in order to limit the effects of global warming for future generations. [21] One promising approach is the use of photochemical systems where sunlight is converted into useful chemical energy. Polyoxometalates have been used in a range of photoreductive reactions with direct environmental impact, i. e. the homogeneous photoreductive activation of CO_2 on POMs and the photoreductive production of hydrogen. In addition, POMs have recently become major candidates in the search for viable light-driven water oxidation catalysts.

13.2.1 Water oxidation by Ru- and Co-polyoxometalates

Over the last years it has been demonstrated that transition metal-substituted POMs can be used as inorganic, homogeneous catalysts for the oxidation of water to molecular oxygen and protons. [22–29] The 4-electron water oxidation represents the critical step of the water splitting reaction which can be used to generate hydrogen from water for use as a secondary energy carrier. Importantly, this system needs to be driven by sunlight to achieve direct conversion of solar to storable chemical energy. The development of stable and economically viable catalysts which can perform the sunlight-driven water oxidation would be a major advance towards carbon-neutral energy systems. Recently, POM-based systems have been reported where the active water oxidation catalyst (e.g. $[Ru_4(O)_4(OH)_2(H_2O)_4(\gamma\text{-}SiW_{10}O_{36})_2]^{10-}$ [25,29] or

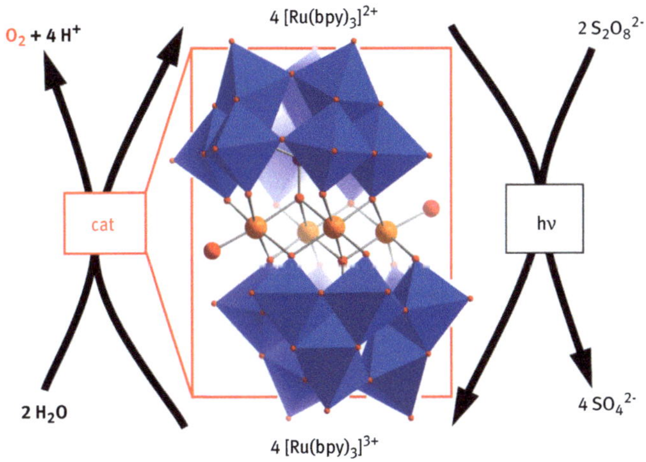

Figure 13.4: Visible-light-driven water oxidation catalyzed by the Co-substituted POM cluster $[Co_4(H_2O)_2(PW_9O_{34})_2]^{10}$. Step 1: Light-induced oxidation of $[Ru(bpy)_3]^{2+}$ in the presence of peroxodisulphate $[S_2O_8]^{2-}$. Step 2: Water oxidation catalyzed by the Co-POM and electron transfer to the oxidised $[Ru(bpy)_3]^{3+}$. Colour scheme: Tungstates: blue polyhedra; Co: orange; O: red. Water ligands on the Co-centres: large red spheres.

$[Co_4(H_2O)_2(PW_9O_{34})_2]^{10-}$ [23,27]) has been coupled with ruthenium-based photosensitizers which allowed the development of stable water oxidation complexes. [22–24, 28] One remarkable example is the cobalt-POM-based system reported by Hill et al. where a $[Co^{II}_4(H_2O)_2(PW_9O_{34})_2]^{10-}$ cluster acts as the water oxidation catalyst and a visible light-driven $[Ru(bpy)_3]^{2+/3+}$ redox couple (bpy = 2,2′-bipyridine) acts as an electron shuttle which transfers electrons from the water-oxidising Co-POM to the stoichiometric electron acceptor, see Figure 13.4. Under optimised conditions, the systems features turnover numbers > 220 and high quantum yields of up to $\phi = 0.30$. [23]

13.2.2 Polyoxoniobate water oxidation

Very recently, Peng et al. demonstrated that water oxidation is also possible using niobium-based POM clusters such as $[\{Nb_2(O)_2(H_2O)_2\}\{SiNb_{12}O_{40}\}]^{10-}$ which features Niobium-based Keggin clusters linked into infinite 1D chains by dinuclear $\{Nb_2(O)_2(H_2O)_2\}$ moieties, see Figure 13.5. [30] The $\{Nb_2(O_2(H_2O)_2\}$ units possess binding sites which allow the coordination of water to the cluster and represent a vital prerequisite for efficient water oxidation. Initial studies show that the Nb–POM can be combined with a hydrogen-evolving catalyst, thus forming a heterogeneous system for the splitting of water into oxygen and hydrogen. When a suspension of the Nb–POM in pure water is irradiated with UV-light in the presence of the co-catalyst NiO, the formation of oxygen and hydrogen was detected, albeit at relatively low turnover

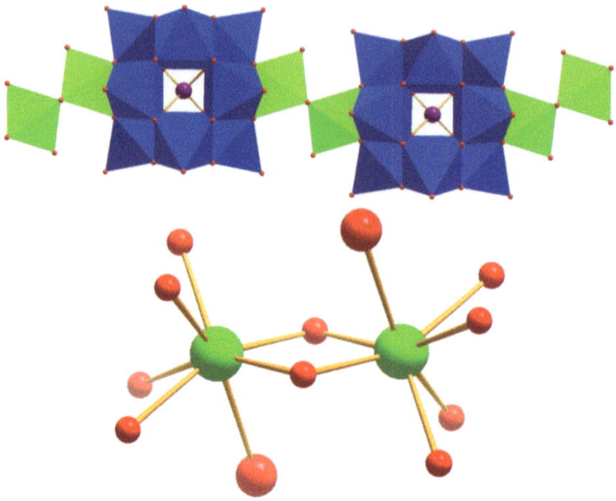

Figure 13.5: Polyoxoniobate water splitting catalyst [30]. Top: Illustration of the $\{Nb_2(O)_2(H_2O)\}$ (green)-linked chains of $\{SiNb_{12}O_{40}\}$ clusters (blue). Bottom: Detailed illustration of the dinuclear $\{Nb_2(O)_2(H_2O)\}$ linking units which feature seven-coordinate niobium(V) centres, each binding one water ligand (large red spheres). Colour scheme: Niobate: blue and green polyhedra; Nb: green; Si: purple; O: red.

numbers. The authors suggest that the system might be combined with Ru-based photosensitizers to allow photoactivity in the visible range. In addition, higher turnover numbers need to be obtained to achieve truly photocatalytic water splitting.

13.2.3 Water oxidation by Dawson anions in ionic liquids

A completely different concept for POM-based water oxidation has recently been reported by Bond et al. In their work, they demonstrated that UV-irradiation of classical Dawson clusters $[P_2W_{18}O_{62}]^{6-}$ in water-containing ionic liquids such as 1-butyl-3-methylimidazolium tetrafluoroborate or diethanolamine hydrogen sulfate results in the oxidation of water, yielding molecular oxygen and protons. [31] This reactivity is remarkable since the Dawson anion $[P_2W_{18}O_{62}]^{6-}$ does not show this type of reactivity in purely aqueous phase or in other traditional solvents such as acetonitrile or dichloromethane. [31]

The authors attribute this unique reactivity to two effects which are both related to the use of ionic liquids. First, the formal reversible redox potentials $E_F{}^0$ of the cluster are much more positive in ionic liquids as compared with other organic solvents and shifts of more than 500 mV are observed. In addition, it has been reported that the structure of water aggregates in ionic liquids is very different from the well-known hydrogen-bonded structure of bulk water and might therefore facilitate water oxidation. [32] A drawback of the reaction is the use of UV-light instead of visible light, and photosensitisation might be employed as one strategy to allow the use of visible light.

13.2.4 Photoreductive CO_2-activation

In order to activate small molecules such as CO_2 on POM clusters, a binding site needs to be incorporated into the cluster shell, as interactions between CO_2 and the negatively polarized oxo-shell of the cluster anion are electrostatically non-favoured and it was suggested that metal-substituted Keggin anions of the type $[TM(L)XW_{11}O_{39}]^{n-}$ (TM = transition metal; L = ligand, X = heteroatom) could be used as CO_2 binding sites. [33]

The concept involves coordination of CO_2 to a transition metal centre such as Co or Ru. [34] Subsequent reductive activation of the carbon dioxide by irradiation of the photoactive LMCT bands of the cluster ion in the presence of a sacrificial electron donor was proposed. Recently, this concept was employed by Neumann et al. and resulted in the development of a homogeneous system for the photochemical reduction of CO_2 by C=O bond activation. [35] A Ru-substituted lacunary Keggin anion, $[Ru^{III}(H_2O)SiW_{11}O_{39}]^{5-}$ (Figure 13.6) was employed in a toluene solution in the presence of CO_2 using triethyl amine as the sacrificial electron donor. After irradiation with UV-light, CO was identified as the only CO_2-based reaction product. The computational study of the proposed reaction intermediates showed that triethyl amine does not only act as a sacrificial electron donor (reducing agent) but, in addition, acts as a supramolecular stabiliser in the cluster-based CO_2-activation, see Figure 13.6.

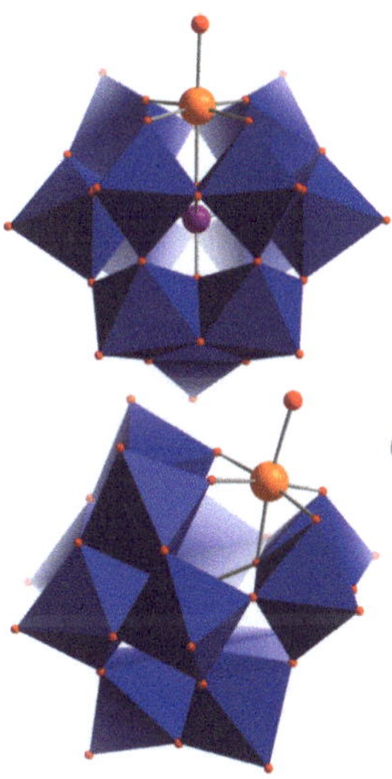

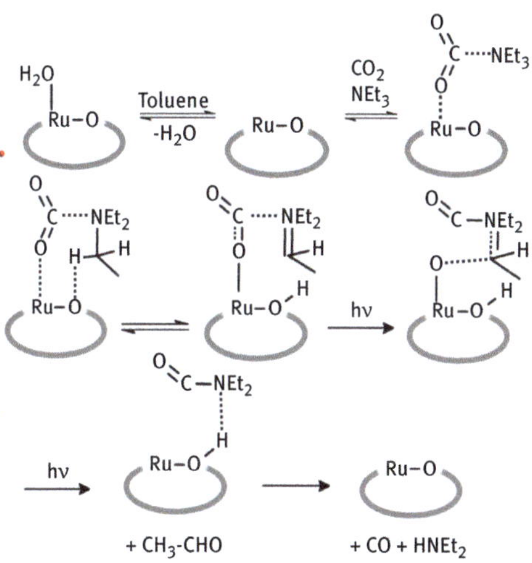

Figure 13.6: Left: Polyhedral representation of the $[Ru^{III}(H_2O)SiW_{11}O_{39}]^{5-}$ cluster in front and side view. Colour scheme: tungstates: blue, Ru: orange, O: red. H_2O: large red sphere. Right: Proposed scheme of the triethyl amine-assisted activation of CO_2 on the Ru-substituted cluster anion $[Ru(CO_2)SiW_{11}O_{39}]^{5-}$.

Based on these analyses, Neumann et al. proposed a reaction mechanism where a CO_2-binding complex, {$Ru(CO_2\cdots NEt_3)SiW_{11}O_{39}$} is formed and subsequent reductive C–O bond activation is achieved photochemically in two consecutive steps, resulting in the formation of carbon monoxide, acetaldehyde and diethyl amine. [35]

13.2.5 Photoreductive H_2 generation

One remarkable feature which has often been observed during the photooxidation of organic substrates was the ability of the photoreduced cluster species to spontaneously reduce protons to molecular hydrogen under anaerobic conditions where re-oxidation of the reduced cluster species can occur by electron transfer to protons (facilitated in the presence of colloidal Pt or other hydrogen formation catalysts). It was shown by several groups that it is crucial to use a cluster which can undergo multiple electron reductions while preventing cluster decomposition. [15, 36–38]

Typical examples for this behaviour are the Keggin and Dawson anions as well as the decatungstate cluster [$W_{10}O_{32}$]. [4–15, 36–38] It should be noted though that to date, this scheme still requires the use of a sacrificial electron (and proton) donor such as primary or secondary alcohols. Therefore, the system can only work economically as a hydrogen generating catalyst if it is further developed to use readily available substrates such as water as the electron/proton donor.

Recently it has been demonstrated that POM-based hydrogen-production systems can be up-scaled from the laboratory scale towards small, portable pilot-plant dimensions. [39] The authors developed a flow-reactor where a reaction mixture consisting of the photocatalyst ($[W_{10}O_{32}]^{4-}$ or $[SiW_{12}O_{40}]^{4-}$), 2-propanol, water and colloidal Pt was pumped through fused-silica tubing equipped with solar concentrators and exposed to sunlight to test the ability of hydrogen generation on a preparative scale. Under typical operative conditions (radiation flux 800–1000 W/m^2), the authors were able to obtain hydrogen production rates of up to 180 ml/h and the system remained operational for several days.

13.3 Optimizing photocatalytic performance of polyoxometalates

13.3.1 Structurally adaptive systems

Traditional POM photocatalysis has been mainly focussed on the reactivity of molybdate and tungstate clusters whereas vanadium-based clusters were often considered to be too labile or too unreactive to show useful photochemistry. Recently, it was demonstrated, that purely vanadium oxide based cluster systems can be employed as photoactive systems for the selective oxidation of organic substrates. [40] The systems illustrate some of the key advantages of vanadates over molybdate-

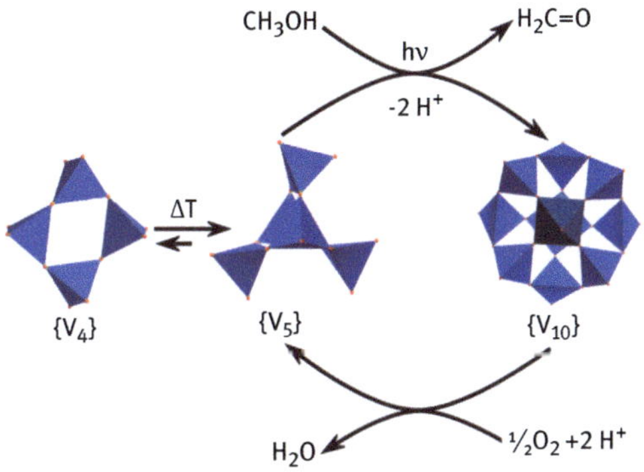

Figure 13.7: Vanadate-based photoredox-cycle. Left: Thermally activated conversion of the inactive {V_4} to the active {V_5} species. Centre: Selective photooxidation of methanol to formaldehyde and reductive cluster aggregation yielding the 2-electron reduced species. Subsequent re-oxidation can be achieved by molecular oxygen (slow) or by H_2O_2 (fast). {V_4} = $[V_4O_{12}]^{4-}$; {V_5} = $[V_5O_{14}]^{3-}$; {V_{10}} = $[V^{IV}{}_2V^{V}{}_8O_{26}]^{4-}$.

or tungstate-based systems: in the study it was shown that visible light photoexcitation can be achieved by the thermal activation of an inactive precursor, $[V_4O_{12}]^{4-}$ (= {V_4}) which at elevated temperatures undergoes a reversible re-arrangement to the active cluster, $[V_5O_{14}]^{3-}$ (= {V_5}), see Figure 13.7. The structural change is accompanied by a bathochromic shift of the LMCT absorption band, resulting in visible light absorption by {V_5}. [41]

Irradiation of a solution of {V_5} in the presence of primary or secondary alcohols using visible light ($\lambda > 380$ nm) leads to the photooxidation of the alcohol and results in the formation of the corresponding carbonyl compound. It was demonstrated that in the case of methanol, this 2-electron oxidation is selective and results in the formation of formaldehyde only. No higher oxidation products (formic acid, CO_x) were observed, see Figure 13.7.

Upon reduction, the {V_5} cluster undergoes an aggregation reaction as the {V_5} unit is not stable under reductive conditions. The resulting aggregation can formally be described as a dimerisation where two reduced {V_5} clusters aggregate and form a 2-electron reduced decavanadate species, $[V^{IV}{}_2V^{V}{}_8O_{26}]^{4-}$ (= {V_{10}}). The reduced {V_{10}} species can be re-oxidised using molecular oxygen or aqueous hydrogen peroxide. It should be noted that the re-oxidation using O_2 is slow and requires a prolonged period of time for complete re-oxidation.

13.3.2 Optimized photoactivity by metal substitution

Recently, it was put forward that functionalization of inert metal oxo clusters with metal-oxo units having visible light photoactivity, such as the vanadium species

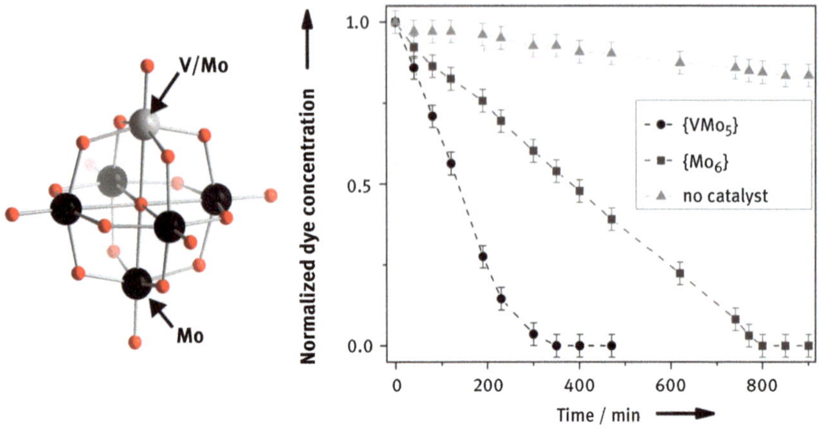

Figure 13.8: Left: Ball-and-stick representation of the general structure of the Lindqvist anion $[M'Mo_5O_{19}]^{n-}$ (M′ = V (n = 3) or Mo (n = 2)). Right: Photocatalytic dye decomposition under UV-irradiation catalyzed by $[VMo_5O_{19}]^{3-}$ ({VMo_5}) and by $[Mo_6O_{19}]^{2-}$ ({Mo_6}).

$[V{=}O]^{3+}$ might be a route for the assembly of highly stable, homogeneous visible-light photocatalysts. In an initial study, the well-known Lindqvist cluster family $[M_6O_{19}]^{n-}$ (M = Mo, W (n = 2); Ta, Nb (n = 8)) was used as a particularly promising model system as the chemical modification of Lindqvist clusters by incorporation of various metal centres has been studied intensely over the last decade, leading to the discovery of a range of heterometallic Lindqvist clusters $[M'(L)M_5O_{18}]^{n-}$ (M = Mo, W; M′ = transition metal, e.g. Co, V, Zr, Hf, L = ligand, e.g. RO^-, O^{2-}, etc.), see Figure 13.8. [42–46]

For the proposed photochemical study, the mono-vanadyl substituted Lindqvist cluster $[VMo_5O_{19}]^{3-}$ was compared with the native $[Mo_6O_{19}]^{2-}$ compound. [47] UV-Vis-spectroscopy showed that the vanadium-substituted unit features several low-energy LMCT transitions in the near-visible range around 400 nm, whereas the molybdate cluster absorbs light in the UV-region only. This was confirmed by time-dependent density functional theory (TD-DFT) calculations which showed that $[VMo_5O_{19}]^{3-}$ features a number of V–O-based LMCT transitions in the region around 400 nm. Vanadyl substitution can therefore in principle be used to photosensitize polyoxomolybdate-based cluster systems. A number of photochemical dye degradation reactions were carried out to understand the photochemical activity of both clusters and it was demonstrated that $[VMo_5O_{19}]^{3-}$ (TON = 1606, TOF = 5.35 min^{-1}) shows higher turnover numbers (TON) and higher turnover frequencies (TOF) for the degradation of a triphenylmethane dye (patent blue V) as compared with $[Mo_6O_{19}]^{2-}$ (TON = 1308, TOF = 1.92 min^{-1}) when irradiated with UV-light. In addition, only $[VMo_5O_{19}]^{3-}$ showed photoactivity when irradiated with monochromatic visible light (λ = 470 nm) whereas $[Mo_6O_{19}]^{2-}$ did not show any residual photoactivity. Metal substitution can therefore be used to improve both the general photoactivity and the photoactivity in the visible region of molecular metal oxide clusters.

13.3.3 Inspiration from the solid-state world

It is noticeable that solid-state metal oxides have attracted considerably more interest as visible-light photocatalysts as compared with their molecular counterparts, the polyoxometalates. It was therefore considered that based on well-known solid-state metal oxide photocatalysts, the synthesis of their molecular analogues might be a promising way for the development of homogeneous, molecular photocatalysts. Bismuth vanadium oxide, $BiVO_4$, is a prime example as the solid-state oxide has been known to be a promising photocatalyst [48–50] whereas no molecular analogues have been known until recently. Access to molecular bismuth vanadium oxide compounds was gained recently using a fragmentation-and-reassembly route where vanadium oxide precursors were fragmented in organic solution and reassembled in the presence of Bi^{III} to give the first molecular bismuth vanadium oxide cluster, $[H_3(Bi(dmso)_3)_4V_{13}O_{40}]$ (= $\{Bi_4V_{13}\}$), see Figure 13.9. [51] The cluster is based on the so-called ε-Keggin architecture $[\varepsilon\text{-}V_{12}O_{36}(VO_4)]^{15-}$ and is stabilized by four Bi^{III} centres arranged on the cluster shell in a tetrahedral fashion. The compound shows visible-light absorption up to ca. 560 nm and photooxidative dye decomposition test reactions show that the cluster is stable as a homogeneous photocatalyst with turnover numbers TON = 1201 and turnover frequency TOF = 1.29 min^{-1}). In addition, the cluster features three acidic protons which might make it interesting as a bifunctional acid and photooxidation catalyst.

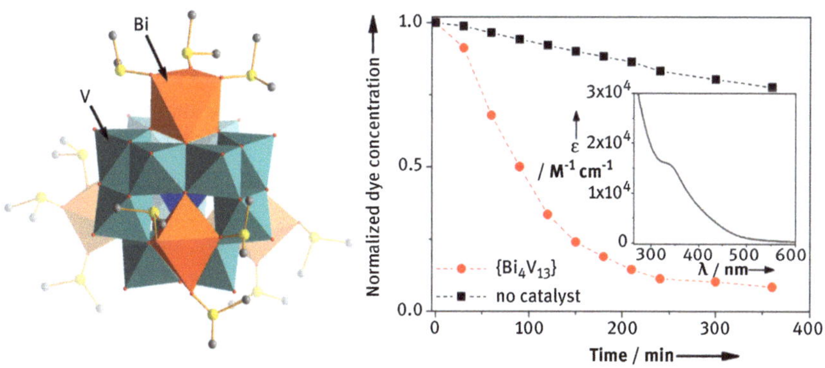

Figure 13.9: Left: Polyhedral representation of the first molecular bismuth vanadium oxide cluster $[H_3(Bi(dmso)_3)_4V_{13}O_{40}]$. Colour scheme: Bi: orange, V: green, C: grey, S: yellow, O: red. Right: Visible-light induced photooxidative dye decomposition by $\{Bi_4V_{13}\}$ under irradiation with monochromatic visible light (λ = 470 nm).

13.4 Conclusion

Over the last decade, the focal point of polyoxometalate photochemistry has shifted dramatically and the main focus has moved from synthetic applications in oxidative organic synthesis to the development of photoactive systems for energy conversion and storage. A particularly thriving field is the development of POM-based water oxidation catalysts which often combine a high oxidative stability with a unique reactivity in light-induced multi-electron transfer processes. In addition, it has been shown that POMs not only offer exceptional properties as photooxidation catalysts, but can be chemically modified to provide reactive sites to allow the development of photoreductive systems.

Several key issues are still impeding the further development of POM photocatalysts, first and foremost the combination of POMs with suitable photosensitizers. Various routes have been proposed in the literature, ranging from purely electrostatic coupling between POM and photosensitizer through to hydrogen-bonded or covalently functionalised systems. In addition, it has to be noted that thus far, the majority of investigations were carried out using traditional POM prototypes such as Keggin- and Dawson-derived structures. However, as synthetic POM chemistry has also changed dramatically over the last decade, there are several cluster families whose photochemical properties have not been investigated yet. The challenge ahead is therefore not only the optimisation of existing systems but also the investigation and development of novel, recently discovered POMs and their application in photoredox catalysis.

13.5 Acknowledgments

This work is supported by the Fonds der Chemischen Industrie (FCI) through a Liebig-Fellowship and by the Deutsche Bundesstiftung Umwelt (DBU) through a doctoral fellowship. Support by the Institute of Inorganic Chemistry II is gratefully acknowledged. Prof Horst Kisch is gratefully acknowledged for many helpful discussions. This article is based on a Perspective Article published in Dalton Transactions, see reference 9.

13.6 References

[1] Jackson SD, Hargreaves JSJ, editors. Metal oxide catalysis. Weinheim: Wiley-VCH; 2008.
[2] Pope MT. Heteropoly and isopoly oxometalates. Heidelberg: Springer-Verlag; 1983.
[3] Long DL, Burkholder E, Cronin L, Polyoxometalate clusters, nanostructures and materials: From self assembly to designer materials and devices, Chemical Society Reviews, 2007, 36, 105–21.

[4] Long DL, Tsunashima R, Cronin L, Polyoxometalates: Building blocks for functional nanoscale systems, Angewandte Chemie-International Edition, 2010, 49, 1736–58.
[5] Pope MT, Müller A, Polyoxometalate chemistry – an old field with new dimensions in several disciplines, Angewandte Chemie-International Edition in English, 1991, 30, 34–48.
[6] Müller A, Beckmann E, Bögge H, Schmidtmann M, Dress A, Inorganic chemistry goes protein size: A Mo-368 nano-hedgehog initiating nanochemistry by symmetry breaking, Angewandte Chemie-International Edition, 2002, 41, 1162–7.
[7] Hayashi Y, Hetero and lacunary polyoxovanadate chemistry, Coordination Chemistry Reviews, 2011, 255, 2270–80.
[8] Yamase T, Photo- and electrochromism of polyoxometalates and related materials, Chemical Reviews, 1998, 98, 307–25.
[9] Streb C, New trends in polyoxometalate photoredox chemistry: From photosensitisation to water oxidation catalysis, Dalton Transactions, 2012, 41, 1651–9.
[10] Hill CL, Prosser-McCartha CM. Photocatalytic and photoredox properties of polyoxometalate systems. In: Grätzel M, Kalyanasundaram K, editors. Photosensitization and photocatalysis using inorganic and organometallic compounds. Dordrecht: Kluwer Academic Publishers; 1993. 307–26.
[11] Papaconstantinou E, Photochemistry of Polyoxometallates of Molybdenum and Tungsten and-or Vanadium, Chemical Society Reviews, 1989, 18, 1–31.
[12] Duncan DC, Netzel TL, Hill CL, Early-time dynamics and reactivity of polyoxometalate excited-states – Identification of a short-lived LMCT excited-state and a reactive long-lived charge-Transfer intermediate following picosecond flash excitation of $[W_{10}O_{32}]^{4-}$ in acetonitrile, Inorganic Chemistry, 1995, 34, 4640–6.
[13] Papaconstantinou E, Hiskia A. Photochemistry and photocatalysis of polyoxometalates. In: Borras-Almenar JJ, Coronado E, Müller A, Pope MT, editors. Polyoxometalate Molecular Science. Dordrecht: Kluwer Academic Publishers; 2003. 381–416.
[14] Tanielian C, Decatungstate photocatalysis, Coordination Chemistry Reviews, 1998, 178, 1165–81.
[15] Hill CL, Polyoxometalates: Reactivity, Comprehensive Coordination Chemistry II, 2003, 4, 679–759.
[16] Jaynes BS, Hill CL, Radical carbonylation of alkanes via polyoxotungstate photocatalysis, Journal of the American Chemical Society, 1995, 117, 4704–5.
[17] Yamase T, Usami T, Photocatalytic dimerization of olefins by decatungstate(VI), $[W_{10}O_{32}]^{4-}$, in acetonitrile and magnetic-resonance studies of photoreduced species, Journal of the Chemical Society-Dalton Transactions, 1988, 183–90.
[18] Mylonas A, Papaconstantinou E, On the mechanism of photocatalytic degradation of chlorinated phenols to CO_2 and HCl by polyoxometalates, Journal of Photochemistry and Photobiology a-Chemistry, 1996, 94, 77–82.
[19] Fox MA, Cardona R, Gaillard E, Photoactivation of metal-oxide surfaces – Photocatalyzed oxidation of alcohols by heteropolytungstates, Journal of the American Chemical Society, 1987, 109, 6347–54.
[20] Tanielian C, Duffy K, Jones A, Kinetic and mechanistic aspects of photocatalysis by polyoxotungstates: A laser flash photolysis, pulse radiolysis, and continuous photolysis study, Journal of Physical Chemistry B, 1997, 101, 4276–82.
[21] Lewis NS, Nocera DG, Powering the planet, Proc Natl Acad Sci USA, 2006, 103, 15729–35.
[22] Geletii YV, Huang ZQ, Hou Y, Musaev DG, Lian TQ, Hill CL, Homogeneous light-driven water oxidation catalyzed by a tetraruthenium complex with all inorganic ligands, Journal of the American Chemical Society, 2009, 131, 7522–3.

[23] Huang ZQ, Luo Z, Geletii YV, Vickers JW, Yin QS, Wu D, et al., Efficient light-driven carbon-free cobalt-based molecular catalyst for water oxidation, Journal of the American Chemical Society, 2011, 133, 2068–71.
[24] Puntoriero F, La Ganga G, Sartorel A, Carraro M, Scorrano G, Bonchio M, et al., Photo-induced water oxidation with tetra-nuclear ruthenium sensitizer and catalyst: A unique 4 × 4 ruthenium interplay triggering high efficiency with low-energy visible light, Chemical Communications, 2010, 46, 4725–7.
[25] Botar B, Geletii YV, Kögerler P, Hillesheim DA, Musaev DG, Hill CL, An all-inorganic, stable, and highly active tetraruthenium homogeneous catalyst for water oxidation, Angewandte Chemie-International Edition, 2008, 47, 3896–9.
[26] Kuznetsov AE, Geletii YV, Hill CL, Morokuma K, Musaev DG, Dioxygen and water activation processes on multi-ru-substituted polyoxometalates: Comparison with the "blue-dimer" water oxidation catalyst, Journal of the American Chemical Society, 2009, 131, 6844–54.
[27] Yin QS, Tan JM, Besson C, Geletii YV, Musaev DG, Kuznetsov AE, et al., A fast soluble carbon-free molecular water oxidation catalyst based on abundant metals, Science, 2010, 328, 342–5.
[28] Orlandi M, Argazzi R, Sartorel A, Carraro M, Scorrano G, Bonchio M, et al., Ruthenium polyoxometalate water splitting catalyst: very fast hole scavenging from photogenerated oxidants, Chemical Communications, 2010, 46, 3152–4.
[29] Sartorel A, Carraro M, Scorrano G, De Zorzi R, Geremia S, McDaniel ND, et al., Polyoxometalate embedding of a tetraruthenium(IV)-oxo-core by template-directed metalation of $[\gamma\text{-}SiW_{10}O_{36}]^{8-}$. A totally inorganic oxygen-evolving catalyst, Journal of the American Chemical Society, 2008, 130, 5006–7.
[30] Zhang ZY, Lin QP, Kurunthu D, Wu T, Zuo F, Zheng ST, et al., Synthesis and photocatalytic properties of a new heteropolyoxoniobate compound: $K_{10}[Nb_2O_2(H_2O)_2][SiNb_{12}O_{40}] \bullet 12\ H_2O$, Journal of the American Chemical Society, 2011, 133, 6934–7.
[31] Bernadini G, Zhao C, Wedd AG, Bond AM, Ionic-liquid enhance water photooxidation by $P_2W_{18}O_{62}$, Inorganic Chemistry, 2011, 50, 5899–909.
[32] Zhao C, Bond AM, Photoinduced oxidation of water to oxygen in the ionic liquid $BMIMBF_4$ as the counter reaction in the fabrication of exceptionally long semiconducting silver-tetracyanoquinodimethane nanowires, Journal of the American Chemical Society, 2009, 131, 4279–87.
[33] Szczepankiewicz SH, Ippolito CM, Santora BP, Van de Ven TJ, Ippolito GA, Fronckowiak L, et al., Interaction of carbon dioxide with transition-metal-substituted heteropolyanions in nonpolar solvents. Spectroscopic evidence for complex formation, Inorganic Chemistry, 1998, 37, 4344–52.
[34] Xu L, Gao GG, Li FY, Liu XZ, Yang YY, CO_2 coordination by inorganic polyoxoanion in water, Journal of the American Chemical Society, 2008, 130, 10838–9.
[35] Khenkin AM, Efremenko I, Weiner L, Martin JML, Neumann R, Photochemical reduction of carbon dioxide catalyzed by a ruthenium-substituted polyoxometalate, Chemistry-a European Journal, 2010, 16, 1356–64.
[36] Argitis P, Papaconstantinou E, Photocatalytic multielectron photoreduction of 18-tungstodiphosphate in the presence of organic-compounds – production of hydrogen, Journal of Photochemistry, 1985, 30, 445–51.
[37] Ioannidis A, Papaconstantinou E, Photocatalytic generation of hydrogen by 1–12 heteropolytungstates with concomitant oxidation of organic-compounds, Inorganic Chemistry, 1985, 24, 439–41.
[38] Papaconstantinou E, Photocatalytic oxidation of organic-compounds using heteropoly electrolytes of molybdenum and tungsten, Journal of the Chemical Society-Chemical Communications, 1982, 12–3.

[39] Muradov N, T-Raissi A, Solar production of hydrogen using "self-assembled" polyoxometalate photocatalysts, Journal of Solar Energy Engineering, 2006, 128, 326–30.
[40] Bouchard DA, Hill CL, Catalytic photochemical dehydrogenation of organic substrates by polyoxometalates, Journal of the American Chemical Society, 1985, 107, 5148–57.
[41] Forster J, Rösner B, Khusniyarov MM, Streb C, Tuning the light absorption of a molecular vanadium oxide system for enhanced photooxidation performance, Chemical Communications, 2011, 47, 3114–6.
[42] Errington RJ, Coyle L, Middleton PS, Murphy CJ, Clegg W, Harrington RW, Synthesis and structure of the alkoxido-titanium pentamolybdate $(^nBu_4N)_3[(^iPrO)TiMo_5O_{18}]$: an entry into systematic $TiMo_5$ reactivity, Journal of Cluster Science, 2010, 21, 503–14.
[43] Errington RJ, Harle G, Clegg W, Harrington RW, Extending the lindqvist family to late 3d transition metals: a rational entry to CoW_5 hexametalate chemistry, European Journal of Inorganic Chemistry, 2009, 2009, 5240–6.
[44] Errington RJ, Petkar SS, Horrocks BR, Houlton A, Lie LH, Patole SN, Covalent immobilization of a TiW_5 polyoxometalate on derivatized silicon surfaces, Angewandte Chemie International Edition, 2005, 117, 1280–3.
[45] Errington RJ, Petkar SS, Middleton PS, McFarlane W, Synthesis and reactivity of the methoxozirconium pentatungstate $(^nBu_{4N})_6[\{(\mu\text{-}MeO)ZrW_5O_{18}\}_2]$: Insights into proton-transfer reactions, solution dynamics, and assembly of $\{ZrW_5O_{18}\}^{2-}$ building blocks, Journal of the American Chemical Society, 2007, 129, 12181–96.
[46] Filowitz M, Ho RKC, Klemperer WG, Shum W, Oxygen-17 nuclear magnetic-resonance spectroscopy of polyoxometalates .1. Sensitivity and resolution, Inorganic Chemistry, 1979, 18, 93–103.
[47] Tucher J, Wu Y, Nye LC, Ivanovic-Burmazovic I, Khusniyarov MM, Streb C, Metal substitution in a Lindqvist polyoxometalate leads to improved photocatalytic performance, Dalton Transactions, 2012, 41.
[48] Kudo A, Omori K, Kato H, A novel aqueous process for preparation of crystal form-controlled and highly crystalline $BiVO_4$ powder from layered vanadates at room temperature and its photocatalytic and photophysical properties, Journal of the American Chemical Society, 1999, 121, 11459–67.
[49] Li GS, Zhang DQ, Yu JC, Ordered mesoporous $BiVO_4$ through nanocasting: A superior visible light-driven photocatalyst, Chemistry of Materials, 2008, 20, 3983–92.
[50] Xi GC, Ye JH, Synthesis of bismuth vanadate nanoplates with exposed {001} facets and enhanced visible-light photocatalytic properties, Chemical Communications, 2010, 46, 1893–5.
[51] Tucher J, Nye LC, Ivanovic-Burmazovic I, Notarnicola A, Streb C, Chemical and Photochemical Functionality of the first molecular bismuth vanadium oxide, Chemistry – A European Journal, 2012, 17, 10949–53.

Thomas Merz, Martin Schütz

14 Description of excited states in photocatalysis with theoretical methods

14.1 Introduction

The theoretical treatment of molecules in their electronic ground state (GS, S_0, respectively) near the equilibrium geometry is nowadays rather straightforward. Efficient and reliable methods are at hand to calculate energies and their derivatives w.r. to various external perturbations, including analytical gradients w.r. to displacements of the nuclear positions as usually used in geometry optimizations. More problematic is the description of the dissociation of covalent bonds where static correlation effects start to play a role (the wavefunction is no longer dominated by a single determinant or configuration state function, but by several or many). Here, considerably more complicated multireference methods have to be used.

The theoretical treatment of molecules in electronically excited states, however, is much more complicated and remains a challenge. In contrast to the GS, electronically excited states can hardly be treated by a single determinant or configuration state function, not even near equilibrium geometry. This calls for multireference methods, or, alternatively, for time-dependent response methods, such as time dependent density functional theory (TD-DFT), or time dependent coupled cluster response theory (TD-CC response).

An accurate quantum chemical treatment also requires a proper treatment of intermolecular interactions (hydrogen bonds, π-stacking). Furthermore, solvent effects may have to be considered, which can have a massive effect, particularly so for charge-separated charge transfer (CT) states. Possible ways to deal with solvation effects are reaction field methods (surrounding the molecule in a cavity by a continuum) or an explicit treatment of solvent molecules via classical model potentials in a quantum mechanics/molecular mechanics (QM/MM) hybrid approach.

All this makes almost every careful theoretical study complicated and time-consuming. Unfortunately, photochemical reactions occur naturally on potential energy surfaces related to excited states: after absorbing one or more photons (in most cases a single photon) the wave packet representing the GS probability distribution of the nuclei, is energetically lifted onto an excited state surface. The GS equilibrium geometry (Franck–Condon point) of course no longer corresponds to a minimum on the excited state surface. The wave packet, therefore, propagates on the excited state surface towards a minimum of the excited state surface. On its way, it may cross conical intersection seams. It may be reflected on a repulsive wall, etc. Propagating such wave packets explicitly, i. e. performing quantum dynamics calculations is computationally very expensive and only possible in a few dimensions.

Fortunately, already from the topologies of ground and the electronically excited state surfaces involved, i. e. from the relevant minimum energy geometries, transition states, and points on the conical intersection seams, a lot of the qualitative photophysics is already revealed.

Theoretical electronic structure methods and experimental techniques complement one another. This is particularly true for ab initio methods, which are entirely independent from any experimental input in contrast to empirical or semi-empirical methods. If the two “dance together”, often new insight and understanding for the posed question can evolve.

In this chapter, we want to present a short description of the concept of potential energy surfaces and possible processes therein. Some theoretical methods, relevant in the present context, are described next. We focus here on time-dependent coupled cluster response methods on the ab initio side, and provide a very short overview on time dependent density functional theory. Applicability, weaknesses and limitations of these methods are also discussed. Subsequently, we discuss the description of solvation effects on the basis of QM/MM. In the last section we present two examples in the context of chemical photocatalysis, in order to illustrate the application of theory in photochemistry and the conjunction with experimental results. For more detailed and profound information, we recommend for general aspects of the QM methods for photochemistry the reviews of Schmidt et al. [1], Gonzalez et al. [2] and Bernardi et al. [3]. A summary of the also described MM methods can be found in Ponder and Case [4]. For a detailed discussion of the QM/MM approach we refer to the review of H. Senn and W. Thiel [5].

14.2 The concept of potential energy surfaces

For the understanding of the photophysics of a system, the information about the shape and the topological properties of the potential energy surfaces (PES) for the ground state and the relevant excited states is crucial. These surfaces describe the energy of the corresponding state of the molecule as a function of the positions of its nuclei. Each surface is 3N-6 dimensional (3N-5 for linear molecules), where N is the number of nuclei. The concept of potential energy surfaces directly derives from the Born–Oppenheimer approximation [6], which separates nuclear and electronic motion on the basis of the relative mass of nucleus vs. electron. Nuclear motion in this framework is driven by forces generated due to the potential energy surfaces, which in turn are determined by the motion of the electrons. The topology of the PES directly maps onto the chemical properties of the system. In general, global minima represent stable, local minima metastable structures, and first-order saddle points are transition states. A chemical reaction corresponds to a path linking a minimum to another minimum via one (or several) first-order saddle points, and eventually other local minima in between. As an example, in Figure 14.1 a three

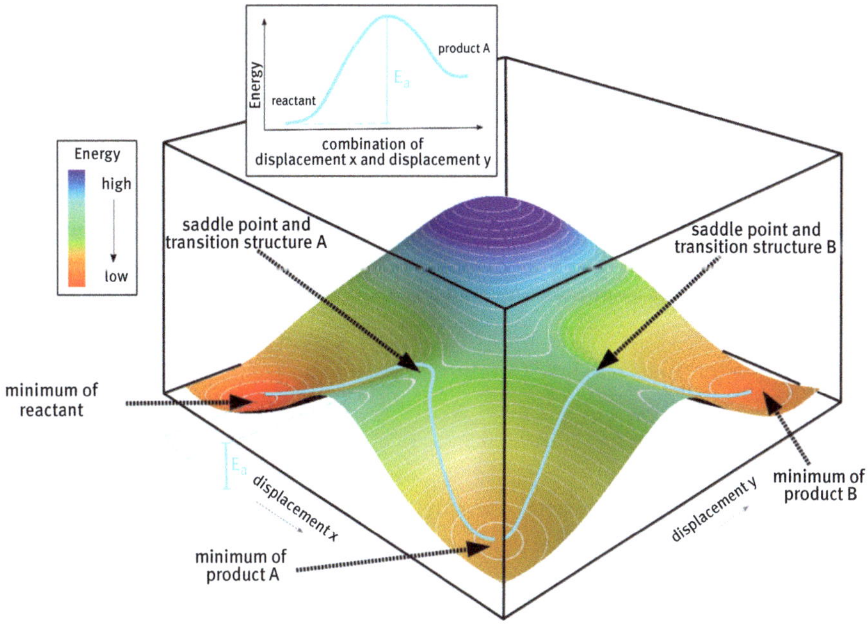

Figure 14.1: Three dimensional potential energy surface (PES), with two dimensional chemical reaction pathway from the reactant, which is here the global minimum, to the product A, which is here a local minimum. In both cases, the reaction energy E_a is depicted. The reactions follow the minimum energy path via the transition structures on top of the saddle points.

dimensional PES is shown. One can see three minima, one global, which represents the reactant of the given system and two local minima representing products A and B. The reaction path connecting these minima via two first-order saddle points is indicated as a blue line. The reaction path linking reactant and product A is also displayed in the more familiar two-dimensional plot, for comparison.

The Born Oppenheimer approximation neglects the non-adiabatic coupling elements, which contain first and second derivatives w.r. to the nuclear coordinates of the electronic wave functions on the ket side. These become large when the wave functions strongly change for small displacements in the nuclear coordinates. Therefore, the Born Oppenheimer approximation only works for insulated states, i. e. for sufficiently separated PES (adiabatic regime). At points however, where two potential energy surfaces intersect, i. e. at the conical intersection seam, the energies of two states obviously are degenerate. Consequently, the two related wave functions describing the molecule at the conical intersection are arbitrary mixtures of the wave functions of the separated pure states. Obviously, when approaching a conical intersection seam the wave functions grossly change their character over a tiny change in geometry. The non-adiabatic coupling elements thus become large and the Born Oppenheimer approximation breaks down. At conical intersections the motion of the electrons and nuclei thus is strongly coupled and the system easily, efficiently, and very quickly crosses from one (adiabatic) PES to the other (non-adiabatic regime).

Conical intersections between electronically excited states and the GS are, for example, used by nature to rapidly shed energy and to avoid photochemical reactions in photoexcited nucleobases. One compelling reason for nature to have the four particular nucleobases as building blocks for DNA is their photostability, which is due to the existence of a conical intersection seam near the Franck–Condon point. [7]

Conical intersections, postulated already in the 1930s [8, 3], have become ubiquitous in photophysics during the past two decades. For a non-linear three-atom molecule like ozone [9] the PES of two states touch in a single point. Due to this topology, it is often illustrated as a double inverse cone [10]. For bigger molecules the space of degeneracy corresponds to a 3N-8 dimensional hypersurface (there are two additional conditions to be fulfilled such that the two by two CI matrix involving the two relevant states yields two degenerate eigenvalues). For a more detailed discussion of this subject we refer to the article of David R. Yorkony [11].

In Figure 14.2 a photochemical reaction along one path is presented as an example. The main adiabatic and non-adiabatic reactions are included in this picture. The first process is the absorption of a photon (in the time scale of *fs*), where the ground state S_0 is lifted energetically to the first excited singlet state S_1. After the relaxation to

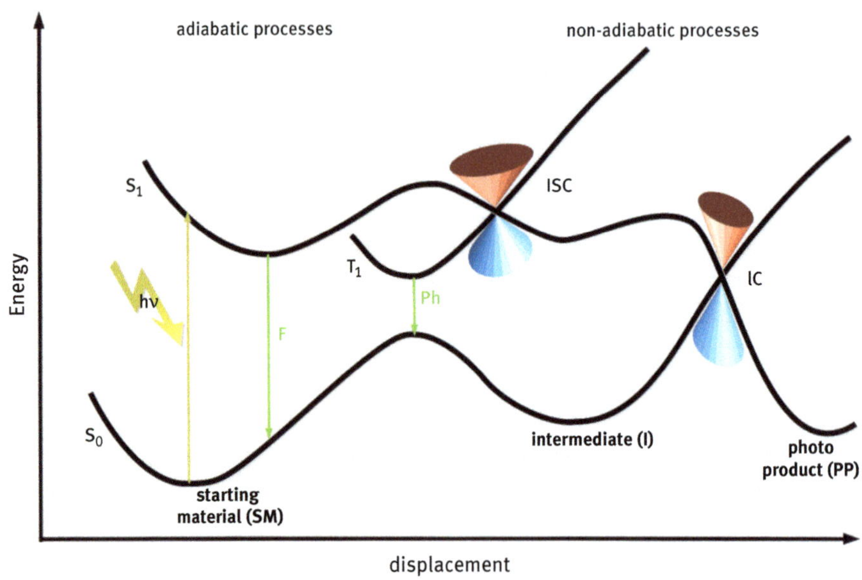

Figure 14.2: Two dimensional PES for a possible photochemical processes. In the green shaded area, the adiabatic processes and photophysics are shown. F is fluorescence, Ph phosphorescence. In the blue shaded area all non-adiabatic processes are shown. The intersystem crossing (ISC) as well as the internal conversion (IC) in this example occurring via a conical intersection (depicted by the typical double inverse cone shape, if plotted in a three dimensional PES).

the lowest vibrational level, the molecule can get back to the ground state by emitting a photon. This is called fluorescence (F), which is happening in the *ns* range.

Radiationless processes as well as the population of other spin states (e. g. triplet states), which are often important for photochemistry or photocatalysis are governed by non-adiabatic processes. The emission of a photon out of the triplet state, which is per se an adiabatic process, is called phosphorescence (Ph). This process takes place in the *ms* time range. As stated before, surface crossings are non-adiabatic processes. These crossings can occur between different spin symmetries. If the crossing involves states with different spin symmetry, one speaks of intersystem crossing (ISC), otherwise of internal conversion (IC). IC is normally based on vibronic coupling, which is then not a surface crossing and therefore an adiabatic process. This process decreases the fluorescent quantum yield. But it is also possible, that an ultrafast conversion, for instance to the photoproduct, can be seen. This ultrafast conversion then is mediated by a conical intersection between the two PES.

In Figure 14.2 the following photochemical reaction should be in the centre of attention. Schematically it can be reduced to the following very simple chemical reaction formula:

$$SM \xrightarrow{h\nu} I + PP\,. \tag{14.1}$$

This can be either a photochemical reaction of a monomer or a complex reaction of a lot of starting materials to a photoproduct. Displacement in this case would then be the classical known reaction coordinate. In our example, due to the knowledge of the shape of the PES, one can go into details for this reaction.

The starting material is excited in the first singlet state. From there, one possible relaxation would be the fluorescence back to the ground state. For some extent also ISC to the triplet state can occur. In this case, the process would be non-adiabatic, mediated by a conical intersection. Deactivation by phosphorescence back to the starting material or an intermediate for the photoreaction would be the outcome. In the example also an IC by a conical intersection from the first excited singlet state to the photo product can be observed.

This simple example, which can be summarized by a very simple chemical formula (Equation 14.1), has to be extended to yield a valid "photocycle" that can be seen as a schematic view in Figure 14.3 and as extended formulas in Equation 14.2:

$$\begin{aligned} SM &\xrightarrow{h\nu} {}^{1}(SM)^{*} \\ {}^{1}(SM)^{*} &\xrightarrow{F} SM \\ {}^{1}(SM)^{*} &\xrightarrow{ISC} {}^{3}(SM)^{*} \\ {}^{3}(SM)^{*} &\xrightarrow{Ph} I + SM \\ {}^{1}(SM)^{*} &\xrightarrow{IC} I + PP \end{aligned} \tag{14.2}$$

Expanded chemical formulas for example.

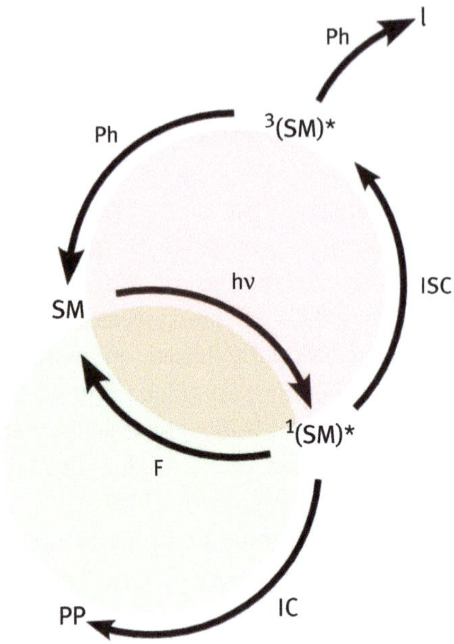

Figure 14.3: Corresponding photocyle to Equation 14.2 of example. The green cycle is productive, the red one is destructive.

Of course, all reactions depend still on the energetic barriers, which have to be passed during the pathway or displacement needed for the forthcoming reaction. But one can see, that the knowledge of the character of the involved PES as well as the conditions for non-adiabatic processes in the photochemical reaction is of benefit for the tuning of reactions towards the desired photoproduct. Theoretical chemists can help explaining the photochemical reactions with quantum mechanical methods.

In the following two classes of methods, which can be used to describe properties of excited states, are presented.

14.3 Computational methods for excited states

14.3.1 QM-Methods

Ab initio wavefunction methods for electronically excited states can be divided into single reference and multi reference methods. The latter are difficult to use in the context of extended molecular systems and are not further discussed here, and we refer to the references [2] and [12] for a detailed discussion of these methods.

Single reference methods include methods like Configuration Interaction, e. g. the configuration interaction singles CIS [13] method, propagator methods like RPA and ADC [14, 15], and time-dependent coupled cluster response TD-CC methods. Another commonly used method for excited states of extended systems is time-dependent density functional theory TD-DFT. TD-DFT and TD-CC both are response methods. Excitation energies are calculated as the poles of the time- or frequency-dependent polarizability by first-order perturbation theory (linear response). This is based on a time- or frequency-dependent external perturbation. Excitation energies thus can be regarded as a time or frequency dependent ground state property of the molecule.

In the following we focus on TD-CC and TD-DFT since these methods are inexpensive enough to describe issues in photocatalysis. Special attention will be given to potential pitfalls.

14.3.1.1 Time-dependent coupled cluster response

The coupled cluster wavefunction $|CC\rangle$ is specified by an exponential excitation operator acting on the Hartree Fock reference wavefunction) $|0\rangle$, e. g. for coupled cluster singles and doubles we have

$$|CCSD\rangle = \exp(T_1 + T_2)|0\rangle, \tag{14.3}$$

where T_1 and T_2 each generate a linear combination of all possible singly and doubly substituted determinants, respectively, from $|0\rangle$. Singly substituted means that a certain molecular orbital occupied in the $|0\rangle$ is being replaced by a virtual molecular orbital.

Due to the exponential ansatz, the wave function is always (at each truncation level) product separable, the energies thus size extensive. Furthermore, all possible excitations of full CI are generated even when the cluster operator is restricted to only singles and doubles as above. The coefficients of higher excitation levels (say quadruples) then factorize as products involving singles and doubles coefficients. Coupled cluster theory as described so far is a highly accurate ground state method, which provides a large fraction of the correlation energy; for a review we refer to R. Bartlett and Musiał [16].

Commonly applied methods for excited states within the couple cluster framework are the equation of motion couple cluster (EOM-CC) methods [17], and time-dependent coupled cluster (TD-CC) linear response [18, 19] based on the hierarchy of the coupled cluster models CC2 [20], CCSD, CC3 [21], and CCSDT. As already mentioned above, TD linear response approaches aim at the frequency dependent polarizability of the system by first-order perturbation theory. Excitation energies are calculated as the poles, transition strengths as the residues of this time-dependent ground state property. The cheapest method still including dynamic correlation effects is TD-CC2 response. Similar to second order Møller–Plesset perturbation theory (MP2), doubles substitutions are just treated to first order w.r.t. to the fluctuation potential. The TD-CC2 is quite a robust method as long as the excited states of interest are dominated by singles substitutions (which is usually the case for low-energy excitations as those considered here). In such situations it provides a qualitatively correct picture of these states with an accuracy of about 0.2–0.3 eV in the excitation energies. Near conical intersections, the excitation energies may become complex, though there are correction schemes to cope with that problem [22]. Note that any response theory breaks down for conical intersections with the ground state, since there the ground state reference no longer is properly described by a single determinant or configuration state function.

The restriction to rather small molecules is a major drawback of TD-CC response. Even TD-CC2 is considerably more expensive than TD-DFT, particularly so for energy gradients w.r. to nuclear displacements, as needed for geometry optimizations on excited state surfaces. A practical approach often employed by us is to utilize TD-DFT for geometry optimizations, and TD-CC2 for excitation energies and other properties at these TD-DFT geometries.

14.3.1.2 Time-dependent density functional theory

Kohn–Sham density functional theory (DFT) is one of the most frequently used theoretical approaches in almost every field of chemistry. According to the Hohenberg–Kohn theorem [23] there exists a functional $E\left[\rho(r)\right]$ mapping the exact electron density to the exact energy. In other words, the energy can in principle be calculated from a much simpler object than the wave function, namely the (one-particle) density function depending only on three degrees of freedom. Yet the Hohenberg–Kohn theorem just proves the existence of this functional, and nobody knows what it actually looks like. The unknown part of $E\left[\rho(r)\right]$ can be condensed in the exchange-correlation part $E_{xc}\left[\rho(r)\right]$ of it, which has to be modelled. DFT with present functionals, thus, in practice, can be considered as a semi-empirical approach. There is a plethora of different $E_{xc}\left[\rho(r)\right]$ which have been proposed over the last decades, ranging from simple local density approximations (LDAs), over gradient corrected functionals (GGAs), hybrid functionals like the famous B3LYP [24], to range-separated hybrids, orbital dependent functionals, and random-phase approximations for correlation. Note that the Kohn–Sham DFT Hamiltonian is an effective one-particle Hamiltonian of a fictitious non-interacting many-electron system, and thus different from the exact Hamiltonian as used e. g. for evaluating the Hartree–Fock energy.

TD-DFT is mostly used as a linear response theory on the basis of DFT. Its equations (also known as Casida's) equations are very similar to those of time-dependent Hartree–Fock theory. As for TD-CC response, the excitation energies correspond to the poles of the frequency dependent polarizability. A recent review of TD-DFT can be found in reference [25].

Like DFT for the ground state, TD-DFT is a very efficient method for calculating excited states of extended systems. Analytic gradients, and thus geometry optimizations, are inexpensive in comparison to ab initio wavefunction methods. Efficiency and simplicity in its usage surely is the key to its success. DFT, and to a lesser degree TD-DFT, are available in most ab initio electronic structure program packages.

On the other hand, both DFT and TD-DFT lack the possibility of a systematic improvement of the method, since there is no such hierarchy of methods as for ab initio wavefunction methods. There are also some nasty artefacts plaguing DFT, like a non-vanishing correlation energy of a single electron, or the self-repulsion of electrons. The latter has severe consequences for TD-DFT, in so far, as excitation energies of charge transfer (CT) excited states are grossly underestimated [26–29] due to the absence of the excitonic hole-particle attraction in local or semi-local functionals [29]. Excitation energies of CT excited states with large separation of hole and particle are easily underestimated by several eV, eventually mixing with and contaminating other low-lying valence states. The use of hybrid functionals with a large fraction of non-local exchange, or range-separated hybrids [30–32] shift CT states back to higher energies and, hence, repair this problem to some extent.

Of course, TD-DFT, as any response method based on a single determinant, is unable to treat conical intersections with the ground state. Knowledge of limitations of TD-DFT is necessary when applying TD-DFT. In particular, low-lying CT states are often meaningless and may even contaminate other valence states. TD-DFT then provides an entirely wrong picture of the photophysics of the system. It is, therefore, recommended to verify TD-DFT results by calculations with other methods like TD-CC2 on the same or a related system.

14.3.2 Solvent description via the QM/MM approach

The combination of QM-methods as those described in the previous section with the less expensive molecular mechanic (MM) method is known as the QM/MM hybrid approach. In the 1970s of the last century, Warshel and Levitt first suggested to combine these two methodologies [33]. Their trendsetting idea had already been all in hand, which is still used nowadays. Field, Bash, and Karplus were the first who combined the CHARMM force field, a widely used model potential for MM calculations, with QM methods [34]. Fourteen years after these first ideas came the breakthrough for the QM/MM approach.

In this approach that is used for very large biological systems as well as for applications in organic chemistry, the whole system is divided into two parts (regions). The first one, which is in almost every case, very tiny with respect to the rest, is treated with an appropriate QM method. This first part, called the QM-region, comprises that part of the system, in which the actual reaction takes place. The second, large part, denoted as the MM-region, comprises the rest of the system.

The main strength of the QM/MM approach is the possibility of investigating truly large systems with quite accurate QM methods. This is possible, because only a small part is treated at the expensive QM level. Inclusion of only the interesting parts in the QM-region, e. g. the enzymatic active region of a protein, or, if the method is utilized for the description of solvation effects, one single solute molecule, or alternatively, a cluster of the solute with its first solvation shell, reduces the computation time dramatically relative to a full QM treatment of the whole system, which would be entirely unfeasible.

The QM/MM approach can straightforwardly be employed to describe solvation effects. Solvation can severely affect the photophysical properties of molecules. These can significantly differ for different solvent environments or the gas phase. The force field can be used to mimic the natural environment, which in biological systems is water. There are a lot of different model potentials for water; the most prominent is the TIP3P model [35]. This is a 3-site water molecule parameterized to reproduce the behaviour of liquid water. A lot of other models have been developed to improve this very simple model. For organic reactions water often is not sufficient to solvate the starting materials for the reaction. However, in principle every solvent can be modelled in the MM scheme.

One should mention, that there are other possibilities to describe solvent environments, such as the reaction field methods, where the solute molecule in a cavity is enclosed by a dielectric continuum. The conductor-like screening model (COSMO) [36] perhaps is one of the most prominent approaches along these lines. However, it is presently unclear if, and how well, such approaches would work for describing solvated-excited states. In the present contribution we therefore focus on the QM/MM approach to describe solvation effects.

Before discussing the coupling of the MM methods with QM and the problems and traps related to this, it is appropriate to provide a short overview of the model potentials or force fields as employed in MM methods.

14.3.2.1 MM methods

MM methods are widely used in the context of biology. The energy of the system in such calculations depends on its atomic coordinates. The energy and the forces on the atoms are related to the gradient of the potential function. This function sums harmonic contributions originating from bond and bond angle displacements, cosine terms reflecting displacement of dihedral angles (torsions), as well as electrostatics (modelled by partial charges or higher moments), van der Waals dispersion, and exchange repulsion terms.

$$\begin{aligned} V(r) = &\sum_{bonds} k_b(b-b_0)^2 + \sum_{angles} k_\theta(\theta-\theta_0)^2 + \sum_{torsions} k_\phi[\cos(n\phi+\delta)-1]) + \\ &+ \sum_{elec}\left[\frac{q_i q_j}{r_{ij}}\right] + \sum_{\substack{van\,der\\ Waals}}\left[\frac{A_{ij}}{r_{ij}^{12}} + \frac{C_{ij}}{r_{ij}^{6}}\right] \end{aligned} \tag{14.4}$$

In Equation 14.4 an example of a simple potential is given. For the following discussions the description above should be sufficient; for a detailed review of force fields and all related issues we refer to an article of Ponder and Case [4].

As is evident from Equation 14.4, quite a lot of parameters enter the specification of such a model potential (for example k_b, b_0, k_θ, θ_0 etc.). They constitute the force field. Different parameterizations distinguish different force fields. A lot of knowledge has to go into the development of the needed parameters. Extending a specific force field to new molecules not covered so far thus has to be done very carefully.

On the MM side, almost every force field can be used. The established ones are AMBER [37], CHARMM [38, 39], optimized for biological systems like enzymes, proteins and DNA, and other systems. It is of course also possible to develop force fields for organic molecules. Such force fields have to be more versatile due to the higher diversity of organic compounds. Specialized force fields like MMx [40–42], UFF [43] can be used here. Gromacs [44] also provides this alternative. On the other hand, the aforementioned AMBER force field also offers the possibility to parameterize organic molecules within its framework by using the GAFF (general AMBER force field) parameters [44]. These force field parameters are designed for compatibility with the original AMBER force field developed for biological systems. GAFF includes parameters for almost every organic or pharmaceutical molecule composed of hydrogen, carbon, nitrogen, oxygen, sulphur and phosphorus. A common procedure to import molecules is available and described by Bayly et al. in reference [46].

14.3.2.2 QM/MM coupling

In the QM/MM hybrid approach the MM and the QM methods are combined. The coupling between these two parts determines the accuracy of the QM/MM calculations. Essentially, there are two distinct coupling schemes, (i) the subtractive (see Equation 14.5), and (ii) the additive (see Equation 14.6) coupling. Both schemes are well developed in the meantime and discussed in detail in a lot of articles and reviews [1, 5, 47, 48].

For (i) the whole system is calculated within the force field. Afterwards a separate MM and QM calculation for the QM-region is carried out. From the sum of the energies of the MM calculation of the whole system, and the QM calculation of the QM region the energy of the MM calculation of the QM-region is subtracted (to avoid double counting). The resulting energy is the QM/MM energy of the system. This subtractive coupling is very simple to implement, but the polarisation of the QM part by the MM part is not taken into account.

$$E_{QM/MM} = E_{MM}(all) + E_{QM}(QM) - E_{MM}(QM) \tag{14.5}$$

On the other hand, in the additive scheme (ii) the MM calculation is only performed for the MM part. Here the coupling is described by an additional term, the so-called coupling term. The sum of the MM energy (only the MM-region), the QM energy (only the QM-region), and the coupling term of the two subsystems, yields the QM/MM energy:

$$E_{QM/MM} = E_{QM}(QM) + E_{MM}(MM) + E_{iact}(QM + MM). \tag{14.6}$$

In order to compute the coupling term $E_{iact}(QM+MM)$, two different procedures are generally used, (i) the mechanical, and (ii) the electrostatic embedding. Mechanical embedding is based on MM-MM electrostatics, while electrostatic embedding implies inclusion of the MM charges in the QM calculation: the MM point charges are included in the one-electron Hamiltonian of the QM region as sources of an external electrostatic field. Here, the polarisation of the QM part by the MM part is taken into account.

Provided that QM- and MM-regions are not connected to each other via chemical bonds, such QM/MM calculations are straightforward and no further problems arise. However, in most cases it is not possible or appropriate to divide the total system in QM and MM parts without artificial bond breaking.

In both regions a cut bond has to be capped somehow. Homolytically or heterolytically cleaved molecules in the QM-region would either lead to artificial radicals or charges, which have little to do with the original system. Also the MM terms involving cut bonds have to be reorganized. For this problematic issue, once again different solutions were developed. The first variant is to introduce an additional atom on the QM region, which is covalently bound to the last QM atom. This scheme is called the Link-atom scheme. As capping atoms hydrogen or fluorine is usually chosen. The second is to add localized orbitals at the boundary, which then are kept frozen in the QM calculation.

There are also other variants to cap the QM-region not discussed any further here [49].

In particular for the link-atom scheme one has to be careful to prevent overpolarisation of the QM density. For a discussion of these problems we refer to references [49] and [50], where these problems are discussed in detail. In the following examples we employ the link-atom approach due to its simplicity.

14.4 Procedure

It is quite difficult to give a general recipe for QM/MM investigations of photochemical or photocatalytical processes, including all aspects from the proper choice of the QM method to the final interpretation of the results. A lot of things have to be considered. But in principle two steps have to be taken to get reasonable results. If one is interested in the conical intersection seam, further and not straightforward calculations have to be done. These procedures are not discussed here.

1. The starting point of every theoretical investigation of a molecule or system is the vertical excitation spectrum for the ground state minimum. This is only one first jigsaw piece for the complete investigation. At that point, the assignment of the different types of electronically excited states has to be done. Different methods should be compared and calibrated at that stage, e. g. TD-DFT with different functionals should be compared to TD-CC2 reference calculations. This provides the needed information on which method or functional to apply for the target problem. Next, the relevant low-energy minima on the potential energy surfaces of the excited states have to be determined. With the geometries related to the reactants, intermediates, and products of a photochemical reaction at hand, one already has a qualitative picture of the mechanism at work.
2. To determine the minimum energy path between reactants and products is more intricate: Provided that the reaction coordinate is a priori known, i. e., reflected in one or just a few internal coordinates, then geometries along the reaction coordinate can be generated by hand. Otherwise more general and costly approaches must be utilized. For example, the NEB (nudged elastic band) method [51], not described here, could be used. For a detailed discussion about optimization methods to determine minimum energy paths we refer to reference [52].

In the examples presented below, TD-DFT was used for geometry optimisations in most cases. In order to identify potential problems (like contamination of relevant valence state by low-lying charge transfer states) TD-CC2 single point calculations were always performed. Comparing the singles substitution vectors between the two methods, as well as dipole moments and transition strengths is very helpful in that context. Furthermore, only the TD-DFT geometries are employed, excitation energies at these geometries are always calculated at the more robust and qualitatively correct TD-CC2 level.

For the presented examples, also solvent effects were taken into account. Parameterisation of the molecular force field was done very carefully. Solvated geometries were generated on the basis of minimum energy structures of the isolated solute (optimized by using QM methods) embedded in a large enough solvent shell, by performing MD simulations after an initial equilibration phase.Typical MD trajectories range between 500 ps and 10 ns, depending on the size of the system.

Subsequently, different snap shots representative of the relevant geometries encountered during the MD simulation were chosen. For a recent review about this issue, we refer to Klenin et al. [53]. These snap shots were then used as starting points for subsequent QM/MM optimisations on the ground and excited state surfaces.

14.5 Examples

In the following, two examples are presented which adopt the procedure described above. In the first example, a flavin analogue, roseoflavin (see Figure 14.4) was investigated in the context of flavin chemistry and biological photochemistry in a protein. Flavin is frequently used as a photo catalyst. As redox cofactor, it is capable of two subsequent electron transfers [54]. The photophysical and photochemical steps for alcohol oxidation were recently studied by Megerle et al. [55]. Furthermore, theoretical investigations on flavin, especially on excited states properties, were also carried out [56–58]. Here, we explore the photophysics of the modified chromophore, roseoflavin, in three different environments.

Introduction of one dimethylamino group to flavin leads to a slightly different absorption behaviour [59]. The usage of ab initio and QM/MM methods to explain and verify experimental results can be shown quite well in this example. Some intramolecular charge transfer process is responsible for inactivating the molecule for photocatalysis in gas phase and in water. Zirak et al. [60] observed a significant fluorescence quenching of roseoflavin in water. For the chromophore in the BLUF protein environment, there is also a charge transfer state between the roseoflavin and one tyrosine moiety of the protein that can be seen. This charge transfer state plays the central role in the photocycle of BLUF domains and their functioning as blue light photoreceptors [61, 62]. It drives the reaction cascade leading from the inactive to the signalling (or light adapted) geometry of the protein. Via a low-lying conical intersection between the charge transfer state (which itself has a very low transition strength), and a locally excited state (with large oscillator strength) the former is populated efficiently after photoexcitation. If the flavin chromophore of the wild type is being replaced by a roseoflavin, the conical intersection is not accessible and the charge transfer state remains unpopulated [63]. The modified BLUF domain then features no photocycle, and the protein is dysfunctional. For a detailed discussion see reference [64].

In the second example, also charge transfer processes play an important role. The chromophore in this case is benzophenone (see Figure 14.5, shortened BP). The molecule is well investigated for biological photochemistry and photocatalysis. ISC to the first triplet state is very efficient [65–75], which leads to better access

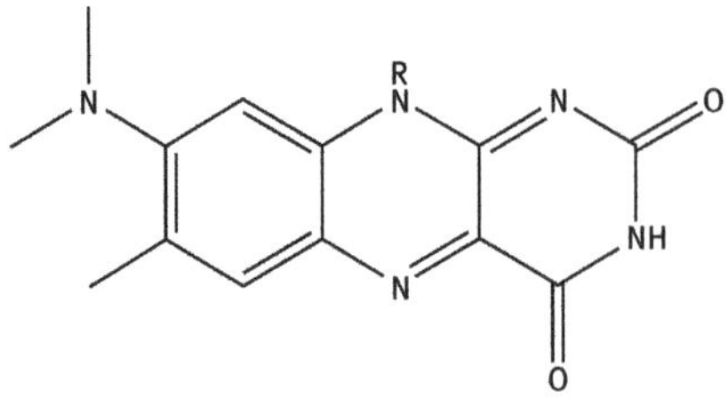

Figure 14.4: Chemical structure of roseoflavin, if R = H.

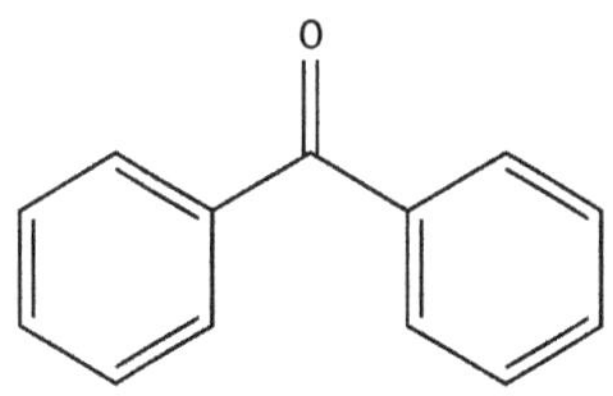

Figure 14.5: Chemical structure of Benzophenone.

for chemistry with the chromophore, due an increase of the life time. For investigations in the framework of DNA-zymes [76, 77], the benzophenone was synthesized as a artificial DNA-nucleobase to be incorporated into DNA [78]. Dimers of one standard DNA-base and the new "benzophenone" DNA-base were the basis for this research.

By employing ultra fast transient absorption spectroscopy it was found that the lifetime of the excited ($n\pi^*$) singlet state is primarily determined by fast intersystem crossing (ISC) to the lowest triplet state.

Yet in methanol and water, different life times of the singlet state for the combination of the artificial "benzophenone" base and guanosine were observed. In methanol the lifetime is shortened about 2 ps compared to water, a substantial reduction of the lifetimes that could not be observed for combinations with other bases apart from guanosine, at least not by such a substantial amount. The experimental an theoretical results will be published elsewhere.

In the following we provide a short account on these two examples.

14.5.1 Roseoflavin

In the investigation of Zirak et al. [63], some well known deactivation processes were postulated as deactivation paths for the photoexcited roseoflavin. So-called TICT (twisted intramolecular charge transfer) for polar solvents or PICT (planar intramolecular charge transfer) for apolar solvents mechanisms, respectively, [79, 80], were postulated to be responsible for it. In order to verify these conjectures, calculations in the gas phase, in water, and in the protein environment (BlrB [81]) were performed.

For the gas phase the procedure described in Section 14.4 was straightforwardly followed. The optimization of the ground state (CC2 in aug-cc-pVDZ AO basis [82]) showed, that the angle between the dimethylamino group and the isoalloxazine skeleton, is tilted. Calculations of the vertical excitation energies at the Frank–Condon point with TD-CC2 and def-SVP and aug-cc-pVDZ basis followed. One can see, that the first state and the second state have ($\pi\pi^*$) character. The third state features ($n\pi^*$) character. The corresponding density difference plots are depicted in Figure 14.6.

The vertical excitations energies at the Franck–Condon point QM methods were applied; were also calculated with TD-DFT (with BP86 [83], B3LYP, and BHLYP [84]

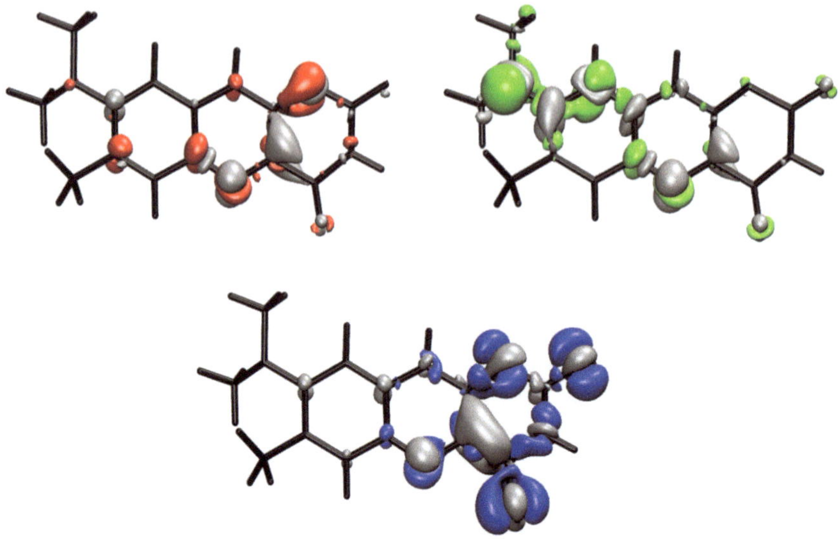

Figure 14.6: Vertical excitations of roseoflavin. TD-CC2 density differences (w.r. to the ground state, aug-cc-pVDZ basis) of the ($\pi\pi^*$) (red and green) and ($n\pi^*$) (blue) excited states. The blue, red, and green isosurfaces show the regions with loss, the grey ones the regions with gain of electron density upon photo-excitation.

functionals, in the def-SVP AO basis [85]). Here, the TD-DFT with local or semi-local functionals like BP86 is bound to fail due to the inherent self-interaction error [29]. The energetic orders of the three different states vary with the used methods. With the exception of the BP86 functional, the lowest singlet state is always the red ($\pi\pi^*$) state.

TD-DFT overstabilizes the charge transfer states. For the hybrid functionals, especially BHLYP, this effect is not that pronounced. For an schematically view see also Figure 14.7.

The red ($\pi\pi^*$) state was optimized at the TD-CC2 level in the gas phase afterwards. Since the system is sufficiently small a geometry optimisation at the TD-CC2 response level was possible.

Comparing the geometries of ground and optimized excited states reveals that the bond distances within the isoalloxazine skeleton exhibit the biggest changes on going from the Franck–Condon point downhill to the related excited state minima. The angle between the dimethylamino group and the ring system remains unaffected. The changes in the isoalloxazine skeleton are quite well described by a single linear combination of the strongly altering bond distance changes. We use this coordinate Q_R to explore the individual potential energy surfaces in more detail.

Figure 14.8 depicts the potential energy surfaces of the ground, and the three lowest excited states along Q_R. The individual geometries along this path starting from the Franck–Condon point ($Q_R = 0.36$) were fully relaxed on the surface of the

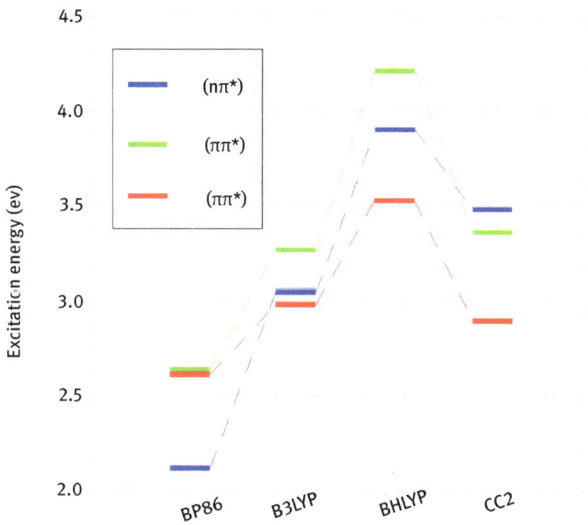

Figure 14.7: Vertical excitation energies for roseoflavin with different methods as schematically pictured.

lowest state (red curve), i. e. all coordinates but Q_R were optimized from this state at the level of TD-CC2 in the def-SVP AO basis. Subsequently at all geometries so obtained (with different Q_R values), single point TD-CC2 calculations in the bigger aug-cc-pVDZ basis were performed.

Evidently, the trace along Q_R leads to a conical intersection between the (ππ*) and the (nπ*) state. Moreover, a second conical intersection is encountered later between the (nπ*) state and the ground state. Apparently, there is a fast deactivation channel from the photoexcited (ππ*) state via two conical intersections back to the ground state. The in-plane ring mode related to Q_R appears to be driving this deactivation process, which would fit the picture of a planar intramolecular charge transfer (PICT) model.

These calculations so far refer to roseoflavin in the gas phase. For roseoflavin solvated in water a TICT mechanism was postulated [60]. In the present work QM/MM calculations for this system were performed. The water solvent was mimicked by the CHARMM force field using the TIP3P model. Roseoflavin was solvated in a preequilibrated water droplet for obtaining representative geometry snapshots. The complete procedure of this is not described in detail here and we instead refer to reference [64]. At the individual snapshot geometries picked from the MD trajectory the vertical excitations were again calculated, now in the framework of QM/MM. The calculations at the TD-CC2 response level show that the lowest excited state now is the (green) (ππ*) state. Its charge transfer character leads in the polar water environment to a substantial stabilization relative to the gas phase. The second state is

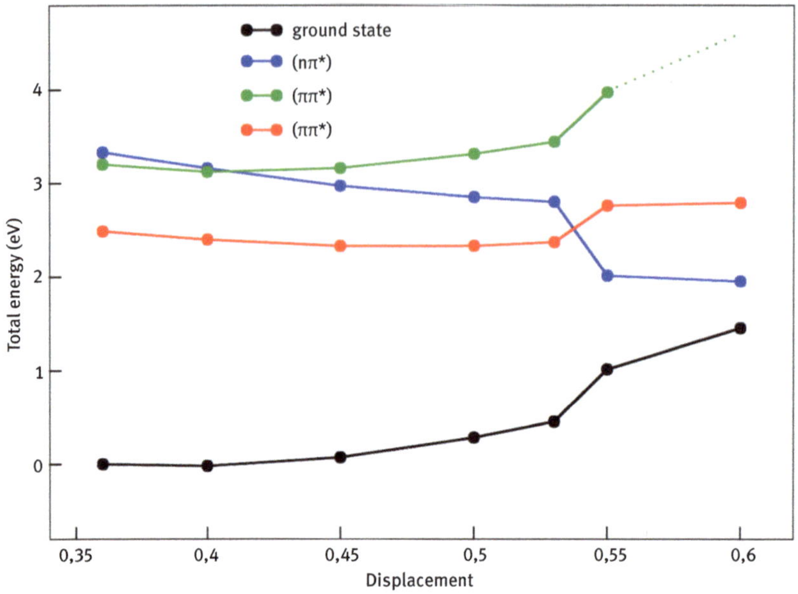

Figure 14.8: Energies (in eV) of the relevant excited states relative to the ground state minimum energy at the relaxed structures along linear combination of bond distances (Displacement or Q_R). The restrained geometry optimizations were performed with TD-CC2 response in the def-SVP basis on the energy surface of the lowest state, i. e. on the (ππ*) surface and on the (nπ*) after the conical intersection. The plotted energies correspond to TD-CC2 response single point calculations at these relaxed structures in the bigger aug-cc-pVDZ basis.

another (ππ*) state, which however is not shown here, because it is irrelevant for the further discussion. It turned out that a twisting angle between the dimethylamino group and the isoalloxazine ring system (denoted as τ) is a relevant downhill coordinate on the surface of the lowest (green) (ππ*) state. Figure 14.9 shows the (on the green ππ* state surface) fully relaxed curves of ground- and the two lowest excited states along this coordinate. Evidently, rotation of the dimethylamino group rapidly leads downhill from the Franck–Condon point on the S_1 surface towards a conical intersection with the ground state. A perpendicular orientation of the dimethylamino group is responsible for this fast relaxation back to the ground state. The postulated TICT mechanism thus could be verified by our QM/MM calculations.

Next, the question regarding the orientation of the dimethylamino group in the protein environment has to be answered. In analogy to the work of Sadeghian et al. [61] the BLUF-domain protein BlrB was investigated. The native flavin was replaced by roseoflavin. Solvation and molecular dynamics simulations to equilibrate the modified BLUF domain were carried out. For the subsequent QM/MM calculations, the very same QM region as in previous work was used to have exactly the same conditions as in the reference calculations on the native system [61].

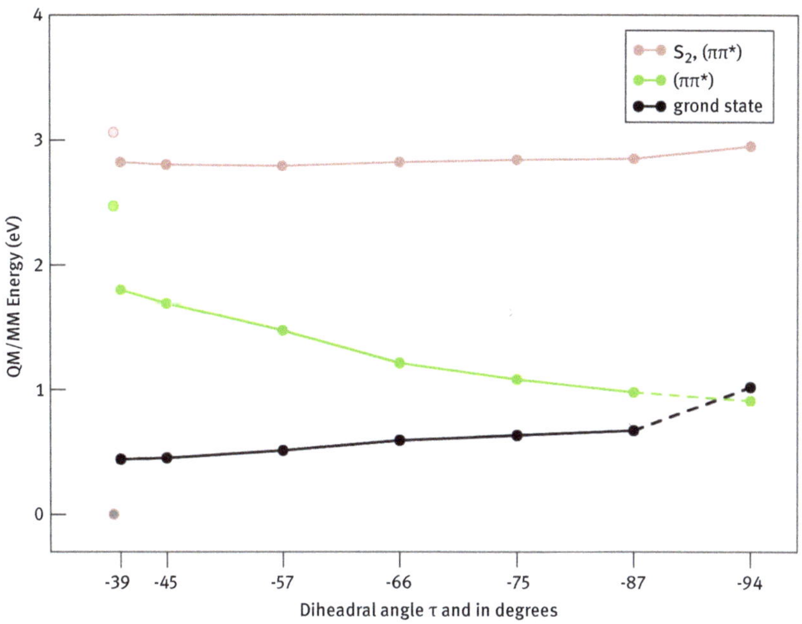

Figure 14.9: QM/MM (TD-CC2/Charmm) energy (in eV) of the ground (S_0) and S_1 state plotted along the path obtained via constrained geometry optimization on the S_1 state. The dihedral angle τ are also shown.

Two possible ground state optimized (DFT, BP functional with def-SVP AO basis) conformations were identified within the binding pocket of the BLUF domain containing now the roseoflavin: a planar structure, which is a little more favourable, and a twisted one.

Figure 14.10 displays density difference plots of the lowest three excited states of the roseoflavin-BLUF domain. The crucial Tyr→roseoflavin charge transfer state, which is driving the reaction cascade of the photocycle, corresponds to the third state.

Comparison of the excitation energies (TD-CC2 in aug-cc-pVDZ AO basis) at the Franck–Condon point, and at the minimum energy point on the lowest excited state surface reveals no crossing of any states, which is in strong contrast to the situation in the gas phase, and in the water environment. Furthermore, in strong contrast to the native BLUF domain, there is also no switching between the Tyr→roseoflavin charge transfer state and the low-lying local excited state. While in the native BLUF domain, the gap between locally excited and charge transfer state closes on going downhill from the Franck–Condon point on the surface of the former, the opposite is true for the roseoflavin-BLUF domain, where the gap opens. The charge transfer state thus cannot be populated via the locally excited state with large oscillator strength due to the lack of a low-lying conical intersection between these two states, rendering the modified BLUF as biological dysfunctional.

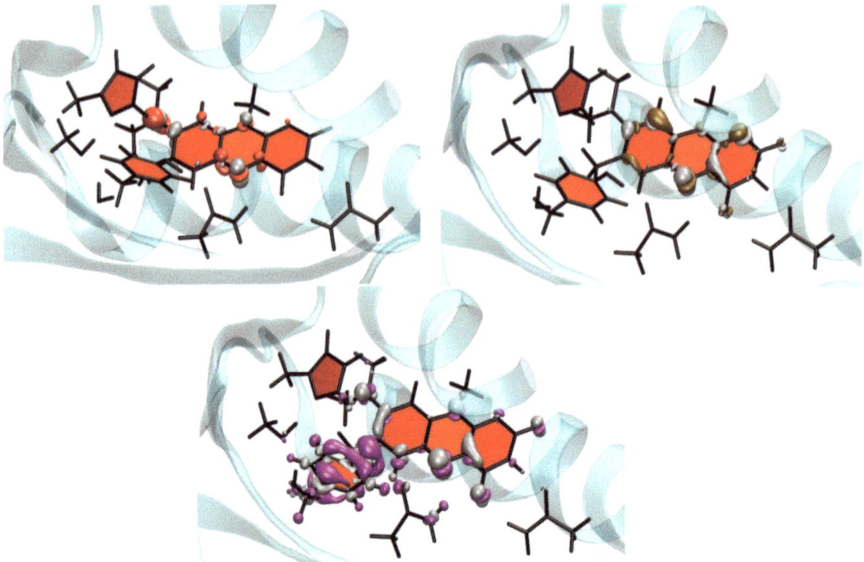

Figure 14.10: TD-CC2 electron density plots for the green (ππ*) state (S_1) brown (ππ*) state (S_2) and the charge transfer state (S_3) at the optimized ground state structure. Grey and coloured areas show the regions with gain and loss of electron density relative to the ground state, respectively. Black indicated atoms show the QM-region The rings were coloured using the Cremer–Pople pucker amplitude [86].

14.5.2 Benzophenone in dinucleotides

As already pointed out above, benzophenone (BP) was attached to 2′-desoxyribofuranoside, to obtain an artificial DNA-base. This artificial DNA-base was then connected to another natural nucleobase. Such a combination of artificial and natural DNA-bases (in this case guanosine) is denoted in this context as a dinucleotide. The base scaffold is depicted in Figure 14.11. In the following the influence of BP on the adjacent guanosine is investigated.

As also mentioned above, the singlet state (with (nπ*) character) of the dinucleotide in water has a drasticallly shortened lifetime. In order to understand this change of lifetime a theoretical study on this particular dinucleotide was conducted.

As before in Section 14.5.1, the system was first studied in the gas phase. Geometry optimisations in the electronic ground state yield two different minimum energy geometries. Using DFT with BP86 functional in the def-SVP basis set a folded structure undergoing π-stacking, and an unfolded structure was located. At the level of CC2 in the same basis set, the unfolded structure is not stable and folds to the compact geometry. The two different geometries are depicted in Figure 14.12 (bottom). In order to get an overview over the relevant excited states and the performance of different methods, vertical excitation energies were calculated.

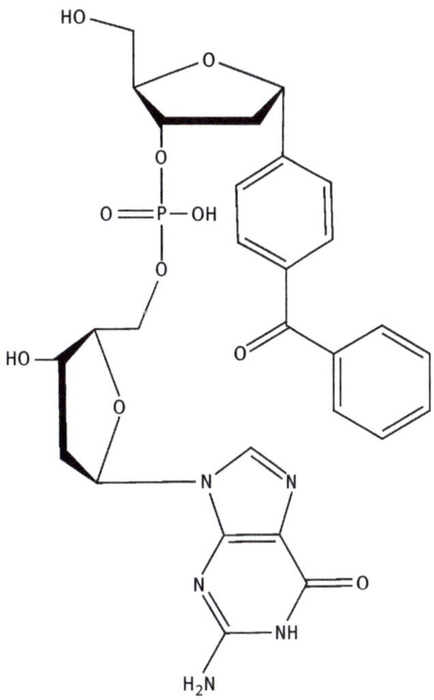

Figure 14.11: Investigated dinucleotide with guanosine as nucleobase attached at the α-position.

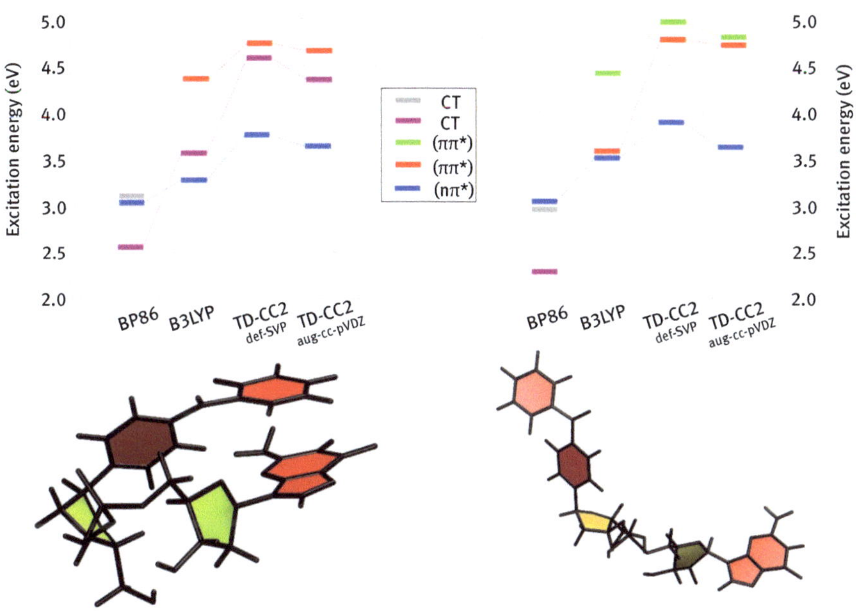

Figure 14.12: Vertical excitation energies for dinucleotide with different methods as schematically picture. Left for the folded and right for the unfolded geometry. Corresponding excitation energies can be found in Table 14.1. The minimum geometries for the groundstate are depicted below. The rings were coloured using the Cremer–Pople pucker amplitude.

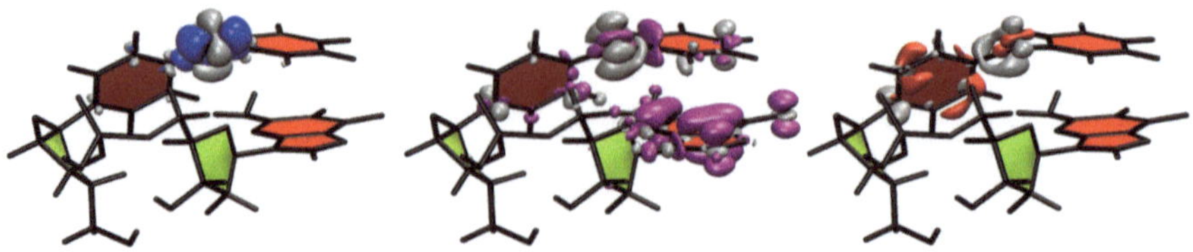

Figure 14.13: Electron density difference plots for the first three excited states of the folded geometry, calculated with TD-CC2/def-SVP basis. Isosurfaces are plotted in blue/purple/red (refers to a decrease in the density upon excitation) and grey (an increase in the density upon excitation). The rings were coloured using the Cremer–Pople pucker amplitude.

The three lowest electronically excited states of the folded geometry are the following: the first is an ($n\pi^*$) excitation, the second a charge transfer state, and the third is a ($\pi\pi^*$) type excitation (see also Figure 14.12). These results were obtained at the level of TD-CC2 in the def-SVP, and the aug-cc-pVDZ basis sets, respectively, as well as with TD-DFT in conjunction with the B3LYP functional and def-SVP basis. If one uses the non-hybrid GGA functional BP86, the order of the states changes; the charge transfer state now is the lowest state, which again is an example for the underestimation of charge transfer states by TD-DFT due to electron self repulsion.

Focussing now on the unfolded geometry, for TD-CC2 with both basis sets, as well as for TD-DFT/B3LYP, the charge transfer state vanishes, which is not too surprising since the distance between the G subunit (donor) and the BP (acceptor) is much larger than in the previous case. On the other hand, TD-DFT/BP86 as above still (wrongly) predicts the charge transfer state as the S_1 state.

From these calculations one can conclude that: (i) the distance between the BP and the G subsystems intimately affects the character of the three relevant, lowest excited states (in particular there is a low-lying CT state present for the folded geometry, which is absent for the unfolded one), and (ii) that all these excitations are either localized on G, or BP, or both, as is evident from Figure 14.13. The sugar and phosphate groups linking these two subunits do not play a role and can therefore be safely omitted from the QM part in the subsequent QM/MM studies of the system in different solvent environments.

The benzophenone dinucleotide was studied thereafter in the water and methanol solvent environments, in which the spectroscopic measurements were carried out. The generation of the starting structures was done as usual via classical molecular dynamics (MD) simulations. Starting from either the folded, or the unfolded gas phase structure, MD simulations were performed in water and in methanol.

Figure 14.14 shows the two MD trajectories obtained with folded starting geometry for methanol and unfolded starting geometry for water. The two MD trajectories with the corresponding other starting geometries are not shown. The blue region indicates an unfolded geometry. Evidently, the unfolded dinucleotide in water folds and stays folded most of the time. In contrast, the folded dinucleotide in methanol

Distance for dinucleotide in used solvents

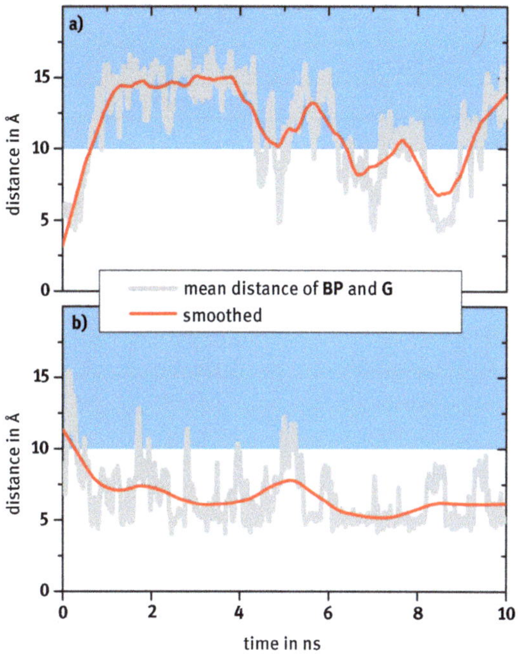

Figure 14.14: Trajectories of MD run. a) shows the dinucleotide in methanol environment and b) in water; in the blue areas the molecule is unfolded. The mean distance of BP and G is plotted against the simulation time.

unfolds and stays in that form most of the time. In water, π-stacking of the hydrophobic ring systems of benzophenone and guanosine is responsible for the folding, whereas in the less polar methanol, the ring systems are individually solvated by the MeOH solvent molecules. For the subsequent QM/MM study, we selected from each of the four trajectories one representative snapshot, so that there were two snapshots from each solvent (A, and B, for each solvent, respectively). The G and BP subunits were included in the QM region; the linking sugar rings were just treated at the MM level.

For each of these snapshots a geometry optimisation in the ground state at the level of DFT/B3LYP-D (B3LYP with empirical dispersion correction [87] to model π-stacking) in the def-SVP basis was performed. Vertical excitation energies were calculated thereafter by single point calculations at these geometries at the level of TD-CC2 in the aug-cc-pVDZ basis. For all snapshots the S_1, S_2, and S_3 states are of ($n\pi^*$), ($\pi\pi^*$), and ($\pi\pi^*$) type, respectively, no matter what solvent is used. The low-lying charge transfer state for the folded form in water was not found among the first four excited states in any of the two snapshots. Apparently, only in the gas phase the charge transfer state is sufficiently stabilized, to find it among the lowest excited states.

Table 14.1: Vertical QM/MM excitation energies in eV for the four chosen dinucleotide snap shots solvated in water and methanol with different methods calculated for the optimized (nπ*).

		BHLYP	TD-CC2	
structure	**state**	**def-SVP**		**aug-cc-pVDZ**
A H_2O	S_1	3.45 (nπ*)	3.13 (nπ*)	3.02 (nπ*)
	S_2	4.63 CT	4.73 (ππ*)	4.57 (ππ*)
	S_3	4.91 (ππ*)	4.95 (ππ*)	4.71 (ππ*)
	S_4	4.96 (ππ*)	5.03 CT	4.74 CT
B H_2O	S_1	3.61 (nπ*)	3.23 (nπ*)	3.12 (nπ*)
	S_2	4.62 CT	4.72 (ππ*)	4.51 (ππ*)
	S_3	4.82 (ππ*)	4.80 (ππ*)	4.57 (ππ*)
	S_4	4.96 (ππ*)	5.06 CT	4.76 CT
A MeOH	S_1	3.58 (nπ*)	3.26 (nπ*)	3.15 (nπ*)
	S_2	4.98 (ππ*)	4.84 (ππ*)	4.66 (ππ*)
	S_3	5.00 (ππ*)	4.88 (ππ*)	4.71 (ππ*)
B MeOH	S_1	3.62 (nπ*)	3.24 (nπ*)	3.13 (nπ*)
	S_2	4.74 (ππ*)	4.65 (ππ*)	4.43 (ππ*)
	S_3	4.83 (ππ*)	4.81 (ππ*)	4.61 (ππ*)

The next step is once again the optimisation of the S_1 state with (nπ*) character. On the route downhill from the Franck–Condon point, perhaps a crossing with the charge transfer state might occur. Hence, at the four different snapshots (A and B for both solvents), geometry optimizations were carried out on the surface related to the lowest excited state (with (nπ*) character). As QM method TD-DFT/BHLYP was employed for the geometry optimisations of the first excited state. At the so obtained structures single pont TD-CC2 calculations (with aug-cc-pVDZ basis) were carried out and the corresponding results are shown in Table 14.1. The first three excited states for these single point calculations at the S_1 state minimum have (nπ*), (ππ*) and (ππ*) character in both solvents. In the folded molecule, which predominantly exists in MeOH environment, the CT state is absent. However, for the folded case in water now the CT state appears as the S_4.

Since the CT state for TD-DFT/BHLYP pops up as S_2 (see also Table 14.1), it was easy to optimise this geometry. In the course of its optimisation at the level of TD-DFT/BHLYP and def-SVP basis the CT state becomes the lowest excited state S_1. Single point calculations with TD-CC2 and aug-cc-pVDZ basis at the so obtained geometry showed, that the CT state is indeed now energetically deeper (also S_1 for this level of theory) as the local excitaion (nπ*). This situation is depicted in Figure 14.15.

From the QM/MM results discussed in this section, the following conclusions can be drawn:

(i) For the dinucleotide solvated in methanol, which exists predominantly in the unfolded form, no low-lying CT state is found, which could quench the fluorescence of the excited (nπ*) singlet state. This suggest a lifetime, which is primarily determined by intersystem crossing to the triplet state.

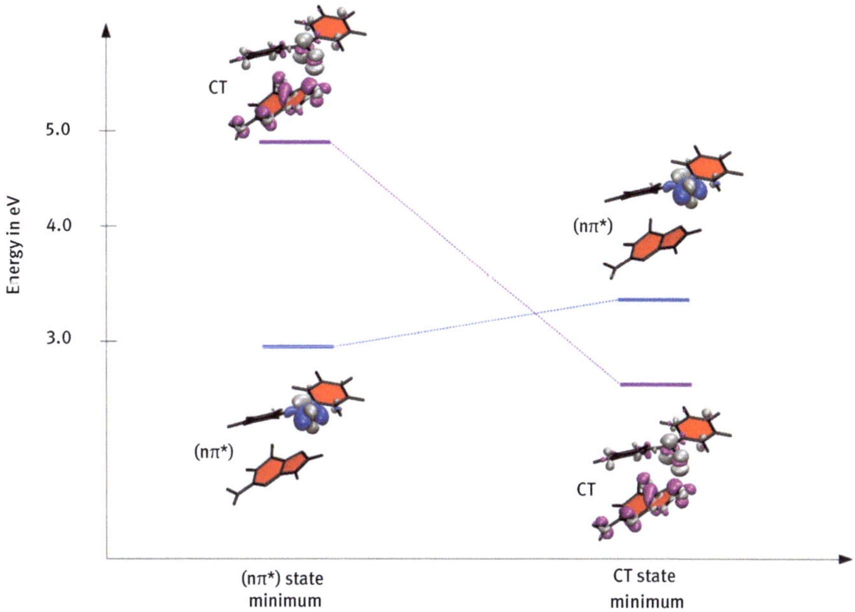

Figure 14.15: Schematic view of solvated dinucleotide (snap shot A) and the vertical QM/MM excitations at the optimized (nπ*) state and the CT state, respectively, calculated with TD-CC2/ aug-cc-pVDZ. Electron density difference plots for the snap shot A of the water trajectory calculated with TD-CC2/ aug-cc-pVDZ basis. Isosurfaces are plotted in blue/purple (refers to a decrease in the density upon excitation) and grey (an increase in the density upon excitation). The rings were coloured using the Cremer–Pople pucker amplitude.

(ii) On the other hand, for the artificial DNA-base solvated in water, which exists predominantly in the folded form, there indeed exists a low-lying CT state in some region of the configuration space, which could quench the fluorescence of the excited (nπ*) singlet state. However, this quenching is probably not very efficient, since no conical intersection between the (nπ*) and the CT state is found in the vicinity of the Franck–Condon point, or on the downhill path to the (nπ*) minimum. Nevertheless, the lifetime of the (nπ*) singlet state may not be solely governed by the intersystem crossing.

14.6 Conclusion

In this contribution we discussed the application of ab initio quantum chemical methods in the context of photochemistry or photocatalysis. Response theories like TD-DFT or TD-CC2 are sufficiently efficient to be used also for larger systems these days, the former primarily for geometry optimizations, and the latter, more robust method, for single point calculations at certain geometries. TD-DFT has to be utilized with some care due to inherent deficiencies of the method (like electron self-repulsion in local or semi-local functionals), and, in particular, low-lying charge transfer states are suspicious. TD-DFT should therefore be combined or verified with a more reliable method (like TD-CC2).

Environmental effects, i. e. of a solvent or a protein environment can be taken into account by applying the QM/MM approach, which divides the system into a large part treated by classical molecular dynamics with an empirical force field, and a small part treated accurately with quantum chemical methods. QM/MM is a non-trivial approach. It is important to choose QM and MM parts properly. The coupling between MM and QM parts is delicate and it is advisable to have the border between QM and MM region as far as possible from the reaction centre in the QM region. Properly setting up a QM/MM calculation is not black box-like and requires care and time.

In this contribution we provide two examples of recent work, where we applied QM/MM methods successfully to two different systems. It has been demonstrated that the combination of such calculations with experiment can lead to new insight into the underlying mechanism of photochemical or photophysical processes.

14.7 References

[1] Schmidt, Thomas et al. 2012. "QM/MM Investigations Of Organic Chemistry Oriented Questions." Pp. 1–77 in. Springer Berlin/Heidelberg.

[2] González, Leticia, Daniel Escudero, and Luis Serrano-Andrés. 2012. "Progress and challenges in the calculation of electronic excited states." *Chemphyschem: a European journal of chemical physics and physical chemistry* 13(1):28–51.

[3] Bernardi, Fernando, Massimo Olivucci, and Michael A. Robb. 1996. "Potential energy surface crossings in organic photochemistry." *Chemical Society Reviews* 25(5):321–328.

[4] Ponder, Jay W, and David A Case. 2003. "Force fields for protein simulations." *Advances in protein chemistry* 66:27–85.

[5] Senn, Hans, and Walter Thiel. 2007. "QM/MM Methods for Biological Systems." Pp. 173–290 in, vol. 268, edited by Markus Reiher. Springer Berlin/Heidelberg.

[6] Born, Max, and Robert Oppenheimer. 1927. "Zur Quantentheorie der Molekeln." *Annalen der Physik* 389(20):457–484.

[7] Luis Serrano-Andrés, Manuela Merchán. 2009. "Are the five natural DNA/RNA base monomers a good choice from natural selection?: A photochemical perspective." *Journal of Photochemistry and Photobiology C: Photochemistry Reviews* 10(1):21–32

[8] Teller, Edward 1937. "The Crossing of Potential Surfaces." *The Journal of Physical Chemistry* 41(1):109–116.

[9] Domcke, Wolfgang, and Clemens Woywod. 1993. "Direct construction of diabatic states in the CASSCF approach. Application to the conical intersection of the 1A2 and 1B1 excited states of ozone." *Chemical Physics Letters* 216(3–6):362–368. Retrieved November 5, 2012 (http://dx.doi.org/10.1016/0009-2614(93)90110-M).

[10] Berry, Michael V., and Michael Wilkinson. 1984. "Diabolical Points in the Spectra of Triangles." *Proceedings of the Royal Society A: Mathematical, Physical and Engineering Sciences* 392(1802):15–43. Retrieved October 26, 2012 (http://rspa.royalsocietypublishing.org/cgi/doi/10.1098/rspa.1984.0022).
[11] Yarkony, David. 1996. "Diabolical conical intersections." *Reviews of Modern Physics* 68(4):985–1013.
[12] Roos, Björn O., Peter R. Taylor, and Per E. M. Siegbahn. 1980. "A complete active space SCF method (CASSCF) using a density matrix formulated super-CI approach." *Chemical Physics* 48(2):157–173.
[13] Foresman, James B., Martin Head-Gordon, John A. Pople, and Michael J. Frisch. 1992. "Toward a systematic molecular orbital theory for excited states." *The Journal of Physical Chemistry* 96(1):135–149.
[14] Oddershede, Jens. 2007. "Propagator Methods." Pp. 201–239 in *Advances in Chemical Physics*. John Wiley & Sons, Inc.
[15] Trofimov, Boris A., Leonid I. Krivdina, J. Weller, and Jochen Schirmer. 2006. "Algebraic-diagrammatic construction propagator approach to molecular response properties." *Chemical Physics* 329(1–3):1–10.
[16] Bartlett, Rodney, and Monika Musiał. 2007. "Coupled-cluster theory in quantum chemistry." *Reviews of Modern Physics* 79(1):291–352.
[17] Sekino, Hideo, and Rodney J Bartlett. 1984. "A linear response, coupled-cluster theory for excitation energy." *International Journal of Quantum Chemistry* 26(S18):255–265.
[18] Koch, Henrik, and Poul Jørgensen. 1990. "Coupled cluster response functions." *The Journal of Chemical Physics* 93(5):3333–3344.
[19] Koch, Henrik, Ove Christiansen, Poul Jørgensen, and Jeppe Olsen. 1995. "Excitation energies of BH, CH2 and Ne in full configuration interaction and the hierarchy CCS, CC2, CCSD and CC3 of coupled cluster models." *Chemical Physics Letters* 244(1–2):75–82.
[20] Christiansen, Ove, Henrik Koch, and Poul Jørgensen. 1995. "The second-order approxlmate coupled cluster singles and doubles model CC2." *Chemical Physics Letters* 243(5–6):409–418.
[21] Christiansen, Ove, Henrik Koch, and Poul Jørgensen. 1995. "Response functions in the CC3 iterative triple excitation model." *The Journal of Chemical Physics* 103(17):7429–7441.
[22] Köhn, Andreas, and Attila Tajti. 2007. "Can coupled-cluster theory treat conical intersections?" *The Journal of chemical physics* 127(4):044105.
[23] Hohenberg, Pierre. 1964. "Inhomogeneous Electron Gas." *Physical Review* 136(3B):B864–B871.
[24] Becke, Axel D. 1993b. "Density-functional thermochemistry. III. The role of exact exchange." *The Journal of Chemical Physics* 98(7):5648.
[25] Casida, Mark E., and Miquel Huix-Rotllant. 2012. "Progress in time-dependent density-functional theory." *Annual review of physical chemistry* 63:287–323.
[26] Grimme, Stefan, and Maja Parac. 2003. "Substantial errors from time-dependent density functional theory for the calculation of excited states of large pi systems." *Chemphyschem: a European journal of chemical physics and physical chemistry* 4(3):292–5.
[27] Parac, Maja, and Stefan Grimme. 2002. "Comparison of Multireference Møller–Plesset Theory and Time-Dependent Methods for the Calculation of Vertical Excitation Energies of Molecules." *The Journal of Physical Chemistry A* 106(29):6844–6850.
[28] Tozer, David. 2003. "Relationship between long-range charge-transfer excitation energy error and integer discontinuity in Kohn–Sham theory." *Journal of Chemical Physics* 119:12697–12699.
[29] Dreuw, Andreas, Jennifer L. Weisman, and Martin Head-Gordon. 2003. "Long-range charge-transfer excited states in time-dependent density functional theory require non-local exchange." *Journal of Chemical Physics* 119:2943–2947.

[30] Yanai, Takeshi, David P. Tew, and Nicholas C. Handy. 2004. "A new hybrid exchange–correlation functional using the Coulomb-attenuating method (CAM-B3LYP)." *Chemical Physics Letters* 393(1–3):51–57.
[31] Peach, Michael J. G., Peter Benfield, Trygve Helgaker, and David J. Tozer. 2008. "Excitation energies in density functional theory: An evaluation and a diagnostic test." *The Journal of Chemical Physics* 128(4):44118.
[32] Kronik, Leeor, Tamar Stein, Sivan Refaely-Abramson, and Roi Baer. 2012. "Excitation Gaps of Finite-Sized Systems from Optimally-Tuned Range-Separated Hybrid Functionals." *Journal of Chemical Theory and Computation* 8:1515–1531.
[33] Warshel, Arieh, and Michael Levitt. 1976. "Theoretical studies of enzymatic reactions: Dielectric, Electrostatic and Steric Stabilisation of the Carbonium Ion in the Reaction of Lysozyme." *Journal of Molecular Biology* 103:227–249.
[34] Field, Martin J., Paul A. Bash, and Martin Karplus. 1990. "A Combined Quantum Mechanical and Molecular Mechanical Potential for Molecular Dynamics Simulations." *Journal of Computational Chemistry* 11(6):700–733.
[35] Jorgensen, William L., Jayaraman Chandrasekhar, Jeffry D. Madura, Roger W. Impey, and Michael L. Klein. 1983. "Comparison of simple potential functions for simulating liquid water." *The Journal of Chemical Physics* 79(2):926–935.
[36] Klamt, Andreas, and G. Schuurmann. 1993. "COSMO: a new approach to dielectric screening in solvents with explicit expressions for the screening energy and its gradient." *Journal of the Chemical Society, Perkin Transactions 2* (5):799–805.
[37] Case, David A. et al. 2005. "The Amber biomolecular simulation programs." *Journal of Computational Chemistry* 26(16):1668–1688.
[38] Brooks, Bernard R. et al. 1983. "CHARMM: A program for macromolecular energy, minimization, and dynamics calculations." *Journal of Computational Chemistry* 4:187.
[39] Brooks, Bernard R. et al. 2009. "CHARMM: the biomolecular simulation program." *Journal of Computational Chemistry* 30(10):1545–614.
[40] Allinger, Norman L., Randall A. Kok, and Mita R. Imam. 1988. "Hydrogen bonding in MM2." *Journal of Computational Chemistry* 9(6):591–595.
[41] Allinger, Norman L., Young H. Yuh, and Jenn Huei Lii. 1989. "Molecular mechanics. The MM3 force field for hydrocarbons. 1." *Journal of the American Chemical Society* 111(23):8551–8566.
[42] Allinger, Norman L., Kuohsiang Chen, and Jenn-Huei Lii. 1996. "An improved force field (MM4) for saturated hydrocarbons." *Journal of Computational Chemistry* 17(5–6):642–668.
[43] Rappe, Anthony K., Carla J. Casewit, K. S. Colwell, William A. Goddard, and W. M. Skiff. 1992. "UFF, a full periodic table force field for molecular mechanics and molecular dynamics simulations." *Journal of the American Chemical Society* 114(25):10024–10035.
[44] Hess, Berk, Carsten Kutzner, David van der Spoel, and Erik Lindahl. 2008. "GROMACS 4: Algorithms for Highly Efficient, Load-Balanced, and Scalable Molecular Simulation." *Journal of Chemical Theory and Computation* 4(3):435–447.
[45] Wang, Junmei, Romain M. Wolf, James W. Caldwell, Peter A. Kollman, and David A. Case. 2004. "Development and testing of a general amber force field." *Journal of computational chemistry* 25(9):1157–74.
[46] Bayly, Christopher I., Piotr Cieplak, Wendy Cornell, and Peter A. Kollman. 1993. "A well-behaved electrostatic potential based method using charge restraints for deriving atomic charges: the RESP model." *The Journal of Physical Chemistry* 97(40):10269–10280.
[47] Hu, Hao, and Weitao Yang. 2009. "Development and application of ab initio QM/MM methods for mechanistic simulation of reactions in solution and in enzymes." *Journal of Molecular Structure: THEOCHEM* 898(1–3):17–30.

[48] Acevedo, Orlando, and William L. Jorgensen. 2010. "Advances in quantum and molecular mechanical (QM/MM) simulations for organic and enzymatic reactions." *Accounts of Chemical Research* 43(1):142–151.
[49] Senn, Hans, and Walter Thiel. 2009. "QM/MM methods for biomolecular systems." *Angewandte Chemie (International ed. in English)* 48(7):1198–229.
[50] Hu, LiHong, Pär Söderhjelm, and Ulf Ryde. 2011. "On the Convergence of QM/MM Energies." *Journal of Chemical Theory and Computation* 7(3):761–777.
[51] Mills, Greg, and Hannes Jónsson. 1994. "Quantum and thermal effects in H2 dissociative adsorption: Evaluation of free energy barriers in multidimensional quantum systems." *Physical Review Letters* 72(7):1124–1127.
[52] Sheppard, Daniel, Rye Terrell, and Graeme Henkelman. 2008. "Optimization methods for finding minimum energy paths." *The Journal of chemical physics* 128(13):134106.
[53] Klenin, Konstantin, Birgit Strodel, David J Wales, and Wolfgang Wenzel. 2011. "Modelling proteins: conformational sampling and reconstruction of folding kinetics." *Biochimica et biophysica acta* 1814(8):977–1000.
[54] Lechner, Robert, Susanne Kümmel, and Burkhard König. 2010. "Visible light flavin photo-oxidation of methylbenzenes, styrenes and phenylacetic acids." *Photochemical & photobiological sciences: Official journal of the European Photochemistry Association and the European Society for Photobiology* 9(10):1367–77.
[55] Megerle, Uwe et al. 2011. "Unraveling the flavin-catalyzed photooxidation of benzylic alcohol with transient absorption spectroscopy from sub-pico- to microseconds." *Physical chemistry chemical physics: PCCP* 13(19):8869–80.
[56] Zheng, Ya-Jun, and Rick L. Ornstein. 1996. "A Theoretical Study of the Structures of Flavin in Different Oxidation and Protonation States." *Journal of the American Chemical Society* 118(39):9402–9408.
[57] Sikorska, Ewa, Igor V. Khmelinskii, Jacek Koput, Jose L. Bourdelande, and Marek Sikorski. 2004. "Electronic structure of isoalloxazines in their ground and excited states." *Journal of Molecular Structure* 697(1–3):137–141.
[58] Salzmann, Susanne, and Christel M. Marian. 2009. "The photophysics of alloxazine: a quantum chemical investigation in vacuum and solution." *Photochemical & photobiological sciences: Official journal of the European Photochemistry Association and the European Society for Photobiology* 8(12):1655–66.
[59] Mathes, Tilo, Christian Vogl, Jürgen Stolz, and Peter Hegemann. 2009. "In vivo generation of flavoproteins with modified cofactors." *Journal of Molecular Biology* 385(5):1511–8.
[60] Zirak, Peyman, Alfons Penzkofer, Tilo Mathes, and Peter Hegeman. 2009. "Photo-dynamics of roseoflavin and riboflavin in aqueous and organic solvents." *Chemical Physics* 358:111–122.
[61] Sadeghian, Keyarash, Marco Bocola, and Martin Schütz. 2008. "A conclusive mechanism of the photoinduced reaction cascade in blue light using flavin photoreceptors." *Journal of the American Chemical Society* 130:12501–12513.
[62] Sadeghian, Keyarash, Marco Bocola, and Martin Schütz. 2010. "A QM/MM study on the fast photocycle of blue light using flavin photoreceptors in their light-adapted/active form." *Physical chemistry chemical physics: PCCP* 12:8840–8846.
[63] Zirak, Peyman, Alfons Penzkofer, Tilo Mathes, and Peter Hegemann. 2009. "Absorption and emission spectroscopic characterization of BLUF protein Slr1694 from Synechocystis sp. PCC6803 with roseoflavin cofactor." *Journal of Photochemistry and Photobiology B: Biology* 97:61–70.
[64] Merz, Thomas, Keyarash Sadeghian, and Martin Schütz. 2011. "Why BLUF photoreceptors with roseoflavin cofactor loose their biological functionality." *Physical chemistry chemical physics: PCCP* 13:14775–14783.

[65] Fleming, Steven A. 1995. "Chemical Reagents in Photoaffinity Labeling." *Tetrahedron* 51(46):12479–12520.
[66] Dorman, Gyorgy, and Glenn D. Prestwich. 1994. "Benzophenone Photophores in Biochemistry." *Biochemistry* 33(19):5661–5673.
[67] Galardy, Richard E., Lyman C. Craig, James D. Jamieson, and Morton P. Printz. 1974. "Photoaffinity Labeling of Peptide Hormone Binding Sites." *Journal of Biological Chemistry* 249(11):3510–3518.
[68] Fagnoni, Maurizio, Daniele Dondi, Davide Ravelli, and Angelo Albini. 2007. "Photocatalysis for the formation of the C–C bond." *Chemical reviews* 107(6):2725–56. Retrieved October 26, 2012 (http://www.ncbi.nlm.nih.gov/pubmed/17530909).
[69] Ravelli, Davide, Daniele Dondi, Maurizio Fagnoni, and Angelo Albini. 2009. "Photocatalysis. A multi-faceted concept for green chemistry." *Chemical Society reviews* 38(7):1999–2011.
[70] Svoboda, Jirí, and Burkhard König. 2006. "Templated photochemistry: toward catalysts enhancing the efficiency and selectivity of photoreactions in homogeneous solutions." *Chemical reviews* 106(12):5413–30.
[71] Wessig, Pablo. 2006. "Organocatalytic enantioselective photoreactions." *Angewandte Chemie (International ed. in English)* 45(14):2168–71.
[72] Müller, Christiane, Andreas Bauer, and Thorsten Bach. 2009. "Light-Driven Enantioselective Organocatalysis." *Angewandte Chemie (International ed. in English)* 48(36):6640–6642.
[73] Müller, Christiane et al. 2011. "Enantioselective intramolecular [2+2]-photocycloaddition reactions of 4-substituted quinolones catalyzed by a chiral sensitizer with a hydrogen-bonding motif." *Journal of the American Chemical Society* 133(41):16689–97.
[74] Cauble, David F., Vincent Lynch, and Michael J. Krische. 2003. "Studies on the enantioselective catalysis of photochemically promoted transformations: 'sensitizing receptors' as chiral catalysts." *The Journal of Crganic chemistry* 68(1):15–21.
[75] Bauer, Andreas, Felix Westkamper, Stefan Grimme, and Thorsten Bach. 2005. "Catalytic enantioselective reactions driven by photoinduced electron transfer." *Nature* 436:1139–1140.
[76] Li, Yingfu, and Ronald R. Breaker. 1999. "Deoxyribozymes: new players in the ancient game of biocatalysis." *Current opinion in structural biology* 9(3):315–23.
[77] Achenbach, J. C., William Chiuman, Rani P. G. Cruz, and Yingfu Li. 2004. "DNAzymes: From Creation In Vitro to Application In Vivo." *Current Pharmaceutical Biotechnology* 5(4):16.
[78] Weinberger, Michael, Hans-Achim Wagenknecht. 2012. "Synthesis of a Benzophenone C-Nucleoside as Potential Triplet Energy and Charge Donor in Nucleic Acids." *Synthesis* 4:648–652.
[79] Yoshihara, Toshitada, Sergey I. Druzhinin, and Klaas A. Zachariasse. 2004. "Fast intramolecular charge transfer with a planar rigidized electron donor/acceptor molecule." *Journal of the American Chemical Society* 126:8535–8539.
[80] Grabowski, Zbigniew R., Krystyna Rotkiewicz, and Wolfgang Rettig. 2003. "Structural changes accompanying intramolecular electron transfer: focus on twisted intramolecular charge-transfer states and structures." *Chemical reviews* 103:3899–4032.
[81] Jung, Astrid et al. 2005. "Structure of a bacterial BLUF photoreceptor: insights into blue-mediated signal transduction." *Proceedings of the National Academy of Sciences* 101:12306–12311.
[82] Woon, David E., and Thon H. Dunning Jr. 1993. "Gaussian Basis Sets for Use in Correlated Molecular Calculations. III. The second row atoms, Al-Ar." *Journal of Chemical Physics* 98:1358–1371.
[83] Perdew, John. 1986. "Density-functional approximation for the correlation energy of the inhomogeneous electron gas." *Physical Review B* 33(12):8822–8824.

[84] Becke, Axel D. 1993a. “A new mixing of Hartree–Fock and local density-functional theories.” *The Journal of Chemical Physics* 98(2):1372.
[85] Schäfer, Ansgar, Hans Horn, and Reinhart Ahlrichs. 1992. “Fully Optimized Contracted Gaussian Basis Sets for Atoms Li to Kr.” *Journal of Chemical Physics* 97(4):2571–2578.
[86] Cremer, Dieter. 1984. “On the correct usage of the Cremer–Pople puckering parameters as quantitative descriptors of ring shapes – a reply to recent criticism by Petit, Dillen and Geise.” *Acta Crystallographica Section B Structural Science* 40(5):498–500.
[87] Grimme, Stefan, and Mirko Waletzke. 1999. “A combination of Kohn–Sham density functional theory and multi-reference configuration interaction methods.” *Journal of Chemical Physics* 111:5645–5655.

Bernhard Dick, Uwe Kensy, Roger-Jan Kutta

15 Transient Absorption

15.1 Introduction

In this chapter[1] we describe the setup and performance of an apparatus for simultaneous time-resolved and spectrally resolved measurements of transient absorption (TA). The key component in this apparatus is a streak camera, yielding, for every excitation of the sample, a two-dimensional data array with 512×512 data points. The apparatus covers the spectral range 350–750 nm and a time range from ca. 500 ns to 10 ms. Because of the large dynamic range of the streak camera of 10000:1 we obtain a multiplex factor of more than 100 compared to sequential measurements at individual wavelengths. This makes it possible to extract the maximum amount of information from small amounts of sample, for example light-sensitive proteins. We show that already a single measurement in a 10-µs time range can yield useful spectra and reliable time constants from a global lifetime analysis. An iterative method is presented for the treatment of artifacts due to scattered excitation laser light or strong fluorescence. As an alternative to a global lifetime analysis we propose a maximum entropy-based inverse Laplace transform for analysis of the data. This results in a wavelength-dependent distribution of amplitudes $p(k, \lambda)$ for all rate constants accessible within a given time window. This analysis is model free and yields a direct visual evaluation of the uncertainties in the rate constants.

Transient absorption spectroscopy is an important tool for the study of the mechanisms of photochemical reactions. Since a photoreaction is initiated by absorption of light, a short light pulse can be used to synchronize the reactions in an ensemble of molecules. Subsequently, the optical absorption of the sample is measured as a function of the time delay t after the excitation pulse and as a function of the wavelength λ of the probe light. The result is a two-dimensional data array $A(t, \lambda)$ that contains information on the spectra of the various intermediates involved in the photoreaction, and their kinetics. In most cases the sample before initiation of the photoreaction is used as the reference, and the data correspond to the difference absorption $\Delta A(t, \lambda) = A(t, \lambda) - A(-\infty, \lambda)$. The highest time resolution can be achieved with ultrashort probe pulses, and the highest wavelength resolution is then obtained with transform-limited pulses. Measurements with transform-limited probe pulses require one excitation of the sample for each time delay t_i and wavelength λ_j. When M measurements are averaged for each data point, the sample has to be excited $N_{ex} = M \times N_T \times N_S$ times, where N_T is the number of time points and N_S is the number of spectral points in the data array. The data array is usually analyzed in terms of the

1 Part of this chapter has been published as an article in Applied Physics [1].

kinetics of several species, often by application of multiexponential fits. A reasonable fit requires a good signal-to-noise (S/N) ratio in the data which requires in turn the accumulation of typically $M \approx 100$ measurements for each data point. Hence, filling a data array with $N_T = 500$ time steps and $N_S = 500$ spectral elements requires $N_{ex} = 2.5 \times 10^7$ excitations of the sample with intense laser pulses.

If the excited molecules undergo a photocycle with very high yield of recovery, such large numbers of excitations can be applied. However, typical organic photochromic molecules sustain only ≈ 1000 excitation cycles, and in many cases the photoreaction is irreversible. Then the sample must be replaced by a fresh sample after every excitation, i. e. the available amount of sample limits the number of measurements that can be performed. In such cases it is advantageous to have a method that measures several data points in the data array following a single excitation of the sample. One option is to use monochromatic probe light of long duration and measure the transmitted light intensity as a function of time. In this case the time resolution is given by that of the detector, which is usually a photomultiplier. Each measurement yields a column of the data array, and the number of excitations required to fill the data matrix is $N_{ex} = M \times N_S$. Alternatively, the data array can be measured one row at a time by using a broad-band short light pulse as a probe in combination with a spectrograph and a detector array, which requires $N_{ex} = M \times N_T$ excitations of the sample.

The ultimate efficiency with respect to the actinic stress applied to the sample is obtained if a broad-band (white) light pulse of long duration is passed through the sample and is subsequently analyzed both with respect to time and wavelength. This can be achieved by the combination of a spectrograph and a streak camera. Streak cameras have been developed as time analyzers for light pulses with a time resolution of a few ps. Their function is schematically shown in Figure 15.1. The light pulse to be analyzed is directed onto a photocathode K where photoelectrons are ejected. These are accelerated towards a detector consisting of a multichannel plate and a phosphor screen. A voltage ramp applied along the *x*-direction deflects the electrons so that photoelectrons generated at later times hit the detector at increasing values of *x*. In this way the temporal structure of the light pulse is transformed into a spatial image on the phosphor screen. This image is recorded by a CCD (charge coupled device) camera, digitized, and processed by a computer. If the photocathode extends into the *y*-direction perpendicular to *x* and *z*, many light pulses applied along the *y*-direction are analyzed simultaneously. In our setup these light pulses correspond to different wavelengths, produced by spectral dispersion of the input light pulse along the *y*-direction by a spectrograph. Among the first applications of streak cameras were the diagnostics of ultrashort laser pulses and other light emission events of short duration. If the light to be analyzed is first dispersed in one direction by a spectrograph and subsequently dispersed temporally in the orthogonal direction by the streak camera, an image of the spectro-temporal characteristics can be recorded for a single light event (e.g. from an explosion).

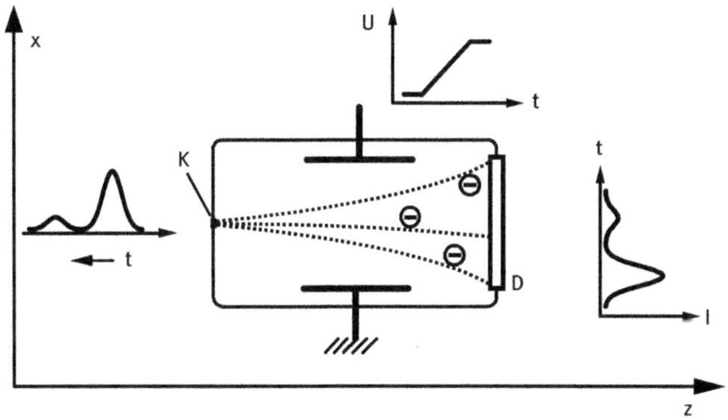

Figure 15.1: Principle of a streak camera.

The first report of the use of a streak camera for transient absorption measurements that we could find in the literature is a communication by A. Müller from 1976 [2]. He reports on the transient absorption spectra of dyes. A Xe flashlamp of 5 µs duration was used as the probe light, and the streak image was recorded by a photographic plate and analyzed by a computer-controlled microdensitometer. Peak positions assigned to $S_1 \rightarrow S_n$ transitions are given, but no time information, and no original data are shown. Three years later Yoshihara et al. [3] reported time profiles from the transient absorption of *trans* stilbene with a time resolution of ca. 10 ps. The fluorescence of a dye was used as the probe light. Other groups reported similar setups in the 1980s [4–7]. The first report of a measurement using wavelength dispersion in combination with time dispersion is by Ito et al. in 1991 [8]. However, only data averaged over all wavelengths or over a certain time interval were discussed. This is also true for several later papers from the same group [9–13]. Whereas most of the papers we have just mentioned report a streak camera measurement mostly as a proof of principle, the groups of D. F. Kelley [14–16] and Du-Jeon Jang [17–24] apparently used their setup for further studies. In all of these studies, the signal-to-noise ratio is in the range $S/N \approx 3$–5, or the standard deviation in the transient absorption is of the order of 0.3 optical density units. Hence, only monoexponential decays of a single species can be analyzed. The main source of fluctuations in these experiments is the low reproducibility of the temporal profile of the probe light, which is either fluorescence from a dye or the emission of a plasma produced by focusing a strong laser pulse into compressed Xe gas. Another limiting factor is the low dynamic range of ca. 100:1 of the streak camera.

Our project follows a quite different approach. In the last decades powerful pump-probe methods have been developed for the study of transient absorption in the time range of a few fs up to 10 ns. If sufficient sample is available for the large number of excitation cycles required, these methods are far superior to a streak camera measurement. Our interest is in transient absorption measurements in the time range of 10 ns up to 10 ms on molecular systems which are available only in small quantities and cannot sustain many excitation cycles. Our aim is to extract a maximum of information on the spectral and temporal dynamics for each excitation of the sample. In contrast to all previous studies on transient absorption with streak cameras we use a streak camera with a particularly large dynamic range of ca. 10^4:1. For the probe light we use a pulsed Xe lamp which provides a flat time profile over a time period of several ms with fluctuations less than 0.1 %. As far as we know, our setup is the first that analyzes the complete two-dimensional data array in terms of multimodal kinetics of several species. Two applications of the setup to the study of a light-sensing protein [25] and a photocatalytic reaction [26] have already been published.

15.2 Experimental Setup

The setup is schematically shown in Figure 15.2. The sample is excited by a light pulse which is generated either directly by a Nd:YAG laser (532 or 355 nm), or by an optical parametric oscillator (OPO) pumped by the third harmonic of the Nd:YAG (Surelite II and Surelite OPO PLUS, Continuum). The pulse duration is typically 8 ns, and the energy typically 10 mJ. For stable operation the laser runs with a constant repetition rate of 20 Hz and a single pulse is selected with a high energy shutter

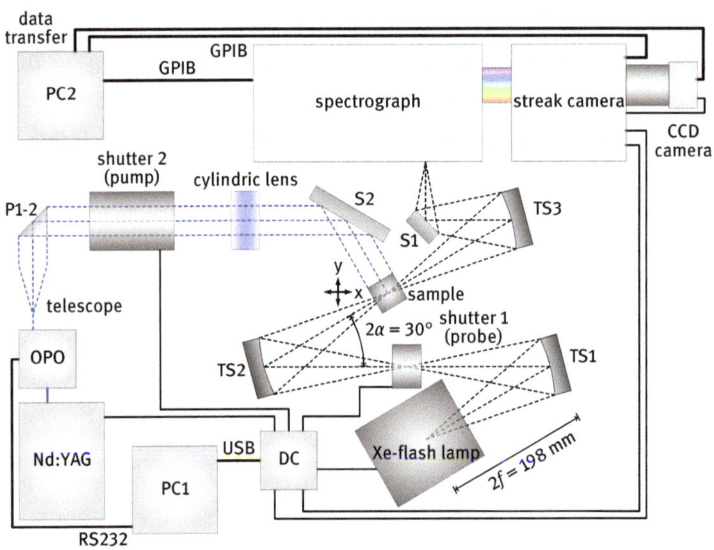

Figure 15.2: Schematic of the setup.

(LS055, nmLaser). The probe light is generated by a pulsed Xenon flash lamp (MSP-05, Müller Elektronik-Optik) which produces a white light pulse with a very flat intensity profile of ca. 1 ms duration. In order to make the light path achromatic, the probe light is focused three times by toroidal mirrors (aluminum-coated blanks of eyeglass lenses, supplied by Rodenstock): At the first focus a shutter (LS6ZM, Uniblitz) is placed which blocks access of the continuous background light of the pulsed Xenon light source to the sample. The second focus is placed at the sample position, the third on the entrance slit of the spectrograph (Bruker 200is) equipped with a grating having 100 lines/mm. After spectral dispersion by the spectrograph the light is temporally analyzed by the streak camera (C7700, Hamamatsu Photonics). The latter consists of two parts: the streak tube which produces an image on a phosphor screen, and a CCD camera (ORCA-CR, Hamamatsu Photonics) which digitizes this image.

A transient absorption measurement requires four streak images:

$$\Delta A = -\log\left(\frac{I_{LP} - I_L}{I_P - I_D}\right). \tag{15.1}$$

In this expression I_{LP} is the intensity image obtained when both laser and probe light are on, I_L and I_P are the intensities when only the laser or the probe are on, respectively, and I_D is the dark image. The measurement of these four images is performed in the sequence I_{LP}, I_D, I_L, I_P. This sequence is repeated M times. When only small amounts of sample are available, i. e. when consecutive measurements have to be performed with the same sample, the cycle time t_C between two consecutive I_{LP} measurements must be comparable to the recovery time of the photochemical reaction. For some biological photosensor proteins like LOV domains t_C can be several minutes. It is then important that the measurement of I_L occurs at $t_C/2$ so that the sample can recover for the same time after each laser excitation.

When the fluorescence generated by the sample is very intense, subtraction of an image with the laser alone may not result in complete cancelation. In this case we found it easier to omit the measurement of I_L and replace I_L with I_D in the numerator of Equation 15.1. The fluorescence then appears as an apparent negative transient absorption which can in many cases be well fitted by the instrument response function. As a benefit, the fluorescence spectrum is obtained within the TA measurement. This should be compared to the situation of one-dimensional TA measurements at fixed wavelengths, which are usually performed with a photomultiplier having an active voltage divider (i. e. one with transistors instead of resistors). When such a multiplier is hit by strong fluorescence, it is “blinded” for several milliseconds. Thus, in our earlier (one-dimensional) measurements of TA in flavin-containing protein domains [27–29], measurements were not possible in the spectral region of the fluorescence. As will be shown below, this is no longer the case with the streak camera.

15.3 Data Analysis

The data sets produced by the streak camera are transient absorption differences as functions of time and wavelength, $\Delta A(t, \lambda)$. For a set of discrete N_T time points t_i and N_S wavelengths λ_j these data form a rectangular matrix $\mathbf{\Delta A}$ of dimension $N_T \times N_S$ with matrix elements $\Delta A_{ij} = \Delta A(t_i, \lambda_j)$. Every column of the matrix is a time trace for a fixed wavelength, every row is a spectrum at a given delay time. Many techniques have been developed for the analysis of data matrices of this kind, and recent reviews can be found in [30–33]. In the following we give a short account of the techniques that we usually apply.

We wish to decompose the data matrix into a sum of products of one-dimensional functions,

$$\Delta A_{ij} = \sum_{k=1}^{N} c_k(t_i) S_k(\lambda_j) = \sum_{k=1}^{N} c_{ik} S_{kj}. \tag{15.2}$$

Such a decomposition can in principle be done exactly, and in an infinite number of ways, when N is equal to the full rank of the data matrix, $N = \mathrm{Min}(N_T, N_S)$. One particularly useful technique is singular value decomposition (SVD). However, we are interested in a decomposition with only a small number of components $N_C \ll N$ corresponding to the number of spectrally distinct chemical species involved in the photoprocess. The k-th column of the matrix $\mathbf{c}$ then contains the concentration time profile of species k, whereas the k-th row of the matrix $\mathbf{S}$ contains the spectrum of species k. Obviously, truncation of the sum in Equation 15.2 results in a model data matrix $\mathbf{D}$ which only approximately reconstructs the original data $\mathbf{\Delta A}$.

$$D_{ij} = \sum_{k=1}^{N_C} c_{ik} S_{kj}. \tag{15.3}$$

Finding the “best” solution constitutes an optimization problem that can, for example, be solved by minimizing the sum of squares

$$\chi^2 = \left\| \mathbf{\Delta A} - \mathbf{D} \right\|^2 = \mathrm{Min}. \tag{15.4}$$

Ideally one would do the decomposition with no or only a small number of assumptions. We perform the analysis in several steps. The first step is model independent and extracts significant information from the data. The effective rank of the matrix is determined by SVD, setting an upper limit to the number N_C of distinguishable species involved. Further processing can then be done either with the original data set or with a reduced data set truncated to the first N_C most important components with the largest singular values. One option is solving Equation 15.4 with a global fit, employing a suitable model either for the concentration time profiles or the spectra.

The other option is performing an inverse Laplace transform along the time axis, leading to a spectral distribution of rate constants $p(k, \lambda)$. Sometimes the data matrix contains invalid data, for example strong scattered laser light, or strong fluorescence that cannot be canceled to the degree required for the application of the fit or the Laplace transform. In these cases the part of the matrix containing invalid data must be modified in an appropriate way.

In the following sections we will first describe the SVD and rank analysis, followed by a short outline of the fit procedure. The procedure developed by us for correction for scattered laser light and/or fluorescence is described next. Although it would appear logical to describe this before the fit, we use the reverse order here since the technique is integrated into the fit procedure. The last section discusses the shortcomings inherently present in a multiexponential fit and describes the method used for the inverse Laplace transform. The latter is based on a maximum entropy concept.

15.3.1 SVD and rank analysis

Singular value decomposition decomposes a matrix $\mathbf{\Delta A}$ with N_r rows and N_c columns into the product of three matrices **U**, **W**, and **V** as

$$\mathbf{\Delta A} = \mathbf{UWV}^t, \tag{15.5}$$

where the superscript t represents the transposed matrix. We assume that $N_r \geq N_c$, otherwise the decomposition is done with the transposed matrix $\mathbf{\Delta A}^t$. The matrix **U** has the same dimension as $\mathbf{\Delta A}$, and the columns of **U** are mutually orthogonal. **W** and **V** are square matrices of dimension N_c. **V** is an orthogonal matrix, and **W** is diagonal with diagonal elements $w_j \geq 0$ usually ordered in descending order. The reconstruction of the matrix $\mathbf{\Delta A}$ with the first K components is then the best representation of the original matrix by a sum of K bilinear products:

$$\mathbf{\Delta A}^{(K)} = \sum_{k=1}^{K} \mathbf{u}(k) w_k \mathbf{v}^t(k). \tag{15.6}$$

In this expression $\mathbf{u}(k)$ refers to the k-th column of the matrix **U**. Thus, the standard deviation $\sigma^{(K)}$ defined as

$$\sigma^{(K)} = \sqrt{\left\| \mathbf{\Delta A}^{(K)} - \mathbf{\Delta A} \right\|^2} \tag{15.7}$$

is the smallest standard deviation that can be expected for a fit to $\mathbf{\Delta A}$ by a model of the form of Equation 15.3 with K components. A plot of $\sigma^{(K)}$ versus K yields a monotonously decaying curve. Usually, this curve shows a steep slope for small values of K, until $\sigma^{(K)}$ comes close to the experimental noise level. The first component that does not lead to a significant improvement of $\sigma^{(K)}$, as well as all further components, can then be regarded as representing noise. Truncation of the reconstruction of Equation 15.6 to only the significant components yields a data matrix which has been effectively filtered for noise. For example, reconstruction of a 512×512 matrix to 4 components is equivalent to averaging over 128 columns and 128 rows, but without any loss in temporal or spectral resolution. Representing the data as a 3D surface plot usually requires such a filtering.

15.3.2 Global lifetime analysis

The global lifetime analysis assumes that the data can be modeled by a factorization into a sum of products of spectra $S_k(\lambda)$ and concentration time profiles $c_k(t)$ as defined in Equation 15.2, where the time profiles can be written as a linear combination of known analytic functions $f_j(t)$,

$$D_{ij} = \sum_{k=1}^{N_C} c_k(t_i) S_k(\lambda_j) \tag{15.8}$$

$$c_k(t) = \sum_{l=1}^{N_F} f_l(t) X_{lk}. \tag{15.9}$$

Here, N_C is the number of distinct spectral species, and N_F the number of analytic time functions. Defining $\mathbf{C}$ as a $N_T \times N_C$ matrix with elements $C_{ik} = c_k(t_i)$, and $\mathbf{F}$ as a $N_T \times N_F$ matrix with matrix elements $F_{il} = f_l(t_i)$, the model data matrix $\mathbf{D}$ can be written in the form

$$\mathbf{D} = \mathbf{CS} = \mathbf{FXS} = \mathbf{FB}. \tag{15.10}$$

The k-th row of the matrix $\mathbf{B}$ with elements $B_{kj} = b_k(\lambda_j)$ corresponds to the spectral changes associated with the time function $f_k(t)$. When these time functions are exponential decays (convoluted with the instrument response), the corresponding spectra are called decay-associated difference spectra (DADS). Optimization corresponds to solving the least-squares optimization problem

$$\chi^2 = \left\| \mathbf{\Delta A} - \mathbf{FB} \right\|^2 = \text{Min}. \tag{15.11}$$

Efficient algorithms exist that find $\mathbf{B}$ by solving Equation 15.11 for fixed matrices $\mathbf{\Delta A}$ and $\mathbf{F}$. The value of χ^2 optimized in this way is further minimized by a nonlinear least squares algorithm by variation of the rate constants in $\mathbf{F}$. The DADS and the corresponding rate constants are the unique result of this global lifetime analysis.

This treatment does not require any model for the kinetics involved in the transient processes. The details of such a model will be entirely defined in the matrix **X** that relates the actual species kinetics to the elementary function $f_k(t)$. Once a model and thus the appropriate matrix **X** has been chosen, the species-associated difference spectra in matrix **S** can be calculated by

$$\mathbf{S} = \mathbf{X}^{-1}\mathbf{B}. \tag{15.12}$$

15.3.3 Eliminating invalid data

Efficient algorithms exist that solve the linear least squares problem of Equation 15.11. These require the full rectangular data matrix $\mathbf{\Delta A}$, or a SVD decomposition of it. Scattered laser light or a strong fluorescence signal will, however, make small spots on the data image invalid for this analysis. Hence these data must be excluded. Since the matrix may not have "holes", this means that the invalid data must be replaced with something else. We tested two strategies.

In the first strategy, the invalid data area is "patched" by replacing all data points in a certain area with values obtained by interpolation of the values along the border of that region. The second strategy replaces the invalid data with the values of the fit according to Equation 15.11. This is done in every iteration. Upon convergence, these data hence do not contribute to the sum of squares. To the best of our knowledge this second technique seems to be new. When the second method is used, the patching will yield a global fit to the full data matrix as a by-product.

15.3.4 Maximum entropy analysis

A global fit with exponential functions yields a set of rate constants k_j and the DADS $P_j(\lambda)$:

$$A(t,\lambda) = \sum_{j=1}^{N} P_j(\lambda) f(t,k_j) \tag{15.13}$$

$$f(t,k) = \int_{-\infty}^{t} dt' \exp(-k(t-t'))g(t'). \tag{15.14}$$

That is, $f_j(t)$ is the exponential decay function with rate constant k_j convoluted with the instrument response function $g(t)$. The latter is approximated by a Gaussian centered at the peak of the excitation laser pulse. Multiexponential fits are notoriously ill-posed problems. As a result, small variations in the noise (e.g. from different measurements on the same system) can lead to substantial variations in the fitted parameters. This can be understood by considering the limit of infinite N. In this limit the set of rate constants becomes a distribution function $p(k)$, i. e. the signal $A(t)$ is the transform of $p(k)$, as shown in Equation 15.15 for a fixed wavelength:

$$A(t) = \int_0^\infty dk\, f(t,k)\, p(k). \tag{15.15}$$

In the limit that the apparatus function approaches the Dirac delta function, $g(t) = \delta(t)$, this is the Laplace transform. In practical application we found it more convenient to use $q = \ln k$ as the variable, i. e. defining

$$A(t) = \int_0^\infty dq\, R(t,q)\, p(q) \tag{15.16}$$

$$R(t,q) = f(t,\exp(q)). \tag{15.17}$$

For the numerical implementation the variables t and q are taken as discrete values in equidistant steps, i. e. $A_i = A(t_i)$, $p_j = p(q_j)$, and $R_{ij} = R(t_i, q_j)$. The transformation is then accomplished by the matrix multiplication

$$\mathbf{A} = \mathbf{Rp}. \tag{15.18}$$

An example is shown in Figure 15.3. In the upper part, a double exponential decay has been simulated. The ratio of the two rate constants is 1:3, with equal amplitudes. The lower part shows two distributions $p(k)$ that fit the experimental data equally well. One corresponds to a double exponential fit, i. e. it consists of two sharp spikes. The other shows two broad bands which indicates that quite an amount of uncertainty in the rate constants can be tolerated.

The dotted curve in Figure 15.3 was obtained with the maximum entropy method [34, 35]. This method optimizes the distribution function $p(k)$ with respect to two criteria: first, the fit must be compatible with the data. This is achieved by fixing the average square deviation to the square of the standard deviation of the measurement,

$$\chi^2 = \sum_j^{M_D} \left(\frac{A_j - A_j^{exp}}{\sigma_j} \right)^2 = M_D. \tag{15.19}$$

In this expression, A_j^{exp} is an experimental data point, σ_j the corresponding standard deviation, A_j the calculated data point according to Equation 15.15, and M_D is the number of data points. The second constraint is that the distribution $p(k)$ should not contain more information than required to fit the data. This is achieved by maximizing the configurational entropy S of the distribution.

$$S = -\sum_{l=1}^{M_L} p_l \ln p_l. \tag{15.20}$$

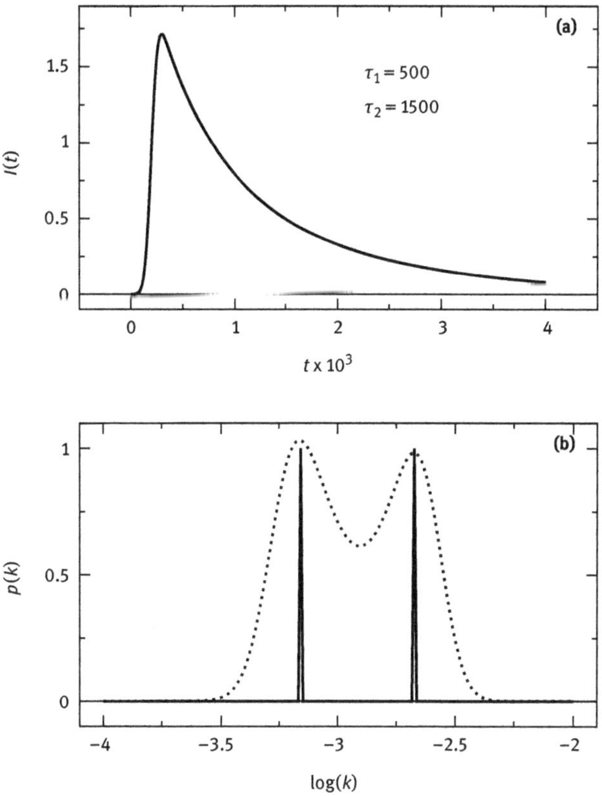

Figure 15.3: Top: A decay curve simulated as a sum of two exponential decays with equal amplitudes convoluted with a Gaussian instrument response function. The ratio of the rate constants is 1:3. Bottom: Distribution of rate constants $p(k)$ obtained by a double exponential fit (full line) and by the maximum entropy method (dotted line). Both distributions yield the same quality of fit to the data set.

That is, the distribution function $p(k)$ is represented by M_L discrete values $p_l = p(k_l)$. In this way the maximum entropy method yields the distribution $p(k)$ with the smallest information content [36]. In other words, every other distribution contains information for which no evidence exists in the data [35]. As far as we know, the first use of the maximum entropy method for solving the inverse Laplace problem with a positive amplitude distribution was given by Livesey et al. [37]. Later, other functions were proposed as entropy functions that have no sign restriction on the distribution. A review and critical comparison can be found in [38]. We use a subroutine, kindly provided to us by Geoff Daniell [34] in 1995, for solving the maximum entropy problem with the definition of Equation 15.20 since this has a strict justification in terms of information theory.

A moderate amount of noise can already significantly lower the true information content of a data set. This is exemplified in Figure 15.4. In the upper panel the same double exponential decay as in Figure 15.3 is shown, with an increasing

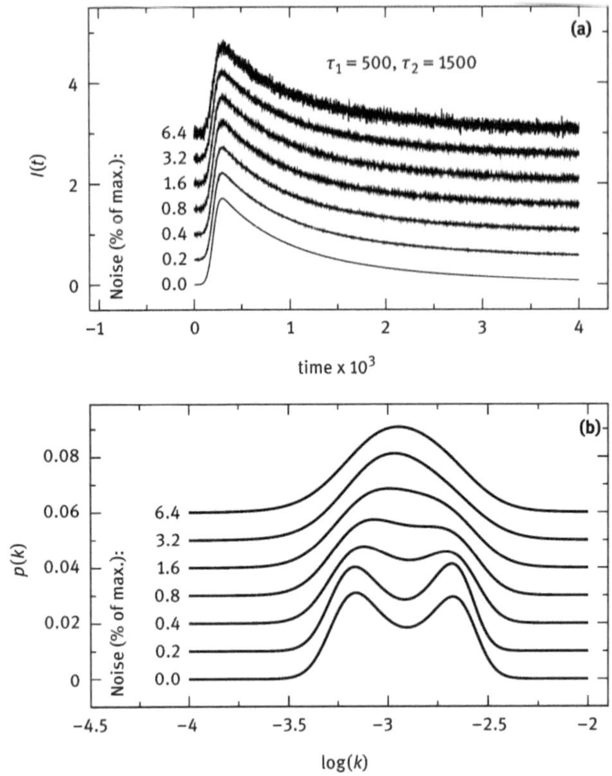

Figure 15.4: Top: Simulated double exponential decay, convoluted with a Gaussian apparatus function, and with increasing amount of Gaussian noise added. Bottom: Distribution of rate constants $p(k)$ obtained by the maximum entropy method for each of the decay curves of the upper panel. When the noise exceeds 1.6 % of the peak intensity, no evidence for two distinct rate constants is left in the data.

amount of noise added. The corresponding maximum entropy distributions are shown in the lower panel. If the noise is larger than 1.6 % of the peak intensity, two decay rates which differ by a factor of three can apparently no longer be distinguished. Application of the maximum entropy technique to the analysis of streak camera images requires two modifications: these data contain decays with positive and negative amplitudes, whereas the entropy expression (Equation 15.20) allows only for positive values. This can be accounted for by representing the distribution $\mathbf{p}$ as the difference between two distributions,

$$\mathbf{p} = \mathbf{p}^{+} - \mathbf{p}^{-}. \tag{15.21}$$

The transformation then takes the form

$$\mathbf{A} = \left(\mathbf{R}, -\mathbf{R}\right)\begin{pmatrix}\mathbf{p}^{+}\\ \mathbf{p}^{-}\end{pmatrix}. \tag{15.22}$$

Every column $\mathbf{r}(q)$ of the matrix $\mathbf{R}$ contains the time profile expected for a species that decays with the corresponding rate constant $k=\exp(q)$. The fastest rate constant is taken as the inverse of the time interval of one pixel of the streak image, the slowest rate constant is set to the inverse of ten times the total time window of the measurement. Between these two limits, 510 equidistant steps are taken on a logarithmic scale. An additional column of $\mathbf{R}$ is filled with the time profile corresponding to $k=0$, and one column contains the constant value 1. The corresponding two values of the distribution $\mathbf{p}$ account for all decay processes that show no significant decay within the measurement time window, and for a nonzero baseline of the whole image.

Since the numerical solution of Equation 15.22 is rather time consuming, the algorithm has been implemented to solve for the first N_C significant columns of the matrix $\mathbf{U}$ from the singular value decomposition of the matrix $\mathbf{A}$ instead of the whole matrix $\mathbf{A}$, resulting in a distribution $\hat{\mathbf{p}}$ in the space of SVD components. The true distribution $\mathbf{p}$ is then obtained by SVD back transform,

$$\mathrm{U}=\left(\mathbf{R},-\mathbf{R}\right)\begin{pmatrix}\hat{\mathbf{p}}^{+}\\ \hat{\mathbf{p}}^{-}\end{pmatrix} \tag{15.23}$$

$$\mathrm{p}=(\hat{\mathbf{p}}^{+}-\hat{\mathbf{p}}^{-})\mathbf{W}\mathbf{V}^{t}. \tag{15.24}$$

The trick of using Equation 15.21 for dealing with negative amplitudes has been applied by others before (e.g. [39, 40]).

15.4 Performance

We demonstrate the performance of the new technique with measurements on the photocatalytic oxidation of p-methoxy benzylalcohol (MBA) with riboflavin tetraacetate (RFTA) as the photocatalyst. Measurements on RFTA show how the data develop with increasing number M of averaged measurements and demonstrate global lifetime analysis and determination of the number of significant components with SVD. Using the photocatalytic system of RFTA with MBA we show the use of the patching procedure and compare analyses of the data by global lifetime fitting and maximum entropy.

15.4.1 RFTA alone

We begin with measurements on the photocatalyst RFTA in acetonitrile/water (50:50-v/v) in the absence of the substrate MBA. Figure 15.5 shows false color representations of data sets obtained with increasing number of accumulated excitation cycles $M=1$, 10, and 100. The graph at the top of Figure 15.5 shows the result of

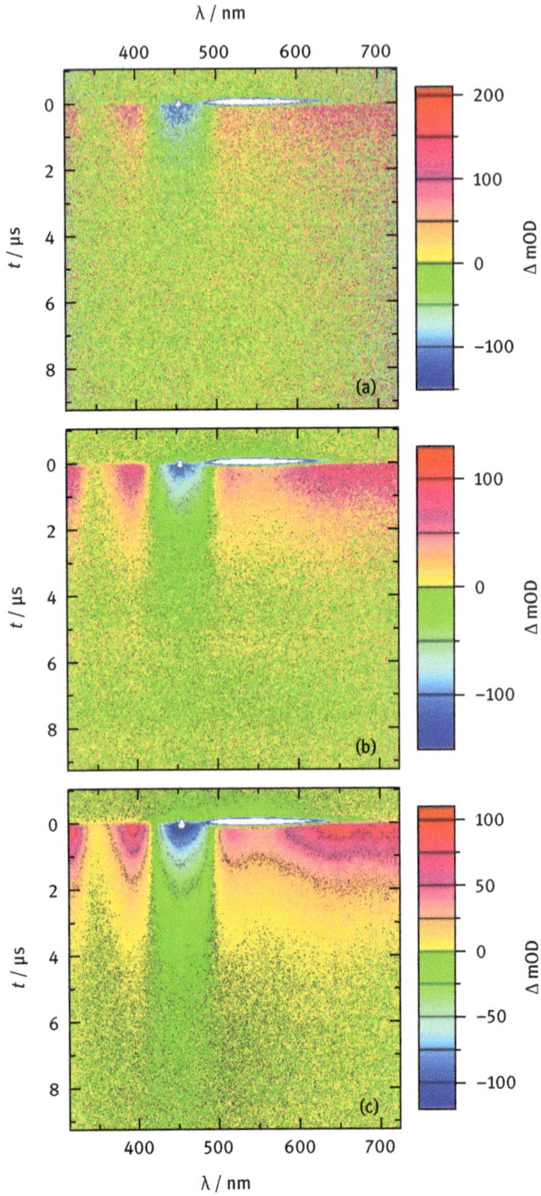

Figure 15.5: False color map of the 2D transient absorption matrix measured after photoexcitation of RFTA dissolved in acetonitrile/water (50:50-v/v). The three panels correspond to the accumulation of 1, 10, and 100 excitation cycles from top to bottom. Yellow and red indicate a positive signal which is assigned to the flavin triplet state absorption. Green and blue data regions correspond to bleaching of the flavin ground state and fluorescence.

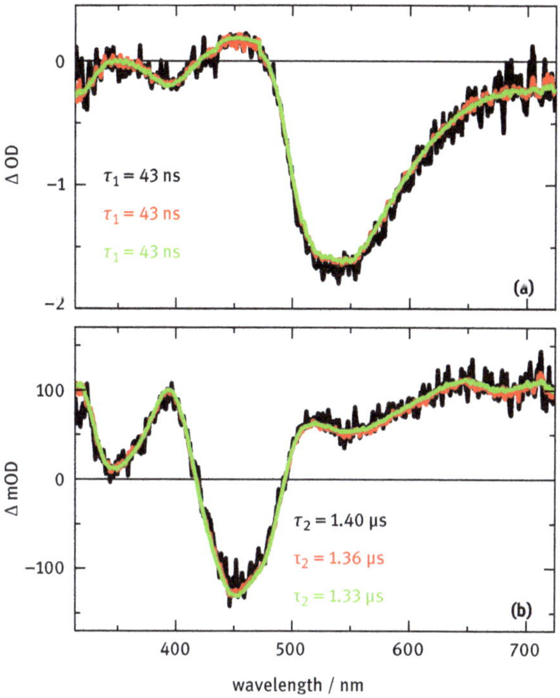

Figure 15.6: Decay-associated difference spectra of the fast and the slow component of a global fit to the data sets measured with $M = 1$ (black curves), $M = 10$ (red curves), and $M = 100$ (green curves). The fast component is assigned to fluorescence, and the fitted lifetime of 43 ns is limited by the time resolution for this time window of the streak camera. The fitted lifetime for the slow component, assigned to triplet–triplet absorption of flavin, is 1.36 μs.

a measurement with only a single excitation cycle. The fluorescence of flavin is clearly visible as the narrow white horizontal stripe. With $M = 10$ (middle panel) the bleaching of the flavin ground state is clearly visible as a green vertical stripe centered at 450 nm. Red stripes on both sides indicate transient absorption by the triplet state. Accumulation of 100 excitation cycles (bottom panel) results in a large improvement of the signal-to-noise ratio.

All three data sets show a small white spot at $t = 0$, $\lambda = 450$ nm. This is an artifact caused by scattered light of the excitation laser. When this is removed by the patching procedure (see below), a global lifetime analysis of these data sets reveals two lifetime components. One is short with a fitted decay time of 43 ns. In the 10-μs time window, a single pixel corresponds to 20 ns, and the resolution of the apparatus is ca. 2.9 pixels. Hence the true lifetime of this component is not resolved. The corresponding spectrum, shown in the upper part of Figure 15.6, is negative throughout the whole wavelength range with a broad peak at 535 nm. It is assigned to the fluorescence of RFTA which shows up in the transient absorption as an increase in

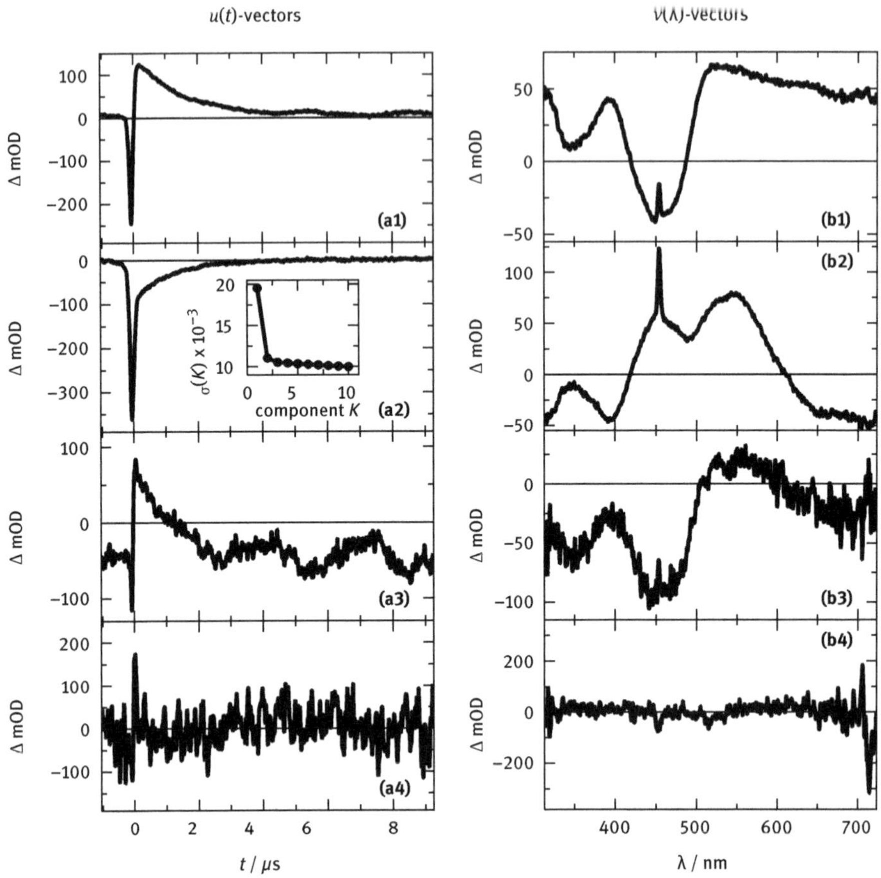

Figure 15.7: The first four temporal (panel a) and spectral (panel b) components $\mathbf{u}_k$ and $\mathbf{v}_k$ from SVD of the data set measured with an accumulation of ten excitation cycles [panel b of Figure 15.5]. The plot of $\sigma^{(K)}$ vs. K is shown as an inset. The third component apparently does not improve the standard deviation, but the corresponding vectors $\mathbf{u}_3$ and $\mathbf{v}_3$ show some structure. The component $K = 4$ is the first that contains only noise.

transmission, i. e. with a negative amplitude. The second component is fitted by a monoexponential decay with a lifetime of 1.4 µs. The spectrum shows the characteristic bands of the flavin triplet state near 650 and 700 nm. The lifetime of the triplet is limited by quenching with oxygen since the solution is saturated by air. We note that a single measurement already yields the correct spectra and decay times. Accumulating more measurements only improves the signal-to-noise ratio.

The effective rank of the data matrix can be found by SVD. As an example we show in Figure 15.7 the first four columns of the matrices **U** and **V** from the SVD analysis of the (unpatched) data set of Figure 15.5b. The plot of the residual error $\sigma^{(K)}$ versus the number of components K is shown in the inset of Figure 15.7. The first two components apparently lead to significant improvement in the residual error, whereas the third component shows only a small effect. The first three spectral and temporal components show structure that is significantly above the noise, but the fourth component is dominated by noise. After patching for the scattered laser light, the number of significant components is reduced to two (data not shown). Hence, we may interpret $\sigma^{(2)}$ as the noise in the data sets after patching for the laser scatter. A plot of $\sigma^{(2)}$ versus $\sqrt{M}$ yields a perfect straight line, indicating that this noise scales with the square root of the number M of averaged excitation cycles as expected for quantum noise. The average number of photons of the probe light that are accumulated on every pixel of the streak image is proportional to the product of M and the width of the time window. Hence, a reliable measurement within the 1-μs time window should require either $M = 10$ accumulated excitation cycles, or an increase of the intensity of the probe light by a factor of 10.

15.4.2 Photooxidation of MBA with RFTA

A detailed study of the mechanism of this photocatalytic reaction by our streak camera technique together with fs transient absorption measurements was presented in [26]. Here we present a more recent data set for the ns–μs time range on the same system, measured with improved signal-to-noise ratio. The results of the previous report were confirmed, but more details became visible.

The data matrix shown as the false color plot in panel a of Figure 15.8 was obtained by accumulation of 100 excitation cycles of 30 mM MBA and 40 μM RFTA dissolved in acetonitrile/water (50:50-v/v). The data matrix covers a time range of 20 μs and the spectral range of 335–755 nm. Red and yellow colors indicate absorption, green and blue colors correspond to bleaching. The white spots at $t = 0$ near 445 nm and in the range 490–610 nm indicate signals exceeding the color-coding scale. They are caused by scattered light of the excitation laser and the flavin fluorescence, respectively. In a first step these two signals were removed by patching the data matrix as described in the section on data analysis. Panels a1 and a2 in Figure 15.8 show a magnified view of this section of the data matrix before and after patching, respectively. The plot of $\sigma^{(K)}$ versus K indicates four significant components before patching (black squares), which are reduced to three components after patching (red circles).

A subsequent global lifetime analysis with three exponential functions yields lifetimes of 0.93, 6.0, and ≫ 20 μs. The three corresponding decay-associated difference spectra (DADS) are shown in panel b of Figure 15.8 as the black, green, and red curves, respectively. They contain the spectral signatures of the flavin triplet state ^{3}RFTA at 700 nm, the radical anion ^{2}RFTA$^{\bullet-}$ at 360 and 480 nm, and the protonated neutral radical ^{2}RFTA-H$^{\bullet}$ at 600 nm, as well as a negative band at 445 nm corresponding to bleaching of the RFTA ground state. The DADS are linear combinations

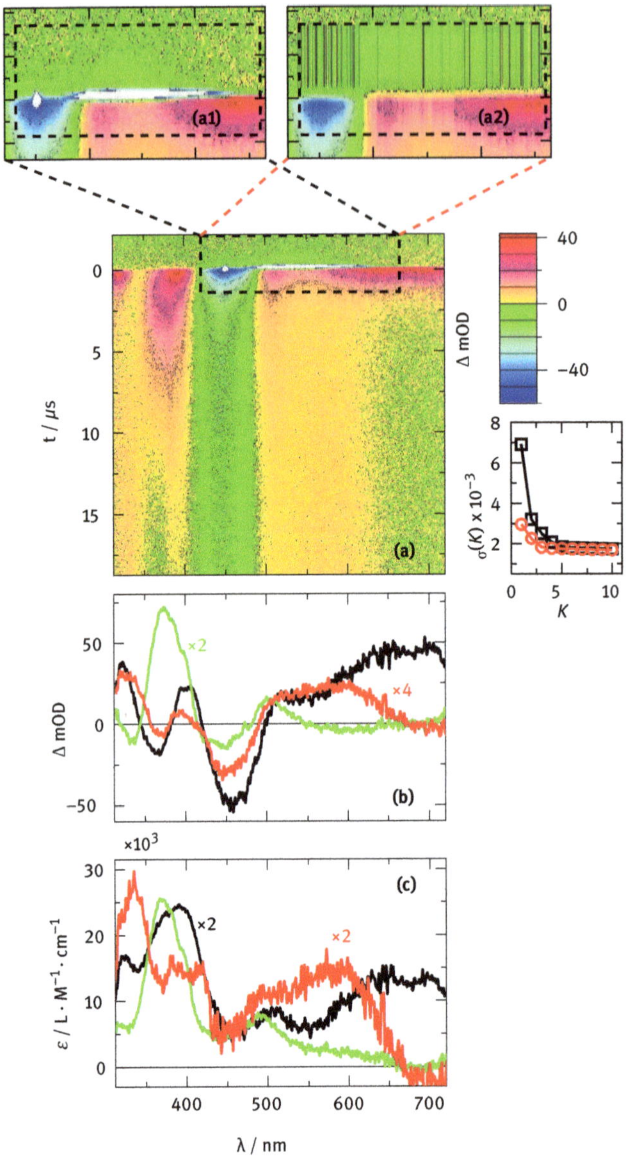

Figure 15.8: (a) False color representation of the 2D transient absorption data measured for a solution of riboflavin tetraacetate (40 µM) and p-methoxy benzylalcohol (30 mM) in acetonitrile/water (50:50-v/v). The magnified sections a1 and a2 show the regions before and after patching for the invalid data caused by strong laser scattering and fluorescence. (b) DADS obtained by a global fit with three exponential functions. (c) Species-associated spectra obtained by transforming the DADS according to the mechanism depicted in Figure 15.9.

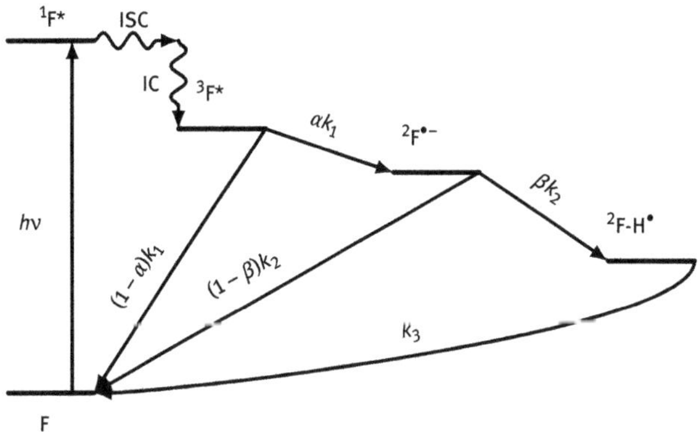

Figure 15.9: Mechanism proposed for the initial steps of the flavin-based photocatalytic oxidation of p-methoxy benzyl alcohol.

of the species-associated spectra (SAS), and the linear coefficients reflect the reaction mechanism. The best agreement with published spectra for the three flavin species is obtained with the mechanism depicted in Figure 15.9: following excitation to the excited singlet state, ^{1}RFTA decays either by fluorescence to the ground state or by intersystem crossing to the triplet state ^{3}RFTA. This process occurs within 3 ns and is not resolved in our measurement. A fraction of $\alpha = 25\,\%$ of the triplet state reacts with MBA and forms the radical anion ^{2}RFTA$^{\bullet-}$, whereas the other 75 % decay back to the ground state. The sum of both decays results in a lifetime of 0.93 μs for the triplet state. A fraction of ca. 46 % of the radical anion falls back to the ground state, presumably via back electron transfer to MBA. The remaining fraction of ca. 54 % is protonated to yield the neutral radical ^{2}RFTA-H$^\bullet$. Transformation of the DADS with this model yields the SAS shown in panel c of Figure 15.8. They agree well with published spectra of these species.

Instead of a global fit the data matrix can also be analyzed by the maximum entropy method. The resulting matrix of the distribution $p(\log k, \lambda)$ is shown in Figure 15.10a. The false color plot reveals three regions along the $\log k$ axis where spectral changes occur: one near $\log k \approx -0.06$, corresponding to lifetimes in the range $\tau = 1.16 \pm 0.46$ μs, a second region near $\log k \approx -0.92$, corresponding to lifetimes in the range $\tau = 8.3 \pm 2.6$ μs, and a third region $\log k < -2.32$, corresponding to lifetimes in the range $\tau > 210$ μs, i. e. outside of the time window covered by the experiment. Integration of $p(\log k, \lambda)$ over the corresponding ranges along the $\log k$ axis, as indicated by the dashed lines in Figure 15.10a, results in the spectra shown as full lines in panels b, c, and d of Figure 15.10. Apparently, these are very similar to the DADS obtained by the global fit, shown again for comparison as dotted lines in the respective panels. The fit corresponds to a distribution $p(\log k, \lambda)$ that consists of three δ-functions along the $\log k$ axis. The maximum entropy method yields, in addition to the "peak"-lifetimes, also a measure of the uncertainty in these lifetimes. In the present case, the characteristic rate constants of the system differ sufficiently so that the peaks along the log k axis are well separated. Thus, the maximum entropy technique yields an *a posteriori* justification for the use of a double exponential fit.

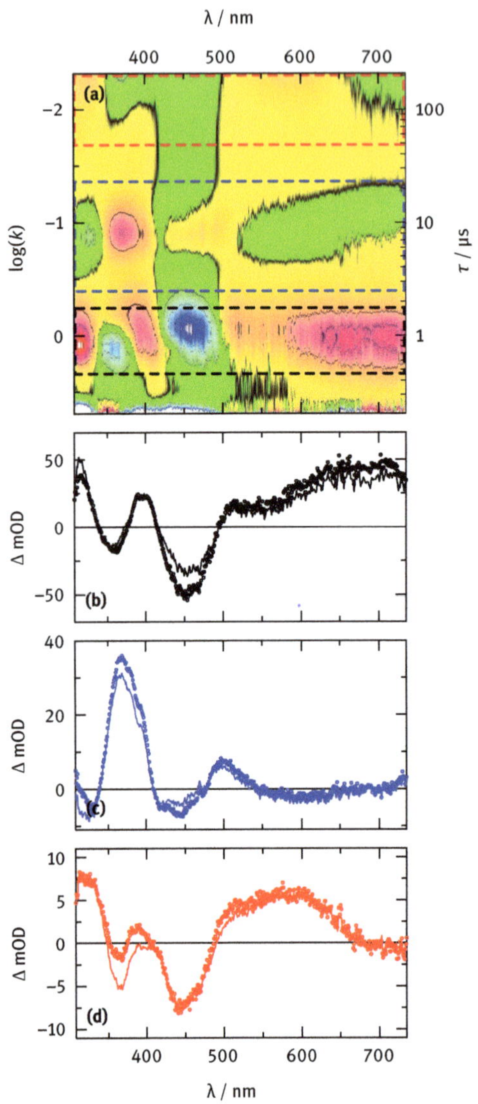

Figure 15.10: (a) 2D representation of the distribution of rate constants $p(k, \lambda)$ obtained by maximum entropy inverse Laplace transformation of the data matrix in Figure 15.8 after patching for scattered laser light and fluorescence. (b–d) Comparison of the DADS (dotted lines) from global lifetime analysis (Figure 15.8) and the spectra obtained by integration of $p(k, \lambda)$ over the rate constant intervals indicated. For a discussion of these results see the text.

15.5 Discussion

We conclude the assessment of the performance of the new transient absorption machine with a discussion of sensitivity. We expect that the noise measured at any individual pixel of the detector scales with the inverse of the square root of the number of photons collected by this particular detector element. This number of photons is proportional to the number M of events averaged, the size of the time window T selected for the streak image, and the width d of the slits of the spectrograph. The standard deviation should be a function of the wavelength since the intensity of the probe light is lower in the red region ($\lambda > 650$ nm) and highest in the visible range between 400 and 600 nm. Hence, it should be possible to model the standard deviation according to

$$\sigma(\lambda) = \sigma_0(\lambda)\left(\frac{T_0 d_0 M_0}{TdM}\right)^{1/2}, \tag{15.25}$$

where T_0 is the time window, d_0 the slit width of the spectrograph, and M_0 the number of averaged events in a typical measurement taken as the reference. Figure 15.11 shows, in panel a, the measured function $\sigma_0(\lambda)$ for $T_0 = 10$ µs, $d_0 = 60$ µm, and $M_0 = 50$. The function was obtained by a measurement on a cuvette filled with pure water and blocking the excitation laser. The standard deviation was calculated for each column of the data matrix, i. e. for all pixels belonging to the same wavelength. In the limit of zero slit widths the resolution of the streak camera CCD combination is ca. 2.5 pixels both in the time and the wavelength direction. We use a slit width of 30 µm for the streak camera, resulting in a width of 2.9 pixels for an infinitely fast (i. e. unresolved) event. For a given time window T the temporal resolution is hence $\delta t = 0.00566$ T. The spectral resolution increases linearly with the slit width d of the spectrograph according to $\delta\lambda/[\mathrm{nm}] = 2.1 + 0.0124\, d/[\mu\mathrm{m}]$. With the settings mentioned above the time resolution is 57 ns and the spectral resolution is 2.9 nm. In this "standard experiment", the noise is ca. 3.5 mOD on each pixel in the spectral range 350–500 nm. It should be noted that this noise can be further reduced by SVD analysis – without any loss in spectral or temporal resolution – by a factor of $\sqrt{K/512}$, where K is the number of significant components.

Panel b in Figure 15.11 shows that the noise indeed scales with the inverse square root of T, d, and M. The standard deviation σ averaged over the wavelength interval 350–500 nm is plotted versus $\sqrt{T_0/T}$ (triangles), $\sqrt{d_0/d}$ (squares), and $\sqrt{M_0/M}$ (circles). Error bars indicate the standard deviation of these errors in that spectral interval. All data points fall nicely on the same straight line. Hence, Equation 15.25 can be used in combination with the curve in Figure 15.11a to estimate the noise for other settings of the apparatus. All these values refer to the same setting of the entrance slit of the streak camera (30 µm), and to the light intensity of our present pulsed Xe flash lamp. Even lower noise levels could be achieved by increasing the

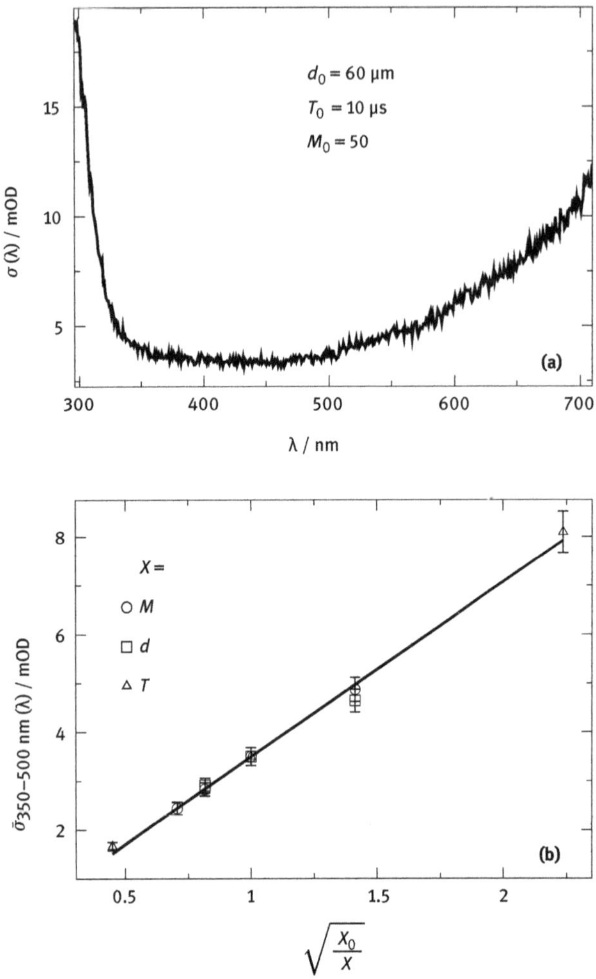

Figure 15.11: (a) Standard deviation σ_0 for an individual data point (pixel) in the streak image for a measurement in a time window of $T_0 = 10\ \mu s$, with a monochromator slit of $d_0 = 60\ \mu m$, and averaging over $M_0 = 50$ excitation cycles. (b) The standard deviation scales with the inverse square root of T (triangles), d (squares), and M (circles).

slit width of the streak camera (thereby sacrificing time resolution), or by employing a brighter flash lamp. We plan to use the latter option, in combination with a faster excitation laser, in order to access shorter time windows in the range 1–100 ns.

15.6 Conclusion

In summary, we have shown that the streak camera apparatus is capable of measuring fluorescence spectra and transient absorption spectra in a single excitation event for time windows $T > 1\,\mu s$ with a time resolution of $\delta t \approx T/200$. Our experience with one-dimensional measurements at a fixed wavelength is that the number of excitation cycles required to obtain a reasonable monoexponential fit is typically $M \approx 100$. Furthermore, measurements are not possible at wavelengths that overlap with the fluorescence. With the streak camera, we can measure over the full wavelength range and obtain the decay characteristics of the system for all these wavelengths with less excitation cycles than required for a single one-dimensional measurement. The fluorescence spectrum is obtained as a by-product. This is possible for two reasons:

1. The streak camera has a much larger dynamic range (10000:1) compared to the typical dynamic range of a photomultiplier.
2. The streak camera data are highly redundant.

This more than compensates for the comparatively low S/N-ratio of each individual pixel. For example, filtering of the data matrix by singular value decomposition to the five most significant components is equivalent to averaging over 100 columns and 100 rows of the matrix without any loss in temporal or spectral resolution. We conclude that the streak camera setup yields transient absorption data with a multiplex factor of more than 100 compared to one-dimensional measurements of time traces at fixed wavelengths. In addition, we can measure transient absorption over the full wavelength range 350–750 nm also for systems with strong fluorescence.

15.7 Acknowledgments

The streak camera was obtained by a grant (Di-299/6) from the Deutsche Forschungsgemeinschaft (DFG) within the Forschergruppe FOR526 “Blue Light Photoreceptors”. The application to flavin photocatalysis was supported by the research training group GRK1626 “Chemical Photocatalysis”. R.J.K. acknowledges a stipend from the DFG within the research training group GRK640 “Sensory Photoreceptors in Natural and Artificial Systems”. The toroidal mirrors were a kind gift from Rodenstock.

15.8 References

[1] R.-J. Kutta, Th. Langenbacher, U. Kensy, B. Dick, Appl. Phys., DOI 10.1007/s00340-012-5320-2
[2] A. Müller, J. Schulz-Hennig, H. Tashiro, Opt. Commun., 18, 152 (1976).
[3] K. Yoshihara, A. Namiki, M. Sumitani, N. Nakashima, J. Chem. Phys., 71, 2892 (1979).
[4] Y. Liang, D. Negus, R. Hochstrasser, M. Gunner, P. Dutton, Chem. Phys. Lett., 84, 236 (1981).
[5] D.-J. Jang, D. F. Kelley, Rev. Sci. Instrum., 56, 2205 (1985).
[6] M. C. Sauer Jr., C. Romero, K. H. Schmidt, Int. J. Rad. Appl. Instrum. C, 29, 261 (1987).
[7] J. A. Schmidt, E. F. Hilinski, Rev. Sci. Instrum., 60, 2902 (1989).

[8] T. Ito, M. Hiramatsu, M. Hosoda, Y. Tsuchiya, Rev. Sci. Instrum., 62, 1415 (1991).
[9] T. Nagamura, H. Sakaguchi, S. Muta, T. Ito, Appl. Phys. Lett., 63, 2762 (1993).
[10] T. Nagamura, H. Sakaguchi, T. Ito, S. Muta, Mol. Cryst. Liq. Cryst. A, 247, 39 (1994).
[11] T. Hayashi, T. Takimura, H. Ogoshi, J. Am. Chem. Soc., 117, 11606 (1995).
[12] A. Watanabe, O. Ito, M. Watanabe, H. Saito, M. Koishi, J. Phys. Chem., 100, 10518 (1996).
[13] T. Kodaira, A. Watanabe, O. Ito, M. Watanabe, H. Saito, M. Koishi, J. Phys. Chem., 100, 15309 (1996).
[14] G. A. Brucker, D. F. Kelley, J. Phys. Chem., 91, 2856 (1987).
[15] N. A. Abul-Haj, D. F. Kelley, J. Phys. Chem., 91, 5903 (1987).
[16] L. F. Cooley, C. E. L. Headford, C. M. Elliott, D. F. Kelley, J. Am. Chem. Soc., 110, 6673 (1988).
[17] D.-J. Jang, Bull. Korean Chem. Soc., 12, 441 (1991).
[18] D.-J. Jang, P. Kim, Bull. Korean Chem. Soc., 16, 1184 (1995).
[19] D.-J. Jang, C. S. Ah, Bull. Korean Chem. Soc., 19, 1063 (1998).
[20] C. S. Ah, H. S. Han, K. Kim, D.-J. Jang, J. Phys. Chem. B, 104, 8153 (2000).
[21] C. S. Ah, H. S. Han, K. Kim, D.-J. Jang, Pure Appl. Chem., 72, 91 (2000).
[22] J. H. Chung, C. S. Ah, D.-J. Jang, J. Phys. Chem. B, 105, 4128 (2001).
[23] H. Kim, T. G. Kim, J. Hahn, D.-J. Jang, D. J. Chang, B. S. Park, J. Phys. Chem. A, 105, 3555 (2001).
[24] S. J. Kim, T. G. Kim, C. S. Ah, K. Kim, D.-J. Jang, J. Phys. Chem. B, 108, 880 (2004).
[25] T. Langenbacher, D. Immeln, B. Dick, T. Kottke, J. Am. Chem. Soc., 131, 14274 (2009).
[26] U. Megerle, M. Wenninger, R.-J. Kutta, R. Lechner, B. König, B. Dick, E. Riedle, Phys. Chem. Chem. Phys., 13, 8869 (2011).
[27] T. Kottke, J. Heberle, D. Hehn, B. Dick, P. Hegemann, Biophys. J., 84, 1192 (2003).
[28] T. Kottke, B. Dick, R. Fedorov, I. Schlichting, R. Deutzmann, P. Hegemann, Biochemistry, 42, 9854 (2003).
[29] H. M. Guo, T. Kottke, P. Hegemann, B. Dick, Biophys. J., 89, 402 (2005).
[30] R. Bonneau, J. Wirz, A. D. Zuberbühler, Pure Appl. Chem., 69, 979 (1997).
[31] N. P. Ernsting, S. A. Kovalenko, T. Senyushkina, J. Saam, V. Farztdinov, J. Phys. Chem. A, 105, 3443 (2001).
[32] I. H. M. van Stokkum, D. S. Larsen, R. van Grondelle, Biochim. Biophys. Acta – Bioenergetics, 1657, 82 (2004).
[33] C. Ruckebusch, M. Sliwa, P. Pernot, A. de Juan, R. Tauler, J. Photochem. Photobiol. C, 13, 1 (2012).
[34] G. J. Daniell in Maximum Entropy in Action, eds. B. Buck, V. A. Macaulay; Oxford University Press, 1992; pp. 1–18.
[35] J. Skilling in Maximum Entropy in Action, eds. B. Buck, V. A. Macaulay; Oxford University Press, 1992; pp. 19–40.
[36] E. T. Jaynes, Proc. IEEE, 70, 939 (1982).
[37] A. K. Livesey, P. Licinio, M. Delaye, J. Chem. Phys., 84, 5102 (1986).
[38] V. A. Lorenz-Fonfria, H. Kandori, Appl. Spectrosc., 61, 74 (2007).
[39] A. T. N. Kumar, L. Zhu, J. F. Christian, A. A. Demidov, P. M. Champion, J. Phys. Chem. B, 105, 7847 (2001).
[40] Z. Ablonczy, A. Lukacs, E. Papp, Biophys. Chem., 104, 249 (2003).

Eberhard Riedle, Matthias Wenninger

16 Time resolved spectroscopy in photocatalysis

16.1 UV/Vis absorption spectroscopy: More than just ε!

The basic principles of optical spectroscopy were discovered in the nineteenth century. The German optician Joseph von Fraunhofer (1787–1826) was the first one, who observed effects of atomic absorption processes in the spectrum of the sun. Narrow black absorption lines, which are now named Fraunhofer lines, correspond to optical transitions of the elements in the solar atmosphere. The spectral position of these lines allows conclusions on the composition of the atmosphere. Later, in the middle of the nineteenth century, the physicist Gustav Robert Kirchhoff and the chemist Robert Wilhelm Bunsen were the first who discovered previously unknown elements (Caesium and Rubidium) with the help of optical spectroscopy. The significant development of spectroscopic techniques since then has pushed spectroscopy to become one of the most important scientific tools. As a recent example, the Curiosity rover, which was landed by NASA on Mars in 2012, carries a spectrometer for the visible and UV spectral range for the optical analysis of matter.

Since there are a huge number of different spectroscopic techniques which are already described in detail elsewhere, we will only give a brief introduction on UV/Vis absorption spectroscopy in the context of time-resolved spectroscopy and photocatalysis. In general, UV/Vis absorption spectroscopy provides a wavelength dependent optical density $OD(\lambda)$. The Beer–Lambert-Law

$$T(\lambda)=\frac{I}{I_0}=10^{-\varepsilon(\lambda)cd}=10^{-OD(\lambda)} \tag{16.1}$$

allows calculating the molar absorption coefficient $\varepsilon(\lambda)$, a quantity describing molecular properties, from the measured optical density $OD(\lambda)$, the concentration c of the sample and the optical path length d. However, the UV/Vis absorption spectrum can provide much more information than just ε values, which arise from the spectral distribution of possible optical absorption and thereby the energy differences. In particular, effects caused either by self aggregation or by the complexation of catalyst and substrate are often well distinguishable in UV/Vis spectroscopy by encountering additional absorption bands with increasing concentration. Also a change of the dipole moment of the excited chromophore between the ground state and its excited state can be often seen with the help of absorption spectroscopy in a variety of polar solvents.

Two different types of self-aggregation of dyes in solution are discussed in the literature: H-aggregates and J-aggregates. J-aggregates, named after their inventor Jelley and H-aggregates, are named for the hypsochromic shift (to shorter wavelengths) of their UV/Vis absorption spectrum when they are aggregated. In the 1930s Günter Scheibe from the Technical University of Munich and Edwin E. Jelley from the Kodak Research Laboratory in Harrow, UK observed independently an unusual behavior of pseudoisocyanine (PIC) chloride in aqueous solution and in ethanol. The absorption maximum was shifted to higher wavelengths and became more intense and sharp upon increasing the dye concentration. Dye aggregates with an absorption band that is bathochromically shifted (to longer wavelengths) with respect to the monomer absorption band are generally termed Scheibe aggregates or J-aggregates. [1–3] Substituted perylene diimides which are also used as a light absorbing chromophore in photocatalysis [4] show – depending on their substituents – very strong effects due to J-aggregation in their absorption spectrum. [5, 6]

Self-aggregation does not necessarily need to affect the energy levels of electronic states but it still can result in changes of the vibronic structure of the electronic absorption. Figure 16.1 shows for instance how self-aggregation of 4-(3′-butenyloxy)quinolone, a pro-chiral target in enantioselective photocatalysis [7, 8], changes the shape of the UV/Vis absorption properties with increasing concentration. It can be clearly seen, that the spectral position of the absorption band is not influenced by aggregation whereas the shape of the absorption band broadens with increasing concentration.

The ability of a chromophore to change its absorption and fluorescence properties due to a change in solvent polarity is called solvatochromism. A hypsochromic shift (to shorter wavelengths) of a transition with increasing solvent polarity corresponds to negative solvatochromism whereas a bathochromic shift (to longer wavelengths) with increasing solvent polarity corresponds to positive solvatochromism. Since polar solvents energetically stabilize states with larger dipole moment, the sign of the solvatochromism generally depends on the change in dipole moment

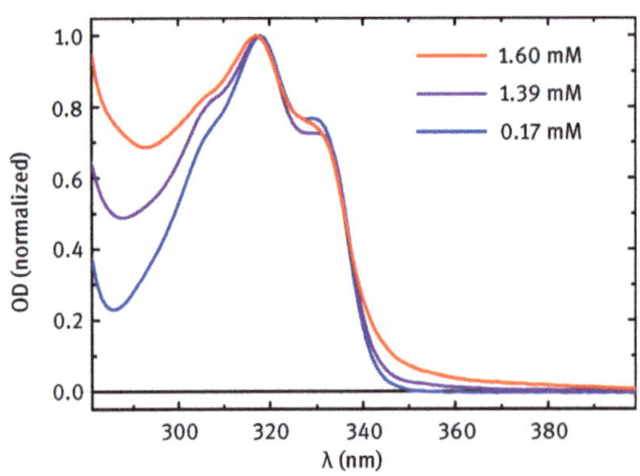

Figure 16.1: Normalized absorption of 4-(3′-butenyloxy)quinolone in trifluorotoluene at different concentrations.

between the ground state and the excited state. The effects of solvatochromism do not necessarily have to affect all ground state transitions of the chromophore. For example, only the second optical accessible transition of riboflavin tetraacetate (RFTA) shows solvatochromism whereas the first accessible transition is nearly unaffected by a change of solvent polarity (Figure 16.2, right panel). The solvatochromism of RFTA is not only caused by a change of polarity. A major part of the solvatochromism of RFTA originates from a change in hydrogen bonding between solvent and chromophore. [9, 10]

The second example to demonstrate the effects of solvatochromism we chose from the group of fluorescin dyes. The dianion of eosin Y (2′,4′,5′,7′-tetrabromo-fluorescin), which has now been reported to provide photocatalytic capability [11], shows a strong negative solvatochromism of its energetically lowest absorption band (Figure 16.2, left panel). A solvent dependent change of the hydrogen bonding is suggested to strongly affect the ground state absorption properties. [12] Also the monomeric and the dimeric form of eosin Y are distinguishable by their absorption properties. [13] It is reported in the literature that the solvent polarity not only affects the ground state absorption spectra but also the excited state dynamics and absorption properties of eosin dyes. The lifetime of the S_1 state, for instance, strongly decreases when diluted in water and in other protic solvents. The occurrence of hydrogen-bond assisted non-radiative deactivation competes with fluorescence and, therefore, leads to an ultrafast relaxation of the excited singlet state back to the ground state. [14]

Intramolecular charge transfer states play a major role in photoredox catalysis. Excitation of these states generally leads to a strong change in the dipole moment of the chromophore with respect to its ground state. Especially coordination complexes, which are often used as photoredox catalysts, exhibit intramolecular charge

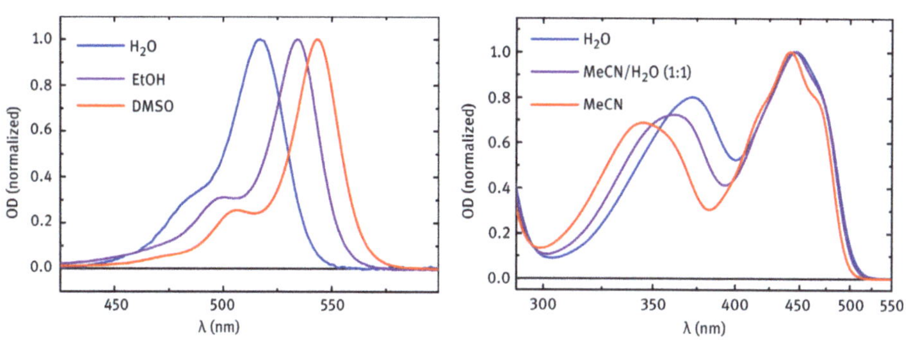

Figure 16.2: Negative solvatochromism of eosin Y (2′,4′,5′,7′-tetrabromo-fluorescin-dianion, left) and positive solvatochromism of riboflavin tetraacetate (right). The empiric values of the polarity of the solvents on the Reichardt scale are 1.00 for H_2O, 0.65 for ethanol (EtOH), 0.46 for acetonitrile (MeCN), and 0.44 for dimethyl sulfoxide (DMSO) [15].

transfer states, so-called metal-to-ligand charge transfer states (MLCT). The excitation of a MLCT transition causes a significant charge rearrangement within the coordination complex in its excited state and, hence, strongly affects the ground state absorption properties with increasing solvent polarity.

Ruthenium centered metal complexes are a well investigated group of coordination complexes due to their versatile application in photoredox catalysis. In reference [16] and reference [17] the solvatochromism of different ruthenium centered coordination complexes is described and a hypsochromic shift of the ground state absorption spectrum with increasing solvent polarity was found which corresponds to a negative solvatochromism. According to reference [17] the excitation of a MLCT transition of the ruthenium centered coordination complex leads to a decrease in dipole moment of the investigated asymmetric coordination complex. The polar solvent environment stabilizes the polar ground state of the chromophore and, hence, enlarges the energetic gap between the ground state and the excited state. This causes the blue shift of the ground state absorption spectrum.

When considering the influence of polar solvent molecules on excited states, it has to be mentioned that immediately after excitation both the nuclear configuration of the chromophore and the solvent environment remain in the ground state configuration. Then, on the time scale of hundreds of femtoseconds to a few ps, the energy is redistributed among intrachromophore modes. The fastest solvation processes take place on the same sub-ps time scale, with a strong dependence on the particular solvent, and diffusive solvent motion occurs on a somewhat longer ps time scale. [18–20] The similarity in time scale for vibrational relaxation and solvation makes their interplay highly complicated. After the two processes have concluded, the excited chromophore and the solvent environment are in equilibrium. This causes, at the same time, an increase of the energy of the electronic ground state. As a consequence, a rather large shift of the emission spectrum to longer wavelengths is found.

To describe the effects of the solvent polarity on optical properties and chemical reactivity as well in a simple model is not sufficient. The effective polarity of a solvent is influenced by a huge number of effects: On the one hand the dielectric constant (ε) and the permanent dipole moment (μ) are contributing to the intensity of the solvent polarity. On the other hand, effects between solvent and solute caused, for instance, by nonspecific Coulombic, directional, inductive and dispersion interactions are contributing to the effective solvent polarity as well as hydrogen bond, electron pair donor/acceptor and solvophobic interactions. [21] The empiric scale of Christian Reichardt reduces all contributing effects on a one-dimensional scale describing the effective polarity of a solvent and can be found in reference [15]. On this scale the highest empiric value of polarity of one is assigned to water whereas the lowest value of 0.006 is attributed to cyclohexane. A fundamental overview of solvatochromism, solvatochromic dyes and their use as solvent polarity indicator is presented by Reichardt in reference [22].

16.2 Time-Resolved spectroscopic methods from fs to μs to elucidate photocatalytic processes

16.2.1 Transient absorption spectroscopy: Signals, time scales, and data processing

In the previous section, an introduction to UV/Vis absorption spectroscopy was given and it was exemplified how optical spectroscopy is able to provide more than only static information of the ground state absorption properties of a molecule. However, to understand the mechanism of a photocatalytic reaction, it is not sufficient only to rely on steady state spectra. Even bimolecular photocatalytic reactions include multiple intermediate steps, which may occur on a huge range of different time scales. Whereas intramolecular electron transfer processes, and the closure or breaking of covalent bonds occur on time scales down to the fs regime (10^{-15}s), diffusion limited bimolecular reactions or triplet chemistry happen on time scales up to the μs regime (10^{-6}s) or even beyond. For this reason it would be a great benefit for the understanding of photocatalysis to observe the individual steps of the photocatalytic reaction in slow motion.

Since many of these processes cause changes in the optical properties of the sample, time resolved spectroscopic methods come into play. There are numerous different spectroscopic techniques, which can be used for the investigation of excited state dynamics. The most common methods provide time resolved emission or absorption properties from the near UV to the IR range of the electromagnetic spectrum. Common methods for the time resolved observation of emission are streak lifetime imaging, time-correlated single-photon counting (TCSPC) and fluorescence upconversion, to mention only the most common ones. Since these techniques only allow the investigation of electronic states which emit photons, the accessible insight into the excited state dynamics is quite limited. Much more information is provided by time resolved absorption spectroscopy. There are also various spectroscopic techniques to access largely different time scales and spectral windows: laser flash photolysis, time resolved infrared spectroscopy, femtosecond stimulated Raman spectroscopy [25], multi-photon absorption spectroscopy and 2D spectroscopy to mention again only a few of them.

The modern version of ultrafast transient absorption (TA) spectroscopy is a direct successor of the flash photolysis technique developed by R. G. W. Norrish and G. Porter around 1950. [23, 24] In their experiments, which earned them the Nobel prize in chemistry in 1967, they used light bursts with millisecond duration from discharge lamps to trigger fast chemical reactions. The course of the reaction was monitored by taking photographs of the absorption of the sample at different delay times after the excitation. Thereby, the sample was exposed to a second light flash with a continuous spectrum that was dispersed in a polychromator before illuminating the photo plates. The basic idea behind this is that different intermediate states of the photoinduced reaction have different absorption properties.

Here we will only focus on time resolved absorption spectroscopy which covers the visible, the near UV and near IR spectral region. The following chapter describes how transient absorption (TA) spectroscopy on largely different time scales can be applied to photocatalytic reactions to gain deep insights into the underlying reaction mechanism and to obtain a fundamental understanding of the intermediate steps. In combination with a suitable data analysis, one is enabled to cover a huge field of questions. With regards to the investigation of photocatalytic processes, especially the following questions are of specific interest:

- Which are the intermediate steps of the photocatalytic reaction?
- Is the catalytic reaction singlet or triplet chemistry?
- How does diffusion contribute to the reaction dynamics?
- Are intramolecular charge transfer processes involved?
- Are catalyst-substrate-complexes formed and which impact has the complex formation on the electronic properties of a molecule?

Time resolved spectroscopic methods provide the quantitative and qualitative determination of the kinetics of intermediate states in light-induced chemical reactions within the experimental limits such as time resolution, maximum pump-probe delay and spectral probe detection bandwidth. Since the fastest molecular electronic processes occur on the fs time scale, ultrafast spectroscopic techniques are necessary to resolve these processes. There are many different approaches for the measurement of photo-induced reaction kinetics. In this chapter we present two spectroscopic techniques allowing the determination of time resolved absorption spectra on timescales from tens of fs to hundreds of μs when they are combined.

In general, time resolved absorption spectroscopy is a pump-probe technique: The sample is first optically excited by a spectrally defined pump pulse tuned into resonance with a selected electronic transition or vibronic feature and then probed with a broadband white light pulse after a chosen delay time Δt. This procedure is repeated for selected values of Δt until the whole temporal window is covered with the desired temporal resolution. The chosen method which determines the pump probe delay Δt depends on the observed time window: On the one hand, for Δt values from tens of fs to few ns, a computer controlled mechanical delay stage to control the optical path length of pump- or probe-pulse is reasonable. An increase of the optical pathway by 3 μm corresponds to a temporal delay of ~ 10 fs between pump and probe pulse whereas an increase of 30 cm corresponds to a ~ 1 ns delay. To resolve ultrashort processes on this time scale, it is necessary that the duration of pump and probe pulse also have to match these temporal conditions. On the other hand, to cover time scales larger than some ns or even in the μs or ms range, a mechanical delay is no longer feasible. Here electronic delay triggering provides the desired temporal accuracy and time window.

One aspect which all methods of transient absorption spectroscopy have in common is the kind of signals that contribute to the observed transient spectrum. Since all common pump-probe techniques measure the change of absorption with respect to the ground state absorption and not the pure absorption spectrum of the excited state or product states, the more correct term for the presented methods would be "transient differential absorption spectroscopy", which is sometimes used. In general there are four major contributions to the transient absorption spectrum. The change of absorption of the excited state with respect to the ground state absorption can be either positive ($\Delta OD > 0$) or negative ($\Delta OD < 0$):

- Excited state absorption (ESA), $\Delta OD > 0$
  Absorption of excited state > absorption of ground state
- Ground state bleach (GSB), $\Delta OD < 0$
  Absorption of excited state < absorption of ground state
- Stimulated emission (SE), $\Delta OD < 0$
  Stimulated emission amplifies probe light within the spectral region of the fluorescence
- Product absorption (PA), $\Delta OD > 0$
  Newly formed species will absorb in general and their absorption will differ from the absorption spectra of the reactants

The signal of the stimulated emission can be clearly assigned to an excited singlet state – usually the S_1 state. Stimulated phosphorescence of the triplet state is generally not often observable in transient absorption spectroscopy since it usually involves a spin forbidden transition from the triplet state back to the ground state, which has singlet character. The ground state bleach corresponds to a lack of ground state absorption due to the excitation of a fraction of molecules within the probed sample volume. The ground state bleach is specifically suited to quantify the fraction of molecules which populate an intermediate or product state of the photochemical reaction: Its persistence indicates that not all of the excited molecules have relaxed back to the ground state. Figure 16.3 a) shows the ground state absorption spectrum and the fluorescence spectrum of riboflavin tetraacetate (RFTA), a versatile investigated organic photocatalyst. The transient differential absorption spectrum of RFTA immediately after excitation at 450 nm is shown in Figure 16.3. b) Since the majority of common pure organic chromophores does not exhibit sup-ps excited state dynamics after excitation into the S_1 state, only the signal of the first excited singlet state is contributing to the measured overall differential transient absorption spectrum at a pump-probe delay of 1 ps. One exception for instance is the organic chromophore xanthone, which undergoes inter-system crossing on the time scale of 1.5 ps. [26] The comparison of Figure 16.3 a) and Figure 16.3 b) clearly shows how the signals of ground state bleach and stimulated emission in the transient differential absorption spectrum correspond to ground state absorption and emission spectra.

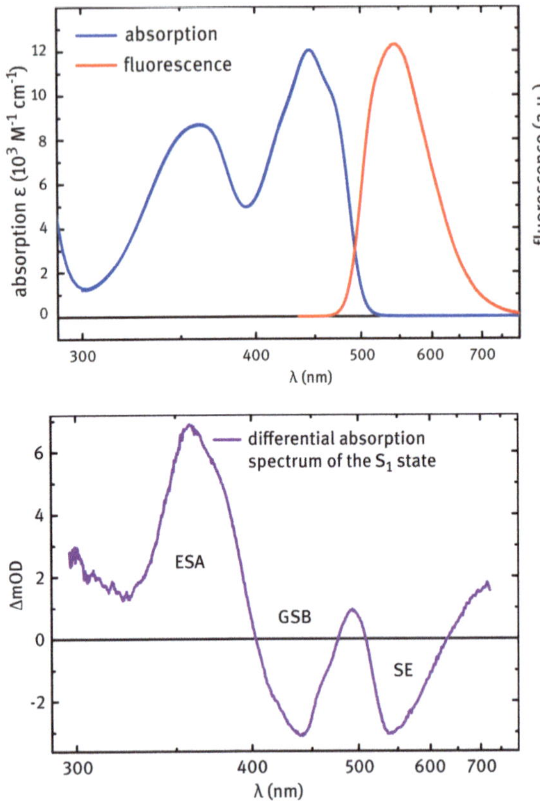

Figure 16.3: UV/Vis absorption and fluorescence spectra of riboflavin tetraacetate (top). Difference spectrum of the S_1 state of riboflavin tetraacetate (bottom) 1 ps after excitation at $\lambda = 480$ nm.

16.2.2 Spectroscopy on the fs to ps time scale

Several technical approaches for time resolved absorption spectroscopy are available to cover a huge range of time scales. The speed of light is the determining physical quantity when choosing the adequate experimental technique for the measurement of time resolved absorption spectra on different time scales. In the 1980s of the last century the first femtosecond lasers based on dye technology became available and allowed seminal investigation into fundamental dynamical processes. Only with the discovery of the modelocked Ti:sapphire, utilizing a solid-state laser material with the necessary broad gain bandwidth, laser pulses with fs duration became readily and reliably accessible. [27, 28] This discovery made the investigation of complex chemical reaction dynamics on the fs time scale possible for many laboratories. The initial work that was relying on the more complex CPM technology was rewarded with a Nobel Prize in chemistry for Prof. Ahmed Zewail in 1999.

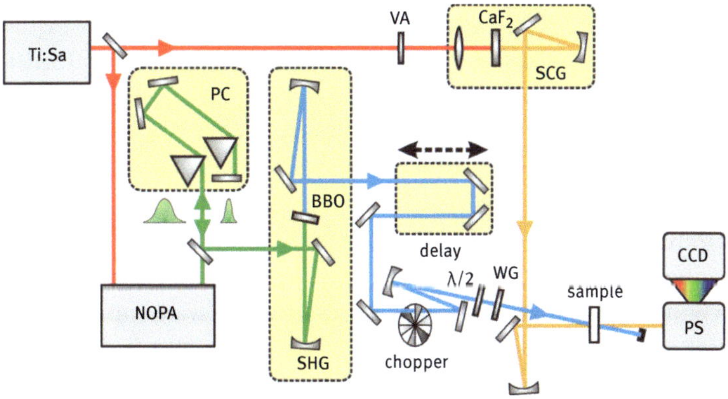

Figure 16.4: Scheme of a state-of-the-art ultrafast pump-probe spectrometer. [29] Tunability of the pump pulse is provided by a noncollinear optical parametric amplifier (NOPA) and second harmonic generation (SHG) in a BBO crystal. The prism compressor (PC) compresses the NOPA output pulse in the time domain. The variable attenuator (VA) adjusts the energy for the super continuum generation (SCG) in the calcium fluoride crystal (CaF_2). The prism spectrometer (PS) detects the spectrally dispersed probe pulse after passing the sample with the help of a CCD array. The combined half-wave plate (λ/2) and wire-grid polarizer (WG) allow adjusting of the intensity of the pump beam.

As mentioned before, time resolved absorption spectroscopy is a pump-probe technique. It does not only require ultrashort laser pulses, also an accurate control of the temporal delay between pump and probe pulse within the sample cell is essential. For the measurement of reaction kinetics on time scales from tens of fs (10^{-15}s) up to a couple of ns (10^{-9}s) the temporal delay can be easily adjusted by changing the optical pathway of pump or probe beam by moving a reflector on a mechanical delay stage. Mechanical delay stages provide accuracy on the µm scale, which is sufficient ($c \cdot 10\,\text{fs} = 3\,\mu\text{m}$) to obtain a time resolution of tens of fs. At the other end, mainly the limitations given by diffractive optics inhibit continuously tunable temporal delays, which are larger than a couple of ns ($c \cdot 10\,\text{ns} = 3\,\text{m}$).

Megerle et al. [29] provide an overview over a state-of-the-art setup for an accurate measurement of transient absorption (TA) spectra on time scales of fs to ns, which is schematically drawn in Figure 16.4. This section only gives a brief introduction into the ultrafast TA spectroscopy.

Modern amplified femtosecond laser sources, which are applied in femtosecond transient absorption spectroscopy, provide ultra short laser pulses centered around 800 nm with a duration of 35–150 fs and pulse energies of about 1 mJ at a repetition rate of a few kHz. Due to the very short duration of the laser pulses, the peak intensities are in the gigawatt regime. By focusing a small part of the pulses into a transparent solid – usually a crystal – a supercontinuum is generated. [31, 32, 39] The spectral shape and the stability of the generated supercontinuum strongly depend on the crystal material. The parametric amplification of a selected region of this supercontinuum allows the direct generation of ultrashort pulses with new wavelengths in the spectral range between 450 nm and the central wavelength of the laser source. Further mixing processes such as second harmonic generation (SHG), sum frequency generation (SFG) and difference frequency generation (DFG) even enlarge the accessible wavelength region down to 189 nm. [33–35] These pulses usually have durations down to below 20 fs, a spectral bandwidth of tenths of THz and a pulse energy of hundreds of nJ to many µJ. They can be used for excitation of a photocatalytic active chromophore at any desired spectral region.

Another part of the laser is focused into a further crystal for supercontinuum generation. This second supercontinuum is used for probing the transient response of the sample. Therefore, an extremely broad and stable white light is desired to gain detailed spectral information within the complete spectral window at once. White light generated in a high quality calcium fluoride (CaF_2) crystal provides a stable probe spectrum within the wavelength region from 290 to 720 nm, when pumped with the fundamental of a Ti:Sa laser at 775 nm. [36] To prevent damage during supercontinuum generation, the crystal has to be kept constantly in motion. Care has to be taken to not negatively affect the shot-to-shot stability of the generated white light by the permanent movement. A translation motion of the CaF_2 crystal within the focal plane provides the best supercontinuum properties for transient absorption spectroscopy. This motion avoids detrimental effects that would be caused by rotation of the crystal. Using a sapphire crystal instead of the CaF_2 crystal provides the advantage that the sapphire crystal is more stable against photo damage and does not need to be translated. However, the generated supercontinuum extends only weakly into the blue (~ 440 nm).

The spectral position of the CaF_2 white light can be shifted to adjacent spectral regions when pumping the supercontinuum generation with the second harmonic or even with an arbitrary chosen wavelength, which is generated by an additional optical parametric amplifier, instead of the fundamental. [32] Figure 16.5 shows

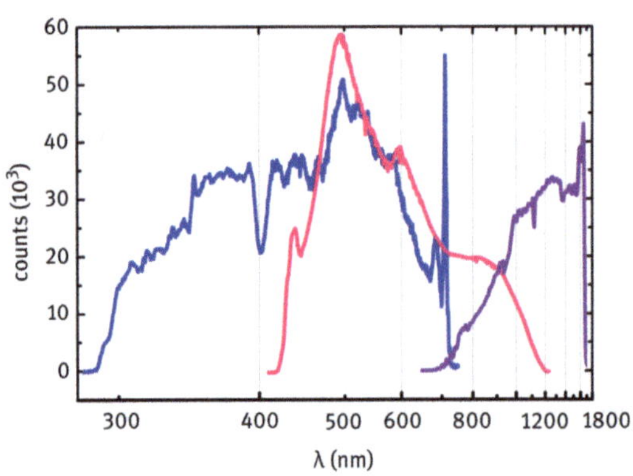

Figure 16.5: Supercontinuum, generated with 775 nm (blue line), with 1100 nm (purple line), and with 2200 nm (violet line) pump pulse in CaF_2. The long wavelength cutoffs are given by spectral filters that suppress the pump pulses.

supercontinua, which are generated in a CaF_2 crystal pumped with femtosecond pulses of different wavelengths. As an alternative to the supercontinuum generation, also a second tunable parametric amplifier can be used for probing the excited sample. The probe pulses, which are generated by a parametric amplifier, are spectrally much narrower than a supercontinuum and, hence, provide less spectral information of the excited state. But in terms of pulse-to-pulse stability and achievable time resolution, this method provides advantages. For recording of broadband transient spectra with this method, it is necessary to repeat the measurement several times with different probe wavelengths to cover the whole spectral range of interest.

Usually in photocatalysis the excited state dynamics in the sub-ps regime only play a minor role, since photocatalysis is a bimolecular reaction and diffusion processes usually occur on the time scale of hundreds of picoseconds or slower. But for the characterization of the underlying sensitizing mechanism, sometimes it becomes inevitable to understand the ultrafast dynamics on the fs time scale. Especially if there is pre-association of catalyst and substrate, dynamics on the femtosecond time scale may be interesting. For instance, electron transfer between the excited state of the photocatalyst flavin and the amino acid tryptophane, both covalently bound to DNA and, hence, in close proximity, occurs on the 100 fs time scale. [38] The achievable time resolution of a transient absorption measurement strongly depends on the duration of the pump pulse, which is used for excitation. The duration of the probe pulse is not significantly affecting the achievable time resolution if the spectral resolution of the detection is accurate enough. [39] Hence, a compressed pump pulse is desired for a high time resolution. The output of a parametric amplifier can be easily compressed by passing a prism compressor, which allows pulse durations in the sub-20-fs regime in the visible and even in the near UV spectral range. To keep the pulse duration short, the use of reflective optics instead of bulk glass material is recommended to avoid additional chirp due to effects of dispersion. [29] Also the use of ultrathin windows for the sample cell minimizes the effect of additional chirp on the pump pulse. Furthermore, effects caused by overlap of pump and probe pulse within the window material, which also negatively affect the achievable time resolution, are minimized.

Time resolved spectroscopic pump-probe techniques require multiple excitation cycles of the probed sample volume to obtain a low-noise transient absorption spectrum with a sufficient density of data points to obtain a high time resolution. When using laser sources with a kHz repetition rate, the duration of a reasonable standard measurement ranges from 20 minutes to 2 hours depending on the averaging and the desired density of data points. Even if the quantum yield of the photocatalytic conversion of the substrate is in the low percent regime, the use of a flow cell is inevitable to keep the reaction conditions constant during the measurement. In addition, effects of the photo degradation of the photocatalyst can be avoided when measuring in a flow cell. A flow cell, which fulfills both the issues of a thin sample size and thin windows of the cell is depicted in Figure 16.6. [29]

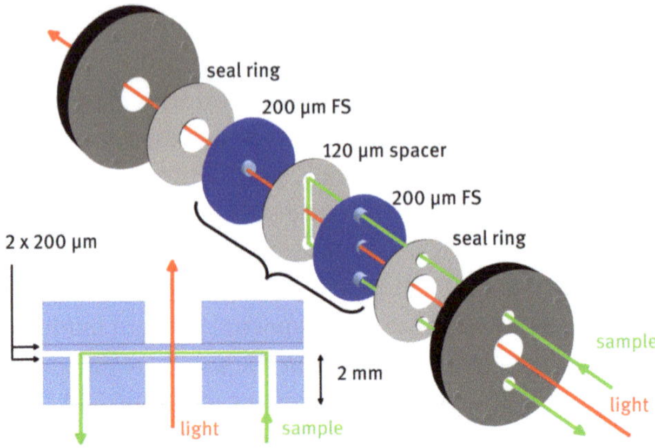

Figure 16.6: Scheme of the custom-made flow cell designed for low dispersion and a small optical path length. Reprinted from reference [29] with permission from Springer.

Another aspect that has to be covered especially when investigating photocatalytic reactions is the total amount of the sample and the effective concentration of the photocatalyst and substrate within the sample solution which is available for the spectroscopic measurement. Due to a limited length of pump-probe overlap within the sample cell, which usually does not exceed ~1 mm, the optical density and; hence, the concentration of the sample solution needs to be sufficiently high. On the one hand, depending on the energy density of the pump pulse, the optical density at the excitation wavelength has to be high enough to generate a reasonable signal of the excited state within the probed spectral region. Also, the optical density within the probed spectrum has to be low enough not to completely absorb any part of the probe spectrum. An optical density of about 0.5 but not higher than 1 is recommended. Especially when product formation or photo degradation of the catalyst plays a major role, one has to be aware of the number of total excitation cycles of one molecule in the sample solution. Performing a transient absorption measurement of a common chromophore ($\varepsilon = 10 \cdot 10^3\,M^{-1}cm^{-1}$) under standard conditions of ultrafast transient absorption spectroscopy ($d = 120\,\mu m$; signal = 10 mOD, spot size of pump pulse: ~100 µm), the amount of excited molecules within a measurement time of one hour is in the range of hundreds of nanomoles. Considering a sample volume in the milliliter regime and a concentration of the chromophore in the mM and sub-mM regime, the excitation probability of one chromophore already comes close to unity.

16.2.3 Spectroscopy on the ns to µs time scale

In the previous section an experimental method for the measurement of photo-induced reaction kinetics on time scales from tens of fs up to a couple of nanoseconds was presented. Since particularly in bimolecular reactions, the relevant intermediate steps may occur on much longer time scales, ultrafast transient absorption spectroscopy on this time scale is no longer sufficient and new experimental approaches are needed. Laser flash photolysis has become the standard tool to gain time resolved spectroscopic data on the ns and μs time scale. But in comparison to the presented ultrafast spectroscopic techniques, it brings some drawbacks: First to our knowledge, common commercially available systems provide a temporal resolution not below 7 ns. Although it is possible to cover the time scale up to ~ 10 ns using a pump-probe setup with a sufficiently long mechanical delay stage, the obtained time constants on the ns time scale are not reliable due to a possible change of overlap of pump and probe beam within the sample cell while moving the delay mirrors. Only a very accurate beam alignment gives access to reliable time constants in the few ns region. Second, in comparison to the presented ultrafast transient absorption setup, the measurement conditions are completely different in classical laser flash photolysis. For instance, the Laser Flash Photolysis Spectrometer LP920 from Edinburgh Photonics [37] works with excitation pulse energies in the millijoule regime whereas the excitation energies in ultrafast transient absorption spectroscopy are many magnitudes smaller and remain in the hundred nanojoule regime. The need for much higher pump energies is due to the lack of spatial coherence of the probe light generated in a pulsed discharge. Therefore a sample volume of about 5 mm diameter and 10 mm length is typically used. The high energy density of the pump pulse and the large optical path length for the probe means the concentration of the chromophore has to be much lower in laser flash photolysis. The concentration is also much lower than typically used for the synthetic investigations. These issues make the transient kinetics obtained by laser flash photolysis difficult to compare directly with transient data measured in ultrafast transient absorption spectroscopy. To avoid all these problems, a transient absorption setup is desirable, which allows the measurement of time resolved reaction kinetics from time scales of femtoseconds up to microseconds within one measurement. In the following chapter an experimental approach is presented which covers all these requirements.

It has been shown that the key element for the access of different time scales is an accurate control of the temporal delay between pump and probe pulse. Whereas an electronic control of the pump-probe delay on time scales below the 100-ps regime is not practical due to physical limitations, in contrast for time scales in the ns regime or moreover in the μs regime, an electronic control of the temporal pump probe delay is easily accessible. An experimental technique for TA spectroscopy on the ns-μs time scale which can be easily integrated into an existing setup for transient absorption measurements on the fs-ps time scale will be presented here.

The basic idea of this approach is to combine two pulsed laser sources. The existing fs laser source is complemented by a second pulsed laser source. To perform pump-probe spectroscopy and to control the temporal delay between pump and probe pulse accurately, both laser sources need to be triggered by a common clock.

Since a mode-locked Ti:sapphire of fiber oscillator cannot be easily triggered, we use the fiber seed laser of our amplified Ti:sapphire system as a master clock. This allows delaying the electronic trigger signal for the second laser source and controlling the temporal delay between the two laser pulses with the help of an electronic delay generator. This is always possible if the

- repetition rates f_1 and f_2 of the two lasers are identical or an integral multiple, and
- Q-switching of at least one laser source can be externally triggered to high precision.

It is also possible to perform pump-probe spectroscopy with two independent pulsed laser sources without synchronization. In this approach the two laser sources, which have to be triggered with slightly different repetition rates, are combined to one pump-probe setup: The first laser generates the pump pulse, the second laser delivers the probe pulses. Depending on the frequency difference $\Delta f = f_1 - f_2$, the pump-probe delay changes from pulse to pulse. Now the pump-probe delay Δt only has to be measured for every excitation cycle and the obtained value has to be assigned to the corresponding measured spectrum. [40] Sorting the recorded spectra according to their assigned pump-probe Δt delay results in a full transient spectrum of the investigated photo-induced reaction. We will not expand on this method, but would like to point out, that it is pursued by others. [41] In this section we will only focus on transient absorption techniques where the temporal pump-probe delay can be directly controlled. In terms of time resolution and duration of a single measurement, this method is clearly favorable.

To allow probing the excited sample over a broad spectral range at once, a fs laser system is suggested for the generation of a spectrally broad white light continuum, which can be used as a probe pulse. Such a femtosecond continuum is spatially coherent and has a sub-ps duration. It is therefore superior to any other broadband light source for transient absorption spectroscopy in our opinion. It allows measurements in an extremely small sample volume and with very low optical load of the sample. Since amplified fs laser sources are anyway available in many laser laboratories due to their use as a fs-ps pump probe spectroscopy and in view of the superior continuum generation with fs laser pulses in bulk materials, the astonishing situation arises, that the femtosecond system is better used for the detection than for the pump, i. e. the triggering of the reaction cycle.

It is highly desirable that the temporal window of the ns-μs transient absorption setup directly attaches to the end of the time windows of the fs-ps pump setup without any gap. This requires two things. The temporal duration of the additional laser source has to be in the low ns-regime and the electronic jitter of the second laser should be sub-ns. Further advantages of the application of a ns laser source are the comparatively low investment costs – as compared to a second fs system – and the availability of easy to handle optical parametric oscillators (OPO) for ns pulses. This provides an easy tunability of the excitation wavelength within a huge spectral

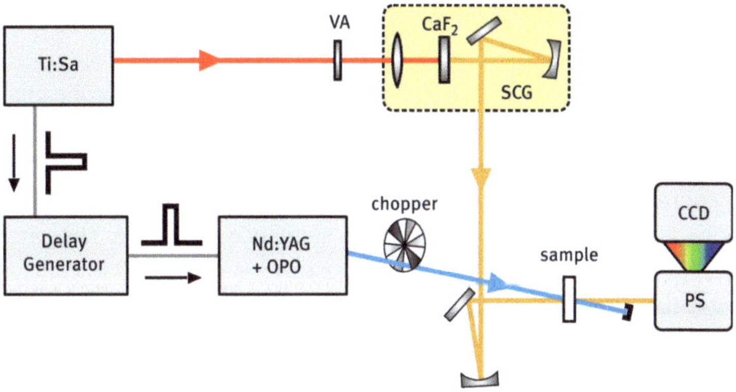

Figure 16.7: Simplified scheme of pump probe setup for the measurement of TA spectra on time scales from ns to hundreds of μs. The fs pulse from the Ti:Sa system is still used for super continuum generation (SCG) in the calcium fluoride crystal (CaF_2). The Nd:YAG laser provides ns pump pulses and is electronically synchronized with the Ti:Sa laser. Parametric amplification in an optical parametric oscillator (OPO) provides tunability of the pump wavelength. The prism spectrometer (PS) detects the spectrally dispersed probe pulse after passing the sample.

range. The diode pumped solid state laser NT242 from the Lithuanian manufacturer EKSPLA [37], which has an easy to handle OPO integrated within the laser housing, fulfills all the above requirements. The following discussion of a suggested setup for ns-μs TA spectroscopy is mainly based on this laser.

By the use of the fs seed laser as a master clock, only the precision of the delay generator and the trigger jitter of the ns system contribute to the achievable stability. Up-to-date electronic delay generators, e.g. the DG535 from Stanford Research Systems [37] provide temporal accuracy of better than 100 ps. The NT242 has a measured trigger jitter of below 200 ps. Both values are well below the measured pulse length of the ns OPO of about 2.5 ns. Since the DG645 delay generator provides an Ethernet and a RS232 interface, it can be controlled by the computer used for data recording. This also allows a simple implementation of the communication software into the existing LabVIEW routines, which are already in use for the acquisition of transient spectra. [29] A simplified scheme of the combined setup is given in Figure 16.7.

The combination of these components allows the measurement of transient absorption spectra up to pump-probe delays of some hundreds of μs with a time resolution below 3 ns. A good estimate of the over-all time resolution, which was achieved during a measurement, is provided by the full width half maximum (FWHM) of the instrument response function. The width of the instruments response function (IRF) can be obtained from fitting the convolution of a Gaussian with the sum of the exponential functions on the measured decay kinetics. If the slowest kinetics of a sample are well below 1 ns, as this is often found for excited state quenching by forward and back electron transfer [42], the IRF can be directly read off the raw data.

Within the presented setup, the fs laser source, which generates the probe continuum, is triggered by the signal of its own seed laser. The ns laser source, which is used for excitation, is triggered by the same laser clock, but the trigger signal first passes the delay generator, which delays the signal for a variable delay time. However, for pump probe spectroscopy, a reversed pulse sequence is needed: The sample first has to be excited and then to be probed. To fix this issue, one has to consider some details regarding the trigger sequence. Consequently, the measurement of one transient spectrum requires two cycles of the laser clock. After passing the delay generator, the first trigger signal releases the ns pump pulse whereas the fs probe pulse is released not until the next trigger pulse. Since two cycles of the laser clock are necessary to generate one transient spectrum, both the jitter of the seed laser and the jitter of the electronic control of the ns laser system contribute to the achievable over-all time resolution. In practice we find that the seed laser stability is so high that it actually does not contribute.

Another issue regarding the triggering sequence has to be solved: When measuring transient absorption spectra, usually transient spectra with pump-probe delays smaller than zero are recorded to obtain a flat base line, which serves as reference. For the measurement of transient spectra with negative pump-probe delays $\Delta t < 0$, the delay generator would need to generate a delay Δt_{DG} of the trigger signal, which is larger than the inverse repetition rate $1/f$ of the signal, since the probe pulse has to be released first. However, the used delay generator does not provide this functionality. To solve this problem and to access negative pump-probe delays, a second delay generator is introduced, which generates a fixed pre-delay Δt_0. This then allows the measurement of transient spectra of negative pump-probe delays at the expense of clipping the temporal pre-delay Δt_0 at the end of the observable time window.

The final issue, which limits the window of accessible pump-probe delays is that the EKSPLA NT 242 requires two trigger pulses: the first one for triggering the Nd:YAG rods for pumping the laser cavity and the second one for Q-switching and outcoupling the laser pulse. For an efficient amplification within the laser medium of the EKSPLA NT 242, a temporal delay of ~ 200 µs between the first and second trigger pulse is required. This presently leads to a clipping of the observable time window by 200 µs. It would also be possible to circumvent this problem by a further computer controlled delay generator.

A more fundamental issue that should be noted is the exchange of the sample solution within the sample cell. Since the primary aim of photocatalysis is the conversion of a substrate into a product, every excitation pulse leads to a certain amount of product formation, as discussed in the previous section. However, the measurement of reliable transient absorption spectra, which requires the acquisition of a huge number of data points, is only feasible under stable conditions within the probed sample volume. Therefore a constant exchange of the excited sample is inevitable. Since the flow cells with ultrathin glass windows, which are necessary to achieve a high time resolution in fs measurements, are not always easy to handle, the use of commercially available flow cells (e.g. Hellma 137-QS [37]) with path lengths of 1 mm or 2 mm are suggested for TA measurements on the ns-μs time scale. This also allows the measurement at somewhat reduced concentrations that are comparable to the ones used in synthetic work.

For recording a transient spectrum of a selected pump-probe delay Δt, at least some hundred excitation cycles are necessary to obtain a low-noise difference absorption spectrum. Due to a shot-to-shot referencing combined with the repetition rate of f = 1 kHz of both the fs and the ns laser source, the temporal delay between two consecutive excitation pulses is Δt = 2 ms. The typical volume current within the circuit of the sample solution, which is pumped by a micro annular gear pump (HNP Mikrosysteme GmbH, model mzr-2921 [37]), is 18 ml/min. Considering the dimensions of the cross-sectional area of the applied flow cell (1 mm × 10 mm) and the volume current, the flow speed of the sample solution within the flow cell can be estimated to be 30 μm/ms. With a spot size of the excitation pulse of ~ 300 μm, roughly 20 % of the excited sample volume is replaced between two consecutive excitation pulses. Assuming a model reaction in which a substrate is converted into an equally absorbing product with a reaction quantum yield of 50 %, the ratio of the signal contribution of the formed product molecules to the over-all signal is 5 % when neglecting diffusion processes, which would lead to a further decrease of the product contribution. The excitation probability was assumed to be 10 %, considering the usual excitation energy densities combined with a typical extinction coefficient of ~ 10,000 $M^{-1}cm^{-1}$ of the chromophore. Our experience is that in the particular flow conditions, pump light focus diameter and possibly a reduced laser repetition rate have to be carefully considered for any measurement that pushes the limits of detection window and sensitivity.

To summarize: For reliable time resolved spectroscopic measurements of photocatalytic processes, performing the measurements in a flow cell is inevitable. If the presented prerequisites are met, commercially available standard components are sufficient for the measurement of time resolved absorption spectra. But if, for example, highly absorbing products in large amounts are produced within the observed photochemical reaction, the presented system has to be modified. One possibility is the use of flow cells with a small cross-section area, either obtained by using a thin optical path length or a narrow flow channel, to increase the flow speed within the flow cell.

To demonstrate the versatility of the presented setup for application in photocatalysis, the flavin sensitized photo-oxidation of methoxy benzyl alcohol to its corresponding aldehyde is chosen as a model reaction. [43] Figure 16.8 shows that the photocatalytic process comprises several intermediate processes and species, which occur on largely different time scales. The time constants, which are applied to perfectly describe the decay dynamics, span five magnitudes of time and range from

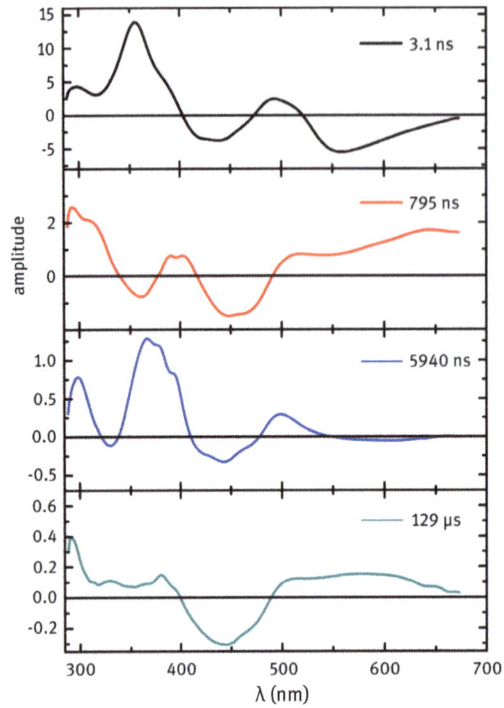

Figure 16.8: Decay associated difference spectra $A_i(\lambda)$ (DADS), obtained from a single TA measurement, which was performed with the presented ns-µs setup of the flavin sensitized oxidation of methoxy benzyl alcohol to the corresponding aldehyde.

few ns to hundreds of µs. All occurring time constants can be obtained within just one measurement. A detailed description of the spectroscopic investigation of flavin catalysis and the underlying intermediate steps can be found in reference [43].

16.2.4 Rate models and the determination of the species associated spectra of the intermediate states

Time resolved differential absorption spectroscopy both on the fs-ps and on the ns-µs time scale generally provides data matrices which describe the change of optical density ΔOD with respect to the ground state, depending on pump-probe delay Δt and spectral position λ. These data matrices $\Delta OD(\lambda, \Delta t)$ contain spectral information of the intermediate states and species. But generally this spectral information is not directly accessible. Data processing is required to gain this spectral information. The species associated spectrum $\varepsilon_i(\lambda)$ (SAS) of an intermediate state or a species allows their identification and the temporal evolution renders the kinetics of the process.

After the measurement the obtained data matrix has to be chirp-corrected in the case of sub-ps time resolution. The chirp correction is necessary since the different colors in the probe pulse arrive at slightly different times in the sample due to group velocity dispersion. This chirp of the broadband probe pulse generates a wavelength dependent time-zero. After chirp correction a sum of time dependent exponential fit functions can be applied to describe the decay dynamics of each individual spectral trace.

$$\Delta OD(\lambda,t)=\sum_i A_i(\lambda)\cdot \exp\left(\frac{-t}{\tau_i(\lambda)}\right) \tag{16.2}$$

Usually the amplitudes $A_i(\lambda)$ and the time constants $\tau_i(\lambda)$, which are both wavelength dependent, are the fit parameters. More sophisticated fit functions can be used to describe more complex processes such as diffusion, which will be discussed in the following section. The data evaluation on the basis of single spectral traces has a number of disadvantages:

- Inferior signal to noise ratio
- Effects of spectral diffusion, e.g. caused by solvation or vibrational cooling result in additional non-exponential contributions to the decay signal
- Contributions of similar time constants cannot be clearly distinguished
- No complete spectral information of intermediate states is obtained

A more powerful method of data evaluation, which avoids these drawbacks, is the so-called global data analysis or global fit. No longer a set of wavelength dependent, individual time constants $\lambda\tau_i(\lambda)$ is applied to the data matrix $\Delta OD(\lambda, t)$ to describe the decay dynamics of each single time trace independently from each other. Moreover, global data analysis tries to apply exponential terms with global wavelength independent time constants τ_i with a wavelength dependent amplitude $A_i(\lambda)$ to describe the decay dynamics of the full spectral window of the data matrix $\Delta OD(\lambda, t)$:

$$\Delta OD(\lambda,t)=\sum_i A_i(\lambda)\cdot \exp\left(\frac{-t}{\tau_i}\right). \tag{16.3}$$

Since the amplitudes $A_i(\lambda)$ describe the spectral composition of each decay with the time constant τ_i, the obtained $A_i(\lambda)$ are called decay associated difference spectra (DADS). Fita et al. [44] suggest a formalism for this kind of data analysis. But before applying the global fit routine on transient data, it is recommended to rescale the wavelength scale from a scale linear in $\Delta\lambda$ to a scale linear in ΔE to avoid the enhancement of the weight of the long-wave components. The number of the applied time constants τ_i is directly linked to the number of processes which occur within the observed time window. Every process which populates an intermediate or final product state leads to an additional time constant τ_i in the global fit. Intermediate states which are not significantly populated, e.g. if the depopulating rate k_{out} of the corresponding state is much higher than the populating rate k_{in}, do not contribute significantly to the global fit. But these states are generally not accessible in TA spectroscopy as well.

Now the challenge is to extract the species associated spectra (SAS) or the species associated difference spectra (SADS) from the DADS. But first, the difference between SADS and SAS has to become clear. The SAS describe the true absorption spectrum of a certain intermediate state or species whereas the SADS only describe the spectral changes of this intermediate state with respect to the ground state absorption spectrum and the fluorescence spectrum of the chromophore. But the calculation of the SAS first requires the extraction of the SADS from the DADS, which are directly obtained as a result of the global fit routine. The conversion from SADS to SAS is very easy: It just requires adding the wavelength dependent absorption spectrum abs(λ) and the wavelength dependent fluorescence spectrum fluo(λ) of the excited chromophore to the SADS as shown in Equation 16.4, where α and β are scaling factors.

$$\varepsilon_1(\lambda) = SAS_1(\lambda) = SADS_1(\lambda) + \alpha \cdot abs(\lambda) + \beta \cdot fluo(\lambda) \cdot \lambda^4 \qquad (16.4)$$

The multiplication of the wavelength dependent fluorescence intensity by λ^4 is required due to the conversion factor of the Einstein coefficient between spontaneous and stimulated emission and due to rescaling the fluorescence intensity per wavelength interval $\Delta\lambda$ to the corresponding quantity per frequency interval $\Delta\nu$. [45] One has to keep in mind that in a fluorometer, the spectral shape of the spontaneous emission is measured whereas in TA spectroscopy the spectral shape of the stimulated emission is detected.

The difference between DADS $A_i(\lambda)$, SADS and SAS $\varepsilon_i(\lambda)$ can be well explained with the following simplified example: A spectrally well characterized chromophore is optically excited. The well confined absorption spectra $\varepsilon_i(\lambda)$ of the ground state, the first excited singlet state (S_1) and the lowest triplet state (T_1) are assumed to be known and are depicted in Figure 16.9. Fluorescence does not play a role for the sake of simplicity.

After excitation of the chromophore into the S_1 state, let there be two competing processes, which depopulate this state: internal conversion with a rate of $k_{IC} = (100\ ps)^{-1}$ and inter-system crossing with a rate of $k_{ISC} = (100\ ps)^{-1}$ as well. The internal conversion process converts the population from the S_1 state back to the ground state and the inter-system crossing process converts the population from the S_1 state to a third state, the triplet state T_1. Due to both processes depopulating the S_1 state, the overall rate k_{S1} which describes the depopulation of the S_1 state can be written as the sum of k_{IC} and k_{ISC}. Therefore, the life time of the S_1 state is $\tau_{S1} = 50$ ps.

$$k_{S1} = (\tau_{S1})^{-1} = k_{IC} + k_{ISC} = (50\ ps)^{-1} \qquad (16.5)$$

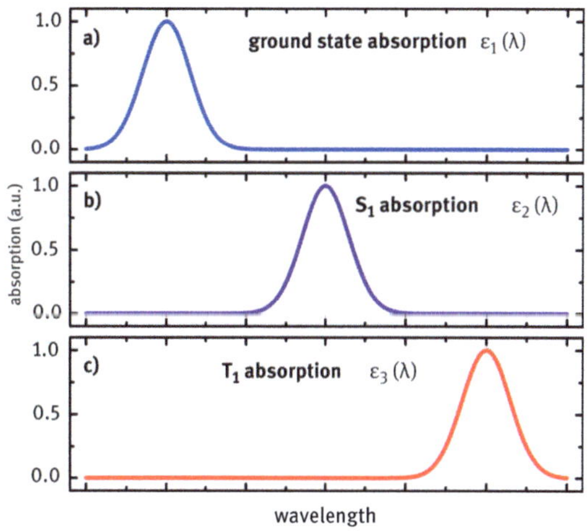

Figure 16.9: Optical absorption spectra of ground state (a), excited singlet state (b) and lowest triplet state (c) of an example chromophore. Since the spectra ε(λ) describe the wavelength dependent absorption properties of each state they are named species associates spectra (SAS).

Since k_{IC} and k_{ISC} are equal, the branching ratio between internal conversion and inter-system crossing is 1:1. This means that for each excited chromophore in the S_1 state both the probability to decay back to the ground state and the probability to decay into the long-lived triplet state T_1 are 50 %. The Jablonski diagram in Figure 16.10 provides a schematic overview of the excited state dynamics.

The species spectra of all populated excited states are contributing in different ways to the transient difference absorption spectrum, and generally, the spectra of these states do not directly appear in the measured data set. Immediately after excitation of the model chromophore, there are only two signal contributions: The depopulation of the ground state leads to the negative signal of the ground state

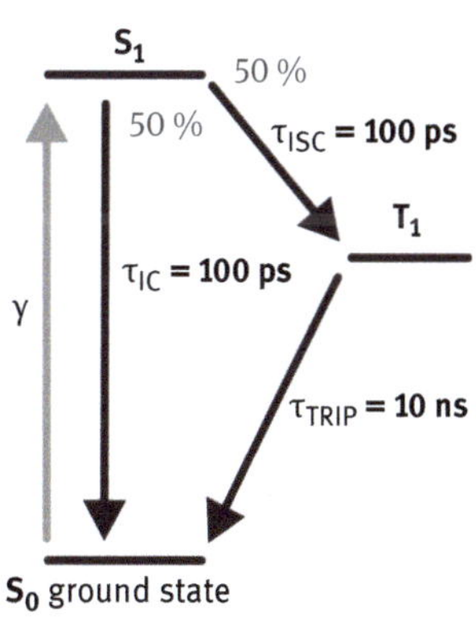

Figure 16.10: Jablonski diagram of the example chromophore

bleach (GSB) whereas the population of the S_1 state results in a positive signal of excited state absorption (ESA) due to the possible excitation into a higher state. The simulated transient signal directly after excitation is shown in Figure 16.11 a).

With the beginning of the inter-system crossing (ISC) process on the time scale of 100 ps, the transient signal now contains contributions of all three states and the spectra of the contributing individual states are no longer directly accessible. Applying the global fit to the full data matrix provides a fit described by two time constants: $\tau_1 = 50$ ps, the life time of the S_1 state and $\tau_2 = 10$ ns, the life time of the T_1 state with the corresponding amplitude spectra $A_1(\lambda)$ and $A_2(\lambda)$. These spectra $A_1(\lambda)$ and $A_2(\lambda)$ are the so-called DADS and are shown in Figure 11 b) and c), respectively. The amplitude spectrum $A_1(\lambda)$ shows, that the S_1 absorption in the middle decays and simultaneously both the ground state and the T_1 state are populated. Since the quantum yield for the triplet state T_1 is only 50 %, its contribution to $A_2(\lambda)$ is only half as intense as the amplitude of the S_1 decay in $A_1(\lambda)$.

The whole population of the T_1 state decays back to the ground state, which can be seen, by comparing the contributions of the amplitudes of the decay of the T_1 ESA and the recovery of the GSB in $A_2(\lambda)$. The full spectral information of the intermediate states now can be extracted from these DADS.

When understanding the relation between DADS, SADS and SAS of the previously presented model reaction, now a more formal treatment can be introduced. It can be seen by comparison of Figure 16.9 and Figure 16.11, that the DADS $A_i(\lambda)$ are composed of a linear combination of the SAS $\varepsilon_i(\lambda)$, including the ground state absorption spectrum $\varepsilon_{S0}(\lambda)$. Consequently, the SAS can be obtained as linear combinations of the DADS. The inverse of the relative yield of the corresponding state deter-

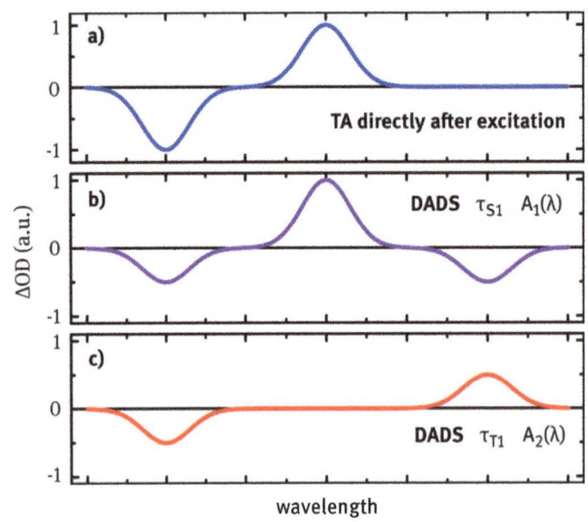

Figure 16.11: Transient difference absorption spectrum directly after excitation (a), $A_1(\lambda)$, DADS of the $\tau = 50$ ps component, that describes the decay of the S_1 state and the population of the T_1 state (b) and $A_2(\lambda)$, DADS of the $\tau = 10$ ns decay of the triplet state (c)

mines the weighting coefficient of the DADS. It can be written as fraction of the involved rates k_i.

$$\varepsilon_{S1}(\lambda) = A_{50\,ps}(\lambda) + A_{10\,ns}(\lambda) + \varepsilon_{S0}(\lambda) \tag{16.6}$$

$$\varepsilon_{T1}(\lambda) = \frac{k_{ISC} + k_{IC}}{k_{ISC}} A_{10\,ns}(\lambda) + \varepsilon_{S0}(\lambda) = 2 \cdot A_{10\,ns}(\lambda) + \varepsilon_{S0}(\lambda) \tag{16.7}$$

A more general treatment of the data analysis is given in the following section. To introduce a mathematical description the use of a rate model is recommended. A rate model is based on the assumption, that an arbitrarily chosen intermediate state *i* is populated from a state *i-1* with rate k_1 and decays with rate k_2 into subsequent intermediate states or the final product state. This allows describing the temporal behavior of the population of the observed state $n_i(t)$ by the following differential equation:

$$\frac{dn_i}{dt} = \dot{n}_i = +k_1 \cdot n_{i-1}(t) - k_2 \cdot n_i(t). \tag{16.8}$$

For the full description of a rate model, involving more than one state, the differential equations for each state, which are no longer independent from each other, can be written in matrix formalism, as shown in Equation 16.9.

$$\begin{pmatrix} \dot{n}_1 \\ \dot{n}_2 \\ \dot{n}_3 \\ \vdots \end{pmatrix} = \begin{pmatrix} -k_{11} & 0 & 0 & \cdots \\ +k_{21} & -k_{22} & 0 & \cdots \\ +k_{31} & +k_{32} & -k_{33} & \cdots \\ \vdots & \vdots & \vdots & \ddots \end{pmatrix} \cdot \begin{pmatrix} n_1 \\ n_2 \\ n_3 \\ \vdots \end{pmatrix}. \tag{16.9}$$

If there is no repopulation from lower to higher states, which means that the matrix A is a lower triangular matrix, the diagonal elements $k_{ii} = (\tau_i)^{-1}$ represent the decay rate of state i and are directly obtained as the inverse of the time constants from the global fit. By solving the differential equation system 16.9 one can obtain mathematical terms for $n_i(t)$ which describe the population dynamics of each state i.

How to solve such a differential equation system is demonstrated by the following example of a sequential model involving three states. A scheme of the process is provided in Figure 16.12.

The population dynamics $n_i(t)$ of each state are described by a differential equation, which relates the temporal behavior of state i $\dot{n}_i(t)$ with its current population $n_i(t)$.

$$\begin{aligned} \dot{n}_1 &= -k_1 \cdot n_1 \\ \dot{n}_2 &= +k_1 \cdot n_1 - k_2 \cdot n_2 \\ \dot{n}_3 &= +k_2 \cdot n_2 \end{aligned} \tag{16.10}$$

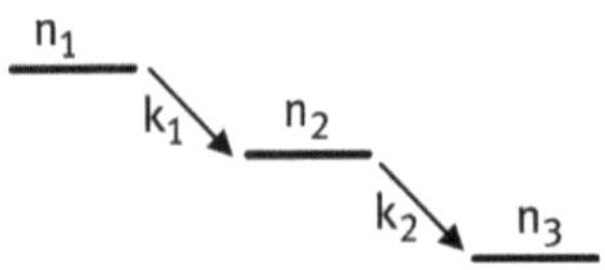

Figure 16.12: Sequential three state model: State 1 with population $n_1(t)$ decays with rate k_1 to state 2. State 2 with population $n_2(t)$ decays back to the ground state with rate k_2.

Now, according to Equation 16.9, a matrix can be introduced for an easier treatment of the differential equation system:

$$\begin{pmatrix} \dot{n}_1 \\ \dot{n}_2 \\ \dot{n}_3 \end{pmatrix} = A \cdot \mathbf{n} = \begin{pmatrix} -k_1 & 0 & 0 \\ +k_1 & -k_2 & 0 \\ 0 & +k_2 & 0 \end{pmatrix} \cdot \begin{pmatrix} n_1 \\ n_2 \\ n_3 \end{pmatrix}. \tag{16.11}$$

Since the introduced matrix A is a triangular matrix, the diagonal entries of A give the multiset of eigenvalues of A.

$$\lambda_1 = -k_1 \qquad \lambda_2 = -k_2 \qquad \lambda_3 = 0 \tag{16.12}$$

The eigenvector $\mathbf{x}_i$ of each eigenvalue λ_i is required for the further treatment and solving of the differential equation system:

$$\mathbf{x}_1 = \begin{pmatrix} 1 \\ \dfrac{k_1}{-k_1 + k_2} \\ \dfrac{-k_2}{-k_1 + k_2} \end{pmatrix} \quad \mathbf{x}_2 = \begin{pmatrix} 0 \\ 1 \\ 1 \end{pmatrix} \quad \mathbf{x}_3 = \begin{pmatrix} 0 \\ 0 \\ 1 \end{pmatrix}. \tag{16.13}$$

Finally, the information, which is needed to solve the differential equation system, is complete. An exponential approach with free parameters c_i solves the differential equation system.

$$\mathbf{n}(t) = c_1 \cdot \mathbf{x}_1 \cdot \exp(\lambda_1 \cdot t) + c_2 \cdot \mathbf{x}_2 \cdot \exp(\lambda_2 \cdot t) + c_3 \cdot \mathbf{x}_3 \cdot \exp(\lambda_3 \cdot t) \tag{16.14}$$

The parameters c_i now only have to be adapted to the boundary conditions. Reasonable boundary conditions for the temporal behavior of the intermediate states of a sequential three state model in transient absorption spectroscopy are:

$$n_1(0) = n_0 \quad n_2(0) = 0 \quad n_3(0) = 0 \quad n_i(t > 0) \geq 0 \quad n_3(t \to \infty) = n_0. \tag{16.15}$$

The solution for the excited state dynamics of the presented sequential three state model, which matches the given boundary conditions, is given in Equation 16.16.

$$\begin{aligned}
n_1(t) &= n_0 \cdot \exp(-k_1 \cdot t) \\
n_2(t) &= n_0 \cdot \left[\frac{k_1}{-k_1 + k_2} \exp(-k_1 \cdot t) - \frac{k_1}{-k_1 + k_2} \exp(-k_2 \cdot t) \right] \\
n_3(t) &= n_0 \cdot \left[\frac{-k_2}{-k_1 + k_2} \exp(-k_1 \cdot t) + \frac{k_1}{-k_1 + k_2} \exp(-k_2 \cdot t) + 1 \right]
\end{aligned} \tag{16.16}$$

One sees readily that the population of the initially excited state 1 decays with a single exponential rate as assumed. State 2 gets populated with k_1 and depopulated with k_2. For the final state 3 both rates are important again, even though there is no direct path from state 1 to state 3. This is only mediated by state 2, but by its involvement, the rate k_1 also becomes important for the population of the final state.

For processes with more than three contributing excited states, which have to be described by a more complex system of rate equations, applying a numeric solution might be more practical to gain reasonable results. Since it now should be clear how to obtain solutions for a rate equation system, the functions $n_i(t)$ describing the excited state dynamics can be used for the interpretation of the results of the global fit. The transient data matrix $\Delta OD(\lambda, t)$ can be written as the time dependent population dynamics $n_i(t)$ of state i multiplied by the corresponding SAS $\varepsilon_i(\lambda)$, which describes the wavelength dependency.

$$\Delta OD(\lambda, t) = n_1(t) \cdot \varepsilon_1(\lambda) + n_2(t) \cdot \varepsilon_2(\lambda) + n_3(t) \cdot \varepsilon_3(\lambda) + \ldots + n_0(t) \cdot \varepsilon_{S0}(\lambda) \tag{16.17}$$

Equation 16.17 contains the population dynamics of the intermediate states $n_i(t)$. The next step towards obtaining the SAS $\varepsilon_i(\lambda)$ of state i, is equalizing the whole expression with the result of the global fit (Equation 16.3).

$$n_1(t) \cdot \varepsilon_1(\lambda) + n_2(t) \cdot \varepsilon_2(\lambda) + \ldots + \varepsilon_{S0}(\lambda) = A_1(\lambda) \cdot e^{-t/\tau_1} + A_2(\lambda) \cdot e^{-t/\tau_2} + \ldots \tag{16.18}$$

Now, the population dynamics $n_i(t)$ of each state, which can be calculated by solving the differential equation system, given in Equation 16.9, has to be inserted into Equation 16.18. The independent population dynamics $n_i(t)$ are written as a sum of exponential terms with individual time constants τ_i. The last step is to sort the different time dependent exponential terms on the left side of Equation 16.18. Now the relation between SAS $\varepsilon_i(\lambda)$ and DADS $A_i(\lambda)$ can be directly seen by comparison of the wavelength dependent pre-factors of the individual time dependent exponential terms of both sides of the equation.

The global fit is one of the most powerful evaluation methods of transient absorption dynamics. But, nevertheless, before it provides confidential results, it requires either a detailed knowledge of the spectra of all occurring intermediate states or a detailed knowledge of the reaction kinetics of each intermediate state. It is not sufficient just to know the decay dynamics of each state – also a majority of

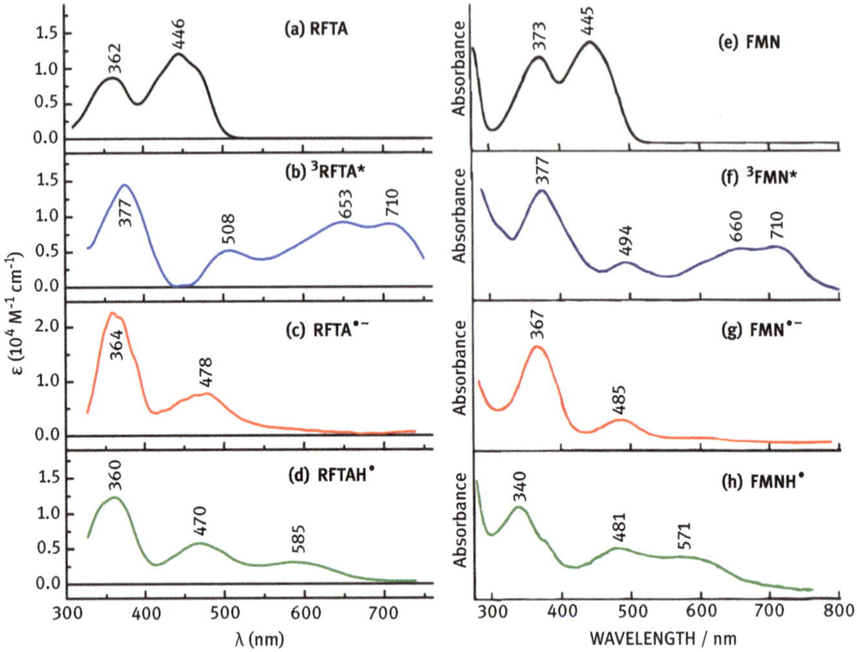

Figure 16.13: Absorption spectra of the ground state RFTA (a) and (e), the triplet state ^{3}RFTA (b) and (f), the radical anion state RFTA$^{\bullet-}$ (c) and (g) and the neutral protonated state RFTAH$^{\bullet}$ (d) and (h) of riboflavin tetraacetate. The spectra on the right side are reprinted from reference [47] with permission from Elsevier, the spectra on the left side are extracted from the DADS, which were obtained by applying a global fit on the data matrices ΔOD(λ, t) of the time-resolved spectroscopic investigation of the photocatalytic oxidation of methoxy benzyl alcohol. [43, 48]

the off-diagonal elements of the matrix in Equation 16.9 has to be identified. Estimating the quantum yield of the intermediate states, for instance with the help of the time dependent intensity of the ground state bleach, may help determining the off-diagonal elements.

Since there have been various further approaches developed over the years to interpret the outcome of the transient spectroscopy, and since the investigated problems vary widely, the interested reader is referred to a more general treatment of data analysis given in a recent review. [46] There it is nicely explained, whichever model assumption is made or not made, one has to input into the various data treatments.

To demonstrate the application of the global fit, the spectroscopic investigation of the flavin sensitized photo-oxidation of methoxy benzyl alcohol (MBA) is used as a model reaction. After excitation and inter-system crossing (ISC) of the riboflavin tetraacetate (RFTA) chromophore, a first electron transfer from the MBA to the triplet state ^{3}RFTA leads to the radical anion RFTA$^{\bullet-}$. The MBA is only a weak chromophore and renders no contribution to the transient absorption signal within the probed

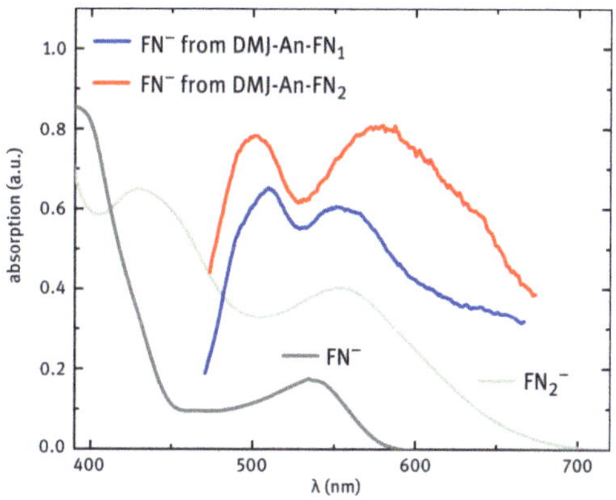

Figure 16.14: Spectra of the fluorenone anion (dark grey line), the anion of two covalently linked fluorenones (bright grey line), and the SAS of an intramolecular charge transfer state of a complex donor-bridge-acceptor molecule (blue and red line), where the electron is temporarily located on the fluorenone moiety of the molecule. [10]

spectral range. The next step of the reaction is the protonation of the radical. [47] The extracted SAS of the occuring intermediate states of the flavin photocatalyst and their assignment to known spectra [47] of the different redox and protonation states of flavin are shown in Figure 16.4.

Another field of application is the combination of global fit analysis with the methods of spectroelectrochemistry. Especially when charge transfer processes between two chromophores, either of intermolecular or intramolecular character, are suspected to occur within the investigated reaction, the combination of these experimental techniques can help to recognize and allocate these intermediate states. One example of the successful combined application of the presented data analysis and spectroelectrochemistry is given in reference [49]. Here, the intramolecular charge transfer of a donor-bridge-acceptor molecule was investigated. The main aspect of the study was to proof an incoherent hopping mechanism, which transports an electron from the donor site of the molecule to its acceptor site. Incoherent hopping includes that the charge transfer state is temporarily located on the bridge molecules, which link electron donor and acceptor. The bridge consists of alternatively one, two or three covalently linked fluorenone molecules. The spectra of the radical anion of a single fluorenone and of two covalently linked fluorenone molecules were determined with the help of spectroelectrochemistry. The spectra of the radical anions exhibit a strong absorption band centered around 550 nm and 575 nm, respectively. The decomposition of the transient spectra with the help of the global data analysis clearly showed a significant spectral contribution of the fluorenone radical anion absorption to the intermediate spectra and proved the proposed reaction mechanism via incoherent hopping. The spectra are shown in Figure 16.14.

16.3 Diffusion limited reactions

16.3.1 Diffusion limited excited state quenching with time dependent reaction rate

For intermolecular photo-sensitization processes, a close proximity between donor and acceptor molecule is inevitable due to a strong distance dependence of the common sensitizing mechanisms. In general, there are two different mechanical approaches to allow photo-induced sensitization: diffusive sensitization after photo-excitation and complex formation before photo-excitation. This chapter will describe the challenges in transient absorption spectroscopy which are related to the investigation of diffusion controlled sensitizing mechanisms.

A very widely accepted principle of the data analysis in transient absorption spectroscopy is the exponential decay modeling. It is based on the assumption, that the number of excited molecules which decay to a lower state within an infinitesimal time period *dt* is proportional to the total number of molecules which are momentarily in the observed excited state. However, to describe the ultrafast population dynamics on the ps (10^{-12}s) time scale of diffusion limited reactions, the exponential decay modeling is inappropriate and a more sophisticated model is required. The reaction rate k, which is a constant in exponential decay modeling, turns into a time dependent quantity k(t) when describing diffusion limited dynamics. The time dependency of k(t) is usually negligible for reaction dynamics on the ns time scale but causes major deviations from strongly monoexponential decay dynamics on the low ps time scale.

The time dependency of the reaction rate can be justified qualitatively: Directly after excitation, there is a certain average distance d_0 between excited sensitizer molecule and quencher molecule, which only depends on the initial quencher concentration $[Q_0]$. The quencher molecules are randomly distributed around the sensitizer molecules if there are no effects of attracting or repulsive forces. This average distance d_0 is the determining quantity of the reaction rate k since, due to the laws of diffusion, the average distance is related to the average time until collision between two molecules occurs. The values of d_0 are typically between 1 and 5 nanometers when using quencher (substrate) concentrations between 10 and 1000 mM which are common for photocatalytic processes. Electron transfer reactions and energy transfer mechanisms require a sub-nm distance to work efficiently.

To quantitatively model the time dependency of the reaction rate k(t) the following ideas are used: After excitation the sensitizer molecules with a quencher molecule in close proximity are quenched first whereas excited sensitizer molecules without a quencher molecule in close proximity remain in the excited state. These excited chromophores are able to decay spontaneously back to the ground state within the life time τ of the excited state. These effects lead to an increase in the average distance d_0 between excited sensitizer molecules and quencher and, hence, to a decrease in reaction rate k. A detailed derivation of an applicable mathematical

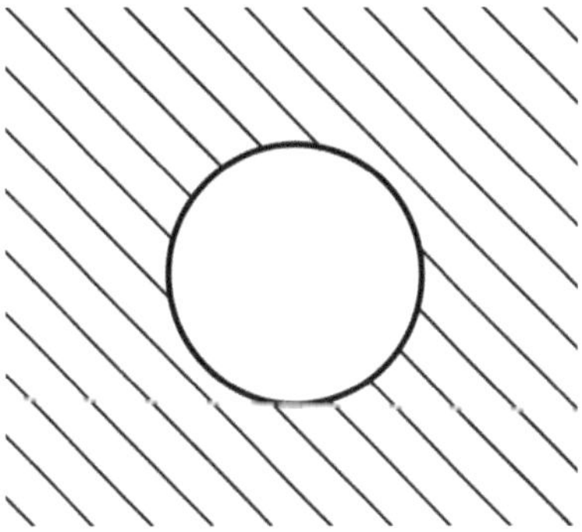

Figure 16.15: Spherical sensitizer with radius R surrounded by the quencher Q which is described as scalar field in spherical coordinates [Q(r)]

model can be found in the book of S. A. Rice, who provides an excellent overview of diffusion limited reaction kinetics. [50] A short summary is presented in the following section.

For the derivation of a mathematical model which describes diffusion limited bimolecular processes, some simplifications are inevitable. The model is based on the Einstein–Smoluchowski-Theory of diffusion and assumes a spherical chromophore with radius R located at the origin of a spherical coordinate system as shown in Figure 16.15. The sensitizer is enclosed by the quencher molecules, which are described as a scalar field instead of discrete molecules. This means that at least in first approximation the distribution function of the quencher is continuous. The concentration [Q] of the quencher is much higher than [S], the concentration of the sensitizer. Inside the spherical chromophore, the quencher concentration equals zero for $t \leq 0$.

The homogeneous initial concentration of the quencher is $[Q_0]$. The following discussion is based on the assumption that the excited chromophore is separated far enough from the next excited chromophore that competing processes can be neglected. For an easy mathematical description, a normalized quencher density $\rho(r)$ which depends on the distance r with respect to the coordinate origin is introduced.

$$\rho(r) = \frac{[Q(r)]}{[Q_0]} \tag{16.19}$$

The assumption, which was applied with the introduction of the model, can be expressed by the following boundary conditions:

$$\begin{aligned} \rho(r,t) &= 0 \quad (t=0 \quad r \leq R) \\ \rho(r,t) &= 1 \quad (t=0 \quad r > R) \\ \rho(r,t) &= 1 \quad (t \geq 0 \quad r \to \infty) \end{aligned} \tag{16.20}$$

This model assumes that the excited chromophore is quenched when the quencher molecule Q is at an encounter distance R. Hence, the quencher concentration at encounter distance R equals zero independently of t, which can be expressed by a fourth boundary condition.

$$\rho(r,t)=0 \quad (t\geq 0 \quad r=R) \tag{16.21}$$

Due to Fick's laws of diffusion the concentration gradient at an arbitrarily chosen distance $r = r_0$ with $r_0 > R$ leads to a particle current density $J(r_0)$ crossing the spherical surface. The particle current density can be written as the product of the diffusion constant D and the concentration gradient at $r = r_0$.

$$J(r_0)=D\left.\frac{\partial c}{\partial r}\right|_{r=r_0} \tag{16.22}$$

The particle current $I(r_0)$ of quencher molecules Q crossing this spherical surface is obtained by multiplication of the particle current density J by the area of the spherical surface.

$$I(r_0)=4\pi r_0^2 D\left.\frac{\partial c}{\partial r}\right|_{r=r_0} \tag{16.23}$$

Now we have to calculate the net loss of quencher Q within a spherical shell with radius r_0 and an infinitesimal thickness dr. First of all, the particle current $I(r_0 + dr)$ through the external spherical surface with radius $r_0 + dr$ has to be formulated.

$$I(r_0+dr)=4\pi(r_0+dr)^2 D\left.\frac{\partial c}{\partial r}\right|_{r=r_0+dr} \tag{16.24}$$

The net loss of quencher Q within this spherical shell now can be calculated by subtraction of the particle current, which is flowing into the shell, by the particle current, which is flowing out.

$$I(r_0)-I(r_0+dr)=4\pi r_0^2 D\left.\frac{\partial c}{\partial r}\right|_{r=r_0}-4\pi(r_0+dr)^2 D\left.\frac{\partial c}{\partial r}\right|_{r=r_0+dr} \tag{16.25}$$

To obtain the rate of loss of concentration of Q per unit time within this spherical shell of infinitesimal thickness one has to divide the net particle loss of this spherical shell by its volume $4\pi r_0^2\,dr$.

$$\frac{\partial c}{\partial t}=\frac{4\pi D}{4\pi r_0^2\,dr}\left[r_0^2\left.\frac{\partial c}{\partial r}\right|_{r=r_0}-\left(r_0+dr\right)^2\left.\frac{\partial c}{\partial r}\right|_{r=r_0+dr}\right]$$
$$\frac{\partial c}{\partial t}=\frac{D}{dr}\left[\left.\frac{\partial c}{\partial r}\right|_{r=r_0}-\left(1+\frac{2\,dr}{r_0}\right)\cdot\left(\left.\frac{\partial c}{\partial r}\right|_{r=r_0}+\left.\frac{\partial^2 c}{\partial r^2}\right|_{r=r_0}dr\right)\right]$$
$$\frac{\partial c}{\partial t}=\frac{D}{dr}\left[-\left.\frac{\partial^2 c}{\partial r^2}\right|_{r=r_0}dr-\frac{2\,dr}{r_0}\cdot\left.\frac{\partial c}{\partial r}\right|_{r=r_0}-\frac{2\,dr}{r_0}\cdot\left.\frac{\partial^2 c}{\partial r^2}\right|_{r=r_0}dr\right]$$
$$\frac{\partial c}{\partial t}=-D\left[\left.\frac{\partial^2 c}{\partial r^2}\right|_{r=r_0}+\frac{2}{r_0}\left.\frac{\partial c}{\partial r}\right|_{r=r_0}\right] \tag{16.26}$$

The concentration c in Equation 16.26 now can be expressed with the normalized quencher density $\rho(r)$ according to Equation 16.19:

$$\frac{\partial \rho}{\partial t}=D\left(\frac{\partial^2 \rho}{\partial r^2}+\frac{2}{r_0}\frac{\partial \rho}{\partial r}\right) \tag{16.27}$$

The right hand side of Equation 16.27 contains two spherically symmetric derivatives of $\rho(r, t)$ which can be replaced by the Laplace operator.

$$\frac{\partial \rho}{\partial t}=D\,\nabla^2\rho \tag{16.28}$$

This equation describes the diffusion of matter as well as the diffusion of heat and its solution has been already studied. In consideration of the boundary conditions in Equations 16.20 and 16.21 the complementary error function erfc(x) solves the differential equation.

$$\rho\left(r,t\right)=1-\frac{R}{r}\,\mathrm{erfc}\left\{\frac{r-R}{\sqrt{4\,D\,t}}\right\} \tag{16.29}$$

The time dependent particle current I(t) of quencher Q diffusing towards the excited chromophore at encounter distance R per infinitesimal time interval represents the time dependent reaction rate k(t) of excited state quenching. The particle current at encounter distance R can be obtained by insertion of $\rho(R,t)$ into Fick's law of diffusion, which is formulated in Equation 16.30. The obtained rate k(t) describes the dynamics of the non-stationary quenching process assuming, that the excited state life time is infinite. The quencher concentration $[Q_0]$ is given in units of particles per unit volume, e.g., m^{-3}.

$$k(t) = I(R,t) = 4\pi R^2 D \left.\frac{\partial \rho}{\partial r}\right|_{r=R} [Q_0] = 4\pi R^2 D \left[\frac{1}{R} + \frac{1}{\sqrt{\pi D t}}\right][Q_0] \tag{16.30}$$

It can be seen that the rate starts at very large values at early times t. However, when we insert the rate into the decay function, and include the intrinsic life time τ of the excited state which is quenched, we see immediately that the t-term in the denominator is more then balanced by the multiplication of the rate by t.

$$n(t) = n_{diff} \cdot \exp(-k(t)t) = n_{diff} \cdot \exp\left\{-\frac{t}{\tau} - 4\pi RD \cdot [Q_0] \cdot \left[1 + \frac{2R}{\sqrt{4Dt}}\right] \cdot t\right\} \tag{16.31}$$

The result can also be written as a product of an exponential term with t in the exponent and an exponential term with the square root of t in the exponent:

$$n(t) = n_{diff} \cdot \exp\left(-\frac{t}{\tau}\right) \cdot \exp\left(-4\pi RD \cdot [Q_0] \cdot t\right) \cdot \exp\left\{-4\pi RD \cdot [Q_0] \cdot \frac{2R\sqrt{t}}{\sqrt{4D}}\right\}. \tag{16.32}$$

This means that the temporal evolution is given by i) the spontaneous decay of the sensitizer excitation, ii) a classical rate term, and iii) a stretched exponential. The last term is the result of the non-equilibrium situation caused by the ultrafast excitation. It deviates strongly from the widely used multi-exponential modeling and can only be poorly compensated in an exponential model by adding ever more time constants.

For $t \to \infty$ the reaction rate k(t) becomes time independent. This time independent rate k_∞ linearly depends on the initial quencher concentration $[Q_0]$ and describes the dynamics of the stationary quenching process.

$$k_\infty = 4\pi RD \cdot [Q_0] \tag{16.33}$$

The proportionality constants have to be multiplied by the product of the Avogadro constant N_A and $1000\,\mathrm{l\,mol^{-1}\,m^{-1}}$ to allow inserting the quencher concentration in units of $\mathrm{mol\,l^{-1}}$ instead of $\mathrm{m^{-3}}$, which was used until now.

$$k_\infty = 4\pi RD \cdot N_A \cdot 10^3 \cdot [Q_0^{mol}] \tag{16.34}$$

Typically all proportionality constants in Equation 16.34 are put together into one factor, usually named K_q, which describes the dependence of the time constant k_∞ of the stationary quenching process on the quencher concentration $[Q_0^{mol}]$. A huge number of quenching constants K_q for different sensitizer quencher combinations

Table 16.1: Rate constants K_q for the stationary regime of diffusive quenching at room temperature for selected sensitizer-quencher pairs.

sensitizer	quencher	solvent	K_q [$10^9\ M^{-1}s^{-1}$]	sensitizing mechanism
eosin Y	2-phenyl-1,2,3,4-tetrahydroisoquinoline	DMSO	0.8	not known
riboflavin tetraacetate	4-methoxybenzylalcohole	98 % MeCN 2 % DMSO	1.9	singlet photo-oxidation
riboflavin tetraacetate	4-methoxybenzylalcohole	50 % MeCN 50 % H_2O	4.0	singlet photo-oxidation
riboflavin tetraacetate	4-methoxybenzylalcohole	50 % MeCN 50 % H_2O	0.02	triplet photo-oxidation
xanthone (derivative)	1,3-pentadiene	trifluorotoluene	2.6	triplet-triplet energy transfer

and a variety of sensitizing mechanisms in different solvents can be found in Montali et. al., "Handbook of Photochemistry". [51] Typical values for K_q range from $K_q = 10^5\ l\ mol^{-1}\ s^{-1}$ for slow triplet quenching processes up to $K_q = 10^{10}\ l\ mol^{-1}\ s^{-1}$ for fast singlet quenching processes.

The last step toward applying the model of diffusion limited reaction kinetics on transient absorption spectra is the derivation of a suitable diffusion fit function, which can be applied to the measured data set. Since signals in TA spectroscopy are generally proportional to the number of molecules n(t) which are momentarily in the observed state, a quantity proportional to Equations 16.31 and 16.32 results. Note that the diffusion fit function includes the intrinsic life time τ of the excited state, which is quenched. Diffusion terms only become significant, if this time is long. However, this just corresponds to the requirement of a long intrinsic excited state lifetime of the sensitizer for efficient catalysis.

16.3.2 Application of the diffusion fit function to experimental data

In the last section the derivation of a fit function, which describes the diffusion controlled reaction kinetics of a photo-sensitizer and a quencher molecule, was shown. The derivation of the model was based on the Einstein–Smoluchowski-Theory of diffusion. It should be noted that we are actually interested in the time dependence of the electron transfer from or toward the substrate, but this is just a different way of looking at the fluorescence quenching of the sensitizer. As the optical signature of the sensitizer is typically more accessible, this focus on the sensitizer is actually closely related with the time resolved spectroscopy. From the synthetic viewpoint of photocatalysis, the substrate is obviously more in the focus.

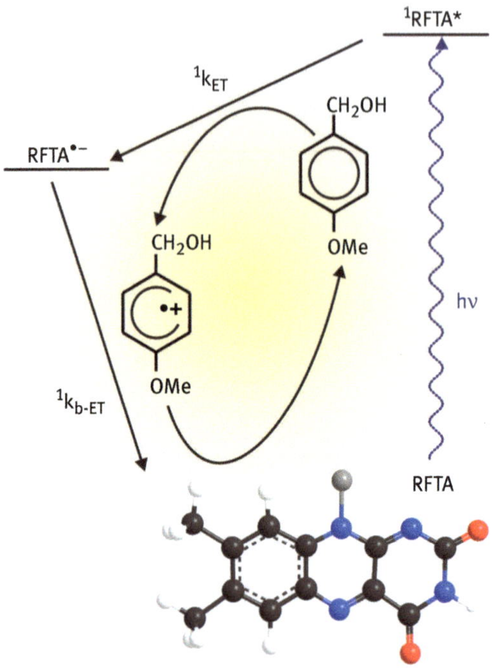

Figure 16.16: Jablonski diagram of the reductive quenching of the S_1 state of riboflavin tetraacetate (RFTA). The rate $^1k_{ET}$ is controlled by diffusion of RFTA and methoxybenzyl alcohol (MBA).

We now apply and test the derived function on photoinduced reaction kinetics, which are controlled by diffusion. As a model reaction the diffusive quenching process of the S_1 state of riboflavin tetraacetate (RFTA) by methoxybenzyl alcohol (MBA) was chosen. The reductive quenching process produces a radical pair comprising the flavin radical anion and the radical cation of the benzyl alcohol. After formation of the radical pair, typically a fast electron back-transfer tends to offset the oxidation of the alcohol. Two neutral ground state molecules result and the energy of the absorbed photon is converted to heat. Since the rate of the charge recombination process $^1k_{b\text{-}ET}$ at common substrate concentrations below 1 mol/l is much faster than the diffusion limited rate of the initial electron transfer step $^1k_{ET}$, no considerable population of the charge separated radical pair state occurs. Hence, the only visible process is the reductive S_1 quenching. Signals of the involved intermediate radicals won't be observable due to the fast charge recombination process. A Jablonski diagram of the model reaction is shown in Figure 16.16.

The transient signal of stimulated emission is a unique sign for excited sensitizers. It clearly show diffusion limited decay dynamics, signaled by non-exponential kinetics and is therefore chosen for the application of the diffusion fit function. Instead of using the signal of only one individual spectral trace within the spectral

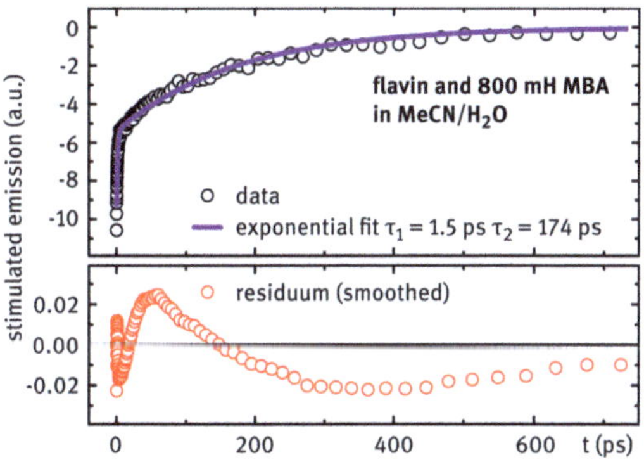

Figure 16.17: Exponential fit applied to the transient signal of the stimulated emission. The main time range of hundreds of ps is augmented by a 1.5 ps component discussed below. Upper panel: experimental data and fit. Lower panel: residuum on an expanded scale.

region of the stimulated emission, the integrated signal of all spectral traces containing major signal contributions of the stimulated emission was selected for the further data processing. This minimizes the influence of spectral diffusion caused for instance by solvation or vibrational cooling and decreases the statistic error of the measurement.

Figure 16.17 shows, that diffusive decay dynamics are not accurately enough described by an exponential fit function. The residuum, i. e. the difference between the data and the fit function, deviates systematically from the measured decay data. In contrast, the application of the diffusion fit function, given in Equation 16.31, results in a very good fit in the range of tens to hundreds of ps (see Figure 16.18 and 16.19). Still, the fit to the measured data points (red line, Figure 16.18) is not yet perfect in the first few ps.

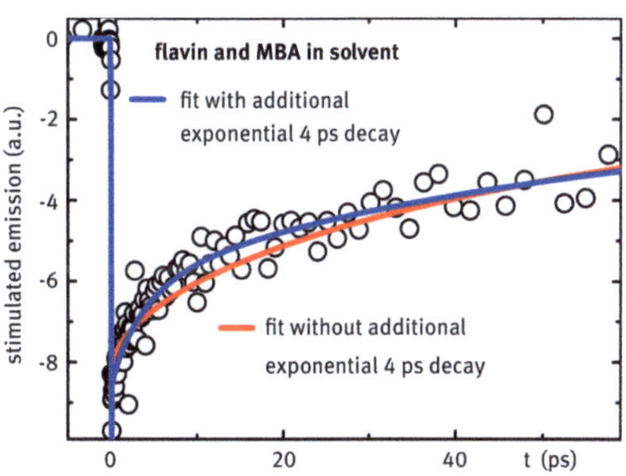

Figure 16.18: Diffusion fit function (Equation 16.34, red line) and extended diffusion fit function (Equation 16.35, blue line) applied to the transient signal of the stimulated emission.

For a full description of the decay dynamics of the observed model reaction an additional term has to be added (blue line, Figure 16.18) to the diffusion fit function in Equation 16.31. From another TA measurement which was performed without using any solvent by diluting RFTA directly in the highly viscous MBA, the intrinsic rate k_{ET}^{0} of the electron transfer from MBA to RFTA could be determined to $k_{ET}^{0} = (4\,ps)^{-1}$. Adding an exponential term, describing the quenching processes with this intrinsic rate k_{ET}^{0}, to the diffusion fit function leads to an extended fit function shown in Equation 16.35. Applying this fit function to the transient data leads to a perfect description of the measured S_1 decay dynamics. The fit parameters are the radius of the chromophore R, the diffusion constant D and the relative amplitudes $n_{nondiff}$ and n_{diff} of both exponential terms.

$$n(t) = n_{nondiff} \cdot \exp\left\{-k_{ET}^{0} \cdot t\right\} + n_{diff} \cdot \exp\left\{-\frac{t}{\tau_{S1}} - 4\pi R D \cdot [Q]_0 \cdot \left[1 + \frac{2R}{\sqrt{4Dt}}\right] \cdot t\right\} \qquad (16.35)$$

The additional term has to be added due to apparent pre-association of chromophore and quencher. The reductive quenching of pre-associated pairs of RFTA and MBA is not a diffusion limited process. It occurs with the constant reaction rate k_{ET}^{0} which was already obtained when both reactants are in closest proximity. The contribution of the constant reaction rate k_{ET}^{0} to the overall decay dynamics becomes more dominant with increasing the quencher concentration. This is in line with a statistical association.

The detailed investigation of flavin photocatalysis has shown that the addition of water to the solvent improves the reaction quantum yield for product formation. For this reason, Figure 16.19 shows the diffusion limited decay dynamics of the integrated

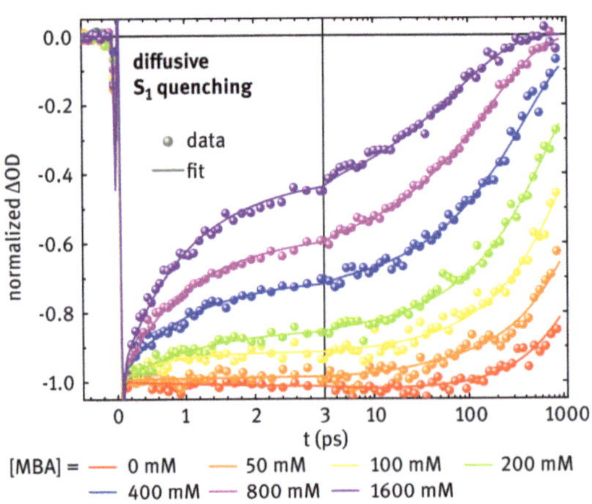

Figure 16.19: Diffusion fit function with an additional exponential decay according to Equation 16.35 applied on TA data of the reductive quenching of excited RFTA by MBA. [43]

stimulated emission of RFTA in a solvent mixture of acetonitrile and water at different MBA concentrations. The extended diffusion fit function from Equation 16.35 has been applied to describe the measured data points. It has to be noted, that in water, the intrinsic time constant for electron transfer changes from $k_{ET}^{0} = (4\,\text{ps})^{-1}$ in pure acetonitrile to $k_{ET}^{0} = (0.9\,\text{ps})^{-1}$ in a mixture of acetonitrile and water.

The simultaneous application of the diffusion fit function to the transient absorption data at varied MBA concentration not only provides an excellent modeling of the diffusion limited kinetics. It provides moreover a detailed microscopic insight into the molecular processes:

For instance, the encounter distance R is a fit constant, which is directly obtained from applying the fit function to the transient data. It provides a rough estimate of the distance at which an electron transfer between substrate and excited photocatalyst becomes efficient. It has to be noted that the underlying assumption that electron transfer occurs at a certain encounter distance is a heavy simplification of the theory of electron transfer. But this simplification seems to be justified due to a strong exponential distance dependency of the common theoretical models of electron transfer dynamics. Usually, exponential damping coefficients for distance dependent electron transfer processes are on the scale of 1 $Å^{-1}$.[10] In contrast, the average distance between photocatalyst and quencher is on the scale of some nm, when working under substrate concentrations, which are usually used in photocatalysis and are in the mM range. Hence, the assumption of instantaneous electron transfer, which occurs at a certain encounter distance R, is legitimate. The resulting interaction radius was determined to $R = 3.8\,Å$ for the observed model reaction in a 1:1 solvent mixture of acetonitrile and water.

Another insight into molecular processes is provided by the fit of the relative amplitudes of diffusion fit function n_{diff} and of the exponential term $n_{nondiff}$, which describes the non-diffusive quenching of pre-associated sensitizer-quencher pairs. In particular, the interesting quantity is the ratio of both amplitudes since its dependency on the substrate concentration allows the direct determination of the association constant of catalyst and substrate. For the model reaction an association constant of $K = 0.63\,M^{-1}$ in a solvent mixture of acetonitrile and water was determined, as shown in Figure 16.20.

Also the last fit parameter, the diffusion constant D, has to be discussed. The Stokes–Einstein equation relates the diffusion constant D to macroscopic accessible physical quantities such as temperature T, viscosity η of the solvent and the effective radius R_{eff} of the diffusing molecule, which is supposed to be spherical.

$$D = \frac{k_B\,T}{6\,\pi\eta\,R_{eff}} \tag{16.36}$$

It has to be mentioned that R_{eff} in the denominator of Equation 16.36 is not directly related with the encounter distance R for electron transfer between photo-sensitizer

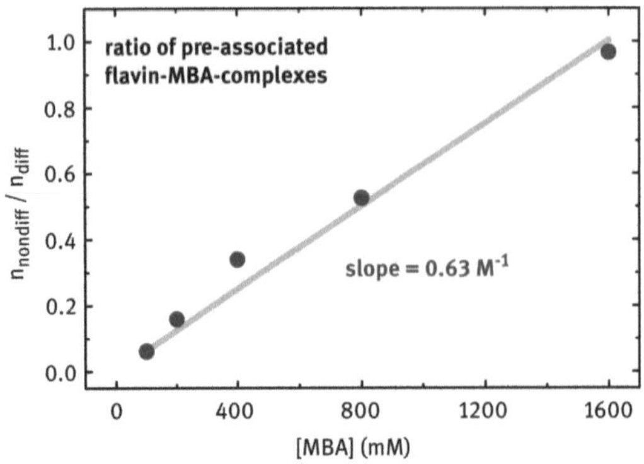

Figure 16.20: Ratio of relative amplitudes of diffusion fit function and of the exponential term, which describes the non-diffusive quenching of pre-associated sensitizer-quencher pairs according to Equation 16.35.

and quencher. The R_{eff} is the effective radius of the molecule, which is relevant for diffusion. Not only the size of a spherical model of the diffusing molecules determines R_{eff} but rather the size of the molecule including the shell of associated solvent molecules can be directly linked to R_{eff}. Since not only the quencher molecules but also the photo-sensitizer are contributing to the diffusion process, R_{eff} gives a description of an effective intermediate radius of both molecules, which is relevant for diffusion.

Figure 16.21 shows the concentration dependency of the fit parameter D resulting from applying the extended fit function from Equation 16.35 to the observed S_1 decay dynamics. The slight decrease of D with higher concentrations of the substrate can be explained by the increase of the solvent viscosity due to adding large amounts of the highly viscous methoxy benzyl alcohol to the solvent with a low viscosity.

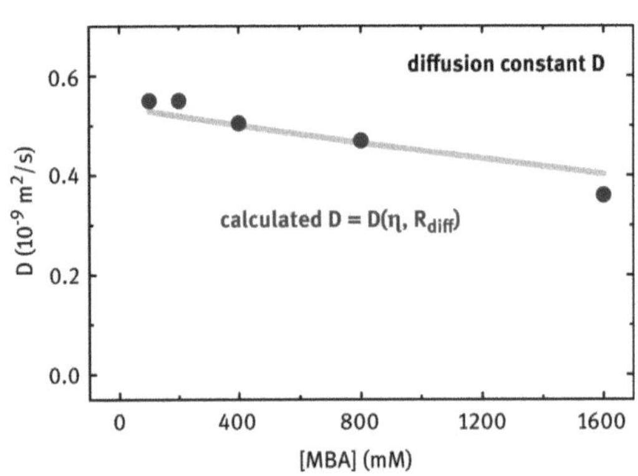

Figure 16.21: Diffusion constant D derived from the diffusion limited S_1 decay dynamics at different concentrations of the highly viscous substrate MBA. D was obtained from applying the fit function, given in Equation 16.35, on the diffusion limited decay dynamics of the S_1 state of RFTA.

For the observed model reaction, which was performed in different solvents, the analysis of the fit of the diffusion constant D offers the following results: R_{eff} increases from $R_{eff} = 4.5$ Å in pure acetonitrile to $R_{eff} = 8.0$ Å in the solvent mixture of acetonitrile and water. This result suggests that a solvent shell of associated water molecules is able to slow down the diffusive movement of the interacting particles.

In summary, it has been shown that the presented model of diffusion limited excitation quenching applied to experimental data provides much more data than only a perfect description of the decay curve. It offers deep insights into molecular processes and extends the previous model of diffusion limited reaction kinetics, which still divides the diffusion limited decay of a signal into three contributions: static quenching, non-stationary quenching and stationary quenching dynamics. [52] Especially for the investigation of photocatalytic processes, a more sophisticated model seems to be a great help to obtain a deep knowledge of the underlying reaction mechanisms.

The literature provides only very few examples where diffusion limited excited state dynamics obtained by time resolved spectroscopic techniques are quantitatively described with the help of suitable fit functions. The application of a very similar model of diffusion controlled decay modelling, which is also based on Einstein–Smoluchowski-Theory, including a Coulomb potential between the reactants, is presented in reference [53]. The model can well describe the diffusion influenced fluorescence quenching reaction of rhodamine B and ferrocyanide. A more general discussion of the model, which describes diffusion controlled fluorescence quenching by electron transfer based on Marcus Theory can be found in reference [54]. It has to be mentioned, that the temporal resolution of the transient spectroscopic techniques, which were applied in this work, was 60 ps. This is somewhat critical, since the decay curves at earlier times are very important for the correct modelling. Some theoretical considerations on distance distribution and photo-induced electron transfer between diluted reactants in solution regarding Marcus theory are presented in reference [55]. Diffusion controlled fluorescence quenching by long range electron transfer in highly viscous non-polar solvents was investigated in reference [56].

16.4 Costs of photocatalysis: The reaction quantum yield

16.4.1 Requirements for an accurate definition of the quantum yield

In photochemistry, the term "quantum yield" has been used for many different quantities over the years. From a physical point of view and in the context of photocatalysis it should describe the ratio of the number of product molecules formed to the number of absorbed photons. Usually a high quantum yield Φ is desired for an efficient photocatalytic process. At this point efficiency is focused onto the best possible use of the available light. As early investigations on the fundamentals of photocatalysis are not so much concerned yet about the use of the photons but rather worry about the endurance of the photocatalyst and the ability to quantitatively convert a substrate, it becomes understandable why other usages have surfaced.

This section will provide a rigid definition of the term "quantum yield" and suggest an easy and cheap method for the accurate determination in homogeneous catalysis. A detailed discussion of the difficulty to properly report photonic efficiencies and quantum yields in heterogeneous photocatalysis can be found in Nick Serpone, "Relative photonic efficiencies and quantum yields in heterogeneous photocatalysis". [57] One of the main aspects for an accurate usage is that the quantum yield Φ has to describe the properties of a molecular reaction independently of the experimental boundary conditions. It is not sufficient to report a quantity which strongly depends on the measurement setup since the result is not reproducible. Whereas the number of incident photons impinging on the external reactor walls is relatively easy to measure, it is more sophisticated to determine the number of the photons actually absorbed by a photocatalyst molecule.

Depending on the sensitizing mechanism and the subsequent reaction mechanism of the photocatalytic reaction, the quantum yield Φ can be affected by a huge number of physical quantities. The solvent environment causes the largest influence on the photo-induced reaction. But also temperature, irradiation wavelength and its bandwidth, and the concentration of catalyst and substrate may affect the efficiency of the catalytic reaction. In particular when intermediate steps of the photocatalytic reaction are limited by diffusion, the concentration of the substrate plays a major role on reaction dynamics and, hence, on the total efficiency of the reaction. [43] Usually the temperature strongly affects the viscosity of the solvent and, therefore, also changes the diffusion dynamics on the microscopic scale. But a change of temperature can also cause major effects in template photocatalysis, where complexation of the catalyst and substrate is required for the photocatalytic reaction. If the catalyst-substrate binding is mediated through hydrogen bonding, where binding energies are close to $k_B \cdot T$, a change of temperature dramatically changes the equilibrium of associated and non-associated catalyst-substrate complexes and therefore affects the quantum yield Φ of the desired template mediated photocatalytic process. [7, 8 ,58] Also, the oxygen concentration in the solvent has to be taken into account. Oxygen is known to efficiently quench triplet states and for this reason strongly reduces the life time of intermediate states with triplet character, which may affect the over-all reaction quantum yield Φ. Another aspect which has to be considered is a wavelength dependent definition of Φ. For example, in organometallic coordination complexes, which are used for photocatalysis, there are examples where it is possible to excite different states depending on the excitation wavelength. [59–61] Only the excitation of metal-to-ligand-charge-transfer transitions (MLCT) directly leads to photocatalytic active states. Also competing absorption of the substrate or the formed product has to be considered when discussing the wavelength dependency of the quantum yield. For the determination of the optimal environment for an efficient photocatalytic reaction, it can be essential to measure Φ under a variety of conditions. Therefore, Φ needs to be easy to measure.

A seminal contribution to this debate is given by Nick Serpone in "Suggested terms and definitions in photocatalysis and radiocatalysis" discussing a large number of terms and definitions for the application in photocatalysis. [62] The following definition of the quantum yield Φ strongly relies on this work. Generally, in chemistry, the quantum yield can be determined for lots of processes such as reactant disappearance, light emission, product formation and various other events occurring as a result of a photochemical process. Since generally in catalysis the disappearance of a substrate molecule does not necessarily lead to formation of the desired product molecule, determining the quantum yield by taking the number of disappearing reactant molecules into account is not appropriate. Rather considering the actually formed product molecules results in a meaningful definition of the quantum yield Φ. Whereas in heterogeneous photochemistry the term quantum yield Φ describes the ratio of the total number of photons incident on the reactor walls and the number of product molecules formed [57, 62], in homogeneous catalysis a more accurate definition is required to describe the actual efficiency of a molecular process.

The definition, given in Equation 16.37, fulfills all the previously discussed conditions. The quantum yield Φ is directly defined as the ratio of the number of photons which are actually absorbed by the photocatalyst N_{abs} and the number of formed product molecules $N_{product}$. In Equation 16.37 the amount of substance $n_{product}$ can be inserted instead of the total number $N_{product}$ of molecules, which were formed within the catalytic process. The corresponding quantity n_{abs} describing the number of moles of absorbed photons can be expressed by the occasionally used physical unit einstein, which is equivalent to one mole of photons.

$$\Phi_\lambda = \frac{N_{abs}}{N_{product}} = \frac{n_{abs}}{n_{product}} \tag{16.37}$$

It should be noted that this definition of the wavelength dependent reaction quantum yield is most precise, if quasi-monochromatic light is used. This can either originate from a laser source, e.g. in basic studies, or from high power light emitting diodes (LEDs) with a spectrum narrower than the typical width of molecular absorption bands. [63] On the other hand, Maschmeyer and Che have argued for the solar spectrum as proper reference. [64] While this is eventually the proper measure for an industrial process, much greater insight can be gained from wavelength selective determinations. If a range of wavelengths has been used, the convolution of the spectrum of a broad light with the catalyst absorption spectrum and the wavelength dependent intrinsic yield will give the overall efficiency.

In terms of efficient photocatalysis, typically the formation of a product molecule is the desired process to occur with a high quantum yield. In enantioselective photocatalysis for instance not only the formation of a product molecule is essential. Only product molecules which exhibit the proper chirality contribute to the number of desired product molecules formed.

Another aspect which has to be considered when determining the quantum yield Φ is the change of reaction environment during photocatalysis since product molecules are formed and substrate molecules are consumed. Upcoming product absorption within the spectral region of irradiance may lead to completely new photochemical processes within the solution. The primary, desired photocatalytic process will be overlaid by another competing photochemical process with unknown products formed. A further contribution to the change of the reaction environment involves the catalytic cycle of the light absorbing photocatalyst. Particularly in photoredox catalysis, the oxidation state of the catalyst has changed after the primary photocatalytic process. Typically the reacted photocatalyst is catalytically inactive and needs to be regenerated in order to retrieve its full photocatalytic capability and to complete the catalytic cycle. The regeneration of the photocatalyst with the help of sacrificial electron donor and acceptor molecules in solution may impact the reaction conditions, since they are consumed with catalyst regeneration. [65] Therefore monitoring the photon absorption for a long period of the photocatalytic reaction is not beneficial for the determination of a meaningful product quantum yield Φ which solely has to describe the yield of the primary photocatalytic reaction. This problem is normally overcome by determination of the quantum yield Φ_λ at small conversions of the reactants, usually less than 10 %, a point not often respected in heterogeneous photocatalysis, where the focus is usually complete transformation of the substrate. [62]

With the discussed requirements on an accurate definition of the product quantum yield Φ and an unambiguous definition, we now focus on experimental techniques which readily allow for the determination of the required quantities. Whereas the denominator in Equation 16.37, the product yield, is relatively easy to measure in a chemistry lab using standard analytical methods, more sophisticated methods are required for the determination of an exact value of the numerator, the amount of the actually absorbed photons. The following summary is based on the article "Laboratory apparatus for the accurate, facile and rapid determination of visible light photoreaction quantum yields". [66] The classical chemical actinometric procedures for the determination of the quantum yield, which are standardized by IUPAC [51, 67, 68] are often too laborious and newly developed ones are in most cases only suitable within the UV spectral region [69, 60, 71]. Efficient solar harvesting does, however, require primarily absorption in the visible.

Our experimental approach for an easy determination of the number of photons, which are actually absorbed by the photocatalyst, is shown in Figure 16.22. The homogeneous photochemical reaction of interest is performed in a standard cuvette with 10 mm optical path length. Since photochemical or photocatalytic active chromophores generally only absorb selected parts of the electromagnetic spec-

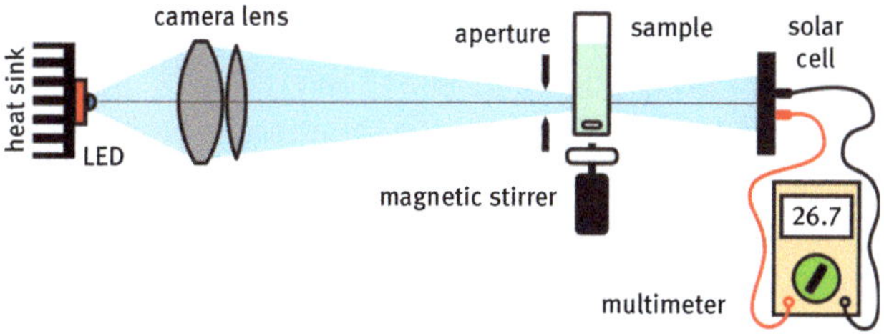

Figure 16.22: Setup for the easy and cheap determination of the product quantum yield Φ of a photo-induced reaction: A camera lens images a high power LED into the sample solution, which is constantly stirred. The power of the transmitted light is measured with the help of a commercially available solar cell as a power meter. Reprinted from reference [66] with permission from the Royal Society of Chemistry.

trum, the illumination spectrum has to be adapted to the absorption properties. Modern high power light emitting diodes (LEDs) provide high photon fluxes at a variety of well defined spectral regions which allows the choice of the optimized light source for a given photocatalytic chromophore. [72, 73] The emitted light from the LED is imaged with the help of a commercially available camera lens into the sample solution to drive the photocatalytic process. In order to facilitate the measurement of the absorbed illumination power and to provide stable conditions to the photo-induced process, the optical density of the sample solution within the illumination spectrum has to be set to values between 0.3 and 1.0. This provides, on the one hand, a reasonable amount of absorption of the irradiating photons but, on the other hand, the absorption takes place within the whole cuvette volume and not in a thin solvent layer at the front side of the cuvette. This would be the case when performing the irradiation under very high optical densities. The transmitted residual light, which passes through the cuvette with the sample solution, is continuously measured by a dedicated and calibrated solar cell. The illumination dependent output current of the solar cell is converted with the help of a low Ohm resistor into a illumination dependent voltage which can be easily measured with a multimeter. The comparison of illumination levels, which were detected with the solar cell, and the values of illumination power, which were measured with a calibrated thermopile powermeter, results in a nearly perfect linearity between output voltage and effective incident illumination power. The use of a solar cell instead of a laser powermeter provides an easy and inexpensive measurement method of the transmitted illumination power.

Finally, for the determination of the number of actually absorbed photons within the sample cell, first the transmitted illumination power has to be measured once with and once without the sample solution within the beam path of the focused LED light. Not only absorption by the photocatalytic chromophore leads to a decrease in illumination power. The back reflection at the terminal air glass interface accounts for a sizable contribution to the illumination power decline as well, which is still in the few percent regime and has to be considered when calculating the absorbed illumination power. A more detailed description of all these issues can be found in reference [66]. With the absorbed power determined, the duration of illumination immediately gives the number of absorbed photons N_{abs}. Together with the total volume of the solution and the independently determined concentration of products, finally the desired value for the yield is found.

We do not want to undervalue the importance to determine other quantum yields like the triplet or photoluminescence yields. They can greatly contribute to the full characterization of photocatalysts or photocatalytic processes. Therefore, we briefly refer to related techniques in the following section.

A completely different technical approach for the determination of the triplet quantum yield of some organic dyes with the help of picosecond laser double-pulse fluorescence excitation and time resolved fluorescence detection [74] is presented in reference [75]. In this two-pulse excitation technique a sequence of picosecond laser pulses is used to determine the yield of triplet state population after photoexcitation. The first pulse strongly excites the molecules to the excited singlet state. From there they decay either back to the ground state or to the triplet state. The second pulse probes the fluorescence reduction by triplet formation, since the fluorescence signal is proportional to the population of the excited singlet state, which has – in contrast to the triplet state – the ability to fluoresce. A modified version of this technique that decouples the triplet yield and cross section determination has been presented recently. [76]

The determination of photoluminescence quantum yields can readily be performed with a commercial spectrofluorometer. A suitable standard, usually a well fluorescing dye solution is needed for absolute values. Relative values and the change with experimental parameters can be determined even more easily. Finally, we want to point the interested reader to a very informative review on light intensity distribution in photocatalytic reactors with some comments on the reaction quantum yield within the focus of heterogeneous photocatalysis in reference [77].

16.4.2 Determination of the quantity of excited molecules in transient absorption measurements

Transient absorption spectroscopy on multiple time scales is in itself a powerful tool for the determination of quantum yields either of the population of an intermediate state or of the formation of a product molecule. Deactivation processes of electronically excited states and recombination of charge separated states are the most important processes which are competing with product formation and, therefore lower the over-all product quantum yield Φ in homogeneous photocatalysis. Both processes result in the formation of ground state molecules of catalyst and substrate. Therefore, the analysis of the ground state bleach (GSB) recovery in transient absorption measurement allows the determination of the number of molecules which are still in an intermediate state or the final product state at selected points of time after excitation. [78] However, the determination of the quantum yield of intermediate or final states using this method requires that the GSB is not significantly overlaid by absorption of the state of interest, which is a prerequisite rather of a purely theoretical nature. If either GSB or the overlaying absorption spectrum of the intermediate state are well structured or exhibit significant characteristic spectral features, the linear combination of spectra can be easily deconvoluted to obtain the contribution of each spectrum and thereby the population of the intermediate states. If in fact the GSB and the excited state absorption of all intermediate states of a photochemical reaction are spectrally sufficiently separated from each other, the quantum yield of each intermediate state and the final product state can be determined readily. For this procedure the intensity of the ground state bleach immediately after excitation of the light absorbing chromophore has to be scaled to 100 % to account for the degree of optical excitation. Then the ratio of the GSB intensity at a certain pump-probe delay Δt to the intensity of the initial GSB directly provides the quantum yield of the observed excited intermediate state of the system. Often transient absorption spectroscopy in the visible and near UV spectral region is reported in the literature, where GSB and excited state absorption are not well enough separated from each other. To still utilize the introduced technique for the direct calculation of the quantum yield for all intermediate states, an extended range of observation and the highest possible fidelity [79] has to be sought. This is the reason why we continuously try to improve the femtosecond spectrometer.

Sometimes, at the beginning of an investigation, only the wavelength dependent extinction coefficient $\varepsilon_1(\lambda)$ of just one intermediate state or of the final product is known beyond the spectrum of the catalyst and substrate. In this case methods of spectroelectrochemistry can provide in addition the wavelength dependent extinction coefficient $\varepsilon_1(\lambda)$ of radical anion and radical cation states of the participating molecules. In case of photoredox catalysis these spectra may directly appear in the transient absorption spectrum. Since transient absorption spectroscopy provides the wavelength dependent change of absorption $OD(\lambda)$ in units of optical density, the application of the Beer–Lambert-Law reveals the effective concentration of molecules, which are momentarily in the observed intermediate state. Now, only the information of the concentration of the initially excited photocatalyst molecules is missing for the calculation of the quantum yield of the observed radical state. Since the extinction coefficient of the ground state absorption spectrum is accessible with UV/Vis absorption spectroscopy and also the excitation energy, focal size and the optical path length within the sample cell is straightforward to measure, the excitation probability of each molecule within the excited sample volume can be easily determined.

A relatively accurate estimation of the excitation probability P_{exc} in transient absorption measurements is shown in Equation 16.38. [28] The P_{exc} usually depends on the excitation energy of the pump pulse E_{pump}, the central excitation wavelength λ, the full width half maximum (FWHM) of the excitation pulse diameter D within the sample cell and the absorption coefficient of the excited molecule at the excitation wavelength ε. The factor ln(2) results from the conversion from a Gaussian beam profile with a FWHM D of the beam diameter to a homogeneous circular beam profile with diameter D. To avoid effects of multi photon excitation it is reasonable to keep the excitation probability below 10 % in transient absorption measurements. An excitation probability of 10 % will typically lead to a contribution of two photon processes of less than 1 % to the transient signal.

$$P_{exc} = n_{ph} \cdot \sigma = \frac{E_{pump}}{\left(hc/\lambda\right) \cdot \pi \cdot \left(D/2\right)^2} \cdot \varepsilon \cdot \frac{\ln(10)}{N_A} \cdot \ln(2) \qquad (16.38)$$

This estimate assumes that the ESA strength is similar to the ground state absorption. Since photocatalysts are expected to be good absorbers, this assumption is most often justified.

Multiplication of the excitation probability P_{exc} with the effective concentration of the photocatalyst in the sample solution results in the concentration of the initially excited molecules and the quantum yield of an observed intermediate state can be easily calculated, if its extinction coefficient $\varepsilon(\lambda)$ is known. Since Equation 16.38 provides an average excitation probability for the whole cross-section area of the excitation beam but usually only a small fraction of this volume – usually the centre area – is within the probed sample volume, there are minor deviations to the actual excitation probability which have to be explicitly modeled if very quantitative results are needed.

16.4.3 Example of the spectroscopic determination of reaction quantum yields: Flavin photocatalysis

The riboflavin tetraacetate (RFTA) sensitized photo-oxidation of methoxy benzyl alcohol (MBA) is a well suited example for the determination of the reaction quantum yield Φ. It allows to prove the picture of the underlying reaction mechanism, which was obtained with the help of time resolved absorption spectroscopy on a large range of time scales. The spectroscopic details of the investigation of the reaction have been described in detail elsewhere. [43] In brief summary: The first electron transfer from MBA to the photo-excited catalyst RFTA is the determining step for the subsequent processes of the photocatalytic reaction. If the singlet state of RFTA is quenched by electron transfer from MBA, fast charge recombination occurs within the formed spin-correlated radical pair on the time scale of 50 ps. This process is fast in comparison to the subsequent processes of product formation and does not significantly contribute to the photocatalytic product formation since it leads back to the

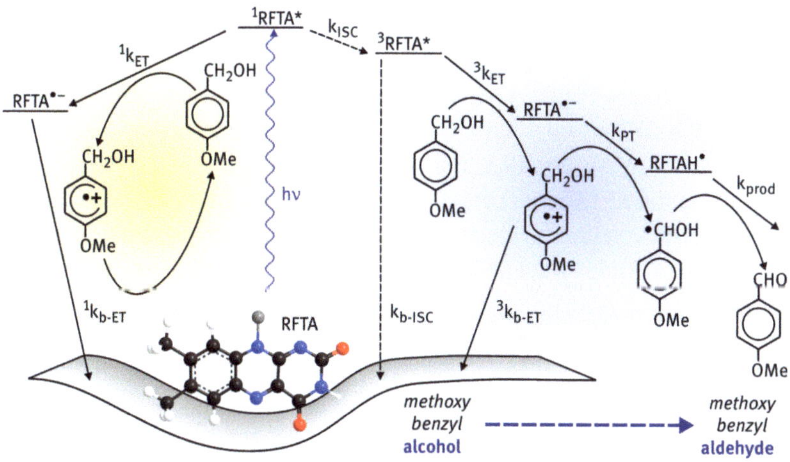

Figure 16.23: Reaction mechanism of flavin sensitized photo-oxidation of methoxy benzyl alcohol to the corresponding aldehyde. [43, 48]

ground state of photocatalyst and substrate. However, the situation is completely different, when the first electron transfer occurs after inter-system crossing of the photocatalyst, i. e. when the RFTA is allowed to reach its triplet state. Charge recombination within a formed spin-correlated radical pair with triplet character is a spin-forbidden process and; therefore, it is significantly slowed down in comparison to the spin-correlated ion pair with singlet character. [80, 81] Now the subsequent reaction steps, which begin with the protonation of the formed radical anion of RFTA, may occur with reasonable yield since competing processes take place on a very slow time scale. A scheme of the full mechanism is depicted in Figure 16.23.

In order to achieve a high reaction quantum yield, the main task is to drive the reaction in the triplet manifold of the photocatalyst. The inter-system crossing of RFTA occurs on the 10 ns time scale with quantum yields up to more than 70 % if there are no competing processes deactivating the excited singlet state. [82, 83, 84] Since diffusive quenching processes of excited states in solvents take place on a comparable time scale and this time scale strongly depends on the quencher concentration, a change of the concentration of the substrate allows an easy and effective control of the reaction pathway. At high concentrations of the substrate, the diffusive quenching of the excited state of RFTA is fast in comparison to the inter-system crossing dynamics and the reaction is driven to the non-productive electron transfer in the singlet manifold with immediately subsequent charge recombination. In contrast, at very low MBA concentrations, the excited singlet state quenching dynamics are slow in comparison to the intersystem-crossing process. From now on the triplet yield of the photocatalyst comes closer to the intrinsic triplet quantum yield since the yield of the singlet radical pair formation has significantly dropped. This provides an effective population of the triplet state of RFTA, from where the productive steps of the photocatalytic reaction start with the drawback that low substrate concentrations include slow diffusive triplet quenching. However, an effective and relatively fast quenching of the RFTA triplet state is required for the next reaction step, the generation of a spin-correlated radical ion pair with triplet character via electron transfer from MBA to the RFTA in its triplet state. But when the substrate concentration is too low, the diffusion limited electron transfer becomes slower than the intrinsic lifetime of the RFTA triplet state and the majority of the RFTA triplet population decays back to the ground state or is quenched by diluted oxygen.

From the point of view of a chemist, a very basic idea to increase the triplet lifetime of RFTA would be to conduct the photocatalysis under oxygen free conditions. However, diluted oxygen is required for this type of reaction since it thermally reoxidizes the photocatalyst which remains in its fully reduced form as a result of the catalytic aldehyde production. So the only remaining parameter to drive the photocatalytic reaction into its efficiency maximum is to adjust the substrate concentration: On the one hand, the concentration has to be low enough for an efficient population of the triplet state but, on the other hand, it has to be high enough for a reasonable generation of the spin-correlated triplet radical pairs, which is a diffusive process. With the help of the time resolved spectroscopic methods, it was possible to quantify each intermediate step of the flavin-sensitized photo-oxidation and especially to understand and to predict the diffusion limited reaction dynamics. Since the data analysis was based on a very complex model describing diffusion controlled reaction dynamics, which was presented in the last section, verification by another independent measurement strongly supports the credibility of the developed model.

Often time resolved spectroscopic methods do not directly give access to the total reaction quantum yield of a photo-induced reaction. This is due to the fact that typically not all intermediate states and species are observed reliably and more importantly, that the reaction quantum yield might be small. In this case the product would have to be detected with very high sensitivity in a sea of other species. But transient absorption spectroscopy reveals a lot of information which can be used to calculate quantum yields of the contributing intermediate steps. The quantum yield of an intermediate step can be determined either, if all contributing reaction rates k_i are known, or if the relative quantum yields of competing reaction pathways can be determined by analyzing the time dependent intensities of the ground state bleach or of other transient signatures with known extinction coefficients. How to calculate quantum yields on the basis of the underlying rate model is shown as follows: For instance, after photo-excitation, an intermediate parent-state is depopulated by three different processes, described by the rate constants k_1, k_2 and k_3 resulting in three different sub-states, named state 1, 2 and 3. The yield for the population of state 1, which leads to the product formation, is given by dividing rate k_1, which populates this state, by the sum of rates $k_1 + k_2 + k_3$, which depopulate the parent-state. This results in the following equation, describing the quantum yield of this intermediate step:

$$\Phi_1 = \frac{k_1}{k_1 + k_2 + k_3}. \tag{16.39}$$

The total reaction quantum yield Φ of a photocatalytic reaction, in case of a sequential reaction, can be written as product of the quantum yields Φ_i of the intermediate steps which are contributing to the product formation.

$$\Phi = \Phi_1 \cdot \Phi_2 \cdot \Phi_3 \cdot \ldots \tag{16.40}$$

Since the flavin sensitized MBA oxidation involves two redox steps, the catalysis comprises two reduction processes of the photocatalyst. Taking the inter-system crossing and the competing non-productive electron transfer in the singlet state of RFTA into account, the product quantum yield is given by Equation 16.41. Since the non-productive singlet quenching and the productive triplet quenching process via electron transfer are both diffusion controlled processes, and, hence, both rates depend on the substrate concentration, also the over-all reaction quantum Φ yield depends on the substrate concentration.

$$\Phi = \Phi([\mathrm{MBA}]) = \Phi_T([\mathrm{MBA}]) \cdot \Phi_{ET\,I}([\mathrm{MBA}]) \cdot \Phi_{PT\,I} \cdot \Phi_{ET\,II} \cdot \Phi_{PT\,II} \tag{16.41}$$

We find that the first proton transfer process and both the second electron and proton transfer do not depend on the substrate concentration. A possible explanation for this finding is the presence of large amounts of water that greatly increases the overall efficiency. As water is dissociated to a sizable degree, protons and eventually the extra electron can be buffered by this reservoir and the yield become independent of the catalyst and substrate concentration. Many rates of the competing processes are known from additional transient absorption measurements which we performed, and we can now write the over-all reaction quantum yield by inserting these rates into Equation 16.41.

$$\Phi = \frac{k_{ISC}}{k_{ISC} + {}^1k_{IC} + k_{RAD} + {}^1k_{ET}([\mathrm{MBA}])} \cdot \frac{{}^3k_{ET}([\mathrm{MBA}])}{{}^3k_{IC} + {}^3k_{ET}([\mathrm{MBA}])} \cdot \Phi_{PT\,I} \cdot \Phi_{ET\,II} \cdot \Phi_{PT\,II} \tag{16.42}$$

Whereas the first factor, which describes the triplet quantum yield, contains the substrate concentration only in the denominator, the second factor, which describes the yield of the formation of the spin-correlated triplet radical pair, contains the substrate concentration both in the numerator and in the denominator. Since the rate of the diffusive triplet quenching by electron transfer ${}^3k_{ET}$ is much slower in comparison to the singlet quenching process ${}^1k_{ET}$, not necessarily the full kinetic diffusion model has to be applied to describe the triplet quenching process. Here a linear dependency between substrate concentration and diffusive reaction rate seems to be sufficient for modeling the quenching process. However, for the calculation of a rate describing diffusion controlled dynamics on the ultrafast time scale, a linear dependency between concentration and reaction rate is not longer sufficient. In the last section we introduced a mathematical model for diffusive decay modeling on the ultrafast time scale, which is used here. According to Equation 16.30 the diffusive singlet quenching rate ${}^1k_{ET}$, which depends on the substrate concentration as well, can be written as

$$ {}^1k_{ET}\left(t,\left[MBA\right]\right) = 4\pi R^2 D\left[\frac{1}{R}+\frac{1}{\sqrt{\pi D t}}\right]\left[MBA\right]. \tag{16.43}$$

Comparable to the diffusive triplet quenching process, which was approximated by a linear correlation, Equation 16.43 also exhibits a strictly monotonic increasing relation between substrate concentration and reaction rate. Under consideration of the monotonic behavior of both singlet and triplet quenching with increasing substrate concentration, Equations 16.41 and 16.42 exhibit an efficiency maximum at a certain substrate concentration with trailing edges both with increasing and decreasing concentration.

For the calculation of the over-all reaction quantum yield, which is generally time independent, the time dependent reaction rate has to be converted into a time independent quantity. By integration from time zero to infinity of an exponential function describing the decay of an excited state by diffusion, an average, time independent reaction rate can be determined.

$$ {}^1k^0_{ET}\left(\left[MBA\right]\right) = \int_{t=0}^{\infty} dt\, \exp\left(-{}^1k_{ET}\left(t,\left[MBA\right]\right)\cdot t\right) \tag{16.44}$$

As shown in the previous section, both diffusive and non-diffusive quenching, caused by pre-association of catalyst and substrate, are contributing to the over-all singlet quenching process and have to be considered when calculating the effective, time independent electron transfer rate. Integration of an exponential function, describing the decay of the excited singlet state only considering the diffusive quenching process, from time zero to infinity results in a time independent average quenching rate, which still depends on the substrate concentration.

$$ {}^1k_{ET}\left(\left[MBA\right]\right) = \int_{t=0}^{\infty} dt \frac{n_{nondiff}}{n_{nondiff}+n_{diff}}\cdot\exp\left\{-{}^1k^0_{ET}\cdot t\right\} + \frac{n_{diff}}{n_{nondiff}+n_{diff}}\cdot\exp\left\{-4\pi R D\cdot[MBA]\cdot\left[1+\frac{2R}{\sqrt{4Dt}}\right]\cdot t\right\} \tag{16.45}$$

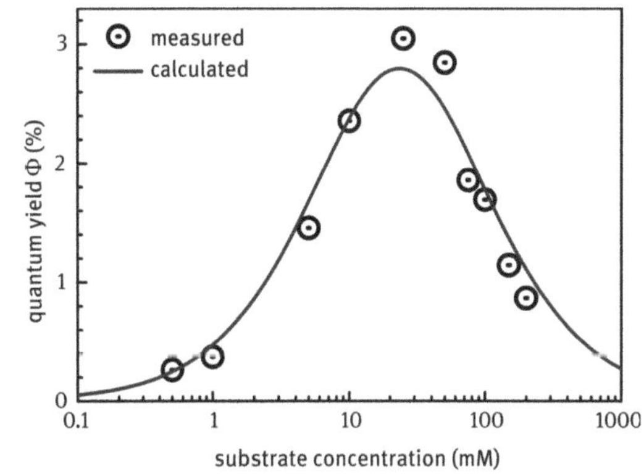

Figure 16.24: Concentration dependent reaction quantum yield of flavin sensitized photo oxidation of methoxy benzyl alcohol on a logarithmic x-coordinate. [43]

Now all contributing contributions of Equation 16.41 are known and the over-all concentration dependent reaction quantum yield can be expressed by quantities which were all determined by time resolved absorption spectroscopy. The result is given as solid line in Figure 16.24. Since the data analysis and especially the diffusive decay modeling is based to some degree on simplifications and assumptions, verification of the predicted substrate concentration dependent reaction quantum yield will strongly support the credibility of the understanding of the photocatalytic reaction.

The comparison of predicted over-all reaction quantum yield in Figure 16.24, which was calculated with the help of the diffusion-based model, and the actual over-all reaction quantum yield determined with the help of the presented setup for the measurement of reaction quantum yields and UV/Vis steady state spectroscopy perfectly confirms the understanding of the photocatalytic process obtained by time resolved spectroscopy. The reaction can be driven in its efficiency maximum, which corresponds to a product quantum yield of 3 %, by adjusting the substrate concentration in the 25 mM range.

16.5 From light absorption to chemistry: Sensitizing mechanisms in homogeneous photocatalysis

16.5.1 Sensitization by excitation energy transfer

Sensitization by excitation energy transfer (EET) processes is sometimes considered for photocatalysis but only plays a minor role in homogeneous photocatalysis. Therefore, we will only comment very briefly on some basic considerations.

Generally there are two different mechanisms to distinguish: Singlet excitation energy transfer via Förster resonance energy transfer (FRET) [85] and triplet triplet energy transfer (TTET) via Dexter [86]. A scheme of both processes is given in Equation 16.46 and 16.47.

$$C+S+h\nu \xrightarrow{\text{abs.}} {}^1C^* + S \xrightarrow{\text{EET}} C + {}^1S^* \quad (16.46)$$

$$C+S+h\nu \xrightarrow{\text{abs.}} {}^1C^* + S \xrightarrow{\text{ISC}} {}^3C^* + S \xrightarrow{\text{EET}} C + {}^3S^* \quad (16.47)$$

Since excited singlet states and triplet states of organic chromophores are usually easily distinguishable with the help of their optical absorption properties, time resolved absorption spectroscopy in the visible is a well suited tool for the investigation of EET sensitization processes both in the singlet and triplet manifold. Whereas triplet sensitization mechanisms are often applied meaningfully, applications for singlet sensitization processes are hardly imaginable in photocatalysis, since the EET usually occurs between two optical accessible excited states. Then in terms of efficient photocatalysis a direct excitation of the substrate is more convenient. A senseful exception might be the case where the lowest optical transition of the substrate leads to the $n\pi^*$ state and consequently the absorption strength is very low. Furthermore, the lifetime of excited singlet states is usually limited by fluorescence decay and inter-system crossing and does commonly not exceed a couple of nanoseconds and; therefore, diffusion limited sensitizing becomes inefficient. If there is no catalyst-substrate pre-association, the expected time scale for triplet EET sensitization is in the ns and μs range, when the reaction dynamics are controlled by diffusion of the reactants. Therefore, long lived triplet states are better suited for photo-sensitization. Especially when the following requirements are matched, applications of TTET can improve the efficiency of a photocatalytic process:

- Triplet sensitization with a photo-sensitizer with high triplet quantum yield Φ_T to substrate with low intrinsic Φ_T to increase the total reaction quantum yield for a triplet reaction
- Triplet sensitization with a photo-sensitizer, which has a lower energy gap between the optically accessible singlet and the triplet state to allow irradiation at longer wavelengths to expand the usable irradiation spectrum far into the visible
- Triplet sensitization of weakly bound catalyst-substrate complexes to allow selective sensitization to provide stereoselective photochemistry. [7, 87]

16.5.2 Photoredox catalysis: Requirements on catalyst and substrate

The general mechanism of a photocatalytic reaction comprises two initial steps: The excitation of the chromophore and the sensitization of the substrate. The sensitization is the use of the excitation energy from the photocatalyst in its excited state for the suitable modification of the substrate in its ground state to start a chemical conversion. The minor importance of the genuine energy transfer processes such as singlet energy transfer via Förster resonance energy transfer [85] or triplet triplet energy transfer via the Dexter mechanism [86] is discussed in the preceding section. The more important sensitization process is photo-induced electron transfer where light serves as energy source for intra- and intermolecular electron transport. Depending on the catalytic system and the sensitizing mechanism, the temporal delay between photo-excitation and sensitization can cover the time range from $1{\cdot}10^{-13}$s, e.g. for intramolecular charge transfer reactions in electronically strongly interacting systems, to $1{\cdot}10^{-6}$s, e. g. for slow diffusion limited intermolecular charge transfer reactions. [43, 51, 88] This section will provide examples of different photo-induced sensitization mechanisms in photocatalysis with strong connection to the application of time resolved spectroscopic methods. The examples are chosen from photoredox catalysis, which is one of the most important sensitizing mechanisms in photocatalysis. [89, 90] Photoredox catalysis relies on the general property of excited states that they are usually more easily reduced or more easily oxidized than the ground state. Therefore, the excited photocatalyst can serve either as highly reactive electron donor or electron acceptor. [90]

From the perspective of time resolved spectroscopy it causes a major difference if the excited state quenching of the photocatalyst occurs in its excited singlet or triplet state. Since charge recombination within the formed spin correlated radical ion pair (RIP) with triplet character is a spin forbidden process, the charge separated state within the triplet RIP is more stable. [80] This allows the subsequent reaction steps to take place with significant yields. In terms of efficient photocatalysis it is, therefore, often advantageous to drive the electron transfer process in the triplet state of the photocatalyst. [43] A lifetime of the charge separated radical ion pair in the microsecond range is considered to be sufficient for efficient catalysis and this has been achieved in many artificial systems as reviewed in various review papers and chapters, although only in a limited number of cases this has been combined with a close to unity total reaction quantum yield. [80, 91, 92]

The sensitization by electron transfer, either photo-induced oxidation or reduction of the substrate by the excited photocatalyst, comprises two spectroscopically accessible steps in the singlet manifold and three steps in the triplet manifold, respectively. All contributing intermediate steps are shown in Equations 16.48 and 16.49, where C is the photocatalyst and S is the substrate and h_ν stands for the incident photon, which is absorbed by the photocatalyst.

$$C+S+h\nu \xrightarrow{\text{abs.}} {}^1C^* + S \xrightarrow{\text{ET}} {}^1\overline{C^{\bullet -/+} + S^{\bullet +/-}} \tag{16.48}$$

$$C+S+h\nu \xrightarrow{\text{abs.}} {}^1C^* + S \xrightarrow{\text{ISC}} {}^3C^* + S \xrightarrow{\text{ET}} {}^3\overline{C^{\bullet -/+} + S^{\bullet +/-}} \tag{16.49}$$

From Equation 16.48 and 16.49 it can be seen, that the photocatalyst remains in its reduced or oxidized state after the photo-induced radical ion pair formation and, therefore, needs to be regenerated in order to retrieve its full photocatalytic capability and to complete the catalytic cycle. On the other hand it can be seen, that the electron transfer process between excited photocatalyst and substrate needs to be thermodynamically allowed.

First we will comment on the latter: The actual sensitizing mechanism of a photocatalytic reaction strongly depends on the electrochemical and photophysical properties of catalyst and substrate. Calculating the change of Gibbs energy ΔG^0 according to Rehm and Weller [94], which is shown in Equation 16.50, helps to predict the sensitizing mechanism or to find a suitable photocatalyst for the desired reaction. The standard oxidation potential of the electron donor in its ground state is $E^0\left(D^{+\cdot}/D\right)$, the standard reduction potential of the electron acceptor in its ground state is $E^0\left(A/A^{-\cdot}\right)$. These can be easily measured by cyclovolametry. The quantity *r* stands for the electron donor-acceptor distance which can be deduced from quantum chemical calculations or knowledge of the molecular structure. The quantity ΔE_{00} represents the energy with respect to the ground state of the excited state, which is quenched by ET. For the excited singlet state ΔE_{00} can be readily read off the absorption spectrum. The change of ΔG^0 serves as driving force for the ET reaction between donor and acceptor and has to be below zero, to thermodynamically allow the ET process.

$$\Delta G^0 = e\left[E^0\left(D^{+\cdot}/D\right) - E^0\left(A/A^{-\cdot}\right)\right] - \frac{e^2}{4\pi\varepsilon_0\varepsilon r} - \Delta E_{00} \qquad (16.50)$$

The corresponding data of common organic photo-sensitizers, which are applied in homogeneous photocatalysis, and of the very useful $Ru(bpy)_3^2$ are listed in Table 16.2. [95] Although these data provide a first idea of the underlying mechanism, an experimental proof of the actual sensitizing mechanism is crucial for a full understanding of the photo-catalytic reaction. Especially time-resolved spectroscopic methods on multiple time scales provide detailed insights into the photo-induced sensitization processes.

The re-oxidation or re-reduction of the photocatalyst after the photo-induced primary ET to its ground state is generally a required process in terms of catalysis. Otherwise the total photo-induced charge transfer process should be discussed as photochemistry, since the light-absorbing chromophore is consumed with the conversion of one substrate molecule. However, the second ET step might seem useless for the desired reaction, but it is necessary for the total catalytic cycle. It is also possible to drive the desired conversion of the substrate with the regeneration step of the photo-formed radical ion of the catalyst. [98] But in this situation, time resolved spectroscopy is not the perfect tool for the investigation of the actual sensitization process of the substrate since the starting point of the reaction is not anymore de-

Table 16.2: Electrochemical and photophysical data of selected organic sensitizers in comparison with the photoredox catalyst $Ru(bpy)_3^{2+}$

sensitizer	E^0 ($D^{+\bullet}$/D) vs. SCE [V]	E^0 (A/$A^{-\bullet}$) vs. SCE [V]	λ_{00} [nm]	E_{S1} [eV]	E_{T1} [eV]	Φ_T
anthracene [a]	1.09 (MeCN)	−1.95 (DMF)	375	3.31	1.84	0.7
benzophenone [a]		−1.83 (MeCN)	384	3.23	3.00	1
eosin Y [a b]		−1.38 (MeCN)	571	2.17	1.83	0.3
riboflavin [d c]		−0.29 [e] (H_2O)	495	2.50	2.17	0.7
xanthone [a]		−1.77 (DMF)	370	3.35	3.21	1
$Ru(bpy)_3^{2+}$ [d]	1.29	−1.33	452	2.74		

(a) reference [51] (b) 2′,4′,5′,7′-tetrabromo-fluorescin-dianion (c) reference 96 (d) reference [97] (e) vs. NHE

fined by the pump pulse. It is the photo-induced generation of the reactive radical ion species of the photocatalyst that is easily and directly accessible. Tracing the second ET step with methods of time resolved spectroscopy is much more challenging, especially if both ET steps are limited by diffusion of the reactants. Then, if both the sacrificial reactant and the substrate are diluted with comparable concentration, the diffusion limited quenching process occurs on equal time scales and, therefore, both processes are difficult to distinguish with the help of time-resolved spectroscopy. The intermediate species might even not be present in high enough concentration for a secure detection.

Another aspect, which has to be considered when comparing photoredox chemistry in the singlet and in the triplet manifold, is the achievable reaction rate for the ET step. It has been found that quenching of excited singlet states by electron transfer is usually fast in comparison to the quenching of a triplet state of the same chromophore by electron transfer. [43, 99] Also the listed diffusive, concentration dependent quenching rates K_q of excited photo-sensitizers by suitable quencher molecules in Montali et al. "Handbook of Photochemistry" [51] strongly support this picture. When discussing rates of inter- and intramolecular ET processes, the Marcus Theory is indispensable for an accurate explanation of the involved processes. [100, 101] The Marcus Equation relates the electron transfer rate k_{ET} to the electronic coupling H_{if} between initial and final state, the reorganization energy λ and the change of Gibbs energy ΔG^0.

$$k_{et} = \frac{2\pi}{\hbar}\left|H_{if}\right|^2 \frac{1}{\sqrt{4\pi\lambda k_B T}} \exp\left\{-\frac{\left(\lambda + \Delta G^0\right)^2}{4\lambda k_B T}\right\} \tag{16.51}$$

A well suited example for this effect is the diffusion limited quenching of excited riboflavin by intermolecular electron transfer from 4-methoxybenzylalcohol, which is 200-fold slower when quenching the triplet state in comparison when quenching the excited singlet state (Table 16.1 and Ref. [43]). Also the quenching of excited states of benzophenone from a covalently bound DNA nucleobase occurs roughly 20 times faster in the excited singlet state than in the triplet state. [102] The Marcus theory provides two explanations for this behavior: Due to the usually energetically lower lying triplet state in comparison to the optical accessible excited singlet states in organic chromophores (Table 16.2), the absolute value of the change of Gibbs energy ΔG^0 is smaller when the excited state quenching by electron transfer occurs in the triplet regime. This leads to a decrease in the electron transfer rate k_{et} and therefore to a decrease in reactivity [99] and clearly can be seen by calculating the ΔG^0 values according to Rehm and Weller (Equation 16.47) and inserting the result into the Marcus equation (Equation 16.51). Another explanation for the decrease in the rate of electron transfer k_{et} can be a change in the coupling term $|H_{if}|^2$ which describes the electronic coupling between initial state and final state of the charge transfer reaction. Generally speaking, on the one hand, in photocatalyst's triplet chemistry provides the advantage of spin-forbidden charge recombination but, on the other hand, it has the disadvantage of a slow electron transfer process in comparison to singlet chemistry.

As a further example, irradiation of organometallic coordination complexes such as $Ru(bpy)_3^{2+}$ which are often used as redox photocatalysts, due to their versatile optical and electrochemical capabilities [91, 103, 104], leads to a species which exhibits catalytic activity. [104] This is a convenient and often used method for the photo-induced activation of a catalyst [105], also adsorbed on a solid [65, 107–109]. Covalently bound to a platin or palladium centered catalytic active site forming a donor-bridge-acceptor-system, ruthenium organometallic coordination complexes can also serve as intramolecular electron donor sites and, therefore, provide photo-induced water splitting capability. [110]

For further information on photosensitization by reversible electron transfer the early but still informative review article from Kavarnos and Turro [111] is strongly recommended.

16.6 Epilogue

In this final chapter we have described, how modern transient spectroscopy can contribute decisively to the understanding of the mechanisms governing chemical photocatalysis. The methods developed in our laboratories over the last years allow the excitation of the catalyst at any selected wavelength. This is important to ensure that indeed the catalyst and not the substrate is excited. In addition it allows one to investigate whether higher electronic states or vibrational excess energy influences the outcome and efficiency of the processes.

We now have a detection range from the UV, as low as 250 nm, and out to NIR, as high as 1700 nm available at a temporal resolution of better than 50 fs. With the help of an auxiliary tunable nanosecond laser we can measure without gaps out to the high μs range. This enormous detection range allows one to identify many of the reaction intermediates. As a consequence, the transient optical spectroscopy does not restrict itself anymore to the determination of specific rates, but it renders a complete picture of a multistep reaction chain.

For this analysis it is far from sufficient to fit exponential models to the signal at an individual wavelength. Such an approach has been used many times in the past, since early experiments only had single detection wavelengths available and the investigators relied on their a priori knowledge where a specific chemical species could be detected. Instead, a global analysis of the whole spectral range has to be performed to arrive at an unambiguous interpretation.

From such an analysis it can readily be determined whether a reaction proceeds via singlet or triplet states. For singlet schemes a pre-association is desired, as the short singlet lifetime does not allow efficient diffusional encounters. We showed, however, that such an approach can easily lead to rapid charge recombination after the initial electron transfer that starts the catalysis. Pre-association is, however, highly desired for stereoselective reactions.

When the initial dynamics is intersystem crossing in the catalyst, the aim has to be to generate a high triplet yield. As this takes typically some nanoseconds, the subsequent step of charge transfer to the substrate cannot be boosted by pre-association but has to be enabled by diffusion. The time resolved spectroscopy should then be repeated for a number of substrate concentrations. What is most important is that the analysis of the resulting set of data has to be performed comprehensively in order to obtain the true system behavior and not just effective values for each value of the concentration. The final set of parameters which truly describes the reaction has to model the complete spectrum and all concentrations simultaneously. From such a comprehensive analysis, a clear picture of the diffusional encounter and separation results. Such information has previously only been used very rarely and as a matter of fact the community of ultrafast spectroscopists has not considered it possible to analyze these rather complex situations. On the other hand, the chemists working on chemical photocatalysis have not dared to apply transient spectroscopy from the femtosecond to the microsecond range to their complex problems.

We hope that the present account and the evolving original work by our efforts will encourage more researchers to use the new methods and capabilities. Not only chemical photocatalysis, but also the sustainable energy source water splitting [112] and a wide range of issues in green chemistry should be able to profit. The detailed understanding of the reaction mechanisms is not only of academic interest, but it can readily lead to strategies for improving the efficiency of the use of the photons. Even so, the solar photons are free of direct cost, and they should be used diligently to keep the demand on collection space low. Only with high reaction quantum yields new routes to solar energy harvesting, in particular the solar backed production of fuels, will gain the necessary technical and industrial acceptance.

16.7 References

[1] Würthner, F.; Kaiser, T. E.; Saha-Müller, C. R. (2011). Angew. Chem. Int. Ed., 50, 3376–3410.
[2] Scheibe, G. (1937). Angew. Chem., 50, 212–219.
[3] Jelley, E. E. (1936). Nature, 138, 1009–1010.
[4] Icli, S.; Demiça, S.; Dindar, B., Doroshenko, A. O.; Timur, C. (2000). J. Photochem. Photobiol. A, 136, 15–24.
[5] Würthner, F.; Thalacker, C.; Diele S.; Tschierske, C. (2001). Chem. Eur. J., 7, 2245–2253.
[6] Würthner, F. (2004). Chem. Commun., 1564–1579.
[7] Müller, C.; Bauer, A.; Bach, T. (2009). Angew. Chem Int. Ed., 48, 6640–6642.
[8] Müller, C., Bauer, A., Maturi, M. M.; Cuquerella, M. C.; Miranda, M. A.; Bach, T. (2011). J. Am. Chem. Soc., 133, 16689–16697.
[9] Sikorska, E.; Khmelinskii, I.; Komasa, A.; Koput, J.; Ferreira, L. F. V.; Herance, J. R.; Bourdelande, J.L.; Williams, S. L.; Worrall, D. R.; Insinska-Rak, M.; Sikorski M. (2005). Chem. Phys., 314, 239–247.
[10] Salzmann, S.; Tatchen, J.; Marian, C. M. (2008). J. Photochem. Photobiol. A, 198, 221–231.
[11] Hari, D. P.; Schroll, P.; König, B. (2012). J. Am. Chem. Soc., 134, 2958–2961.
[12] Klonis, N.; Clayton, A. H. A.; Voss Jr., E. W.; Sawyer, W. H. (1998). Photochem. Photobiol., 67, 500–510.
[13] Chakraborty, M.; Panda, A. K. (2011). Spectrochim. Acta A, 81, 458–465.
[14] Fita, P.; Fedoseeva, M.; Vauthey, E. (2011). J. Phys. Chem A, 115, 2465–2470.
[15] Reichardt, C. (2003). Solvents and Solvent Effects in Organic Chemistry, 3rd edition (Wiley-VCH, Weinheim)
[16] Curtis, J. C.; Sullican, B. P.; Meyer, T. J. (1983). Inorg. Chem., 22, 224–236.
[17] Fantacci, S.; De Angelis, F.; Selloni, A. (2003). J. Am. Chem. Soc., 125, 4381–4387.
[18] Stratt, R. M.; Maroncelli, M. (1996). J. Phys. Chem., 100, 12981–12996.
[19] Fleming, G. R.; Cho, M. (1996). Annu. Rev. Phys. Chem., 47, 109–134.
[20] de Boeij, W. P.; Pshenichnikov, M.S.; Wiersma, D. A. (1998). Annu. Rev. Phys. Chem., 49, 99–123.
[21] Reichardt, C. (1979). Angew. Chem. Int. Ed., 18, 98–110.
[22] Reichardt R. (1994). Chem. Rev., 94, 2319–2358.
[23] Norrish, R. G. W.; Porter, G. (1949). Nature, 164, 658–658.
[24] Porter, G. (1950). Proc. Roy. Soc. Lond. Ser. A, 200, 284–300.
[25] Kukura, P.; McCamant, D. W.; Mathies, R. A. (2007). Annu. Rev. Phys. Chem., 58, 461–488.
[26] Satzger, H.; Schmidt, B.; Root, C.; Zinth, W.; Fierz, B.; Krieger, F.; Kiefhaber, T.; Gilch, P. (2004). J. Phys. Chem A, 108, 10072–10079.
[27] Keller, U. (2003). Nature, 424, 831–838.
[28] Spence, D. E.; Kean, P. N.;Sibbett, W. (1991). Opt. Lett., 16, 42–44.
[29] Megerle, U.; Pugliesi, I.; Schriever, C.; Sailer, C.; Riedle, E. (2009). Appl. Phys. B, 96, 215–231.
[30] Alfano, R. R.; Shapiro, S. L. (1970). Phys. Rev. Lett., 24, 584–587.
[31] Alfano, R. R.; Shapiro, S. L. (1970). Phys. Rev. Lett., 24, 592–594.
[32] Bradler, M.; Baum, P.; Riedle, E. (2009). Appl. Phys. B, 97, 561–574.
[33] Kozma, I. Z.; Baum, P.; Lochbrunner, S.; Riedle, E. (2003). Opt. Express, 11, 3110–3115.
[34] Homann, C.; Lang, P.; Riedle, E. (2012). J. Opt. Soc. Am. B, 29, 2765–2769.
[35] Bradler, M.; Homann, C.; Riedle, E. (2011). Opt. Lett., 36, 4212–4214.
[36] Huber, R.; Satzger, H.; Zinth, W.; Wachtveitl, J. (2001). Opt. Commun., 194, 443–448.
[37] Mention of vendor names and model numbers is for technical communication purposes only and does not necessarily imply recommendation of these units, nor does it imply that comparable units from another vendor would be any less suitable for this application.
[38] Wenninger, M.; Fazio, D.; Megerle, U.; Trindler, C.; Schiesser, S.; Riedle, E.; Carell, T. (2011). ChemBioChem, 12, 703–706.

[39] Polli, D.; Brida, D.; Mukamel, S.; Lanzani, G.; Cerullo, G. (2010). Phys. Rev. A, 82, 053809.
[40] Schriever, C.; Pugliesi, I.; Riedle, E. (2009). Appl. Phys. B, 96, 247–250.
[41] Ganz, T.; von Ribbeck, H. G.; Brehm, M.; Keilmann, F. (2008). Opt. Commun., 281, 3827–3830.
[42] Pugliesi, I.; Krok, P.; Lochbrunner, S.; Blaszczyk, A.; von Hänisch, C.; Mayor, M.; Riedle, E. (2010). J. Phys. Chem. A, 114, 12555–12560.
[43] Megerle, U.; Wenninger, M.; Kutta, R.-J.; Lechner, R.; König, B.; Riedle, E. (2011). Phys. Chem. Chem. Phys., 13, 8869–8880.
[44] Fita, P.; Luzina, E.; Dziembowska, T.; Radzewicz, C.; Grabowska, A. (2006). J. Chem. Phys., 125, 184508.
[45] Birks, J. B. (1970). Photophysics of Aromatic Molecules (Wiley-Interscience, London)
[46] Ruckebusch, C.; Sliwa, M.; Pernot, P.; de Juan, A.; Tauler, R. (2012). J. Photochem. Photobiol. C, 13, 1–27.
[47] Sakai, M.; Takahashi, H. (1996). J. Mol. Struct., 379, 9–18.
[48] Megerle, U. (2011). Photoinduced molecular dynamics in increasingly complex systems: From ultrafast transient absorption spectroscopy to nanoscopic models. (Dissertation, Ludwig-Maximilians-Universität, Fakultät für Physik. München)
[49] Ricks, A. B.; Brown, K. E.; Wenninger, M.; Karlen, S. D.; Berlin, Y. A.; Co, D. T., Wasielewski, M. R. (2012). J. Am. Chem. Soc., 134, 4581–4588.
[50] Rice, S. A. (1985). Diffusion-limited reactions, (Elsevier Science Publishers B.V., Amsterdam)
[51] Murov, S. L.; Carmichael, I.; Hug, G. L. (1993). Handbook of Photochemistry, 2nd edition (M. Decker, New York)
[52] Koch, M.; Rosspeintner, A.; Angulo, G.; Vauthey, E. (2012). J. Am. Chem. Soc., 134, 3729–3736.
[53] Eads, D. D.; Dismer, S. G.; Fleming, G. R. (1990). J. Chem. Phys., 93, 1136–1148.
[54] Murata, S.; Matsuzaki, S. Y.; Tachiya, M. (1995). J. Phys. Chem., 99, 5354–5358.
[55] Murata, S.; Tachiya, M. (1996). J. Phys. Chem., 100, 4064–4070.
[56] Burel, L.; Mostafavi, M.; Murata, S.; Tachiya, M. (1999). J. Phys. Chem. A, 103, 5882–5888.
[57] Serpone, N. J. (1997). Photochem. Photobiol. A, 104, 1–12.
[58] Bakowski, A.; Dressel, M.; Bauer, A.; Bach, T. (2011). Org. Biomol. Chem., 9, 3526–3529.
[59] Vichova, J.; Hartl, F.; Vlcek, A. (1992). J. Am. Chem. Soc., 114, 10903–10910.
[60] Tfouni, E. (2000). Coord. Chem. Rev., 196, 281–305.
[61] Xia, H.-L.; Liu, F.; Ardo, S.; Sarjeant, A. A. N.; Meyer, G. J. (2010). J. Photochem. Photobiol. A, 216, 94–103.
[62] Serpone, N.; Emeline, A. V. (2002). Int. J. Photoenergy, 4, 91–131.
[63] Cherevatskaya, M.; Neumann, M. Füldner, S.; Harlander, C.; Kümmel, S.; Dankesreiter, S.; Pfitzner, A.; Zeitler, K.; König, B. (2012). Angew. Chem., 124, 4138–4142
[64] Maschmeyer, T.; Che, M. (2010). Angew. Chem. Int. Ed., 49, 1536–1539.
[65] Fagnoni, M.; Dondi, D.; Ravelli, D.; Albini, A. (2007). Chem. Rev., 107, 2725–2756.
[66] Megerle, U.; Lechner, R.; König, B.; Riedle, E. (2010). Photochem. Photobiol. Sci., 9, 1400–1406.
[67] Parker, C. A. (1953). Proc. R. Soc. London, Ser. A, 220, 104–116.
[68] Kuhn, H. J.; Braslavsky, S. E.; Schmidt, R. (2004). Pure Appl. Chem., 76, 2105–2146.
[69] Jankowski, J. J.; Kieber, D. J.; Mopper, K. (1999). Photochem. Photobiol., 70, 319–328.
[70] Rahn, R. O. (1997). Photochem. Photobiol., 66, 450–455.
[71] Zepp, R. G.; Gumz, M. M.; Miller, W. L.; Gao, H. (1998). J. Phys. Chem. A, 102, 5716–5723.
[72] Ghosh, J. P.; Sui, R.; Langford, C. H.; Achari, G.; Berlinguette, C. P. (2009). Water Res., 43, 4499–4506.
[73] Schmidhammer, U.; Roth, S.; Riedle, E.; Tishkov, A. A.; H. Mayr, H. (2005). Rev. Sci. Instrum., 76, 093111
[74] Reindl, S.; Penzkofer, A. (1996). Chem. Phys., 211, 431–439.
[75] Reindl, S.; Penzkofer, A. (1996). Chem. Phys., 213, 429–438.

[76] Peceli, D.; Webster, S.; Fishman, D. A.; Cirloganu, C. M.; Hu, H.; Przhonska, O. V.; Kurdyukov, V. V.; Slominsky, Y. L.; Tolmachev, A. I.; Kachkovski, A. D.; Dasari, R. R.; Barlow, S.; Marder, S. R.; Hagan, D. J.; Van Stryland E. W. (2012). J. Phys. Chem. A, 116, 4833–4841.
[77] Pareek, V.; Chong, S.; Tade, M.; Adesina, A. A. (2008). Asia Pac. J. Chem. Eng., 3, 171–201.
[78] Middleton, C. T.; de La Harpe, K.; Su, C.; Law, Y. K.; Crespo-Hernandez, C. E.; Kohler, B. (2009). Annu. Rev. Phys. Chem., 60, 217–239.
[79] Dobryakov, A. L.; Kovalenko, S. A.; Weigel, A.; Pérez-Lustres, J. L.; Lange, J.; Müller, A.; Ernsting, N. P. (2010). Rev. Sci. Instrum., 81, 113106.
[80] Verhoeven, J. W. (2006). J. Photochem. Photobiol. C, 7, 40–60.
[81] Hore, P. J.; Broadhurst, R. W. (1993). Prog. Nucl. Mag. Res. Sp., 25, 345–402.
[82] Heelis, P. F. (1982). Chem. Soc. Rev., 11, 15–39.
[83] Insinska-Rak, M.; Sikorska, E.; Bourdelande, J. L.; Khmelinskii, I. V.; Prukala, W.; Dobek, K.; Karolczak, J.; Machado, I. F.; Ferreira, L. F. V.; Komasa, A.; Worrall, D. R. and Sikorski, M. (2006). J. Mol. Struct., 783, 184–190.
[84] Salzmann, S.; Tatchen, J.; Marian, C. M. (2008). J. Photochem. Photobiol. A, 198, 221–231.
[85] Förster, T. (1948). Ann. Phys., 437, 55–75.
[86] Dexter, D. L. (1953). J. Chem. Phys., 21, 836–850.
[87] Cauble, D. F.; Lynch, V.; Krische, M. J. (2003). J. Org. Chem., 68, 15–21.
[88] Zewail, A. H. J. (2000). Phys. Chem. A., 104, 5660–5694.
[89] Hsu, C.-P. (2009). Acc. Chem. Res., 42, 509–518.
[90] Zeitler, K. (2009). Angew. Chem. Int. Ed., 48, 9785–9789.
[91] Wasielewski, M. R. (1992). Chem. Rev., 92, 435–461.
[92] Gust, D.; Moore, T. A.; Moore, A. L. (2001). Acc. Chem. Res., 34, 40–48.
[93] Lhiaubet, V.; Paillous, N.; Chouini-Lalanne, N. (2001). Photochem. Photobiol., 74, 670–678.
[94] Rehm, D.; Weller, A. (1970). Isr. J. Chem., 8, 259–271.
[95] Ravelli, D.; Fagnoni, M.; Albini, A. (2012). Chem. Soc. Rev.
[96] Meisel, D.; Neta, P. (1975). J. Phys. Chem., 79, 2459–2461.
[97] Narayanam, J. M. R.; Stephenson, C. (2011). Chem. Soc. Rev., 40, 102–113.
[98] Ischay, M. A.; Anzovino, M. E.; Du, J.; Yoon, T. P. (2008). J. Am. Chem. Soc., 130, 12886–12887.
[99] Steiner, U.; Winter, G.; Kramer, H. E. A. (1977). J. Phys. Chem., 81, 1104–1110.
[100] Marcus, R. A. (1956). J. Chem. Phys., 24, 966–978.
[101] Marcus, R. A.; Sutin, N. (1985). Biochim. Biophys. Acta, 811, 265–322.
[102] März, T.; Wenninger, M; Weinberger, M.; Wagenknecht, H.-A.; Riedle, E.; Schütz, M. in preparation
[103] Campagna, S.; Puntoriero, F.; Nastasi, F.; Bergamini, G.; Balzani, V. (2007). Top. Curr. Chem., 280, 117–214.
[104] Juris, A.; Balzani, V.; Barigelletti, F.; Campagna, S.; Belser, P.; von Zelewsky, A. (1988). Coord. Chem. Rev., 84, 85–277.
[105] Hennig, H.; Dietze, F.; Thomas, P.; Rehorek, D. (1978). Z. Chem., 18, 458–460.
[106] Hennig, H. (1999). Coord. Chem. Rev., 182, 101–123.
[107] Wada, Y.; Nakaoka, C.; Morikawa, A. (1988). Chem. Lett., 25–26.
[108] Ballinger, T. H.; Yates Jr., J. T. (1992). J. Phys. Chem., 96, 9979–9983.
[109] Ichikawa, M.; Pan, W.; Imada, Y.; Yamaguchi, M.; Isobe, K.; Shido,T. (1996). J. Mol. Catal. A: Chem., 107, 23.
[110] Tschierlei, S.; Presselt, M.; Kuhnt, C.; Yartsev, A.; Pascher, T.; Sundström, V.; Karnahl, M.; Schwalbe, M.; Schäfer, B.; Rau, S.; Schmitt, M.; Dietzek, B.; Popp, J. (2009). Chem. Eur. J., 15, 7678–7688.
[111] Kavarnos, G. J.; Turro, N. J. (1986). Chem. Rev., 86, 401–449.
[112] Nocera, D. G. (2009). ChemSusChem, 2, 387–390.

Index

Bei Fragen zur Produktsicherheit wenden Sie sich bitte an:
If you have any questions regarding product safety,
please contact:

Walter de Gruyter GmbH
Genthiner Straße 13
10785 Berlin
productsafety@degruyterbrill.com